Revised Printing

Zoology

An Inside View of Animals

Ken Hyde, Ph.D.
Virginia Turner
Craig Stettner

Harper College

Kendall Hunt
publishing company

Cover background image © Corel

Kendall Hunt
publishing company

www.kendallhunt.com
Send all inquiries to:
4050 Westmark Drive
Dubuque, IA 52004-1840

To

Dr. James Barrow,
an incomparable teacher whose passion for zoology
first lit the spark

Contents

An Inside View... v
Preface/Acknowledgements... vi

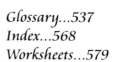

An Inside View

Certain topics, because of their special nature, deserve special treatment. So, scattered throughout the text are a number of discussions under the heading **An Inside View.** These are listed below. Some are given to further illustrate a topic, while others are presented as an in-depth coverage of an important subject; still others are more of a general interest. A few are simply bizarre or outlandish. All, however, are intended to give the student greater insight into the world of animals. (KH)

Preface

The urge to write this book is simple. I want to share some of the pure exhilaration that the study of animals has given me. For a lifetime, I have been intrigued by animals. Their form, behavior, beauty, and complexity have captivated me. If this text can convey a sense of that excitement, then it will have been a worthwhile enterprise.

Any introductory textbook is a terminological minefield. Technical terms are everywhere, thousands of them. Many are necessary, for they lead us through the battlefield of learning. Yet, they can be overwhelming to a student. Thus, I have tried to keep the vocabulary basic; whenever possible emphasizing concepts over minutia. Nonetheless, there are many terms, most new to the student, littered throughout this book. So, to assist students as they traverse this tangled path, all new terms are printed in boldface and are listed separately at the end of each chapter. Also, terms are defined in the glossary section at the end of the book.

Zoology is an immense subject. No single text can be expected to cover it entirely. Thus, authors are challenged to pick and choose presentations carefully. With this in mind I have tried to give a balanced coverage to the major groups of animals. I have striven to write with clarity and enthusiasm without sacrificing content. Above all else, I have tried to write in a way that engages students as they attempt to grasp this marvelous subject. My hope is that this book will cause students to pause from time to time to ask why?...or, how?

This textbook is, in many respects, similar to others that are available. It is descriptive and it is conceptual. It, as with most in the field, also adheres to the phylogenetic perspective.

But in one respect this book is very different. Through the years I have seen students struggle trying to juggle one textbook for the lecture component of a course and another "manual" for the laboratory portion. I believe this is cumbersome and unnecessary. So, I have combined the two. Almost every chapter concludes with a laboratory exercise or demonstration over the material just covered. These exercises are intended to guide students by way of direct observation, analysis, written explanations, and/or dissections. I believe this approach is not only more manageable (and hopefully more comprehensive for the student), but also, more cost effective.

Ken Hyde

Acknowledgements

The number of teachers, colleagues, students, and friends who have contributed, in small ways and large, to this effort are too many to chronicle. Their contributions, however, can be found liberally sprinkled on every page.

There have been a few individuals who have been of special help. A former student, Scott Yoo, has lent his critical and creative touch to a portion of the manuscript. I thank him for his time and valuable effort. Another student, Roberto Blanco, has helped me to understand digital photography and has laughed along with me at my feeble attempts to learn. And my son Chris has walked this dinosaur through the amazing world of computers. Without his artful guidance, I'm afraid this effort would have suffered the same fate as those other intractable reptiles.

A colleague, Prof. Sandy Kreiling, has provided invaluable overall feedback on content. I value her opinion and appreciate her input.

Finally, I thank my wife for her continual support.

Ken Hyde

1 Chapter 1

ANIMAL DIVERSITY

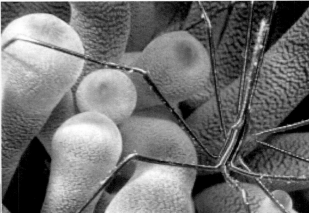

Where order in variety we see,
And where, though all things differ, all agree
- Windsor Forest

Animal Diversity

Introduction

*T*hat animals are spectacular is a conclusion often reached by the most casual observer. Indeed, even experienced zoologists find their study inspiring. And why not? When you consider the many ways animals differ from each other— their enormous size variations, the contrasting palette of colors, their bizarre structures, the changing complexion of behaviors, the cacophony of sounds, and their mosaic habitats— you begin to understand the awe-inspiring nature of animals. This is the world we are about to enter: the world of zoology.

At first glance this world may seem hopelessly confusing: just a hectic collection of mysterious and bewildering creatures running, swimming, crawling, and flying about aimlessly. There seems to be little design...just randomness. However, a closer look reveals not a world of chaos, but one instead interlaced with structure, organization, intricate patterns, and complex relationships. An underlying order exists.

It is this order, this hidden inner sanctum, that zoologists struggle to uncover. Through observation and penetrating investigations, we are slowly unmasking the secret universe of animals. In fact, for some of the more common species we have already amassed a fairly large reservoir of knowledge concerning their structures, needs, habitats, and behavior. For instance, there has been more than one book written about the common housefly or the cockroach. Certain mosquitoes are well known. The common earthworm, the bald eagle, and the white-tailed deer are also some of the better-known species. But for the vast majority of animals we have barely scratched the surface. We know virtually nothing of their anatomy, of how or where they live, of how they interact with each other. For them our encyclopedia of understanding is appallingly empty.

How Many Kinds?

One of the very first questions people ask about animals has to do with their numbers. Animals seem to be all around us, so exactly how many kinds are there? To find that answer we look to **taxonomy**, the science of naming and classifying organisms. When we do, we get a mixed picture. First of all, we learn that about 1.2 million different animal species have been discovered and named so far. However, the real question is not how many animals we have met face-to-face, but instead, how many kinds actually exist? And here the picture gets somewhat muddled. That's because our best estimates tell us that the one million or so thus far identified is just a small fraction of the total. But what that total really is we're just not sure. Oh, we can offer educated guesses, but they are just that...guesses. If you were to ask a flock of zoologists how many species of animals exist, you would surely receive a flutter of answers. On the more conservative side the number may be somewhere in the neighborhood of 10 million species. However, most experts would place that number much higher, somewhere in the 40 to 50 million species range...maybe even twice that number!

Whatever the exact number might be, and we will never know for sure, it at least means that there are millions and millions of animals scrambling about, most unknown to us. Each is surviving in its own little corner, differing from all others in its basic needs, anatomy, physiology, specialization, size, behavior, and so on. Each has found a special place, a special role, a special way in which to thrive. Think about it... animals are everywhere: under rocks, on tree trunks, hidden in decomposing carcasses, clustered amid vegetation, teeming in rivers, lakes, and oceans, swarming in the air, abundant in their secret underground labyrinths. They even persist in deserts and polar barrens. There are thousands in each handful of humus, in each cup of pond water. Swamp muck is rich with them; estuaries are flooded with their heirs; the rain forests are showered with their uniqueness. The earth literally pulsates with their activity. Whether you prefer to use a microscope or a telescope, you will find them everywhere.

How Many Individuals?

Perhaps as important as the number of different species is the number of individual organisms lurking about. Just how many robins are there? How many houseflies, wildebeest, cockroaches, bees, or sidewinder rattlesnakes. How many fleas, lice, or mites? How many soldiers march within the vast legions of army ants? How many sharks police the oceans? How many termite workers toil in their subterranean bivouacs? What is the number of earthworms plowing the soil or seahorses riding the waves? The number of roundworms alone is said to be in the millions in each square meter of soil. A single gulp of planktonic brew by a megamouth shark may engulf billions upon billions of individual, microscopic animals.

So what is the total number? Just how many individuals are alive at any one time? Plainly, it's not possible to know. There are too many to count. They move too fast, are too secretive, hatch too quickly, and die too rapidly: a census is clearly out of the question. Yet we can imagine in our mind's eye what that massive pyramid of animals is like. And from that mental image, we realize that their number must be staggering, their biomass enormous, and their impact immeasurable!

Understanding Diversity

While we cannot be sure of how many different animals exist nor of the total number of animals on earth, we can agree that the sheer volume is mind-numbing. But what do the numbers have to do with biodiversity? To answer this question we must first get a handle on the meaning of the term diversity. In its simplest context, biodiversity refers to the kinds of organisms living in a given area. That area may be as small as a rotting log, a pond, or the canopy of a single tree. On the other hand, the area may encompass ecosystems as vast as deserts, forests, prairies, oceans, or the entire biosphere. Regardless of the area, *diversity is the kinds of different organisms living there.*

Figure 1.1. Each species is adapted to a specific niche where it secures the needed resources for life. For example, a moray eel (above) lies in wait of prey along the ocean bottom while a sea otter (bottom) prefers to bring its food to the surface.

In general, we think of diversity as the way in which species vary from each other: an alligator differs from a woodpecker, which differs from a butterfly, which, in turn, differs from a tapeworm. This is **interspecific diversity**...each species is different from every other kind. There are even different types of woodpeckers and different types of butterflies. But this, too, is interspecific diversity because each kind of woodpecker still belongs to a separate species: downy, red-headed, red-bellied, three-toed, and yellow-bellied sapsuckers...all woodpeckers, but each belonging to a different species. And there are literally thousands of different species of butterflies.

Yet diversity can be expressed in another way. Each member of a given species, especially in higher organisms, also differs from all other members of the same species. Each red-headed woodpecker is different from all other red-headed woodpeckers. Every monarch butterfly varies from all other monarchs. Just think of how we humans are each unique. Unless you are an identical twin, you are different from every other person alive. This is the second level of diversity, **intraspecific diversity**; variety within a given species. Each member of a sexually reproducing species differs genetically from all other members of that same species.

Figure 1.2. All members of a sexually reproducing species differ genetically from each other as seen here in an assemblage of monarch butterflies.

When biologists use the term diversity they are usually referring to the different species that may occur in an area, not to individual raccoons or paper wasps. In other words, they mean *interspecific* rather than intraspecific diversity. And that is the case in this chapter and, indeed, throughout this book. We are, in most cases, exploring how animal species differ from each other in their needs, structure, function, and behavior.

So, What Is a Species?

Both interspecific and intraspecific dynamics are interwoven to give us the tremendous variety of animals on planet earth. One depends upon the other. But before we explore just how this occurs, let's first see what is meant by the term **species**. What criteria are used to differentiate one kind of animal from another? How can we tell them apart? We know from personal experience that sometimes it's quite easy. For example, discriminating between an elephant and a fish is simple, and distinguishing a robin from a hawk is not so difficult either. On the other hand, separating some species from others can be nearly impossible. Beetles, belonging to the insect taxon Coleoptera, are perhaps the most numerous kind of animal on earth. There are literally hundreds of thousands (maybe millions) of different kinds of beetles. How can anyone sort out one beetle type from another? Can anyone tell slugs apart? Can you distinguish between the several kinds of "sea gulls" that inhabit our coasts or lake front?

The task can be daunting. Yet it's highly meaningful to know what a species is and how to tell one from another. Otherwise, we cannot logically discuss animal behavior, habitats, breeding biology, and so on. So we must search for a definition, but finding a good one is difficult. However, one definition that seems to satisfy many biologists is that *a species is a group of organisms sharing a common gene pool*. In other words, if two animals mate and yield fertile offspring, who, in turn, can also successfully reproduce, then both are, according to this **biological-species concept**, members of the same species. If they can't, or don't, they belong to separate ones. So if *animals are reproductively incompatible* (i.e., cannot produce fertile offspring), *they exist in nature as entirely different species*.

Stated in another way, this definition declares a species to be a population of organisms reproductively isolated from all other populations. They are separated in such a way that they cannot or do not reproduce; they are reproductively isolated. And the actual isolating mechanisms can be wide ranging. They may consist of **geographical separation** in which two populations never come into contact with each other. Or isolation may be **behavioral** (i.e., courtship rituals are incompatible), **physiological** (i.e., the egg and sperm cannot mix), or even **anatomical** (i.e., reproductive structures are mismatched). In any event, the result is always the same: mating cannot or does not yield viable offspring. Thus, their genes do not intermingle. The available genes within each population (i.e., the **gene pool**) remain intact.

As reliable as this definition might be, it's not without one limitation. It works quite well for sexually reproducing organisms, those producing eggs and sperm and, thus, usually requiring a male and a female animal. But the definition does not apply to asexually reproducing animals. An asexual animal doesn't require a mate, it doesn't exchange genes with another animal, and therefore the question of reproductive compatibility is pointless. For them...and there are large numbers of asexual animals...a definition of species relies more upon their phenotypes. In other words a definition of species for asexual animals is *a group of similar organisms that are recognizably different from all other groups*.

> **FOR THE MAJORITY OF ANIMALS, A SPECIES IS DEFINED AS "A GROUP OF SIMILAR ORGANISMS REPRODUCTIVELY ISOLATED FROM ALL OTHER GROUPS".**

But there are other, more important, questions for us to consider. For example, if there are indeed 40 or so million different species of animals on earth...why so many? And how did they get to be so numerous? To address these questions, we only have to look to the daily life of individual animals. When we examine them carefully, we discover that each one shares two things with all others. Regardless of the animal in question, it is preoccupied with two simple strategies: *staying alive* and *reproducing*.

Figure 1.3. Animals are consumed by two preoccupations: self-preservation (staying alive) and species perpetuation (reproducing).

Self-Preservation

Animals do whatever they can to stay alive. They struggle daily to do so. In fact, a large portion of their time and energy is spent doing just that. They must eat to survive and at the same time be sure that they do not become someone else's dinner. They need to hunt without becoming the hunted. They use what sensory structures they possess to become aware of their surroundings so that food is found and enemies avoided. They scurry about finding food and shelter and all the while staying out of harms way. This is all part of staying alive.

However, self-preservation goes beyond just food and safety. Staying alive also involves maintaining inner stability. **Homeostasis** is the balancing of internal processes so that an animal functions within a narrow range of efficiency. Such things as proper temperature,

ion and fluid balance, pH level, and scores of other metabolic processes must be continually monitored and adjusted. These homeostatic mechanisms, too, demand their share of an animal's energy budget.

There's no question that an animal spends much time, effort, and talent on just staying alive, on surviving. But, it is also true that, at times, just as much of its energy, perhaps even more, is expended on reproduction—on **self-perpetuation.** In point of fact, we can argue that animals make such a dedicated effort to stay alive just *so* they can reproduce. Reproduction is that strong of a drive. Does that mean that the very "goal" of life then is procreation? That question is one of the most important in biology. Let's see if we can find an answer.

Multiplying

Zoologists realize that animals have a compelling urge to reproduce and can do so with alarming efficiency. In fact, animals, at times, can increase far beyond our imagination. Unchecked, their numbers can skyrocket to unbelievable proportions. Examples can be drawn from a variety of sources: insects, such as grasshoppers, locusts, ants, and flies can produce huge, swarming populations in short order; squid do likewise; mice and rats, too. Flamingos take flight in immense pink clouds numbering well over a million individuals. Passenger pigeons traveled in such huge flocks that they eclipsed the sun. Even penguins occur in enormous breeding colonies. A single female human roundworm (*Ascaris*) located in your small intestine will carry in excess of 27 million eggs in her uterus at any given time. She releases, on average, about 200,000 of these each day, year after year. That's something like 73 million eggs per year. However, a single oyster may hold the record; it can produce over 500 million eggs in a single year!

However, one doesn't need to be a biologist to realize that animals have an unquenchable desire to reproduce. We have all heard of the suffocating plagues of locusts or the swarming, marching columns of army ants. We have all suffered the mass emergence of gnats, mosquitoes, bees, or wasps. We have, indeed, all witnessed the result of this ravenous need of animals to propagate.

Truly, animals are great reproducers and they spend an inordinate amount of energy and time doing just that. In fact, occasionally the whole procreative process appears chaotic, confusing, and even counterproductive. For instance, some animals regularly reproduce at great personal risk. Consider the many males who stake their very lives on the chance to court and mate: scorpions, tarantulas, black widows, and praying mantids, to name a few. Females of these species are cannibalistic and, following the mating act, will devour the male if given the chance. He may, if not careful, end up as his partner's post-nuptial snack!

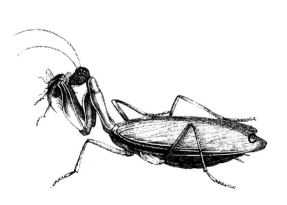

Figure 1.4. Animals often multiply in huge swarms such as the flock of flamingos (**A**) or the cluster of aphids (**B**). Sometimes they reproduce at great personal risk (the female mantid may devour the male shortly after mating (**C**) and the female blacksnake (**D**) is exposed to predators as she guards her eggs).

Even the robin or cardinal in your backyard makes extreme sacrifices to produce next year's yield. They must establish and defend a territory, attract a mate, create pair bonds, build nests, lay and incubate eggs, and then are obliged to feed the relentlessly hungry nestlings. All of this is conducted at a substantial energy cost to the adults. And this scenario is repeated by countless animals during each breeding cycle.

When we think of what it takes for animals to reproduce successfully, it becomes obvious that reproduction is one of the most important undertakings in an animal's life. The process is, at the very least, an energy-demanding one, and at most, a survival risk. If that is so, then we must ask the simple question...why? Why does an individual animal spend so much time, effort, and sacrifice on reproduction? Why would a male face possible death just to multiply? What forces are there that compel them to do so?

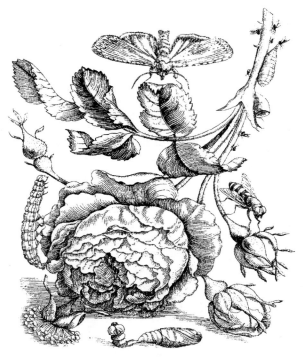

Figure 1.5. All animals are intimately concerned with staying alive and reproducing. Here several different organisms feed and reproduce on a single plant.

The Urge to Reproduce

We know that the urge to reproduce is powerful and universal among animals. If that is so, then we can also assume that there must be some reason, some reward, some payback. But what kind of payback? Perhaps it just feels good; maybe animals reproduce in order to reach a state of procreative bliss. Do they march to the seductive drum of reproduction because of tender feelings? Hardly. It is rather difficult to imagine that organisms such as the blob-like amoebas, the glassy sponges, the gelatinous jellyfish, hungry tapeworms, or blood- thirsty mosquitoes and millions of others reproduce just to reach a state of euphoria. The limited sensory structures of these "lower" forms scarcely support such an idea. And even with more complex species (except, perhaps, humans), it is hard to conceive that "sexual bliss" is the dominating drive. No, we must consider other forces at work here. So let's look at possible reasons a little more closely.

Just why do animals pursue mating with such persistence? If it's not for sexual bliss, then perhaps it's done to satisfy some paternal instinct...the conscious desire to care for offspring. Many animals develop a strong and enduring bond between one or both parents and the offspring. But is that the reason they mate?

Figure 1.6. Many animals develop strong bonds between parent and offspring.

Indeed, close observations of many "higher" animals, such as birds and especially mammals, reveal many patterns of such bonding. However, just because bonds are formed doesn't mean animals do so with any insight or foreknowledge. It doesn't mean they know what they are doing. The ability to form such reasoned, emotional, and reflective decisions would require a highly complex and sophisticated nervous system; one that probably only humans, and maybe a few other mammals, possess. Our current knowledge of the brain tells us that, among animals, only humans can figure out just why they perform certain acts, and only humans know what the probable outcome of these acts will be. Thus, it seems likely that other animals are destined to reproduce without ever knowing why!

Genes...Our Key to Immortality

If that is true, then they must reproduce instinctively, innately. They must do so because of a ready-made, built-in program. And, of course, we all know exactly what that program is. It's our genetic code—our chromosomes, our DNA, our genes, our very inheritance.

So, if animals multiply because of this code, then the most plausible explanation of animal reproduction is that they do so, not out of any desire or understanding, not because of any insight or passion, but rather in response to genetic commands, a set of instructions given to them by their predecessors. And it's only through the unraveling of the DNA code that the universal instructions are rewoven into the fabric of each new generation. It is the legacy of our ancestors, one we most eagerly pass along each time we reproduce. That thread holds the vital information for survival. It is a coded manual for every aspect of an animal's life: its development, daily behavior, body design, defensive activities, food preferences, and, yes, even its reproductive choices. The information interwoven within the DNA molecule is the very blueprint for life itself.

So, again, why do animals reproduce? The answer is clear: to pass along that valuable, timeless, coded DNA message...and they do so without even realizing it. It's the echo of the ancient genetic call. They propagate because the code compels them to.

THE DNA MOLECULE IS A CODED, GENETIC MANUAL FOR DAILY BEHAVIOR, FOR REPRODUCTIVE CHOICES, FOR SURVIVAL ITSELF.

Just Who Reproduces?

This brings us to yet another intriguing question. Who reproduces? Exactly which members of a population get to hand down their genetic message? Is the whole process by chance alone, or is there some order, some kind of design involved?

Who reproduces? Why, survivors do, of course. Those very individuals who possess the capacity to flourish within their own habitats are most likely to pass on survival traits to their offspring, who will, in turn, survive to pass those same traits, and so on. The key point here is not that an animal survives, but that it survives long enough to reproduce, to pass on those survival skills to the next generation. We call this *fitness*.

But is it through luck or by design? Well, there's no question that blind luck can play a part in who survives life's hazards and eventually reproduces. For instance, a nestling bird may fall into a river and drown or a gust of wind may sweep a spider far from its normal habitat before it has had time to meet a proper mate. But, on average, luck has little to do with who reproduces. Instead, propagation is primarily the outcome of an orderly, patterned process: an orchestrated genetic symphony played out on ancient, programmed instruments. *An animal reproduces, not because it's lucky, but because it has the devices and skills to survive long enough to do so.* If these traits provide an advantage for survival, they will eventually become present at a higher frequency in the population. This process is termed *differential reproduction*. Those with less advantageous traits will decrease in the population.

Figure 1.7. Reproduction is no haphazard process but is designed in a way to propel the best genes possible to the next generation. Here a single bull fur seal who has successfully defended his breeding rights guards his large harem.

Even If It Hurts?

We can now return to the question of why some animals reproduce at great sacrifice. Recall that as part of the reproductive scheme, some males may be killed and eaten by the female immediately after they mate. This is true for many spiders, praying mantids, and a variety of other animals. We can envision how powerful the reproductive drive must be if an animal will face near-certain death just for the privilege. And we can further conclude that an animal must, therefore, have no idea of what he is doing. If he did, reproduction in that species would surely come to a screeching halt!

The strange thing is, this is not a rarity. In fact, throughout nature there are many instances of animals rushing headlong toward death in order to reproduce. Numerous species make the most extreme efforts to breed, often traveling hundreds of miles or waiting years to do so, only to die shortly thereafter. From squid to cicadas, from mayflies to salmon, entire populations of animals perish immediately after mating.

A well-known case is the pacific salmon. After spending most of its life at sea, it responds to the reproductive summons by swimming exhaustively upstream until it locates the very spawning grounds of its parents. There, where it hatched years before, a breeding orgy takes place. Following the release of sperm and eggs, all participants, both male and female, die by the thousands. Streams literally reek with the foul stench of their rotting remains.

A less celebrated, yet no less fascinating, example is the 17-year cicada. After almost two decades living underground, the immature stages, called nymphs, emerge in massive numbers and quickly molt into adults. Males then "sing" vigorously and incessantly to attract females. Mating takes place, eggs are laid, and then they all, too, soon die.

What does this tell us? Do you think these animals enter into this contest fully aware of what is about to happen? I hardly think so. So we return to our original question: "Why do animals clamber so to reproduce?" Clearly, the answer is that they simply don't know any better. They do so because of a genetic drive, an instinctive pulse to procreate at all cost. They reproduce because the process is etched within their genetic code. They are just following orders. And once again, who reproduces? It is those who are best suited for their particular nook. Why are they best suited? Because they received the success formula from the genes of their parents who were themselves well suited. The cycle continues through the generations by way of coded, genetic instructions shackled to an overpowering, blind drive to reproduce.

Figure 1.8. Migrating salmon (**A**) gather as a prelude to mass mating after which both males and females die. They must run the gauntlet, including hungry grizzly bears (insert). In **B** above, cicadas emerge from their nymphal case prior to their sessions of mass mating. They, too, all die shortly afterward. A male frog (**C**) grasps a female in a mating embrace called amplexus. This causes her to release an egg mass, which he covers in a cloud of sperm. Of the hundreds of eggs fertilized, only a few will survive long enough to become a reproducing adult.

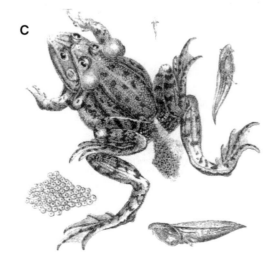

An Inside View 1.1

A Campus Challenge

One April morning as I strolled across campus I found myself in the middle of an embittered battle. About 10 yards in front of me stood a male Canada goose. Nearby was his mate. In front of him was another gander equal in size and, at least to my eyes, equal in appearance. The mated male, upon spying the interloper, immediately charged. The meaning of the attack was inescapable. He was determined to dislodge the intruder from his domain...and that included, above all else, his female. The foray was made *en grande tenue*—in full battle regalia—goose regalia, that is. With his head lowered, beak partially agape, wings spread, and feathers upright, he attacked. All the while he uttered his best rendition of a calvary charge...delivered, of course, goose style. The best I can describe this outcry was as a bellowing, drawn out, raucous, wonk...w-o-n-k, getting louder and more frenzied as he faced the trespasser.

But the intruder was not to be outdone. He also charged—his feathers erect, his head lowered, his wings also held aloft. He bellowed the same battle cry as he lunged forward. They stood face-to-face, not three feet apart, each proclaiming his own grievance. This was obviously something never encountered by our goose. An interloper who did not retreat. But he was determined. He charged again. His rival did likewise. They again rushed each other with equal intensity. They scolded each other with equal fervor. They signaled, postured, strutted, lunged with lusty vigor, each matching the other. The air was filled with their energy. This was true warfare, goose style. A duel in the shadow of ivied halls.

For several minutes the battle raged. Charge after charge was met with amazingly similar counter-charges. Each posture, each move, each eruption was matched perfectly by the challenger. Try as he might, our defender couldn't dislodge the invader; he couldn't even induce a ruffled feather. After an exhausting series of assaults and counter assaults, each finally gave up. Our male turned to his mate, gave a few recognition waggles (geese pairs shake their heads at each other to show their white cheek flags as a sign of recognition) and the two sauntered off. To me, as they disappeared over the hill, our gander's strut seemed less confident, his shoulders were slightly drooped, his "honk" less menacing, his eyes less spirited. In his mind he was not the hero he wanted to be. He had failed in this age-old contest of out-bluffing one's foe.

I turned to view the gate-crasher, but he, too, had mysteriously disappeared. And no wonder. He didn't exist. He wasn't real. He was but a mirror image. You see, this all took place at the entrance to the science building, in front of the glass doors. The male was charging his own reflection. Each act was met with an equal act—his own. But he didn't realize it. He never figured out that he was in mortal combat with himself. (KH)

IT'S A FACT

The majority of animals behave instinctively, not rationally. They court, reproduce, and defend territories but without a clue of what they are doing or why they are doing it. It's one of the strange contradictions of animals: they flourish without knowing how or why.

Maybe the goal of life really is to reproduce, after all. It surely seems that way. Each animal is propelled blindly, unknowingly toward self-perpetuation. They may even do so at great personal sacrifice and risk. We can easily imagine the billions of animals, generation after generation, rushing toward the reproductive prize, all compelled instinctively, genetically to procreate.

But wait, there's a catch here, a hidden agenda, one unrecognized by the individual animal. While the individual organism strives to reproduce, the actual advantage, as we have seen, is not for the individual at all but rather for the entire population. The real objective of reproduction *is not self-perpetuation*, but rather the continuation of the species: *species-perpetuation*. You see, it's not the individual that's important genetically, it's the population. It's the population that really matters. Individuals are just the conduits through which the population's immortal genes pass to the next generation. Stated more succinctly, *the individual reproduces so that the species can endure.*

> ## IT'S A FACT
>
> THE INDIVIDUAL ORGANISM STRIVES TO SURVIVE AND REPRODUCE AND BY DOING SO ASSURES THE CONTINUATION OF THE SPECIES

Figure 1.9. The individual strives to survive and to reproduce so that the species can endure.

The Niche Concept

There's a concept in biology called the ecological niche that's central to our understanding of biodiversity. In fact, trying to comprehend animal diversity without understanding the niche is like studying genetics with no concept of DNA. The concept has two basic ideas. First of all, *niches are unique for each species of animal.* They are occupied by just one species at a time. Each animal is specifically adapted to exist in its own ecological niche and nowhere else. A successful animal possesses all of the attributes needed to survive and prosper there. It's his private domain. He survives there when no one else can because he is better equipped to compete and will do so at the expense of any intruder. We call this the **law of competitive exclusion**. It's how niches are occupied.

So it sounds then like a niche is a place. Well, it is...but it's much more than just a place. It also includes an animal's essential resources, its physiological and "psychological" requirements. It includes its interactions with other organisms, what type of food it needs, and how it acquires that food. The niche truly includes a complete inventory of an animals' living and nonliving environment. A niche is all inclusive. That's the second principle of the ecological niche—*it includes everything that an animal needs.*

Thus, a popular definition among ecologists is that a niche is the *role an animal plays within its environment.* This definition includes the idea of space, place, behavior, resources, and so on. And, though it's a little vague, it is, nonetheless, very useful in unraveling the mystery of animal diversification.

THE **LAW OF COMPETITIVE EXCLUSION** STATES THAT EACH SPECIES OCCUPIES ITS OWN, UNIQUE ECOLOGICAL NICHE

Niche Adaptation

However, as nice as this definition might be, it's of far greater importance for us to determine, not just what a niche is, but rather how a species becomes customized for its special niche. How do animals become "fit" and why do some become better fit than others? How does a vampire bat become well suited to survive within its niche? How does it acquire the necessary devices to thrive there? The same questions can be asked of snails, and of great horned owls, leeches, tiger salamanders, jumping spiders, armadillos, clams, and human head lice. How does each become so well suited to its respective niche? That's really a prime purpose of this chapter...to discover just how animals become adapted to their particular ecological niche.

Keep in mind that an animal's environment is not static. It, instead, is constantly changing. So, for a species to remain tailored to its niche it too must be able to change, to adjust. We call that give-and-take between a species and its environment **adaptation,** developing a degree of harmony within a changing niche. But adaptation is not like a light switch; it cannot be turned on or off at a whim. Instead, it is a natural process, often slow, always opportunistic. So, how does it occur? Well, our best calculation so far is that adaptation is the result of three underlying agents: (1) reproduction; (2) competition; and (3) genetic variations.

(1) **Reproduction:** As we have already seen, each animal has received the necessary survival information from its parents through the genetic act of reproduction. We have also seen that reproduction is no haphazard process but is designed in a way to shower the next generation with a constant sprinkling of the best genes available. Thus, through that genetic thread of DNA each succeeding generation receives the **genotypes** necessary to cope with a changing environment. But it doesn't end here. Instead, to succeed, each individual must also compete with its neighbors for survival rights.

(2) Competition: To survive, thrive, and reproduce within one's niche, an organism needs to secure a number of essential resources. Among these are food, space, shelter, moisture, and reproductive partners. But resources are not always abundant, and so, to survive, organisms must compete with other members of the same species for their share. Those with the most adaptive attributes, such as size, speed, and alertness, are more successful in this struggle. They can compete more effectively because they are better suited to their ecological niche: they are the ones who stand a better chance to survive; they are the ones most likely to reproduce; they are the ones who will pass on that genetic blueprint to the next generation. And again, what is the nature of that blueprint? It's the very code containing the needed information— some would say, the best information— to survive in one's own niche.

There are, however, limits to what an ecosystem can tolerate. For example, a given area can support only so many organisms...a **carrying capacity** of sorts. Yet animals have an inherent ability to reproduce far beyond what the environment can sustain. Thus, in most cases, there are too many animals competing for too few resources.

Because resources are often scarce, competition for them can become intense. As a result, not all members of a population survive. Some will, others won't. **Differential survival** of this sort is based upon the ability among individuals to compete. Just recall the law of competitive exclusion that declares only one species can occupy a niche at a time. And the result can be stark. Of the millions and millions of eggs produced by an oyster, perhaps only a few, maybe only one or two, will survive to reproduce. The same can be said for most animals: sea anemones, roundworms, ladybugs, geese, alligators, jack rabbits, and so on. Large numbers of eggs or young are produced, but due to the hardships of competition (avoiding accidents, eluding predators, securing food, finding concealment, locating reproductive partners, rearing young, etc.) very few reach adulthood.

Not only does competition serve to control population size but it also allows for the population as a whole to become more vigorous, that is, more fit. It does this by giving some individuals, those with the best attributes, a survival advantage while at the same time culling out those less suited. Successful animals survive to reproduce, unsuccessful ones don't.

Competition therefore favors the most fit within a population and dispenses with the less fit. It serves as a weeding out process to produce smaller yet hardier (fitter) populations. So, even though animal numbers can explode at tremendous rates, competition for rare or limited resources helps to keep populations in check and by doing so also produces individuals even better suited to their niche.

These two factors, differential reproduction and resource competition, play an integral role in niche adaptation. But there's another component that's equally significant. That third factor is **genetic variation.**

(3) Genes at Work: We know that animals differ in their genetic makeup. In fact, each sexually reproducing organism, except for identical twins, possesses a different set of genes from every other member of its species. These personalized sets were received from parents through the biological processes of **meiosis** and **fertilization** (see page 61). During the meiotic formation of gametes (eggs and sperm), genes are rearranged into different combinations much like shuffling a deck of cards. In this way, every single mating produces genetically different offspring. This reshuffling of genes is termed **genetic recombination** and is a primary source of population variation. Each sexually reproducing animal differs slightly from all others because it possesses a different deck of genes received from two parents.

Figure 1.10. Competition for food and other limited resources drives adaptive fitness and niche colonization.

Although recombination does result in new arrangements of genes, it does not produce any new genetic material. That, instead, is the work of **mutations**. Unlike recombination, which can only reorganize existing genes, mutations actually alter the nature of the DNA molecule. Through heritable mutations, the nucleotide sequence of DNA is modified and that, in turn, causes genotypic changes in individual organisms. These genetic changes themselves may result in small, incremental evolutionary steps within a population— a process known as **microevolution**.

Mutations, by their nature, are spontaneous and incessant. In fact, their frequency can be so high that, at times, they spatter the genetic landscape in a constant barrage. And since they provide the only new source of genetic information, they are vital to the process of niche invasion. They are, in reality, the very raw material upon which populations depend for adaptations— the foundation upon which the pyramid of diversity rests. There is little doubt that mutations are essential to the fitness scheme.

There are, however, a couple of leaks in the mutation bucket. As essential as they are, mutations are not without their shortcomings. First of all, most are **lethal**. The vast majority disrupt life processes to the point that an organism just cannot survive. Since they result in death, these **maladaptive mutations,** of course, quickly disappear from that population. Thus, the great majority of mutations don't contribute one whit to diversity.

The other problem with mutations is that they are **nondirectional**. Just because a niche requires certain adaptations doesn't mean that a mutation will occur to lead a population in that direction. Mutations cannot anticipate the future; they can't know what sort of adaptation might be needed. Nor do they occur in a logical sequence. Instead, they are the spontaneous, random, and rather unpredictable dinner guests at the genetic banquet.

Fortunately, however, some mutations occur that are **nonlethal**. These may prove to be immediately beneficial and may, therefore, thrust an individual to the competitive forefront within a new or changing niche. Indeed, these favorable mutations provide the building blocks upon which adaptation and niche invasion depends. A beneficial mutation within a population may improve the ability of some individuals to compete and thus, step-by-step, the population becomes increasingly better adapted to its niche. On the other hand, nonlethal mutations may be neutral and, as such, just linger within the population, doing no good, but also doing no harm. These **lingering mutations** may provide a genetic reservoir by which members of a population can later take advantage of a new niche and its myriad requirements.

IT'S A FACT

MUTATIONS ARE SPONTANEOUS, PERMANENT CHANGES IN THE GENETIC CODE. THESE ALTERATIONS IN THE BASE SEQUENCES OF THE DNA MOLECULE PROVIDE THE RAW MATERIAL, THE FUEL UPON WHICH THE FIRE OF FITNESS BURNS.

The Survival Formula

1. RECEIVE GENETIC INSTRUCTIONS FROM YOUR PARENTS.

2. COMPETE FOR NECESSARY RESOURCES.

3. THE MOST FIT SURVIVE AND REPRODUCE.

4. POPULATIONS ACCUMULATE ADAPTIVE MUTATIONS.

5. SURVIVAL GENES ARE TRANSMITTED TO THE NEXT GENERATION.

The Diversity Connection

What does all of this have to do with animal diversity? Well, everything. Diversity is nothing more than the net result of animals striving to fill ecological niches. The more niches, the greater the diversity. If there are indeed 40 million niches, then there can be 40 million different species of animals, each striving to occupy one for itself.

New niches emerge continually through a variety of mechanisms: extinctions, geological forces, climactic instability, diseases, starvation, human interference, and so on. When niches become available, animals seek to occupy them. *Nature abhors a void* is an old expression that communicates this effort. But how does that happen? The best explanation is that it occurs through a process known as **preadaptation**. Animals who already possess the skills, techniques, and structures needed in the new niche will be the ones to colonize and survive there. Those lacking the necessary skills cannot.

The next logical question then is "Just how does an animal become preadapted?" Here, again the answer is quite clear...through mutations. This is how it works. Nonlethal mutations occur constantly within populations. Some of these mutations are not immediately beneficial in the present niche but, since they are nonlethal, they may linger within the population at large. If these lingering (neutral) mutations provide an organism the slightest advantage in a new situation then, it is already adapted to survive there. It is ready to colonize a new niche even before the niche materializes. And this invasion can happen rather swiftly. Why? For the simple reason that the organism is, to some degree, already preadapted to survive there.

However, mutations, as we have also stated, are random and nondirectional. So, in this respect, mutations, unlike reproduction, are very much a product of chance. Does this mean that adaptation and niche fitness relies on genetic randomness? To a great extent, yes. But in case you think trial and error cannot provide the impetus for niche invasion, you need only reflect upon the enormous numbers of individuals we are dealing with, the relatively high rate of mutations, and the millions of years over which these events occur.

But chance alone is not the answer either. In fact, as we shall see, the process, on the whole, is opportunistic. Certain individuals have a distinct advantage over others due to their fitness, due to their ability to compete. Their survival is not due just to their blind luck.

> **SUMMARY:**
>
> EVEN THOUGH MUTATIONS ARE RANDOM AND NONDIRECTIONAL, THEY ARE THE ONLY SOURCE FOR NEW GENETIC INFORMATION. THEY, THEREFORE, PROVIDE THE RAW MATERIAL FOR NATURAL SELECTION TO OCCUR. IN OTHER WORDS, MUTATIONS LEAD TO PREADAPTATION.

Figure 1.11. Certain individuals have an advantage over others in the competition for limited resources. They tend to have a higher rate of survival. Their genes will then be passed to the next generation.

"The Noiseless Foot of Time"

Where Did They All Come From?

Can you just imagine, 40 million different kinds of animals, maybe more? Where did they all come from and how did they get here? Why these particular 40 million and not some other combination? What makes one type prevail and another fail? How long did it take to reach this level of diversity? Where do we go from here? It seems that our questions are endless. So where can we look for the answers? Well, one good place to go when you want answers is to a depository of information... a library. But is there any such reservoir of animal records? Fortunately, there is. The history of some animals has been well recorded, preserved in a stone library. We know it as the fossil record. Throughout time, because of a combination of factors (body design, habitat, internal or external skeletons, etc.), certain organisms have been prone to fossilization. Their past is documented in this rocky record. They have just been waiting through the eons for us to decipher their cryptic messages. As we pick, scrape, brush, and reconstruct the past, we are slowly reading that ancient book. And what we have uncovered is basic to our understanding of animal diversity. The record tells us what was present and what was absent, and if read accurately, it gives us a glimpse into the very lives of those ancient animals and, more importantly, it gives us insight into the very process of species formation. Unlike the Shakespearean quote above, the silent record is really not so noiseless after all; on the contrary, theirs is a small monstrous voice!

Speciation

New species originate along two pathways, and like a junction in a road, these two paths may lead in totally different directions. We find that one route, under proper conditions, can produce a profusion of new species, while the opposite path, produces but one. Similar to a tree, the first path branches out into new niches in a radiating pattern. Here, one ancestral parent species is able to produce a plethora of new genotypes to invade a variety of niches. Each branch of this ancestral tree may ultimately result in the formation of a new species. In other words, many new forms can spring from a common stem, like twigs on the same bush. This form of speciation is called **adaptive radiation** and, much like smoke emanating from a fire, it permeates its surrounding. At times, this radiation may produce only a handful of new species then it stops. But at other times, the process may literally explode into a veritable landslide of new forms. Such explosive radiations are common, for instance, on remote islands as well as in the aftermath of drastic environmental or geologic changes.

The second path to speciation occurs when one population responds to environmental changes and, over time, is transformed into a single new species. Most often in this case of **linear descent**, the original species disappears from the scene leaving only a lone offshoot. Note that, unlike adaptive radiation, linear descent produces no net increase in the number of animals, just the appearance of a single new species.

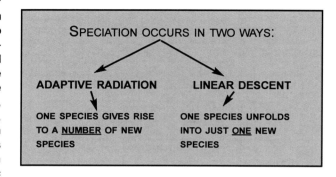

In both cases, however, speciation results from the same force: accumulated mutations, which, in turn, leads to the appearance of new morphological types or **phenotypes**. These phenotypes may, in time, lead to more pronounced changes—a process termed **macroevolution**. These cumulative changes may produce such distinctive phenotypes that eventually an entirely new biological species appears. But the process can be agonizingly slow. In fact, it may take millions of years of accumulated adaptations for a single new species to emerge in this way. Nonetheless, scientists believe that this is exactly how many, if not most, of our animals originated— through a slow-paced mechanism called **gradualism**.

However, gradualism is not the only way that new species may arise. In fact, recent interpretation of the fossil record supports another possibility, **punctuated equilibrium**. There are apparently circumstances in which macroevolution occurs rapidly, in spurts, resulting in an avalanche of new species. These radiations often follow massive extinction episodes when a profusion of new niches become available. Certain preadapted members of the population can quickly spread into these newly formed niches. These radiating spurts are then followed by relatively long periods of evolutionary inactivity or stasis, followed by yet another spurt, and so on. An unsteady, punctuated equilibrium may then be achieved.

Natural Selection

Overview

The appearance of new species occurs through two mechanisms: the usual way seems to be through a slow, lingering, step-by-step process called **gradualism**. However, it also seems that new species may literally erupt in the blink of the geologic eye. These spurts are followed by periods of relative evolutionary calm; a **punctuated equilibrium** dominates. But the actual process is always through *natural selection.*

When a new niche appears, you can bet that an animal will soon follow to occupy it. New niches surface constantly and animals respond by relentlessly filling them. We have seen that it takes a number of steps for this to occur: reproduction plays a big part; competition is essential; and mutations are indispensable. But there is one more step involved, **natural selection**. This is a simple but absolutely crucial step. However, before we discuss it, let's first review what we have learned so far.

(1) We know that mutations are the source of genetic change. They occur spontaneously, randomly, and continually. Although they have no predictive capability, they, nonetheless, provide the very means of adapting to changing niches. We have further learned that there is no known way to produce a particular mutation. No organism, for example, can command a given mutation to produce a desired outcome. No animal can simply push a button and cause a needed genetic change.

Thus, speciation may be slow or rapid, subtle or explosive. Which one predominates in a given situation depends on both internal (mutation rates) and external (extinction rates) forces. When niches become available on a massive scale due, for example, to a catastrophic climactic event, then speciation appears to speed up. Niches are then colonized rather quickly. On the other hand, under more stable environmental conditions, the process is much more gradual and relies on slow, accumulating microevolution.

(2) We have also learned that for mutations to have a genetic influence they must occur within the reproductive tissue, the testes or ovaries. Regardless of what happens to a structure (such as an appendage, antenna, liver, or eyeball) during an individual's lifetime, that change cannot be transmitted to the next generation. Only if a mutation occurred within the reproductive (gonadal) tissue will changes be inherited. Thus, the one-time popular concept of Lamarkian "use and disuse" has been discredited as a player in evolution's creative stage.

(3) We have learned that the only route open to invade a new niche is through preadaptation. The inheritable advantage must be present in the population before invasion of the new niche can occur.

(4) We know that competition for limited resources is a fundamental step in evolution. An animal must survive long enough to reproduce, long enough to pass on the survival traits or mutated genes. Otherwise, adaptation cannot and does not occur. An animal does this by out-competing its neighbors for essential resources. If successful, it will reproduce and pass the successful features to its offspring.

SUMMARY

1. MUTATIONS ARE THE SOURCE OF ALL GENETIC CHANGE
2. MUTATIONS MUST OCCUR IN REPRODUCTIVE TISSUE
3. PREADAPTATION IS THE KEY TO THE EVOLUTIONARY PROCESS
4. ANIMALS COMPETE FOR LIMITED RESOURCES WHICH LEADS TO NATURAL SELECTION

So again, we come to the all-important question, "How does an animal get the necessary features to invade a new ecological niche?" Well, the only possibility is that it must rely upon genetic transformation, recombinations, and random mutations that have already occurred. In other words, to invade a niche an animal must already possess the needed attributes. It must be preadapted for survival in that niche. Then as more preadapted individuals flood into these niches, or as more mutations occur, the population ultimately acquires the necessary skills, structures, and abilities for that particular niche. They accumulate a competitive repertoire of sorts. As a consequence, they are the ones that have a higher probability of surviving and reproducing. Their traits are selected, through differential reproduction, to be passed on. This is the final step in our diversity equation: selection of fit traits, or in other words, **natural selection.** Fit individuals survive (are selected) and pass on their skills to the next generation. The unfit don't.

Natural selection, then, is the ultimate referee in evolution's genetic contest. It serves as a "judge" by which the fit pass muster and the unfit don't. But natural selection is a two-way street. It rewards the fit at the expense of the unfit. In fact, many zoologists refer to natural selection as **nonrandom elimination.** By this they are observing that not only does natural selection choose who survives, but it also chooses who doesn't.

It's important to note here that in natural selection there is a premium placed on those characteristics that lead to fitness, and those sets of characteristics are different for each ecological niche. Note also that, while mutations occur randomly within individual organisms, natural selection itself is *nonrandom.* Chance may be instrumental in the appearance of mutations, but natural selection works through anti-chance; that is, it favors the preadapted individual and thus is not random at all. If natural selection occurred through luck or chance then each individual in a population would have an equal opportunity to pass on its genes. However, we know all too well that is not the case. On the contrary, it's the fit, not the lucky, that tend to reproduce. Natural selection chooses the best set of traits for a given niche; it fine-tunes every population to fit better and better within its particular niche.

One final and important point about evolution. *It's not the individual but the population that evolves.* Individuals sow mutations but the population reaps the evolutionary harvest. From the vast variations within a population, natural selection chooses only those with the means to survive. Whether slow or fast, plodding or explosive, natural selection does, indeed, create new species of animals. And so, through the combined processes of mutations, genetic recombination, preadaptation, competition, and natural selection, biodiversity continues its universal unfolding.

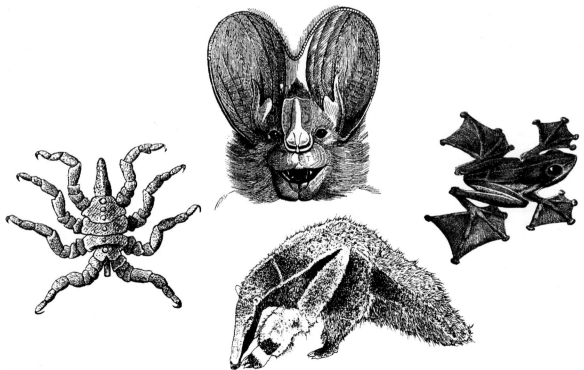

Figure 1.12. At times, natural selection seems to prefer the bizarre or exotic. However, while some features may seem strange to us, keep in mind that whatever offers survival advantage in a given niche will most likely be selected.

The Specialization Paradox

As we have learned, the more specialized a group becomes the better it fits into its niche. So there's an advantage for a species to become highly specialized, right? Well, maybe not. Ironically, there's a potential, but ever so real, danger that if a population goes too far in its race toward specialization it may be drawn into a trap, the specialization paradox. Here's what might happen. On the one hand, the pressure is to become more and more specialized because survival rates increase for those groups that do so. On the other hand, if the specialization is along very narrow lines the group may lose its ability to withstand even a minor environmental change. Indeed, it can easily become entangled in a web of no return. If that happens extinction is sure to result.

Parasites often face such a dilemma. They tend to become restricted to a single host species and will only be found on that one host. If that happens, the parasite's fate becomes irreversibly linked to that of its host. If the host numbers decline, so do the numbers of the parasite. If the host vanishes altogether, so does the parasite. Overspecialization is a clear problem for many parasites as it is for a multitude of animals.

A comparison between two similar species, the Giant panda *(Ailuropoda melanoleuca)* and the American black bear *(Ursus americanus)* further illustrates this dilemma. The panda feeds almost exclusively on bamboo, which quickly passes through its digestive system so that much of the nutrients are lost. Thus, the panda must eat large amounts each day. The problem is that bamboo supplies are becoming scarce as people replace them with agricultural crops, and due to natural bamboo cycles. As a result, the panda, because it's so specialized, faces a bleak future (fewer than 1,000 exist in the wild).

On the other hand, the black bear is much less selective in its feeding habits. As an omnivore it will eat virtually anything from insects, to vegetation, to small game, to human garbage. Thus, it is a much more adaptable animal and the population of black bears is actually on the rise, even in the midst of tremendous human encroachment.

So the trick is to be specialized, but specialized to a broad range of conditions rather than to just one or two. For instance, a parasite might specialize on birds in general rather than on one species of bird. However, there's danger lurking here, too. For if a population becomes overly generalized, it lessens its ability to compete... that's the paradox. It must be specialized but not overspecialized; it must be generalized but not overgeneralized. If the species cannot solve that puzzle, it is destined to travel along that one-way path of no return.

Figure 1.13. The panda and the black bear show the extremes of the specialization paradox. Pandas are specialized to feed only on bamboo whereas black bears accept a wide range of food. As a result, the panda populations are declining as bamboo supplies dwindle but the black bear continues to thrive even in the face of considerable human intrusion.

Solving the Specialization Puzzle

When we consider the long-term survivors, the "living fossils," such as the centipede, the dragonfly, scorpions, alligators, opossums, and sharks, we find they have all been able to solve the mystery of the specialization paradox. They have balanced the equation well, being specialists and generalists at the same time. Through the eons they have withstood the countless pitfalls, hazards, and catastrophes of the ecological landscape. They have survived because of their genetic elasticity. They have persevered and have left the overspecialists and the overgeneralists behind, encased in their fossilized tombs.

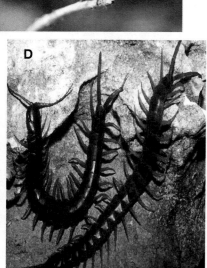

Figure 1.14. Many animals have solved the specialization paradox by balancing their degree of generalization against their level of specialization. Crocodiles (**A**), dragonflies (**B**), sharks (**C**), centipedes (**D**), opossum (**E**), and tarantulas (**F**) all have survived the millennia by solving this paradox.

An Inside View 1.2

Form Follows Function

Let's see how this process might work through a simple, hypothetical example. Suppose that in a large population of mammals some individuals undergo a specific, heritable mutation. The result of this genetic alteration is that these animals inherit a slightly thicker coat of fur. Now suppose that the environment, too, changes slightly: it becomes just a little cooler. It's not unreasonable to assume that those individuals with thicker fur will now have a competitive advantage. As a result, they have a greater probability of surviving and reproducing than their less furry comrades. They will then pass on their "fur" genes to their young, who will do likewise. In time, assuming the climate doesn't change again, those "fur" genes will be broadcast throughout the entire population. Eventually, all members will become furrier.

Now let's suppose that a new player arrives on the scene, a vicious predator. A thick coat of fur might help to some degree but not nearly as much as, say, a coat of armor or even spines. If a mutation had already occurred in the population in which hair was slightly modified into a scale or a spinelike covering, then those animals would now have the advantage and they would more likely survive to reproduce.

We could imagine similar scenarios for an endless number of features: say, an elongated tail, or maybe larger canine teeth, or perhaps larger eyes, louder vocalization, longer legs, improved eyesight, maybe an enlarged cerebral cortex... and on and on, until at some point the mammal we began with no longer exists, but is now something quite different, *a new species*.

But regardless of the outcome, remember that the forces are always the same: heritable mutations, which lead to enhanced survival and reproduction.

Note that the process followed this course:

(1) A specific, inheritable mutation occurred

(2) The environment changed, or a new predator arrived

(3) Those with the mutations had an advantage; they were preadapted to the new situations.

(4) They survived and reproduced at a higher rate than others

(5) The "furrier," "spinier," or "scalier" phenotype was selected and eventually predominated.

Also note that:

✓ The mutation was random; it did not anticipate the change in climate or the arrival of a new predator.

✓ The mutation must occur in reproductive tissue to become part of the genetic program.

✓ Those individuals without the mutation were at a competitive disadvantage.

✓ Natural selection works by filtering out the fit from the unfit within a particular niche.

Individuals sow mutations, but it's the population that reaps the evolutionary harvest

How Do We Know?

All of this sounds fascinating, a rather logical explanation of the animal diversity question. But do we really know that biodiversity originates in the way we have described? Are mutations, genetic recombination, competition, and natural selection the true agents of diversity? The answer is that we are reasonably sure they are, but not absolutely sure. We can surmise and deduce, we can use our educated rationale, and we can make scientific assumptions, but we cannot know absolutely. No one was there during those millions of years to witness and record the emergence of different animals. No one can predict just how, when, or if certain mutations might occur. No one can even measure the degree of fitness required for any particular niche. So, while evidence does exist to support our model, we cannot know with certainty.

What evidence? Well, there is actually an abundance of proof in favor of this process. But because the process takes so long, the evidence is, for the most part, indirect. Indirect...but powerful. So powerful, in fact, that few serious zoologists doubt its veracity. This evidence comes to us through a number of separate, yet interconnected routes: the geological record, fossils, the geographical distribution of animals, comparative anatomy, comparative embryology, artificial selection, and direct observation. Let's examine several of these.

The Fossil Record

We know from literally thousands of discovered and radioactive dated fossils that many life-forms no longer exist. We know that certain animals are prominent or even abundant in fossil form but have completely disappeared as living entities. Just think of some of the more famous extinctions: dinosaurs, archaeopteryx, trilobites, passenger pigeons, hairy mammoths, giant armadillos, saber-toothed tigers, giant sloths, the Irish elk, and an endless stream of others. All are gone. Fossils alone bear their ancient footprints.

In contrast, representatives of the great majority of living animals (those on earth right now) are absent from the fossil record. They apparently did not exist in those prehistoric days. They have evolved since then; they are newcomers to the world's fauna. So we can reasonably conclude that biodiversity is vastly different today from what it was in ancient times. The animal spectrum has, indeed, changed. About that there is little doubt.

Not only are there different kinds of animals alive now, but they occur in a much greater variety than ever before. By comparing living animal populations to the fossil record, it is apparent that diversity is considerably greater now than at any time in the past. New animals continue to appear in radiating fashion. Like lava flowing from a volcano, natural selection has produced an ever-expanding stream of animals, resulting in the 40 million or so species thought to exist today. By listening to their petrified echos, we can learn much of history that fossils are longing to tell.

Trilobite
(extinct)

Sabertooth Cat
(extinct)

Figure 1.15. The fossil record has shown that both the number and kinds of animals have changed drastically throughout time. Most that once existed are gone; others that didn't exist in the past now flourish.

Comparative Anatomy

The essence of diversity is that every species differs from every other one. And while that is correct, the opposite is also true: animals are also very much alike. For example, they all use DNA as the purveyor of genetic information; they all share the same 20 amino acids to construct the magical protein molecules; they all contain cells with a similar molecular makeup; they all require the same biochemical materials (e.g., calcium, phosphate, sodium, potassium). All animals are also heterotrophic, in that they must obtain their energy from external, rather than internal, sources.

Furthermore, many animals have similar body designs. The forelimbs of amphibians, reptiles, birds, and mammals show structural similarities. Even limbs that appear much different from each other are based on a common pattern. For example, legs, wings, and flippers all have a basic anatomical construction.

Many other body parts show correlations: location and action of various muscles; uniformity within digestive systems; similarities in nervous systems and photoreceptors (eyes); even body coverings show interconnections...scales, feathers, and fur have common roots, for example. The list of structural similarities is a very, very long one.

These morphological associations, in fact, form the basis for the important concept of **common ancestry**. When two or more animals show common designs, engineering, or features, we begin to wonder if they have descended from a common source. There is the real likelihood that we are seeing the result of **adaptive radiation.** The closer the resemblance, the closer the possible evolutionary relationship.

Then there's the question of **vestigial** structures. These are features that have lost their function, or their function is drastically reduced, yet they linger in some animals. Are they remnants of ancient usage giving us some insight into what an animal was like in the past? Maybe. Consider the following structures in the human: (1) the *lacrimal caruncle* (the little reddish mass in the corner of your eye considered by some to be a remnant of the nictitating membrane, or third eyelid, of reptiles and birds); (2) muscles in some people that move the external ear; (3) the *coccyx* or tailbone; and (4) the *arrector pili muscles* that cause our body hairs to "stand on end" (erect hairs are important in other mammals, making an individual appear larger and more ferocious and even trapping air for insulation, but they have no known function in humans). All of these structures have either lost their original function or their function is greatly reduced in adult humans, yet each persists.

Vestigial structures are evident in other animals as well. For instance, in some snakes and whales, the internal remnants of a pelvic girdle remain, even though they have long ago lost their hind limbs. Many birds have become flightless, yet reduced, stubby wings remain. Also, cave-dwelling fish and subterranean moles retain their residual, sightless eyes. Perhaps these structures, too, reflect an evolutionary link to the past. Maybe they really do open a window and let us peer, furtively, into our yesterday.

COMMON ANCESTRY

ANIMALS HAVE AN ANCESTRAL CONNECTION--THEY ARE MORE ALIKE THAN THEY ARE DIFFERENT. FOR EXAMPLE, ANIMALS ARE UNITED WITH EACH OTHER THROUGH THEIR MOLECULAR MAKEUP, THEIR CELLULAR MACHINERY, AND THEIR GENETIC THREAD. THEY ALSO HAVE SIMILAR METABOLIC AND NUTRITIVE NEEDS; THEIR REPRODUCTIVE PATTERNS ARE SIMILAR. MANY HAVE COMPARABLE BODY DESIGN. SOME CLAIM A CLOSER AFFINITY WITH EACH OTHER, WHILE OTHERS ARE MORE DISTANTLY RELATED. YET ALL HAVE A SHARED ANCESTRY WOVEN WITHIN THEIR PROTOPLASMIC FABRIC.

ANIMALS ARE TRULY DIFFERENT...BUT RELATED. THEY SHARE A COMMON LINEAGE.
EACH ANIMAL ALIVE TODAY IS CONNECTED TO ITS PREDECESSORS;
EACH IS LINKED, IN ONE WAY OR ANOTHER, TO THAT ENDLESS, ANCIENT EVOLUTIONARY CHAIN.

Comparative Embryology

Embryology is the study of an organism's early growth and development. It traces an individual's progress from conception through each formative stage, to the time of hatching or birth. Through these detailed events we have discovered the wonders of development: gamete formation, fertilization, gene control, tissue formation, organ formation, and birth.

When a variety of embryos are compared, it is clear that there are vast developmental similarities. For example, the actual sequence of development is almost universal. It involves a number of stages including fertilization, cleavage, blastulation, gastrulation, histiogenesis, and organogenesis. These are but a few of the rather technical terms used to describe an animal's development, and the pattern is uniform throughout most animal groups. In other words, the developmental forces seem to be the same for each animal. The same evolutionary mechanisms seem to be at work in the embryo as well as in the adult.

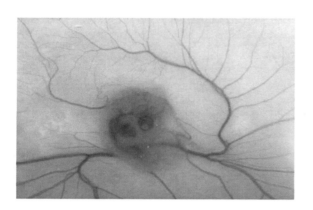

MANY EMBRYOS OF HIGHER ANIMALS PASS THROUGH SIMILAR STAGES OF DEVELOPMENT WHICH TENDS TO SUPPORT THE IDEA OF A COMMON ANCESTRY

Geographical Distribution

Of all the evidence in favor of natural selection, the distribution of animals upon our planet is perhaps the most compelling. A study of zoogeography unveils some rather surprising findings.

First of all, throughout the world, regardless of location, similar habitats are occupied by physically similar animals. Take, for instance, the North American prairie. Animals suited for survival there, such as the pronghorn antelope, the coyote, or the prairie dog, are similar in structure, size, and behavior to animals in prairie ecosystems throughout South America and the African continent (Figures 1.16 and 1.17). The same is true for animals living in deciduous forests, ponds, lakes, swamps, tropical rain forests, mountains, and deserts. Most have their ecological counterparts in other areas of the world. The Alaskan brown bear is similar to the Russian brown bear; the North American golden eagle is similar to the white-tailed eagle of northern Europe and Asia; the Tasmanian wolf is similar to the North American timber wolf; the South American condor is similar to the California condor; New World and Old World monkeys resemble each other; and so on and so on.

Figure 1.16. Even though the Asian male Mandarin duck (top) and the North American wood duck live on different continents, they occupy similar habitats and show remarkably similar body designs and coloration.

When two habitats are alike, their occupants, too, often share a surprising similarity. In fact, the uniformity is sometimes astonishing. Mutations, competition, reproduction, and natural selection have created "look-alike" animals in similar ecosystems throughout the world. This is called **convergent evolution** in which similar habitats produce similar life-forms. Examples are endless, the correlations irrefutable, the conclusion inescapable. The same evolutionary forces are at work regardless of locale.

But let's not get carried away! For it's just as important to recognize that, while occupants of similar habitats are themselves similar, they are not the same. In fact, they differ in virtually every case. An animal observed in a Madagascar rain forest will be comparable to, but different from, an animal occupying a corresponding ecosystem in Central America. Why is that? One would think that if the habitats were the same and if evolutionary process are, indeed, universal, then the animals occupying them would, likewise, be the same.

To think that way, however, is to discount the fact that although habitats seem to be identical, the ecological niches are not; each is unique. Also, evolution, as we have seen, is driven very much by *random, chance* events. Thus, you would not expect any two regions, regardless of how similar their niches might be, to produce identical species. Each niche requires a different set of adaptations. Habitats such as prairies, rain forests, and deserts produce similar, but unique, life-forms because each habitat contains unique ecological niches. For example, antlers may be an advantage to animals living in comparable ecosystems, say a prairie. Thus, prairie-living antelopes develop horns. However, mutations and natural selection will dictate the exact type of horn in each instance. The outcome will be similar, but different, as in the figure below.

I think there are two important considerations here. First, the forces are the same for each niche: mutations, competition, survival, and reproduction work together to produce similar, yet unique animals. Secondly we learn that natural selection favors success, and success can mean different things in different niches. Thus, while animals in parallel niches may be very much alike, they are nonetheless, also very different. Evolution does not have an "ideal" size, shape, or body plan. It just selects those best fit in a given situation.

IT'S A FACT

THE GOAL OF EVOLUTION IS NOT PERFECTION...IT IS SIMPLY TO SELECT THE MOST FIT INDIVIDUALS FROM AMONG THOSE COMPETING FOR ANY GIVEN RESOURCE.

Figure 1.17. Animals occupying similar habitats, even continents apart, show comparable body design. Here a pronghorn (*Antilocarpa americana*) from the North American prairie (left) is contrasted with springbuck (*Antidorcas marsupialis*) of south Africa. Note that they are similar but different.

Island Biogeography

The distribution of animals on islands provides additional insight into the evolutionary process. In fact, this feature probably had as much impact on Darwin as any other as he pieced together his now famous theory.

Some conspicuous and interesting things happen to animals when they invade island ecosystems. Because competition, predation, and diseases are often absent or reduced on islands, a colonizing life-form can experience pronounced radiations in a rather short period of time. Empty niches are often plentiful on islands and relatively easy to invade. In fact, a single colonizing animal can disperse into a variety of available niches like ripples on a glassy pool.

That was the case on the Galapagos Islands visited by Darwin in the mid-1800s. On this small archipelago off the coast of Argentina he discovered several episodes of island radiation. For example, he observed several species of birds that had apparently sprung from a small founding group of finches. Throughout the world finches are small, seed-eating birds. They have stout, conical beaks engineered to crush hard seed coats. Darwin found these basic, seed-eating finches on the islands. But he also recognized a number of finch like birds that had strange, new beak types: a chisel-like woodpecker beak; a second one now razor-sharp and designed to feed, vampire like, on blood; an elongated beak meant to probe into small crevices; yet another one intended to catch insects; and another one devised to feed on flower nectar. He discovered thirteen different designs in all— thirteen different species of finch like birds derived from a single finch like ancestor. He saw this genetic divergence as powerful evidence for his developing theory of natural selection.

Darwin found similar radiations of birds and other animals on each island visited in the chain. It seems that each island had gone through its own version of evolution. This, he realized, was the final piece of the evolution puzzle. These islands were fabulous, living laboratories where natural selection had created, on each, a unique and diverse animal community.

Figure 1.18. The Galapagos Islands played an important role in our understanding of the evolutionary process.

Artificial Selection

Further confirmation of evolution comes from human manipulation of animal genetics. Through the centuries man has domesticated a number of animals for his own use: for food, sport, labor, clothing, companionship, and now, it appears, even for organ implants. Breeders and scientists alike are constantly discovering and manipulating the phenotypes of domestic animals. By doing so they can select the very traits deemed desirable in each species: a full slab of ribs, a swift hawk for falconry, sweet cow's milk, a luxurious fur stole, a work horse, a hairless cat, a miniature poodle, a floppy-eared rabbit, a talking parrot, beasts of burden, and even pigs for heart and liver transplants!

This artificial selection in the laboratory or barnyard mirrors natural selection in nature. The operating forces are the same: mutations occur; some resulting traits are selected by a breeder or a farmer, others are neglected; animals with the desired features are permitted to breed while the undesirables are not. As the coveted traits eventually spread throughout the population, those preferred traits in each animal become the norm.

Figure 1.19. Relying on the principles of organic evolution, humans have "invented" an amazing variety of domesticated plants and animals. From floppy-eared rabbits to wooly sheep, artificial selection has underscored the processes of natural selection.

Direct Observation

One of the earliest and most basic questions asked about the evolutionary theory was, "Is it happening right now, and if so can we see it?" For almost 100 years the answer has been, "Well, yes...and no." But that has changed. Over the last 50 years or so we have been keeping a scorecard, recording each instance of possible evolution. The results are now coming in, and they show, clearly, that we have, indeed, witnessed evolution in progress.

Take houseflies for instance...those ubiquitous nuisances. They have been our constant and detested companions for thousands of years. And for just as many years we have tried everything imaginable to wipe them off the face of the earth. By the middle of the 20th century the best we could come up with was the flyswatter and sticky flypaper. Then along came the insecticide DDT. We had, at long last, discovered a remedy for this torment. Housewives everywhere, armed with their "bug bombs," staged an onslaught. Billions upon billions upon billions of unsuspecting flies fell victim to DDT's lethal mist. We had finally conquered this squalid, disease-carrying, maggot-producing, filth-spreading pest. But, had we?

Unknown to us, the lowly housefly had a trick up one of its six sleeves. The problem was that we didn't strike a fatal blow when we had the chance. In our assault, we didn't know that some flies were less affected by DDT than others and they had survived the purge. *They survived because some were naturally resistant to DDT's poison.* They were preadapted. These survived in small pockets, here and there, and that resistance was gradually transmitted from one generation, then on to the next, and so on. The rest is history. By the early 1970s houseflies were virtually unaffected by DDT exposure. Flies survived, DDT did not. Flies had evolved while DDT is gone.

We have seen many, many similar examples of evolution in progress. Since insects reproduce so rapidly and expansively, they have been the most obvious targets of our observations. But we have seen it elsewhere, in bacteria, plants, and even viruses (the AIDS virus, for example). It seems that evolution, especially microevolution, occurs constantly.

Evidence for Evolution: A Summary

Evidence for the natural selection paradigm of evolution comes from a number of quarters. Collectively, they make a compelling case in favor of this model:

(1) Fossils: The fossil record shows that the kinds and numbers of animals have changed throughout time. It also reveals a number of mass extinctions followed by progressive increases in diversity.

(2) Comparative Anatomy: Animal body design, molecular makeup, and biological machinery show such similarities that common ancestry is uniformly accepted among zoologists.

(3) Comparative Embryology: Developmental patterns among animals indicate strong interconnections. Also, many embryos show features seemingly reflective of their evolutionary past, a concept called biogenesis.

(4) Animal Distribution: The geographic distribution of animals on earth provides strong evidence that the principles of natural selection are universal: similar habitats produce similar animals. On the other hand, animals occupying similar habitats are alike, but different. This indicates that each ecological niche is unique and, in turn, produces unique animals. Island ecosystems are conspicuous in their display of these features.

(5) Artificial Selection: The domestic production of plants and animals parallel evolution in that humans select those traits deemed desirable, just as natural selection works in nature.

(6) Direct Observation: An increasing body of data derived from direct observation (pesticide-resistant insects, resistant pathogens, industrial melanism) have further bolstered the natural selection model.

Now How Does It Work, Again?

The theory of natural selection, as proposed by Charles Darwin and later modified by many others, can be rather confusing at first. After all, it is a concept meant to explain the very mechanism of biodiversity on our planet. It is trying to answer the age-old question: How have all organisms, in their endless variety, come to be?

You could expect, therefore, that such an ambitious, comprehensive, and all-encompassing theory would be complex and difficult to grasp. But that is not the case at all. Indeed, what is surprising about the evolutionary concept is not its complexity, but rather, its fundamental simplicity. The theory of evolution can be summarized in just six, easy-to-understand principles. It's supported by six biotic pillars.

The Six Pillars of Evolution

1. Individuals within a given species have *inherited differences*; they vary from each other in subtle, but important, ways.

2. Animals have a *great reproductive potential*, often beyond that which a given habitat can sustain.

3. But *natural resources* (food, space, etc.) *are limited*.

4. Since population numbers can outstrip a habitat's carrying capacity, organisms must *compete* for these limited environmental resources.

5. *Those individuals better suited* to their niche, because of inherited beneficial traits, compete more effectively, and *stand a greater probability of surviving and reproducing*; they will be the ones to see their genes pass through the filter of natural selection.

6. Because of competition and natural selection, population size over the long term remains relatively stable, fitness is improved (due to differential survival), phenotypes change (due to differential reproduction and accumulated mutations), and eventually *new species are launched*.

The Enchanting Parade:

Biophilia and Alternate Explanations

It seems to me that people are fixated on nature. Some are seduced by it, others are repelled. Think about it: rarely is someone lukewarm when it comes to things wild. For some it's an aversion or a phobia, such as a fear of spiders, snakes, centipedes, or tapeworms. They may prefer the sidewalk to the hiking trail, the motel to the campsite. For others there's a reverence. They are the bird-watchers, the whale enthusiasts, the butterfly chasers, the SCUBA divers, the fishermen. They may be active environmentalists, or may simply relish walking quietly along a wooded trail. Some may even become zoologists.

So, in one way or another, people are involved with nature. Our innate being is somehow tied to the world around us. It's part of the human spirit. Some might go so far as to say it's woven into our genetic fabric. This fascination with nature even has a name: **biophilia**. Perhaps that's why you're in this course. You realize that you're a part of this enchanting parade and you want to learn more about how it came to be.

In this chapter, we have seen that there is a tremendous, even awe-inspiring, diversity to the zoological landscape. We have also seen that science explains this vast biodiversity by way of natural selection through differential reproduction. In other words, through evolution. So it seems that evolution then is the drummer to which animals march in this incredible parade. But is that the only interpretation? Are there other ways to explain the appearance of so many different life-forms? Certainly, more than one. In fact, each culture has its own version of the grand history of life. However, alternative viewpoints are, for the most part, religious in nature and as such lie well beyond the scope of science. These alternate explanations are, therefore, not detailed here. Not because they are unimportant, but rather because they belong to another domain, a very personal, yet separate, domain.

Yet, we cannot simply ignore alternate explanations altogether for, although separate, science and religion have a shared history— one that has, at times, been rather turbulent. As science advanced over the centuries, religion adjusted many of its interpretations to allow for scientific discoveries of the natural world. For instance, the role of miracles to explain unknown phenomena has given way as science uncovered nature's mysteries one after another. On the other hand, science has realized its limitations in explaining moral issues. Science, for instance, does not pretend to deal with such claims as a soul or life after death; those are purely the realm of religion.

Thus, each of us is left to form our own world view that somehow incorporates both science and religion. Over time, three such "world views" have emerged.

The Defenders

Since it was first forwarded some 150 years ago, the theory of evolution has been the focus of considerable debate between the scientific and religious communities— a debate that is alive today in certain arenas. Many people, on both sides of the issue, claim that evolution and religion are incompatible, that they lead to conflicting conclusions. This is the either/or approach: either one or the other explanation is true, but not both. In this interpretation, since both explanations cannot be true, one must therefore deny the other.

This approach has led to the establishment of two opposing camps, each defending its own position while refuting the validity of the other; each entrenched behind its position as battle lines have been drawn, opposing positions taken, and volleys fired. In the United States, for example, creationists recently declared their explanation a "science" and, as such, demanded equal time in science curricula. Biologists, and the wider scientific community, responded with the view that creationism is a religious position, and, as such, cannot be viewed as science— a view the courts have upheld. And so, the two camps remain unyielding and far apart.

The Bridgebuilders

A contrary viewpoint is taken by many others who believe in a connecting link or bridge between evolution and religion. They propose that, in reality, there is confluence between the two; that science and religion are two elements of a Supreme Whole and each has a specific contribution to the understanding of that whole.

In this world view theologians claim that a Creator began the process on earth from which all life has sprung; however, it's up to science to explain the outcome. In other words, it's the evolutionists who will unravel the mechanism by which this creation is fulfilled. Religion then forms the framework for our understanding of life while science fills the gaps. Thus, one discipline supports the other. Here, one overall Truth prevails and both science and religion contribute their special insight to its resolution. According to the bridgebuilders, we can only understand life on earth if the two sides work together toward its true revelation. In the end, the One Truth will be upheld by two supporting buttresses: science *and* religion.

The Separatists

Then there are a number of scientists and theologians alike who have found satisfaction in the very idea that the two disciplines offer separate, but equal, lines of inquiry. In point of fact, say the separatists, both science and religion, seek answers to different, yet fundamentally profound questions. Science seeks to provide factual accounts of *how* we came to be, what is our history, and how we are related to other life-forms on our planet. Religion, on the other hand, addresses the question of *why* we came to be. It searches for the meaning behind our existence; it seeks to place values within the human context.

Both search, but in different realms. Science inspects the physical, tangible, factual universe while religion looks to the metaphysical, the philosophical, the ethical. Science relies upon experimentation, objective reality, direct observation, and reproducible tests to resolve its uncertainties while religion unveils its mysteries through revelation, personal faith, introspection, and moral discourse. These are both important endeavors, but different.

In the opinion of many scientists and theologians alike, science and religion have equally relevant messages. Both are, after all, seeking to answer universal questions, but with different tools, different approaches, different perspectives. Both are integral to the human experience, yet we should be careful not to confuse one with the other, they are not the same. One doesn't necessarily substantiate nor deny the other. If science is successful in its search to answer how we came to be, it doesn't, by default, answer the question of why we are here. Likewise, should religion reveal the why, it doesn't, by extension, answer the how question. As Stephen Gould has so aptly put it, "the causes of life's history cannot solve the riddle of life's meaning."

We cannot deny the fact that both science and religion have had an immeasurable impact on human understanding. But they do so in different ways and for different reasons. However, they do come together in one respect... they both help us to appreciate the wonder of the world around us. So, regardless of the route you might follow in your discovery of the natural world, one thing is certain: animals enrich the earth with their complexity, their matchless beauty, and their fascinating diversity. And in some way the human spirit, your spirit, is linked to that complexity, beauty, and diversity.

...REGARDLESS OF THE ROUTE YOU MIGHT FOLLOW IN YOUR DISCOVERY OF THE NATURAL WORLD, ONE THING IS CERTAIN: ANIMALS ENRICH THE EARTH WITH THEIR COMPLEXITY, THEIR MATCHLESS BEAUTY, AND THEIR FASCINATING DIVERSITY.

An Inside View 1.3

What's in a Name?

A Question of Relationships. We have seen that there are millions of organisms occupying planet earth. And while each is unique, different from all others, there are, nonetheless, definite patterns of similarity. Evolution has produced some animals that are very much alike and yet are quite different from others. It is obvious that animals with feathers are more closely related to each other than they are to animals adorned with fur, or with a scaly covering, or to those with slime-covered bodies. One-celled organisms are more closely aligned to each other than to multi-cellular forms. Roundworms are quite different from jellyfish. It is upon these shared relationships that scientists have developed an organizational scheme...a classification system to help us grasp the order in life's vast diversity. What may seem like a chaotic, disorderly sea of organisms becomes sensible and manageable when we apply a systematic approach.

There are two disciplines that deal with life's vast diversity. **Systematics** is the study of the relationships that occur among organisms and their evolutionary history (phylogeny) while **taxonomy** is the actual naming and classifying of organisms. As a result of these two activities, each known organism is placed into a taxonomic hierarchy that reflects its relationship to all other organisms. The hierarchy is based upon the early work of Karl von Linnaeus (1707–1778) and is known as the Linnaean classification system. He realized that organisms could be placed into groups based upon shared relationships. As a result he devised a ranking system consisting of the seven distinct levels listed below. Note that each level (called a taxon) is more inclusive than the one below it.

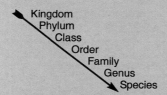

Kingdom
Phylum
Class
Order
Family
Genus
Species

Although early scientists recognized only two kingdoms (plants and animals), we now distinguish five:

Monera*   prokaryotic cells (bacteria and cyanobacteria)

Protista   eukaryotic, single-celled organisms

Fungi   multicellular, saprophytic decomposers (slime molds, yeast, mushrooms)

Plantae   multicellular, photosynthetic organisms (plants)

Animalia   multicellular, heterotrophic organisms (animals)

*Some biologists have divided the monerans into two separate groups, the Archaebacteria and Eubacteria, giving us six kingdoms. We, however, will stay with the more traditional system of five.

The Genius of Linnaeus. In addition to giving us a logical system to categorize all organisms, Linnaeus also devised a way of naming each and every one. It's a brilliant, ingenious approach that gives each animal a name consisting of two parts. The first part is the **genus** name and the second is the **specific** (species) epithet. Together they make up the scientific name of an organism that's unique to that one species. In this way much of the confusion inherent in common names is eliminated. Take, for instance, the largest carnivore in the United States, the mountain lion. In various parts of the country it is known as the cougar, panther, painter, and puma...all the same animal. Also, it may be confused in the southwestern part of its range with the jaguar, or in the West with the bobcat. However, by using its scientific name, *Felis concolor*, biologists from across the country can be sure they are referring to the same animal. That's the beauty of the **binomial system**. It's unique for every species; it creates order out of chaos.

Rules of Binomial Nomenclature. When a new animal is discovered it's eventually given a name. To do this, a number of steps, or guidelines, must be followed. The more common ones are listed below:

(1) Each scientific name has two parts: the genus and the species epithet. For example, *Amphispiza belli* is the sage sparrow. Sometimes a third name, the subspecies, is used for distinct subpopulations. For example, *Amphispiza belli belli*, is the San Clemente Island sage sparrow that occurs only on that island.

(2) The name is to be in Latin or otherwise "latinized," such as *Crotalus horridus*, the timber rattlesnake. Note that the specific name is an adjective that often describes a characteristic of the animal. What feature do you think the spotted skunk, *Spilogale putorius* has, or how about the rabbit *Sylvilagus aquaticus*?

(3) The name is to be set off in different typeface, usually in italics, or underlined.

(4) The first letter of the genus name is to be in uppercase; all others, including the first letter of the species name is written in lowercase.

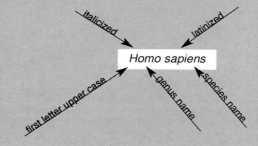

italicized   *latinized*

Homo sapiens

first letter upper case   genus name   species name

Haliaeetus leucocephalus, the American bald eagle
(leuco = white, cephalus = head)

Felis concolor, Mountain lion
(Felis = cat, concolor = one color)

Lumbricus terrestris, the common earthworm
(terrestris = terrestrial)

Ursus horribilis, the grizzly bear
(Ursus = bear, horribilis = to tremble)

IT'S A FACT

MANY SCIENTIFIC NAMES ARE QUITE DESCRIPTIVE. THEY MAY GIVE INFORMATION ABOUT AN ANIMAL'S APPEARANCE, BEHAVIOR, HABITAT, GEOGRAPHIC DISTRIBUTION, OR OTHER FEATURE. SEVERAL EXAMPLES ARE GIVEN ON THIS PAGE.

Anas platyrhynchos, mallard duck
(platy = flat, rhynchos = beak, snout)

Chapter 1 Summary

Key Points

(some common questions about evolution)

1. *Q. How many different animals are there?*

A. Although approximately 2.2 million species of animals have been identified, it is thought that the actual number is many times that...perhaps as many as 40 million or even more.

2. *Q. What causes biodiversity?*

A. There are four forces behind animal variety: (1) mutations; (2) competition; (3) differential survival; and (4) differential reproduction.

3. *Q. What is meant by adaptive fitness?*

A. Fitness is measured by the ability of an animal to compete, survive, and more importantly, to reproduce. The most fit animal stands the best chance of surviving and reproducing. Thus, through the natural selection filter, fitness is rewarded.

4. *Q. Are all mutations beneficial?*

A. No. The majority of mutations that occur are maladaptive. Individuals that undergo these mutations either die very early or are less equipped to handle the vagaries of the competition process. Thus, they fail to maintain the reproductive pace. So, most mutations are perilous to an individual and to the population.

5. *Q. Then how do mutations increase fitness?*

A. Some mutations are beneficial and will immediately improve the survival chances of its possessor. Others are neutral; they neither increase or decrease survival rates. These lingering mutations may eventually become beneficial if niche conditions change.

6. *Q. What is an ecological niche?*

A. A common definition is that a niche is the role that an organism plays within its environment. This includes, but is not limited to, space, food, reproduction, shelter, and resources.

7. *Q. Do new niches appear, and if so, how?*

A. New niches appear constantly. The primary forces appear to be environmental catastrophes (volcanoes, ice flows, climate changes), geologic activity, and human involvement.

8. *Q. How are new niches invaded?*

A. Through a process called preadaptation. This means that an individual already possesses a needed feature before it's required. It can colonize a new niche and will then pass its traits to the next generation, and so on.

9. *Q. How does an animal get a preadapted feature?*

A. By way of either beneficial or lingering (neutral) mutations.

10. *Q. Is there any peril in invading new niches?*

A. Yes. A species may be either too specialized or too generalized in its adaptations. This specialization paradox may lead a species toward extinction.

11. *Q. What is resource competition and how does it contribute to the evolutionary process?*

A. Competition contributes to the vitality of a species by producing smaller yet hardier populations. It does this through favoring the fit and filtering out the unfit.

12. *Q. If evolution leads to increasing fitness, why don't all animals evolve into higher forms such as humans?*

A. Natural selection does not have a goal. For example, it does not lead toward perfection or toward any other objective. Instead, it rewards those individuals who possess the traits needed to survive in a particular ecological niche. Niche adaptation, then, is the outcome of the evolutionary process.

13. *Q. Why do animals possess so many features?*

A. Each color, pattern, shade, or tone; each spine, hair, feather, or scale; every talon, claw, tooth, or fang; each antennae, tentacle, sensory papillus, or receptor; all tissues, organs, or systems are the result of natural selection's precise appointment. Every animal dimension or feature has a specific function. Each has been the living target for natural selection's fitness arrow.

14. *Q. What evidence is there for natural selection?*

A. Support for evolution comes from a variety of sources such as comparative anatomy, comparative embryology, geographical distribution of organisms, artificial selections, and direct observation.

15. *Q. Are there other ways to explain biodiversity besides evolution and natural selection?*

A. Yes, but other explanations are primarily spiritual in their viewpoint. Therefore, although relevant in a religious context, they remain beyond the sphere of science.

Key Terms

adaptation	macroevolution
adaptive radiation	maladaptive mutations
artificial selection	meiosis
binomial nomenclature	microevolution
biological species concept	mutation
biophilia	natural selection
carrying capacity	nonrandom elimination
common ancestry	phenotype
convergent evolution	preadaptation
differential survival	punctuated equilibrium
genetic recombination	self-preservation
genotypes	species perpetuation
gradualism	specialization paradox
homeostasis	systematics
interspecific diversity	taxon
intraspecific diversity	taxonomy
linear descent	vestigial structures

Points to Ponder

1. What is biodiversity?

2. What is the relationship between diversity and ecological niche?

3. Explain what is meant by the idiom "Nature abhors a void." What does this have to do with animal diversity?

4. What is the Specialization Paradox. What may happen to a species caught in such a predicament?

5. What is biophilia? What is your personal status related to nature?

6. Why is reproduction so important in the life of an animal?

7. On average, what decides if an animal reproduces?

8. How does an animal become fit?

9. What role does mutation play in the evolutionary process?

10. What is preadaptation? What is its relationship to natural selection?

11. Contrast gradualism and punctuated equilibrium.

12. Define the term species, then give two ways that speciation may occur.

13. Which of the six lines of evidence in support of evolution do you think is the strongest? the weakest? Support your selections.

14. To create order out of animal diversity, biologists have adopted a taxonomic hierarchy consisting of seven levels from kingdom to species.

15. At the current time, five kingdoms of life are recognized on planet earth.

16. The naming of organisms involves a number of guidelines leading to the <u>binomial</u> <u>system</u> of scientific nomenclature.

Inside Readings

Darwin, C. 1859. The Origin of Species. Cambridge: Harvard University Press.

Gould, S. J. 1999. Rocks of Ages: Science and Religion in the Fullness of Life. New York: Ballantine Publishing.

_____ 1980. The Panda's Thumb. New York: W. W. Norton & Company.

Freeman, S., and J. C. Herron. 1998. Evolutionary Analysis. Upper Saddle River, N J: Prentice Hall.

Grant, P.R. 1991. Natural selection and Darwin's finches. Sci. Am. 265(4):82-87.

Harris, C.L. 1981. Evolution: Genesis and Revelations. Albany, NY: SUNY Press.

Kellert, S. R., and E. O. Wilson (eds). 1993. Biophilia. Washington, D C: Island Press.

Lewin, R. 1982. The Thread of Life. Washington, D C: Smithsonian Books.

Mayr, E. 1997. This Is Biology: The Science of the Living World. Cambridge: Harvard University Press.

Wilson, E. O. 1992. The Diversity of Life. New York: W. W. Norton and Company.

Brusca, R. C., and G. J. Brusca. 1990. Invertebrates. Sunderland, MA: Sinauer Associates.

Pechenik, J. A. 1996. Biology of the Invertebrates. Dubuque, IA: Wm. C. Brown.

Wilson, E. O. 1984. Biophilia. Cambridge: Harvard University Press.

Chapter 1

Laboratory Exercise

Animal Diversity

Laboratory Objectives

✓ to learn to use microscopes properly
✓ to appreciate differences among various animal groups
✓ to use a dichotomous key to identify unknown specimens

Supplies needed:

- dissection microscope
- compound microscope
- selection of prepared specimens

Chapter 1
Worksheet

*The whole secret of the study of nature lies
in learning how to use one's eyes.*
- George Sand

Introduction

We live in two worlds, one visible, the other invisible. The visible universe, that world we all witness, displays a patchwork of colors, sizes, shapes, and complexities. It amazes and dazzles us with its panoramic diversity. This visible world allows us to view the animal kingdom in its vast variety...but not all of it. There are countless animals that exist in a shadowy underworld, imperceptible to the human eye. And this invisible world is no less intriguing than the visible one. It, too, shows an infinite kaleidoscope of shapes, colors, and complex designs. It, too, astonishes and entices. It draws us in. And when we enter this unseen world, we become absorbed in its intricate nuances.

Nature has given you eyes to view the visible world while science has provided a tool, the micro-scope, to unveil the invisible. No other scientific discovery has done more to open new horizons than that of the microscope. It reveals the hidden, the minuscule, the unexplored, as nothing else can. It exposes subtleties and intricacies undetected by the human eye...what we could only once imagine the microscope exposes. It is truly one of our most magnificent instruments.

During this class, we often use the microscope to reveal what we cannot see otherwise. Therefore, understanding its features and limitations is well worth the effort.

I. The Microscope

Microscopes magnify small objects that are often well beyond the resolution of the human eye. To do this a 'scope uses an illuminating and an imaging system. The illuminating system projects light on the specimen while the imaging system enlarges the object for detailed inspection.

In this class, we use two types of microscopes, the compound light microscope and the stereoscopic (dissecting) microscope. An effective compound micro-scope can improve the power of the human eye over 1,000 times. Used properly it can expose details of even the smallest animals. Compare the compound scope at your table to the diagrams on the following page. Be sure to locate each of the items listed below:

✓ eyepiece
✓ base
✓ stage
✓ condenser
✓ revolving nosepiece
✓ aperture diaphragm
✓ objective lenses
✓ coarse and fine adjustments

II. Animal Diversity

To fully appreciate individual animals, we must first get an idea of the range in structure, appearance, and behavior of all animals. This is an activity that is emphasized throughout this course and begins in this first exercise.

Here we use a device called a "dichotomous key" to identify a number of "unknown" animals. In essence, a key of this type walks you through a series of steps in which two statements appear. Because of the way the key is written, one statement is true for a given specimen and the other is false. The steps are followed sequentially until the identity of the unknown animal is revealed. While a key can identify an animal to any taxonomic level, in this exercise we will identify the specimens only to the rank of phylum.

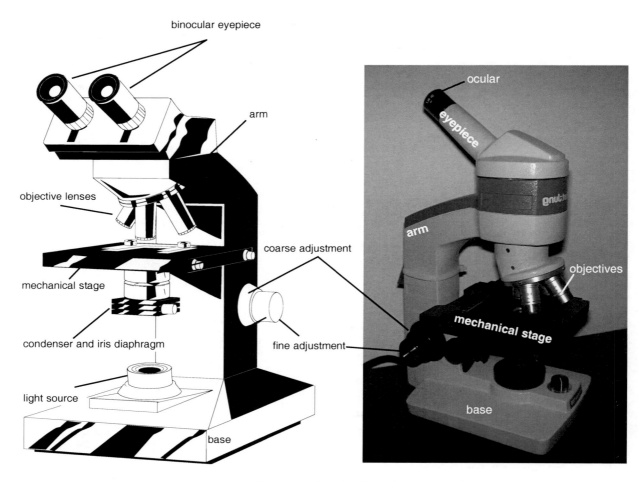

The Compound Light Microscope

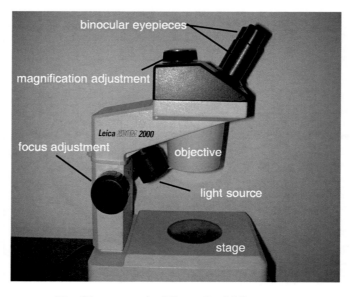

The Stereoscopic (Dissection) Microscope

Chapter 1
Worksheet

(also on page 581)

_____   _____
(name)   (lab section)

Instructions: Use the dichotomous key to identify each "unknown" specimen. Give its phylum and common name in the space provided

Animal Identification

Specimen #   Phylum   Common Name

1. phylum Nematoda _____

2. phylum cinidana _____

3. phylum Mollusca _____

4. phylum Mollusca _____

5. Earthworm _____

6. Phylum chordata _____

7. phylum plateyliminteS _____

8. phylum Echnodermata _____

9. phylum poriteria _____

10. phylum Arthropoda _____

DICHOTOMOUS KEY FOR SELECTED ANIMAL PHYLA

Instructions. There are ten numbered "unknown" specimens located at your lab station. Use this key to determine the phylum for each specimen then place your answer on the preceding page. While you are completing this exercise, take time to carefully observe each specimen as to:

- ❑   general body size and form
- ❑   outer covering (feathers, scales, shell, exoskeleton)
- ❑   the number and types of appendages

Each couplet has two statements. Determine which statement is correct for a given specimen, then follow the key until you have identified each unknown specimen.

1. Body microscopic, consisting of only a single cell.............."Phylum" *Protozoa*
Body not microscopic, consisting of many cells and forming several layers.....**go to couplet #2**

2. Body design is bilaterally symmetrical (equal right and left sides).....**go to 4**
Body design is not bilateral.....**go to 3**

3. Asymmetrical animals, body contains numerous pores.....Phylum *Porifera*
Radially symmetrical animals (body arranged like spokes on a wheel).....**go to 10**

4. Animals with a backbone (i.e., vertebrate animals).....Phylum *Chordata*
Animals lacking a backbone (i.e., invertebrate animals).....**go to 5**

5. Body divided into repeating segments (surface covered with rings).....**go to 8**
Body not segmented.....**go to 6**

6. Body flattened dorso-ventrally (entire body is paper thin).....Phylum *Platyhelminthes*
Body not flattened dorso-ventrally.....**go to 7**

7. Body long and slender, unsegmented; not enclosed within a shell.....Phylum *Nematoda*
Body not long and slender; animal enclosed within one or two shells.....Phylum *Mollusca*

8. Body bearing paired, jointed appendages; some serving as legs.....**go to 9**
Body without paired, jointed appendages.....Phylum *Annelida* (earthworm)

9. Body segmented, appendages short, microscopic animals.....Phylum *Tardigrada*
Body segmented, not microscopic, covered by an exoskeleton; jointed appendages.....Phylum *Arthropoda*

10. Body unsegmented, gel-like with long, slender tentacles.....Phylum *Cnidaria*
Body unsegmented, covered with short, spiny projections.....Phylum *Echinodermata*

2 Chapter 2

FORM & FUNCTION

The harmony of the world is made manifest in form...

- Wentworth Thompson

Form & Function

Form ever follows function.
- Sullivan

Introduction

*T*he forces of evolution that shape animal populations function on two levels: the **genotype** and the **phenotype**. In the last chapter, we discussed evolution, in its broadest sense, as the transmission of an individual's genes (genotype) from one generation to the next. However, I assume by now you have come to recognize that natural selection really acts on the physical characteristics of an organism, its phenotype. As a result of competition, certain characteristics lead to greater survival than others, and it's these "fit traits," expressed in the phenotype, that are selected through reproduction.

Generally speaking, the purpose of this chapter is to gain some insight into animal phenotypes: what animals are and what makes them tick...how they are constructed and how they function. So, I guess we are actually trying to define an animal. A little thought reveals that to do so is not all that simple. In fact, it requires that we first look at the various life-forms on our planet. At the current time biologists have divided living forms into five major categories known as kingdoms. These are:

(1) **Monera** - the bacteria

(2) **Protista** - the single-celled organisms; protozoans and algae

(3) **Fungi** - mushrooms, rust, molds, yeast; many are decomposers

(4) **Plantae** -plants with cell walls and capable of converting sunlight into chemical energy through photosynthesis (autotrophism)

(5) **Animalia** - everything else; nonautotrophic (heterotrophic) organisms; animals

While this scheme may be instructive, it still doesn't do a lot to help us define animals. In fact, if we look at an even broader category, that of cells, the outcome can be even more confusing. According to our current understanding there are only two types of cells on earth. The simplest are the **prokaryotes,** cells lacking a distinct nucleus; only bacteria are prokaryotes. All other organisms have cells of the **eukaryote** type in which DNA is confined within a nucleus.

Well, back to our definition of an animal. From the discussion above we can conclude that an animal is nothing more than *a nonphotosynthetic organism, consisting of eukaryotic cells without walls.* This definition is certainly precise, you've got to give it that, but it's hardly inspiring. Where's the symmetry of a hummingbird in flight? the mystique of a great white shark? the magic of a soaring albatross on her 13-foot wingspan? Animals are certainly more than a heterotroph lacking cell walls...much more. They are the sum of five basic features: *structure, stability, energy, responsiveness, and reproduction.*

Figure 2.1. Animals are more than a collection of eukaryotic cells incapable of carrying out photosynthesis. They have form, function, symmetry, and beauty.

Five Fundamental Features

Since their appearance on earth some 700 million years ago animals have been competing with each other for limited resources. Little by little, like an enormous wind, natural selection has swept away the chaff leaving behind the wondrous diversity we see today. And what has made each survivor in this grand experiment so successful? Let me suggest that it's due to the interaction of just five fundamental features.

1. Structure

To function effectively within their respective niches animals must have a dependable structure, both internally and externally.

Structural Stability. Three major grades of construction are recognizable in animals: *cellular, tissue,* and *organ* levels. There's no doubt that cells are the immortal assembly factories where life processes begin. Regardless of the organism, whether a simple sponge or a complex mammal, it's nothing but the sum total of the action of its individual cells.

If cells are the building blocks of animal life, are there any discernible patterns? The answer is yes, there are three. First of all, the protozoans are single-celled organisms with animal-like characteristics (movement, eukaryotes, heterotrophic nutrition). Thus, they constitute what might be considered the **unicellular grade** of animals (see A in the plate below). They accomplish all life functions, including the stimulus-response pathway, within the confines of a single cell. From feeding, to reproduction, to thermo-reception, one cell is enough for them.

Second, we soon see that life progressed from the single-celled stage of protozoans to **multicellularity** (C below). Indeed, all true members of the animal kingdom exist, not as single cells, but as groups of cells, at times numbering into the trillions. To contrast these animals with the unicellular protozoans, we give them the collective name of **metazoans**. Sponges are the first animals and at the same time, the first metazoans.

The ancestral source of multicellularity has been the subject of much argument. While it is not in our interest to enter into this contracted debate, suffice it to say that the actual precursor of the metazoans long ago disappeared, joining that long frozen line of the extinct. Nonetheless, we can safely assume that something like a modern-day protozoan provided the springboard toward multicellularity.

Whatever the source, multicellularity was an extremely popular design adopted by the many millions of animal species. The advantage of consisting of many cells has to do primarily with size...individual cells are invariably small, real small. Most occur well below the resolution of the naked eye. There are a few exceptions, but very few. In contrast to what the science fiction writers may tell us, there are no giant-sized amoeba...the "glob" did not and could not exist.

But the implications of size are immense. If animals were to become larger, they must vacate the single cell in favor of a multicelled existence. You only need to look around to realize that they did just that. They triumphed beyond any expectation; the elephant, the rhinoceros, the whale-shark, the sperm whale, and the giant of all animals, the blue whale amply attest to the success of being multicellular.

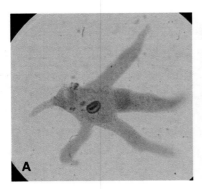
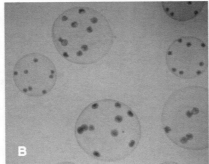
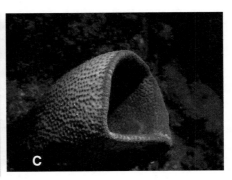

A. unicellular (amoeba)   B. colonial (volvox)   C. multicellular (sponge)

Figure 2.2. The three major grades of animal organization.

One other pattern has emerged from the unicellular lineage. These are not entirely multicellular, nor are they entirely unicellular. Although most protozoans are one-celled organisms, many consist of a number of genetically identical cells attached together; they exist as a **colony.** A *Volvox* (below and B on preceding page), for example, may contain as many as 50,000 cells within its mother colony. Yet, since *Volvox,* and other colonial forms show virtually no trend toward cellular specialization, few would consider them to be true metazoans. As such, colonial protozoans probably form a bridge between unicellular and multicellular organisms.

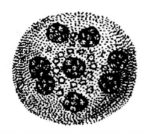

A colonial organism similar to this may have been the ancestor of all multicellular animals.

As multicelled animals gain bulk, they begin to show a division of labor among the various cells. Instead of every cell carrying out every function, it's more efficient if *different* cells perform *different* functions. Thus, instead of expecting one cell type to be responsible for everything from defense, locomotion, movement, feeding, and excretion, to sensory reception, we see a trend toward **cellular specialization.** Along with cellular specialization comes improved efficiency, and greater size results. The first sign of cellular specialization occurred in the simple sponges wherein four different cell types divide the workload among themselves (see chapter 4). The number of different cell types then escalates as different groups of animals appear. By the time we reach *Homo sapiens,* the number of discrete cell populations has blossomed into about 200: muscle cells, nerve cells, skin cells, liver cells, bone cells, lung cells, and so on.

This division of labor is the focal point around which animals are constructed as they became larger and more complex. Cellular specialization, for example, soon led to the next level of organization in which similar kinds of cells group together to form a functional mass...the **tissue.** This first occurred in the jellyfish and their relatives as two distinct tissue layers (chapter 5). Before long a third one is added and those same three occur throughout all higher animals. On the inside, lining their gut, is a layer known as **endoderm.** Covering an animal's exterior is a protective layer, the **ectoderm,** and sandwiched between the two is the third tissue layer, the **mesoderm.**

From these three primary layers come all of the many features common to animals: the simple structures, the not so simple, the complex ones, the **organs,** and finally, the **organ systems.** As animals become larger and more sophisticated, cells or tissues alone cannot provide the needed resources. As a result, organ systems emerge as intricate *combinations* of organs and their tissues designed to carry out specific functions within an organism. To best understand how and why animals are constructed as they are, we need to look briefly at the various systems that exist in a higher animal such as a human.

Animal Organ Systems

(1) **Integumentary System.** Acts as a protective sheet that covers the entire animal.

(2) **Muscular System.** Initiates movement of the entire body or its parts; also helps to maintain posture.

(3) **Skeletal System.** Supports overall body, protects body parts, serves as site for muscular attachment, produces red blood cells.

(4) **Nervous System.** Senses external and internal environment, integrates information and coordinates responses.

(5) **Endocrine System.** Produces hormones to assist the nervous system in the control of body functions.

(6) **Circulatory System.** Transports materials (e.g., nutrients, oxygen, carbon dioxide, and wastes) throughout the body.

(7) **Lymphatic System.** Defends body against infection and aids in the body's fluid balance.

(8) **Digestive System.** Ingests food and water, breaks down food, absorb the products and removes the undigested portion.

(9) **Respiratory System.** Exchanges those respiratory gases (oxygen and carbon dioxide) needed by all cells.

(10) **Excretory System.** Regulates body fluid levels and removes nitrogenous wastes.

(11) **Immune System.** Protects the body from invasion by foreign materials.

(12) **Reproductive System.** Produces gametes for fertilization and may provide a site for embryonic growth.

As animals progressed beyond the relatively rudimentary cellular level of organization, they soon adopted the tissue, and eventually the organ system grade. As they did so, they also adopted distinct shapes and orientations.

Body Shapes. Animals come in many disguises. Each species seems to have its own special stamp, a blueprint that varies from all others. In structure, shape, and format, they are all unique, at least it appears that way. But as diverse as animals seem, they are nonetheless arranged into just three basic body plans. And symmetry is the common theme. The overall arrangement of an animal so that it can be divided into equal halves is seen in virtually all animals...but not all.

Most sponges are **asymmetrical**, they cannot be bisected along any plane to produce two equal or similar parts. Adult snails, too, are asymmetrical, but secondarily so since their larval stages are symmetrical. Except for these few groups, however, all other animals display some form of symmetry. For example, an animal is said to be **radially symmetrical** if there are a number of bisection planes that can yield similar parts, sort of like spokes on a wheel. As long as the plane passes through the center of the wheel, equal halves are produced. All radial animals are aquatic, usually sedentary or slow moving forms. As such, they may receive equal stimuli from any direction. In this setting, a radial pattern works well since an animal so arranged can best exploit its environment. The entire phylum Cnidaria (jellyfish, sea anemones, corals) are radial animals, as are some sponges and adult echinoderms (sea stars).

The final body plan, and the one adopted by virtually all animals, is **bilateral symmetry.** In this design, an animal can only be bisected along one plane, the long axis, to produce equal halves. The result is a right and left side of equal proportions and of equal appearance. In other words, the two sides are mirror images of each other. The overwhelming prevalence of bilateral animals on earth has to do with an active lifestyle; there is a direct relationship between a bilateral design and active, forward locomotion. See chapter 6, pg 117 for a further discussion of the significance of bilateral symmetry.

Figure 2.3. The three basic body designs are asymmetrical as shown in the sponge (**A**); radial symmetry as seen in the corals (**B**); and bilateral symmetry in all other animals and illustrated by the young chimpanzee (**C**).

An Inside View 2.1

Body Directions

In biology we use standardized terms to describe the various regions of the body. So, while we could describe a shark as having a front and a rear or a back and a belly, these terms are so vague as to be of little value. Instead, by using standardized terminology we can avoid confusion and misunderstanding in our discussions of body regions and parts. The following terms are common anatomical references:

◆ The surface of the animal that faces the substrate is **ventral.**
◆ The region that is opposite of the ventral surface is **dorsal.**
◆ The part of the animal that enters a new environment first is the **anterior** end.
◆ The last region to enter a new environment is **posterior.**
◆ A part of the animal away from its midline is termed **lateral.**
◆ A part that is toward the midline is **medial.**
◆ The part of a structure that is away from the midline is **distal.**
◆ The part of a structure that is toward the midline is **proximal.**

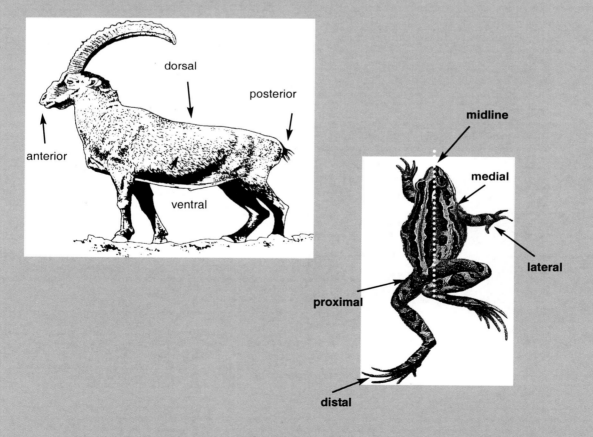

Planes of Section

There are three general ways to section (cut) an animal into two parts. These bisection planes are:

◆ A plane separating an animal into anterior and posterior parts is a **transverse** section, also called a **cross section.**
◆ A plane that divides an animal into right and left halves is a **sagittal** section.
◆ A **frontal** plane separates an animal into dorsal and ventral parts.

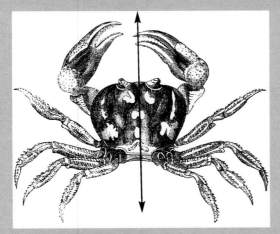

Sagittal Section

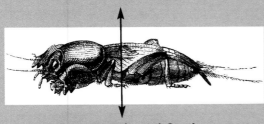

Transverse (cross) Section

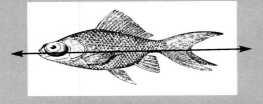

Frontal Section

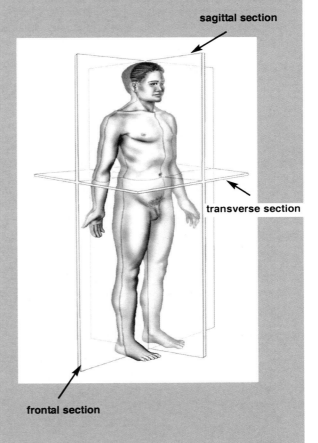

sagittal section

transverse section

frontal section

II. Functional Stability

Beneath the title at the beginning of this chapter is a quote from Sullivan, "Form ever follows function." Is that statement accurate? Does a structure develop before or after it's needed? Does a particular function appear and then the animal adapts a structure to meet that demand, or is it vice versa? Perhaps a structure evolves and then the body finds a function for it. Or maybe a function appears and the animal evolves a structure to meet the function. Which is it? Certainly, there is a relationship between form and function. But what is it?

First of all, we must conclude that the makeup of a feature determines its functions. No matter how much a muscle "wants" to be a lung, it cannot undertake gaseous exchange; it cannot behave as a lung. The structure of muscle limits its function. Perhaps the best way to solve this puzzle is to look, again, at the evolutionary process. We know there is no way for an animal to produce a particular structure upon demand. There's just no mechanism to direct the necessary mutations and then to select the outcome; it's impossible. The structure must already exist for the need to be met. Throughout this text, we have referred to that process as *preadaptation*. So, I guess the best answer to the "egg" first question is that the structure must precede the need; the need then draws out the structure.

This is a prime example of how form and function intertwine. To fully understand animals we must not only look to the various evolutionary forces and pressures that led to their amazing variety, but we must also recognize that each animal is an intricate mixture of cells, tissues, organs, and organ systems. These parts all interact with each other, depend upon each other, and operate only within the realm of the others to produce a given organism.

One of the most important manifestations of this inner-reliance is the concept of **homeostasis**. All organisms must overcome the tendency toward disorder. They must all preserve an inner stability for life to proceed. For example, an animal cannot afford to have a temperature that's too high, or too low. It cannot afford to have its pH too high or too low. Likewise, it cannot afford to allow the buildup of toxic wastes, nor can it permit excessive fluid loss...or gain. In these cases, and countless others, an individual attempts to sustain a relatively narrow range of internal dynamics. We call this tendency to maintain a stable internal environment *homeostasis*. It's a common theme throughout the animal kingdom.

A striking example of homeostasis is a process called **osmoregulation**, the attempt of each cell to maintain a proper water and ion (salt) balance. The process is shown graphically below and includes the following important features:

◆ It is based upon diffusion, the inherent movement of molecules from an area of higher to lower concentration.

◆ All cells attempt to remain in an isotonic state: equal salt levels inside and outside of a cell.

◆ Water will diffuse *into* any cell residing in a hypotonic environment such as freshwater, resulting in cellular enlargement.

◆ Water will diffuse *from* a cell that resides in a hypertonic environment such as in seawater or on land, resulting in cellular shrinkage.

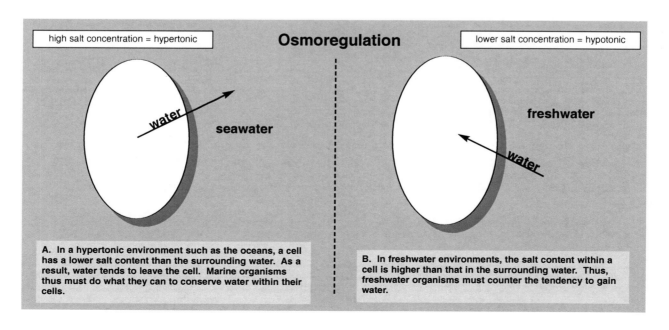

Osmoregulation

high salt concentration = hypertonic

lower salt concentration = hypotonic

water → seawater

freshwater ← water

A. In a hypertonic environment such as the oceans, a cell has a lower salt content than the surrounding water. As a result, water tends to leave the cell. Marine organisms thus must do what they can to conserve water within their cells.

B. In freshwater environments, the salt content within a cell is higher than that in the surrounding water. Thus, freshwater organisms must counter the tendency to gain water.

III. Energy Dynamics

All living cells, without a single exception, require energy. No cell has much of a future without a steady stream of it; all animals must be able to find it, collect it, consume it, and, finally, to assimilate it. Since they are made of cells and since they are *heterotrophic*, animals must rely upon an external source of food for their energy supply. The food, once **ingested**, undergoes either mechanical and/or chemical breakdown in a process called **digestion**. The products of digestion are then **assimilated** by individual cells. Once in the cell, the energy is then converted to **ATP** through the process of **cellular metabolism**.

The energy converting, metabolic process in the cell requires *oxygen* to complete its activity. In the presence of oxygen the energy is captured in the ATP molecule for later use. But during the process three by-products are liberated. First, some *water* is produced and the animal must finely regulate its concentration both inside and outside of the cells. Second, *carbon dioxide* is liberated; and finally, *nitrogen* is produced from the breakdown of amino acids present in digested proteins. Both carbon dioxide and nitrogen are toxic to the cells and must either be quickly removed or transferred into a nontoxic form. The diagram below represents the basic metabolic process within a cell.

Any undigested food is eliminated from the animal as **waste**. The names given to this waste material are many: feces, droppings, offal, excrement, stool, manure, sewage, as well as many not fit to mention here.

Feeding Mechanisms. As heterotrophs, animals have a number of feeding strategies available to them. If they feed primarily on plant material, they are known as **herbivores**. In contrast, animals that feed on other animals are called **carnivores**. However, while many animals prefer a herbivorous *or* a carnivorous diet, it is rare to find one that is exclusively one or the other; instead, most live as **omnivores**, feeding on both plant and animal materials. Animals that feed on dead flesh are **scavengers,** while those that prefer decaying material are **detritus feeders.**

Many invertebrate animals have two special feeding strategies at their disposal, **suspension** or **deposit** feeding. In suspension feeding, animals filter food from their aquatic surroundings. To do that requires water to be brought across a feeding surface (often lined by mucous as a trapping medium) followed by the removal of food particles such as bacteria, plankton, and detritus.   Examples of suspension feeders include certain protozoa, sponges, some annelids, barnacles, and mollusks, among others.   Deposit feeders extract food from a soft substrate such as mud, soil, or sand. Those animals that simply swallow large portions of substrate are known as *direct deposit feeders* (e.g., earthworms and some snails); whereas, animals that carefully remove only the uppermost layer of a substrate are *selective deposit feeders* (e.g., sea cucumbers and many annelids).

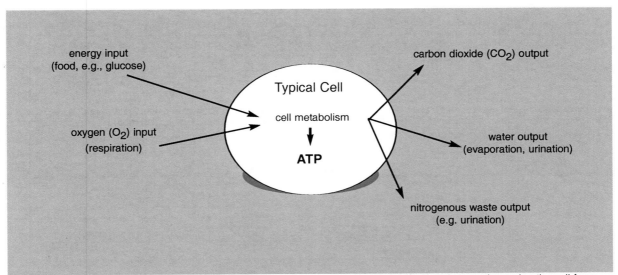

Figure 2.4. This summary diagram of cellular metabolism shows required material (food and oxygen) entering the cell from the left while by-products (water, CO_2, and nitrogenous wastes) exit from the right. The result of metabolism is the production of ATP molecules for use by the cell in its normal functions.

IV. Responsiveness

One of the universal characteristics of living organisms is their ability to respond to their environment. To be responsive, a circuit must be established in which an animal receives information and then acts accordingly. This circuit involves three steps: (1) a **sensory pathway** in which information is received and then forwarded by *receptors* to (2) an **integration center** (e.g., a brain or similar accumulation of nerve cells) where processing of incoming information takes place, and finally (3) the signal passes along a motor pathway to elicit an **appropriate response**. This generalized pathway is shown in the diagram below.

Impulses generated by receptors travel along specialized cells called **neurons** as chemo-electrical signals. Neurons typical of higher nervous systems are shown in Figure 2.6 on the next page. A neuron is composed of three parts: an enlarged *cell body*, and two types of fibers, the *axon* and *dendrite*. The nervous impulse travels toward the cell body on dendrites and away from the cell body along the axon. The interlacing network of neurons gives a nervous system its structural and, more importantly, its functional complexity.

Brains or similar structures developed as animals became more and more complex and needed to process an ever-expanding load of information. These will be discussed in detail as we explore the different groups of animals.

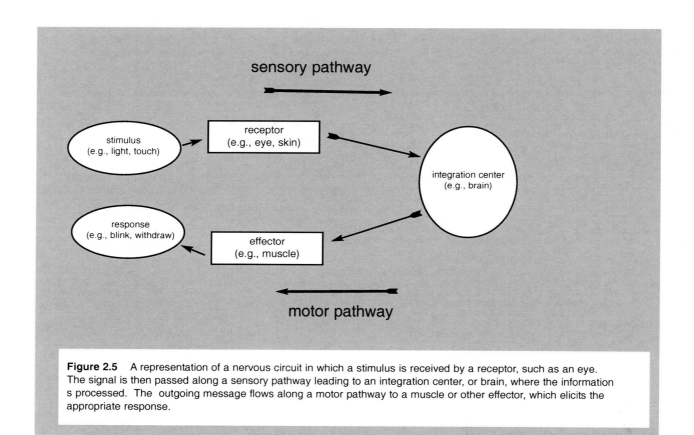

Figure 2.5 A representation of a nervous circuit in which a stimulus is received by a receptor, such as an eye. The signal is then passed along a sensory pathway leading to an integration center, or brain, where the information s processed. The outgoing message flows along a motor pathway to a muscle or other effector, which elicits the appropriate response.

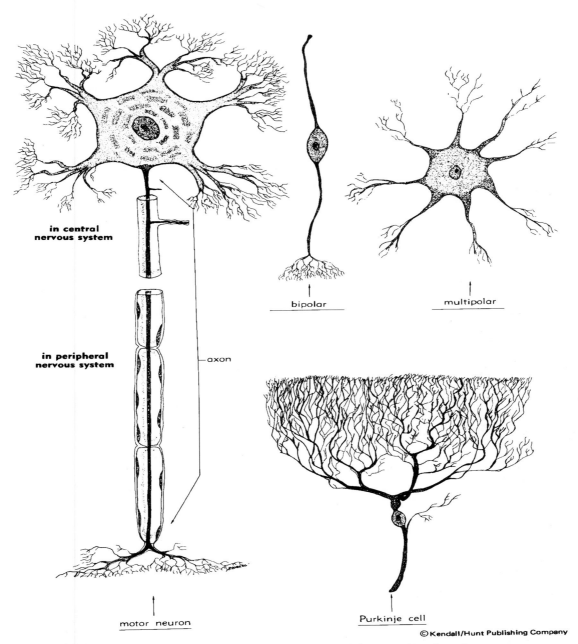

in central
nervous system

in peripheral
nervous system

axon

bipolar

multipolar

motor neuron

Purkinje cell

Figure 24-2.   NERVE TISSUE, Types of Neurons (As shown)

Figure 2.6   A variety of neurons found in higher animals is shown above. The motor neuron carries the bulk of information from the central nervous system (brain and spinal cord) to muscles where responses take place. Bipolar, multipolar, and purkinje neurons are designed to carry specialized signals.

Receptors

Structures that receive stimuli are known as receptors. They are given the responsibility of detecting or sensing an animal's external and internal environment. They fall into a number of categories, five of which are described below.

Thermo-receptors

These are devices able to sense temperature and temperature changes. This allows the organism to locate and remain in habitats of optimum, rather than extreme, temperature. Also, thermo-receptors permit some animals, such as leeches and ticks, to locate their blood meals. In higher animals, such as humans, thermo-receptors are scattered throughout the skin as sensory nerve endings.

Tactile (touch) receptors

Essentially all animals are sensitive to touch. Indeed, the tactile response is one of the most primitive yet one of the most important to the survival and well-being of an animal. Tactile sensors usually occur as projections from the body such as hairs, pegs, bristles, spines, or tubercles. As an animal contacts its environment, these projections send appropriate signals to the nervous system.

Many animals also have an ability to detect vibrations within their surroundings, especially in water. Since vibrations travel better in water than on land, it's not surprising that many aquatic animals are able to detect waves as they travel across the body surface. The lateral line system of sharks is a well-known touch-sensitive device.

Chemo-receptor

Protoplasm, in general, has a basic sensitivity to a variety of chemicals, especially those that are toxic or otherwise offensive. Thus, an animal is constantly monitoring its surroundings for the presence of chemical substances. In addition to sensing toxins, animals may also use these receptors in the location of mates. For example, some moths can detect a chemical "scent" or pheromone of a potential mate over 1/2 mile away. Their sensitivity defies imagination since it's estimated that they can detect one molecule out of a trillion in the surrounding air! Other functions of Chemo-receptor include location of food, monitoring the quality of an aquatic environment, and detecting humidity or pH. Many animals have special chemical-sensing devices such as pits, antennae, tentacles, or hairs.

Geo-receptors

These specialized receptors respond to the pull of gravity and provide vital information related to an animal's orientation. Virtually all animals resist being "upside down." If you turn any animal on its back, it will immediately do what it can to right itself. The reason for that is clear. An upturned animal is extremely vulnerable: it cannot feed properly, it cannot defend itself, it cannot locomote efficiently, nor can it carry out the details of reproduction. In short, it cannot survive. Perception of one's orientation, balance, or equilibrium is thus of utmost importance.

Geo-receptors belong to a class of sensors known as **statocysts**. Their design is extremely simple, consisting of little more than a fluid-filled capsule lined by ciliated epithelial cells. Within the fluid is a floating granule (in some instances the granule is nothing but a grain of sand) called a **statolith**. As the body tilts, the fluid within the capsule shifts, causing the granule to touch the sensitive cilia on one side. This sends information to the nervous system, which allows the animal to right itself. Many invertebrate animals have statocysts or similar structures. In fact, the elements of the human balancing system are quite similar to that of jellyfish!

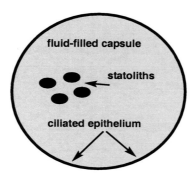

Figure 2.7. A generalized representation of a geo-receptor known as a statocyst. When tilted, a stonelike statolith strikes sensitive epithelial hairs along the statocysts' inner edge. As a result, information regarding body position and orientation is transmitted to the nervous system.

Photoreceptors

A tiny, single-celled, autotrophic organism by the name of *Euglena* is considered by many biologists as a forerunner to plants and animals. Within its cytoplasm is an array of chloroplasts used in photosynthesis. To undertake the amazing feat of transforming sunlight into chemical energy, *Euglena* must be able to orient itself in optimum lighting. So, tucked away beneath the cell membrane, at the base of the flagellum, is a photosensitive organelle, a **stigma** or "eyespot" that enables the organism to do just that. *The stigmatic apparatus is one of the first photoreceptors to appear in the parade of animals.*

All photoreceptors have in common special light-sensitive pigments that trap photons of light. The absorbed light is then used to generate an electrical potential that may affect the organism's behavior. There are four general types of photoreceptors.

(1) *Scattered photosensitive cells* may be distributed over the body's surface or concentrated in strategic areas. Earthworms, for example, have this type of photoreceptor.

(2) *Ocelli* consists of cups containing photosensitive receptors that can detect light direction and intensity. This simple type of eye cannot form distinct images.

(3) *Compound eyes* are found in some annelids, a few mollusks and, especially, arthropods. Their conspicuous eyes are composed of individual photoreceptors called **ommatidia** massed together giving a honey-combed appearance. Each ommatidium of the compound eye is aimed in a slightly different direction leading to an overlapping field of vision and the typical convex shape. Such an eye is well adapted to perceive even the slightest of movements and is likely able to detect color as well as to form some type of image.

(4) *Complex eyes* of squid and other cephalopods are very similar to the vertebrate eye (figure 2.8 C). In both, light passes through an outer membrane (the **cornea**), enters through an opening (the **pupil**), passes through a focusing element (the **lens**), and then strikes a photosensitive field (the **retina**) where light is transformed into a visual pathway carried to the brain. Such an eye is well designed to form distinct three-dimensional images and to focus sharply. Many complex eyes have photosensitive cells of two types, rods and cones, to perceive images in dim light and in color.

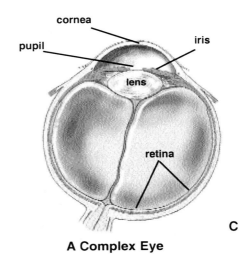

A Complex Eye

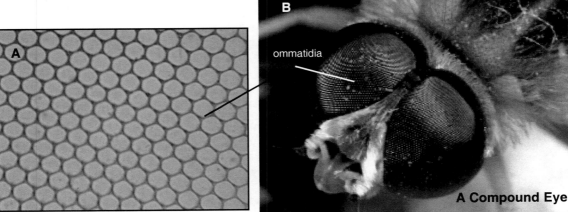

Figure 2.8. The insect compound eye is a concave surface consisting of thousands of individual units called ommatidia. Each unit detects a slightly different field, thus giving a three-dimensional image designed to detect the slightest of movements. A highly magnified surface view is given in plate **A** and is compared with the entire eye in **B**. A complex eye, such as that of cephalopods (squid) and vertebrates (humans), is diagrammed in plate **C**. Note the focusing element (lens), which gives the complex eye the advantage of sharp vision on both near and far objects.

V. Reproduction

As we have seen, reproduction is one of the most important events, if not the most important event, in the life of an animal. In certain respects we can view an individual animal as just a storehouse for the genetic instructions wound within the DNA molecule. Each animal temporarily holds a small share of the total genes (i.e., gene pool) for its particular species to be passed on to the next generation. Members of that next generation then store the message for a short time only to convey it on to their heirs, and so on. Certainly from the perspective of the population (i.e., the species), it is absolutely crucial that an animal passes along that genetic baton; chastity, for all of its supposed merits in the world of humans, goes unrewarded in the rest of the animal kingdom. Reproduction, then, can be thought of as the *measure of success* within a population. Among animals, there are but two methods available—asexual and sexual.

Asexual Reproduction

Asexual reproduction refers to multiplying through mitosis: one cell producing two identical genetic copies of itself (see the Inside View at the end of the chapter for a review of mitosis). Many simple organisms, especially the protists, reproduce primarily through mitotic division. In unicellular organisms (e.g., amoebas, flagellates, etc.), the one cell divides into two, a process termed binary fission. The cleavage plane can occur perpendicular to the long axis **(transverse binary fission)** or the plane can be parallel to the long axis as in **longitudinal binary fission** (see diagrams below). In another form of asexual reproduction, **multiple fission,** the nucleus of a single cell divides simultaneously into a number of individual nuclei, resulting in large numbers of new cells. In some cases of multiple fission, such as **sporogony** and **schizogony**, the number of new cells produced can be utterly astounding as witnessed in the malarial parasite, *Plasmodium.*

Other patterns of asexual reproduction include **budding, fragmentation,** and **gemmule** formation. Many multicellular, and a few unicellular, animals reproduce asexually through a small outgrowth from the body of the adult. In this type of reproduction, called budding, the new growth attains a critical size then drops off to assume an independent life of its own. Budding is widely used by ciliates, sponges, many cnidarians (e.g., *Hydra*, sea anemones, and *Obelia*), and even in some annelid worms. Another form of asexual reproduction is **fragmentation.** Here pieces of the adult break away to continue growth into a new individual. Sponges, sea anemones, and even starfish can reproduce in this manner. The final example of asexual reproduction is illustrated by sponges. Many freshwater sponges cannot withstand freezing or drying conditions and so have developed a strategy to bypass adverse circumstances. Special capsules called **gemmules,** containing amoeboid cells, are highly resistant to low temperature and desiccation (i.e., drying out) and are thus able to overwinter safely. Upon resurgence of favorable conditions, the amoeboid cells emerge to become the next generation of sponges.

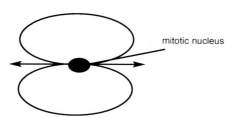

B. Longitudinal Binary Fission - A mother cell splits along its horizontal axis producing two identical daughter cells.

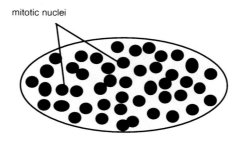

C. Multiple Fission - The nucleus within a single mother cell undergoes multiple divisions yielding a large number of daughter cells.

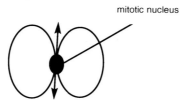

A. Transverse Binary Fission - A mother cell splits along its vertical axis forming two identical daughter cells.

Sexual Reproduction

Although many animals use asexual reproduction in one or more of its forms, the vast majority choose, instead, to multiply through sexual means. Here *meiosis*, not mitosis, is the mechanism (see the Inside View at the end of the chapter for a review of meiosis). As a result, haploid sex cells called gametes (i.e., spermatozoa and ova [eggs]) are produced. Each gamete carries specific genetic information, usually (but not always) from a different diploid parent. Once the two gametes unite during fertilization, a new, genetically unique, individual is formed. In other words, unless the product of identical twinning, each offspring produced sexually possesses a combination of genes that makes it a unique being, different from all others of its race. Although the patterns and structures involved in sexual reproduction are many, there are only two basic pathways.

(1) **Hermaphroditism.** In many animals a single individual has the capacity to produce both sperm and eggs; it's a hermaphrodite. **Monoecious** is another term used to describe this phenomenon. If an animal can form both egg *and* sperm, it only follows then that it might have the ability to fertilize itself. In actuality, some animals do just that, especially when they live a solitary existence. For example, tapeworms, which are monoecious animals, often occur singly within an intestine. If they are to take advantage of sexual reproduction, they have little choice but to self-fertilize. Some sessile or attached animals, too, are self-fertilizers. They may rarely encounter another of their kind, so fertilizing one's own eggs is the only sexual outlet.

However, it seems that if given the opportunity tapeworms and other hermaphrodites always elect to use two separate individuals to reproduce sexually. They apparently prefer mating partners. In such cases, when an egg from one individual is fertilized by the sperm of another, **cross-fertilization** has been achieved. Keep in mind that when this occurs, genetic material (usually sperm) is exchanged between both hermaphroditic individuals, and so each animal is simultaneously acting as both male *and* female. We call this type of synchronized pairing **mutual copulation** or, as it is sometimes referred, **simultaneous hermaphroditism.**

(2) **Separate sexes.** As we move along the parade of animals, we speak more and more about males and females. The reason for this is that hermaphroditism gradually gives way to the **dioecious** condition. Here, an animal can make either eggs *or* sperm, but not both. In general, gametes are produced by specialized organs called **gonads**; males produce sperm in testes, while eggs are produced by females in ovaries. *This strategy is used by the great majority of animals; most invertebrates are dioecious, as are all vertebrates.*

For sexual reproduction to be successful in dioecious animals, sperm and eggs must come together in the act of fertilization. There are only two ways this can occur: **external fertilization** and **internal fertilization.** In many groups (sponges, corals, annelids, various fishes), gametes are merely released into the surrounding water where external fertilization occurs. In **broadcast spawning** of this nature, huge numbers of eggs and sperm are discharged with the prospects that some will unite and survive to adulthood. Parents provide little if any care of the young in this reproductive strategy.

As you can imagine, broadcast spawning has some serious drawbacks. Too much is left to chance; egg and sperm may never collide, the aquatic environment may be hostile, and predators may eliminate most, if not all, of the offspring. The process is just too haphazard. So, to counter the gamble inherent in external fertilization, many aquatic and all terrestrial animals have opted for *internal* fertilization. Here, sperm is actually introduced into the female where fertilization is all but assured. To be effective, however, two things are required. First, the two animals must come together. This, too, has some risk. Qualified, mature mates may be hard to find, females may already be impregnated, or they may just be unreceptive. Thus, we see arise a whole array of behaviors or **courtship rituals** designed to bring the two sexes together. In zoology, there is no amphitheater more bizarre or fascinating than that of the courtship arena. The lengths to which some animals will go to establish mating bonds is truly amazing. From insects to birds and mammals, the time, energy, effort, and evolutionary creativity of courtship can be utterly astounding.

The second problem encountered with internal fertilization is one of **copulation**. Once the two sexes have gotten together, sperm must be introduced into the female. To ensure that this happens in the most effective manner, a number of male structures or devices have evolved. These intromittent organs, all designed for the delivery of sperm, include stylets, modified appendages, palps, tentacles, cirri, gonopods, and penises.

Figure 2.9. The dioecious condition in animals, in which separate sexes emerged, gradually gave rise to internal fertilization. This, in turn, gave rise to courtship rituals and eventually to copulation between the sexes as shown by the mating beetles and leopards.

Special Situations

There are two reproductive schemes in animals, conjugation and parthenogenesis, that defy easy classification. Each seems to lie somewhere between the asexual and sexual strategies.

Conjugation occurs in some ciliates (e.g., *Paramecium*), and although it seems to be a sexual process, it isn't. Separate sexes do not occur, and, in fact, sperm and eggs are not even formed. Instead, two mating types line up side-by-side and remain attached for several hours. During the pairing process each individual undergoes meiosis to produce haploid pro-nuclei, which are exchanged with its partner. Following this exchange of genetic material, the two organisms separate and each undergoes two sessions of binary fission to produce four individuals. Note that in conjugation there are no sperm or eggs *per se,* yet new genetic combinations do appear because of meiosis.

It is relatively common for a number of otherwise sexual animals (flatworms, roundworms, crustacea, and insects) to reproduce without normal fertilization. **Parthenogenesis** (Greek partheno, "virgin"; genesis, "birth") is the term given to a process whereby an embryo develops *without* input from both sperm and egg. There are several species of fish, for example, that reproduce parthenogenetically as well as a number of amphibians, lizards, and snakes. It has even been recorded within domestic turkeys.

Sometimes sperm are marginally involved and serve as a mere stimulus toward egg development but do not contribute their chromosomes to the process. In most instances of parthenogenesis, however, the male is even less involved; he's superfluous and doesn't participate in any way. In fact, in some populations there are no males at all! How is that possible? Well, most female animals have identical sex chromosomes (e.g., the familiar **xx** pattern of the human female). If an **xx** egg develops parthenogenetically (without spermatic input), then the only possible outcome is **xx,** or females. These species, in fact, consist of nothing but females.

As a reproductive pathway, parthenogenesis seems to have elements of both asexual and sexual reproduction. It appears to be asexual in that only one "parent" is involved. Yet, at the same time, it is somewhat sexual in that meiosis is used to form the eggs and so the offspring may be slightly different, genetically, from their mother. Entire generations of certain animals, such as aphids, ants, bees, fish, amphibians, and reptiles, may reproduce exclusively in this rather odd fashion.

Combinations

Many animals hedge their bets when it comes to procreation. They seem to straddle the reproductive fence by combining two or more techniques. For instance, many protists integrate binary and multiple fission, and some, such as the ciliates, combine binary fission with conjugation. Many metazoans, on the other hand, combine fragmentation, budding, and sexual reproduction. Sponges are monoecious (sexual), yet use asexual means to form gemmules as a way to evade inclement conditions. There are even a few species (e.g., *Obelia*) that are asexual during one phase of their life cycle only to take on the sexual mantle later on. Some species of insects, such as aphids, alternate successive generations between normal sexual reproduction and parthenogenesis.

But when viewed from a broad perspective, a definite trend emerges—reproduction evolves from asexual to sexual. The fission of single-celled organisms yields to budding and/or fragmentation. Next emerges those species that use a combination of budding/fragmentation and hermaphroditism. Finally, separate sexes are attained. And once that happens, there's no turning back. In the vast majority of animals, starting with the nematodes (i.e., roundworms), sperm are formed only by males, and eggs, in turn, by females. This is evolution's choice for virtually all species of mollusks, arthropods, and chordates (including fishes, amphibians, reptiles, birds, and mammals).

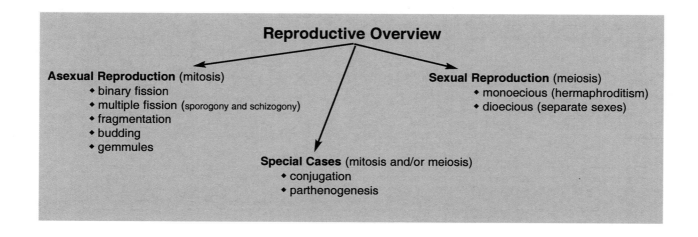

Reproductive Overview

Asexual Reproduction (mitosis)
- binary fission
- multiple fission (sporogony and schizogony)
- fragmentation
- budding
- gemmules

Special Cases (mitosis and/or meiosis)
- conjugation
- parthenogenesis

Sexual Reproduction (meiosis)
- monoecious (hermaphroditism)
- dioecious (separate sexes)

Asexual or Sexual...Which Is Better?

At this point you may very well be wondering why all the different reproductive patterns? Or you may be asking whether there's any reason for an animal to select one pattern over another? Is it better to reproduce asexually or sexually? Are there advantages that one strategy offers in comparison to another? The answer to these questions is that there are advantages to both, depending upon circumstances.

The Asexual Arsenal

When we examine the asexual strategy, we uncover two important factors: (1) asexual reproduction is practiced by the more simple animals, primarily protists and certain "lower" invertebrates, and (2) the vast majority of asexual animals are aquatic, especially marine. Putting these two factors together gives us real insight into the appeal of asexual reproduction. Many niches within aquatic environments are relatively stable; they change little. If a species is already well tuned (adapted) to its aquatic habitat, it's to that population's advantage to pass on those very fitness genes to the next generation. In other words, *genetic uniformity is an advantage in a stable environment.* Since asexual reproduction is based upon the process of mitosis, all offspring are genetically identical to each other. Thus, generation after generation the population remains suited to its aquatic home. In addition, the process of asexual reproduction, since it doesn't require a partner, is easy and quick. This is an important consideration for the more simple animals that may lack the sophistication to form gametes or to locate mates. Thus, energy and time is not wasted seeking a partner, convincing that partner of your desirability, and engaging in copulation.

A final, but very important, advantage of asexual reproduction is that it can produce huge numbers in a relatively short period of time. Just recall the outcome of such asexual practices as sporogony or schizogony in the malarial parasite. The numbers can be simply overwhelming.

> *The allure then of asexual reproduction is that it's genetically uniform, easy, fast, and prolific.*

On the other hand, environments are not static; they are subject to change, say due to drought, a severe fluctuation in temperature, or even change brought on by the invasion of a virulent disease. If that happens, then the very advantage of asexual reproduction may become a population's downfall. If a well-adapted, genetically uniform asexual species suffers such a catastrophe, it most likely lacks the genetic resources to survive the onslaught. Why is that? To understand the problem here, we must recall that natural selection operates upon genetic variety in a population. When variety is lacking, as it is in asexual populations, the entire population may fall prey to a catastrophic change. An asexual population's only source of genetic variation is through mutation. However, mutations are unpredictable and are, in any case, primarily detrimental or lethal. So mutations in asexual populations are just too unreliable to offset a drastically changing circumstance.

> *Asexual reproduction's major disadvantage lies in its very genetic uniformity. It just doesn't work well in a changing environment.*

The Sexual Invasion

The vast majority of animals that inhabit the earth are sexual. This tells us that sexual reproduction must hold a powerful attraction. And, as with asexual reproduction, the reason is linked to the environment. Most animals live in habitats that undergo periodic, and at times, regular change. As indicated in the discussion above, a population's survivability during environmental insults is related to genetic variation. The more diverse the population, the more likely some individuals will be preadapted in favor of a change. As a result, they are the ones who will survive and reproduce to pass along their survival genes. But sexual reproduction has several drawbacks: (1) in many cases, it's an energy-demanding process; (2) it's often metabolically wasteful in that many eggs and sperm go unused; and (3) it may be time-consuming (locating a proper partner, persuading that partner to accept you, and so on). Nonetheless, sexual reproduction has one mighty advantage...**genetic variation.**

Sexual reproduction relies on the formation of gametes produced through meiosis. When fertilization results, a new, genetically unique individual is formed. It is upon this variation that natural selection can exercise her magic wand. Evolution in a changing environment depends mightily upon this genetic variety. Whereas asexual reproduction preserves the genetic *status quo* through mitosis, sexual reproduction, by its very nature of mixing genes from both gametes, dismantles it. But by dismantling the parental genotype, new and unique genotype are created. The effect of this inherited variability is so potent that most animals have opted for dioecious (male-female) reproduction. The clear advantage of separate sexes is the genetic versatility bestowed by the combination of genes from two distinct parents.

IT'S A FACT

THE OVERWHELMING POWER OF SEXUAL REPRODUCTION IS ITS GENETIC VARIETY WHICH PROPELS EVOLUTION WITHIN CHANGING ENVIRONMENTS. THIS MORE THAN COMPENSATES FOR THE ENERGY EXPENDED, THE WASTEFULNESS, AND THE DEMANDS INHERENT IN THE SEXUAL PROCESS.

The Pros and Cons of Reproductive Strategies

Asexual Reproduction

Disadvantages = nonadaptive; cannot readily adapt to changing environments

Advantages = genetically uniform, quick, easy, and prolific; works well in a stable environment

Sexual Reproduction

Disadvantages = time-consuming, energy demanding, may waste gametes

Advantage = genetic variation permits adaptation to changing conditions.

An Inside View 2.2

The Mitotic Marvel

What happens when you cut yourself? How does healing actually occur? How did you grow from a toddler to adulthood? How are the millions upon millions upon millions of cells that you lose each day replaced? What happens to your cells as you become older? Why can't you live forever? How does a primitive organism such as an amoeba reproduce? The answer to each of those questions lies in the fundamental biological process of mitosis, a process common to all known organisms. What is it about this process that makes it such a vital part of life?

Zoologists universally agree that single-celled organisms were the ancestors to multicelled ones...to the rest of us. They have concluded, and rightfully so, that all higher animals came from simple single-celled protozoans. However, for a unicellular organism to become multicellular it must first find a way to make a copy of itself— an exact copy. If you don't quite understand why that's so, just imagine that you have burned your finger, resulting in the destruction of a number of epithelial cells. You want those dead cells replaced—but not by brain cells, not by liver cells, not even by muscle cells. You want them replaced by epithelial cells...you want your finger back the way it was. The events leading to the replacement of your finger cells lies at the very heart of mitosis.

We know that the instructions for life's many processes are wound in the genetic code. It's the DNA in our cells that's responsible for the structure and function of each cell, and by extension, for structure and function of the entire body. So, the answer to the problem of making perfect copies must be found in DNA. The first thing a cell must do to copy itself, therefore, is to make a copy of its DNA. That first step is called **replication**, chromosomal replication, to be exact. If a cell can make an unerring copy of its DNA, then the DNA will, in turn, direct each cell to carry out its specific functions. A skin cell behaves as a skin cell, a brain cell does what brain cells are supposed to do; DNA directs them along their proper paths. Not to belabor this point, but the key here is to make an exact copy of each and every chromosome. Replication is, in fact, the crucial event in the mitotic marvel.

Mitosis proceeds as a series of steps leading to the production of two new daughter cells genetically identical to each other. The first step, replication, occurs during cell interphase and results in the appearance of a double-stranded chromosome as diagrammed below. Each identical strand is a sister chromatid held together by a centromere. The actual process of mitosis, then, simply involves a sequence of events (prophase, metaphase, anaphase, and telophase) during which the sister chromatids are separated from each other and eventually passed on to two new, genetically identical cells (see Figure 2.10 on page 60).

Once the chromosomal material has been allocated to each daughter cell, the new cells can either separate to produce new organisms or they may remain together, continuing the process over and over, to produce a scar, or to replace damaged or lost cells, or to form a growing tissue.

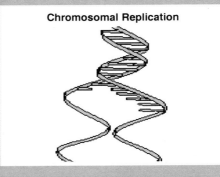

Chromosomal Replication

The process is quite simple and reliable, as long as the chromosomes are replicated exactly. But mitosis is not without its difficulties. First of all, mutations can occur within the DNA molecule, which may lead to failed copies and potential disaster. Also, the process does not go on indefinitely. Quite the contrary. Have you noticed that as you age you take longer to heal? That cut on your finger that took a few days to heal when you were a youngster takes longer and longer to recover as you age. We think the reason for this is that DNA can only be copied so many times. Each replication slightly erodes the process until eventually cells lose their ability to reproduce altogether...a process termed **senescence**. Alas, we cannot live forever; the genetic code has seen to that! But our genes haven't abandoned us entirely. Instead, they have improvised a process that allows a part of us to live on. That process is *sexual reproduction through meiosis* and is the topic of our next *Inside View.*

The important thing to remember about mitosis is that, because of replication, each chromosome is duplicated **exactly,** which makes it possible for each new daughter cell to be genetically identical to each other. Since all multicellular organisms (including you and me) started life as a single cell (a zygote) and then relied upon mitosis to add cell after cell after cell, we can conclude that each cell in our body is genetically identical. That is true. In fact, it's the basis for cloning, for producing exact replicas of a living organism...a frog, a sheep, a human!

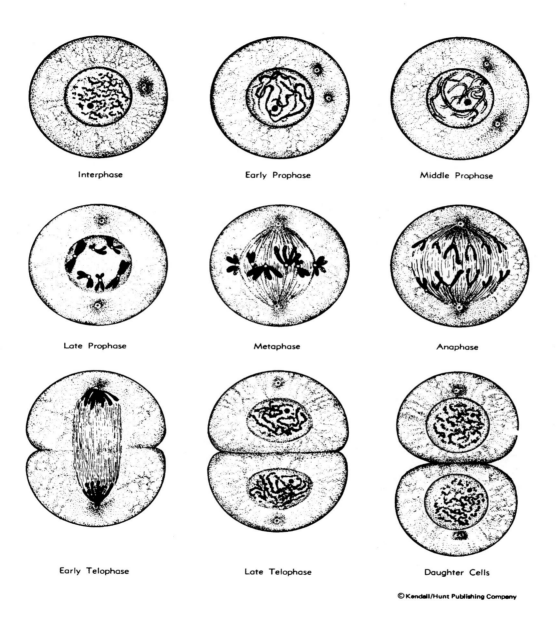

Interphase   Early Prophase   Middle Prophase

Late Prophase   Metaphase   Anaphase

Early Telophase   Late Telophase   Daughter Cells

Figure 2.10. The process of mitosis in an animal cell.

An Inside View 2.3

Meiosis: Evolution's Revolution

It should be apparent from the preceding discussion on mitosis that asexual reproduction leads to genetic uniformity. In other words, at the conclusion of mitosis, two genetically identical daughter cells are produced. Thus, huge populations of identical cells, whether tissues or amoebas, can be issued in a relatively short period of time. But uniformity, as important as it can be, is not the result, nor is it the goal, of sexual reproduction. On the contrary, sexual reproduction leads, not to homogeneity, but to genetic variability. Diversity is its very essence. And keep in mind that it's sexual reproduction that's chosen by the vast majority of animals.

While we may think of sexual reproduction as a process requiring two separate parents, that's not always the case. Indeed, hermaphroditism is alive and well in the animal world. The key to understanding the sexual pattern of reproduction is not in the action of separate parents, but in the production and action of two different sex cells called **gametes**. These two cells, each carrying different genetic information, unite during fertilization to form a new, genetically unique, cell, the **zygote**. *Sexual reproduction then is the union of different sex cells, usually termed sperm and egg, and usually, but not always, derived from a male and a female parent.*

Since the gametes are genetically distinct from each other, so then is the zygote a genetically distinct individual. That's the key to sexual reproduction. Soon after its appearance, the zygote begins to divide into an ever-expanding explosion of cells...it begins to grow and develop. To do so, it relies upon mitosis. But at some stage, let's call it adulthood, the new individual has the capacity to contribute its genes to the production of gametes, sexual reproduction follows, a new (genetically distinct) zygote is formed to eventually reproduce, and so on. You can now see that the secret to sexual reproduction lies squarely in the production of gametes. And how are gametes formed? The answer is not found in mitosis, but through a similar process called meiosis.

Meiosis starts out in the same manner as mitosis in that chromosomes are replicated. But instead of the sister chromatids separating from each other, they stay together and each links with a predetermined partner, its homolog. The diagram on the next page shows this process.

Homologs are defined as a pair of chromosomes carrying genes for the same traits: eye color, skin color, fingernail shape, and so on. These homologous chromosomes come together early in meiosis in an event called synapsis. Following synapsis, the homologs (still consisting of two sister chromatids) are separated from each other and distributed to two individual cells. Each of these cells then undergoes a process reminiscent of mitosis in which the sister chromatids do finally separate into the formation of two new cells. Therefore, as a result of meiosis, four daughter cells (gametes), each containing one-half the number of chromosomes of the parent cell, are produced. In other words, during meiosis a diploid cell undergoes reduction division to produce haploid gametes. Two of these gametes (sperm and egg), each with a different genetic pattern, unite during fertilization to produce a new, genetically distinct individual.

Of utmost importance in sexual reproduction is that each newly produced individual is genetically different from every other member of that species. The lone exception is when identical twins develop. Other than that, all sexually reproducing organisms are distinct.

The importance of this diversity is difficult to overstate since the entire evolutionary process in higher animals depends upon it. Because members of a population differ from each other, some will most likely have a competitive advantage over others of their race; they are better fit for their ecological niche. They will have a higher survival rate. Those that survive, or otherwise out-compete their cohorts, will, of course, be the ones to reproduce. The result is that the survival genes are passed along.

IT'S A FACT

Why do the vast majority of animals "chose" the sexual pathway? The answer is clear. They do so because biodiversity is the route to evolutionary success. And what's responsible for this great diversity? Why, nothing less than the intermingling of genes resulting from meiosis.

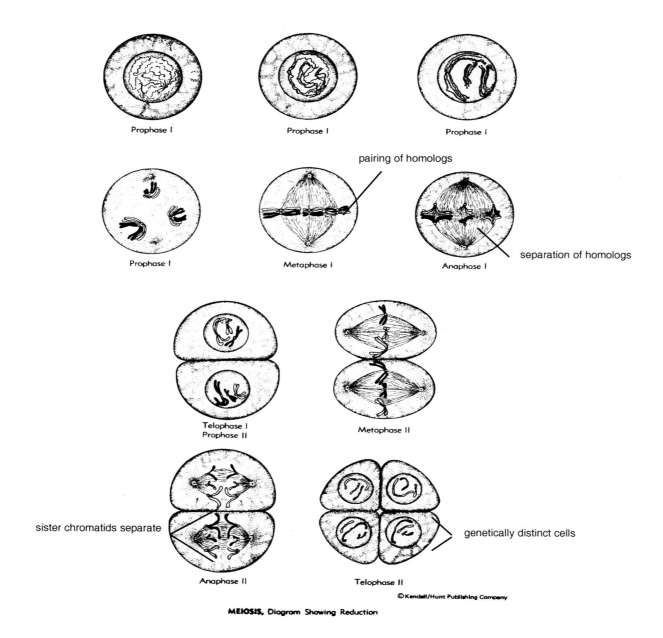

Figure 2.11. A diagrammatic representation of meiosis in an animal cell.

Chapter 2 Summary

Key Points

1. There are five recognized life kingdoms: Monera, Protista, Fungi, Plantae, and Animalia.

2. According to the Cell Theory, all life consists of cells, cells are the fundamental unit of living organisms, and all existing cells originate from pre-existing cells.

3. There are two types of cells. The bacterial are composed of prokaryotic cells while all higher life-forms consist of eukaryotic cells.

4. A definition of an animal is: *a non-photosynthetic organism consisting of eukaryotic cells lacking cell walls.*

5. There are three basic grades of animal structure: cellular, multicellular, and colonial.

6. Multicellular organisms may be arranged into cells, tissues, organs, and organ systems (of which there are 12 in the human). All levels show division of labor in that they are to one degree or another specialized to perform different functions. Division of labor leads to greater efficiency and increased size of the organism.

7. Animal form is related to a great extent to either its habitat or its mode of locomotion. Some animals have an asymmetrical body design, a few show radial symmetry, but most are bilaterally symmetrical.

8. As animals become more mobile, they develop specific body patterns. The most common is arranged along an anterior-posterior axis with the anterior end aimed in the direction of travel. The body portion normally directed toward the substratum is ventral and that portion away from the substrate is dorsal.

9. Any structure farther from the body midline is termed distal or lateral; a structure closer to the midline is proximal or medial.

10. Higher animals can be sectioned into three planes: transverse, frontal, and sagittal.

11. All animals strive to maintain an internal stable condition with respect to a large number of physiological states. Such examples as temperature, pH balance, fluid balance, and ionic balance illustrate this attempt at homeostasis.

12. Osmoregulation is the tendency to maintain proper fluid and ion (esp. salt) balance and is an excellent example of homeostatic mechanisms. If a cell's internal salt level is equal to the surrounding medium, it is in an isotonic state. However, if the salt level surrounding the cell is higher than that inside, the cell is in a hypotonic state. Conversely, if the environmental salt level is lower than that internally, the cell is said to be in a hypertonic state.

13. Getting sufficient energy is a vital matter for animals since all animals are heterotrophic and must rely upon plants or animals for their nourishment. As a result, they have adopted a wide range of feeding strategies: they may be herbivores, carnivores, omnivores, scavengers, suspension feeders, deposit feeders, or detritovores.

14. All tissue is responsive to environmental stimuli.

15. Relative simple animals have developed nervous systems capable of sending impulses along a specialized cell, the neuron. These electrochemical impulses make it possible for animals to sense, integrate, and respond to a variety of stimuli such as light, temperature, touch, gravity, and chemical.

16. Animals have invested a great deal of their evolutionary capital into photosensors. Some have light-sensing organs scattered over the body surface, while others have concentrations of photosensors such as ocelli, compound eyes, or complex eyes.

17. Animals have two basic reproductive strategies: asexual, in which genetically identical offspring are formed, and sexual, which results in genetically unique individuals.

18. Asexual reproduction is quick, easy, and prolific but is limited because of its genetic uniformity.

19. Sexual reproduction may be slower, less prolific, and more wasteful than asexual means, but it has a powerful attraction due to its genetic diversity.

20. Sexual patterns may be monoecious in which individuals produce both eggs and sperm or they may be dioecious, wherein eggs and sperm are formed by separate individuals, usually males and females.

21. Fertilization may be external or internal.

22. Two unusual reproductive patterns (conjugation and parthenogenesis) occur in which elements of both asexual and sexual mechanisms are used.

23. Many animals use two or more reproductive strategies.

24. Mitosis is a biological process whereby one cell divides into two genetically identical cells. It is used for asexual reproduction as well as to replace damaged tissue and for overall growth and development.

25. Sexual reproduction is accomplished wherein the diploid number of chromosomes is reduced by one-half in the production of gametes.

26. Haploid gametes unite during fertilization to reestablish the diploid chromosome number in a genetically unique zygote.

Key Terms

anterior
asymmetrical
bilateral symmetry
binary fission
broadcast spawning
budding
carnivore
centromere
Chemo-receptor
chromosomal replication
colonial
complex eye
compound eye
conjugation
cross-fertilization
deposit feeding
diffusion
dioecious
diploid
distal
dorsal
ectoderm
endoderm
eukaryote
external fertilization
fragmentation
frontal section
gametes
genetic uniformity
genetic variation
genotype
geo-receptors
gonads
haploid
herbivore
hermaphrodite
heterotrophic
homeostasis
homologous chromosomes
hypertonic
hypotonic
integration
internal fertilization
intromittent organs
isotonic
lateral
medial
meiosis
mesoderm
metazoan
midline
mitosis
monoecious
motor pathway
multiple fission
mutual copulation
neuron
ocellus

omnivore
osmoregulation
parthenogenesis
phenotype
photoreceptor
posterior
prokaryote
proximal
radial symmetry
sagittal section
scavenger
schizogony
self-fertilization
senescence
sensory pathway
simultaneous hermaphroditism
sister chromatid
sporogony
sporozoite
statocysts
statolith
stigma
suspension feeding
symmetry
synapsis
tactile receptors
thermo-receptor
transverse section
ventral
zygote

Points to Ponder

1. There are five kingdoms of living organisms. Name them and give an example of each.

2. What is the advantage of multi-cellularity?

3. What are the three tissue layers found in animals?

4. Be able to match the 12 animal organ systems with their functions.

5. Compare form of an asymmetrical, radially symmetrical, and a bilaterally symmetrical animal.

6. What is meant by homeostasis? Give three examples of items that animals regulate through homeostasis.

7. Describe binary fission.

8. Define osmoregulation. Describe the process by including the following terms: isotonic, hypertonic, and hypotonic.

9. Diagram the nervous system pathway including the sensory, integrative, and motor components.

10. Be able to define and give examples of the various types of animal receptors (especially photoreceptors).

11. Compare deposit feeding and suspension feeding.

12. Give examples of animals that are herbivores, carnivores, omnivores, and scavengers. Define the term heterotrophic.

13. What are the advantages and disadvantages of asexual reproduction?

14. List and define the various types of asexual reproduction used by animals.

15. What are the advantages and disadvantages of sexual reproduction?

16. Compare the processes of self-fertilization and cross-fertilization used by hermaphroditic animals. When would each be used?

17. Contrast external and internal fertilization. Under what circumstances would each be preferred? What is broadcast spawning?

18. What two problems are introduced by internal fertilization? What are its advantages?

19. Define the role of mitosis and meiosis in both asexual and sexual reproduction.

20. Why would some organisms use sexual reproduction during part of their lives only to revert to asexual at another time?

21. Can you think of any advantages to the use of parthenogenesis by animals?

22. On the animals below indicate the following:

- anterior
- posterior
- ventral
- dorsal
- proximal
- distal
- medial
- lateral
- transverse section
- frontal section
- sagittal section

Inside Readings

Brusca, R. C., and G. J. Brusca. 1990. Invertebrates. Sunderland, MA: Sinauer Associates.

Cumings, M. R. 1994. Human Heredity. St. Paul: West Publishing Co.

Freeman, S. and J. C. Herron. 1998. Evolutionary Analysis. Upper Saddle River, N J: Prentice Hall.

Hamer, D. and P. Copeland. 1998. Living With Our Genes. New York: Doubleday.

Hauser, M. D. 2000. Wild Minds: What Animals Really Think. New York: Henry Holt and Company.

Mayr, E. 1997. This Is Biology: The Science of the Living World. Cambridge: The Belknap Press.

Moir, A. and D. Jessel. Brainsex. 1991. London: Mandarin Press.

Page, G. 1999. Inside the Animal Mind. New York: Doubleday.

Prochiantz, A. 1989. How the Brain Evolved. New York: McGraw-Hill.

2 Chapter 2

FORM & FUNCTION

Laboratory Exercise

Laboratory Objectives

✓ to differentiate between the various body forms
✓ to understand the function of the three body forms

Supplies needed

● modeling clay or "play dough"

Chapter 2   Worksheet

Body Form & Function

(also on page 583)

I. Body Surfaces

1. Make a model of an animal of your choice out of "Play Dough". Just so we're sure which end is which, use the probe to draw a mouth and an anus on your animal.

2. Use the probe to mark the midline on your model between the anterior and the posterior end. Mark both the dorsal and the ventral surface.

3. Sketch your animal in the space below. Label anterior, posterior, dorsal and ventral.

II. Sections (to view the internal structures of an animal, it is often cut in half or a thin slice is cut from it)

1. Draw a line on each of the animals below to indicate a **transverse** and a **sagittal** cut.

(a transverse cut produces a **cross section** and a sagittal cut produces a **longitudinal section**)

TRANSVERSE

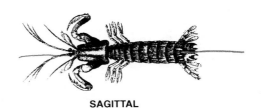

SAGITTAL

2. Using a scalpel, cut your "animal" in half making either a transverse or a sagittal section. Your partner should make the other cut. How would your view of the interior of the animals differ in the two cuts?

III. Radial Symmetry

1. Recycle your animal and use your Play Dough to make a model of a radially symmetrical animal with several appendages.

    a. where would you put the mouth?

    b. is there an anterior and a posterior end?

    c. does it matter where you draw the midline?

octupus, Jellybish, Starbish

IV. Body Design (sac-like vs. tube-within-a-tube designs)

1. Use 1/2 of the Play Dough to make a simple sac-like body with only one opening.

2. Use the other 1/2 of the Play Dough to make a tube-within-a-tube body with an opening at each end and a continuous tube connecting the two openings.

    a. if you were to feed these two animals, where would the food enter? where would wastes exit?

3. In the space below draw each of your body plans. Use arrows to indicated the flow of food and wastes.

Snake have two opening mouth to enter bood, anus to take out waste

Jellybish have one opening where bood & waste exist

3 Chapter 3

THE PROTISTS

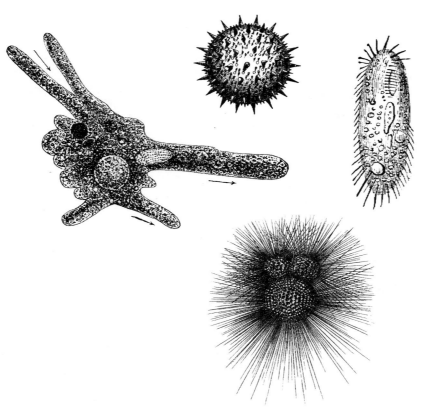

At the interface between plant and animals...
lie single-celled organisms, the protists.

Kingdom Protista

*....from so simple a beginning endless forms most beautiful and
most wonderful...are being evolved.*
-Charles Darwin

Introduction

The single-celled organisms of the Kingdom Protista exist at the crossroads of plant and animal life. Many of these microscopic forms display basic plant characteristics such as the presence of chlorophyll-filled organelles called plastids. Chlorophyll confers a greenish hue to these organisms and, more importantly, gives them the capability of manufacturing their own energy through the complicated process of photosynthesis. Because photosynthetic organisms produce their own food from sunlight they are known as **autotrophs**. Examples of plant-like protists are *Euglena*, *Chlamydomonas*, and *Volvox*.

On the other hand, many Protists seem to be animal-like in their appearance and behavior. Some are highly mobile and others move about sluggishly in their search for food. Many are predators on other protists and may show a variety of specialized "hunting" skills. Still others simply filter their food from the surrounding aquatic environment. However, regardless of their particular food-gathering tactics, none of the animal-like forms are able to manufacture their own energy and must, instead, rely directly upon external sources for their nourishment. This type of energy acquisition is referred to as **heterotrophism**.

What is important to remember about protists is that *they exist as a single cell.* And as such they are able to accomplish all life processes in the microscopic arena of that one cell. Think about it...they are highly successful organisms yet live their lives as tiny, unseen individuals. They accomplish everything necessary from feeding to locomotion, from respiration to osmoregulation, from homeostasis to reproduction as minuscule, one-celled creatures.

Figure 3.1. Two amoebas showing the elongated pseudopodia, the central nucleus, a number of food vacuoles, and a contractile vacuole.

Phylum Sarcomastigophora

Subphylum Sarcodina (the "ameboids")

The amoeboid organisms are among the most common of the protists. They are found in both fresh and marine environments nestled under leaves, stones, plants, and other debris. There they feed on a variety of miniature organisms such as bacteria and other protists.

They are characterized by two major features: a semitransparent cytoplasm and a variable number of elongated **pseudopodia** (Figure 3.1). The cytoplasm is a constantly changing matrix housing a nucleus, numerous food vacuoles, and one or more contractile vacuoles. The outer region of the cytoplasm is a translucent layer, the **ectoplasm,** enclosing a viscous, granular inner region of **endoplasm**.

All amoeba are **asymmetrical**, lacking any semblance of a head, a rear, or right and left sides. They crawl about their tiny microhabitats by steadily extending and retracting their fingerlike pseudopodia— an action aptly called *amoeboid movement*. It is fascinating to watch a living amoeba as it shifts direction, alters it shape, and streams through the channels of its ever-extending pseudopodia in search of food. As they collide with any edible tidbit, it is surrounded and engulfed through **endocytosis**. The food is then converted into a food vacuole where **intracellular digestion** (i.e., inside of the cell) occurs.

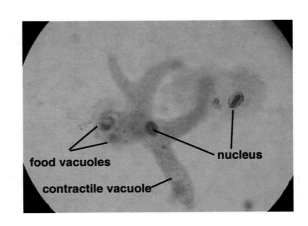

food vacuoles

nucleus

contractile vacuole

All animals are in constant interaction with their environments. Of particular importance is the ability of an organism to maintain an internal balance between water and ions, especially salts, in relation to its surroundings. This universal internal balancing act, **osmoregulation**, is an ongoing process within all animal tissue.

In freshwater animals there is a contrast between ions (salt) inside of the cell and that of their environments. Because molecules move along a gradient from high concentration to a lower concentration, water (which is more highly concentrated outside of the cell) tends to diffuse into the cytoplasm. Without consistent removal of excess water these cells will continue to expand and stand the very real threat of rupture. In amoeba, and most other protists, osmoregulation is accomplished by a **contractile vacuole**, a small, clear organelle that consistently constricts to remove excess water. Indeed, one of the interesting aspects of observing a freshwater amoeba is the disappearance and reappearance of these bubble-like vacuoles as they squeeze out surplus water.

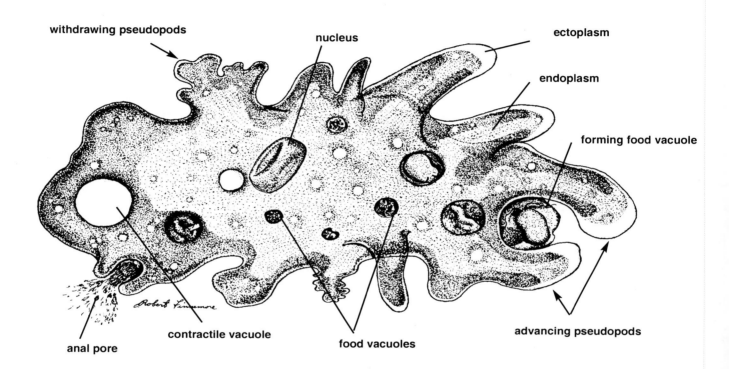

Figure 3.2. Structure of an amoeba.

1. contractile vacuole
2. nucleus
3. advancing pseudopod
4. food vacuole

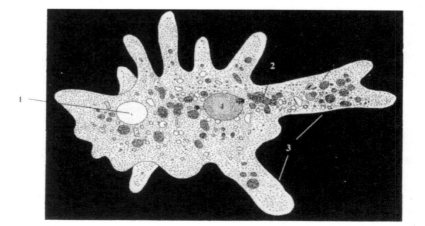

A number of marine sarcodines (amoeba), such as the radiolarians and foraminiferans, form protective outer shells (see below). These external skeletons, called **tests,** consist of calcium carbonate, silica, or other materials. Some species can fill the test with air and just float away from danger. For millions and millions of years these beautifully intricate shells have rained down upon the ocean floor as *sea snow.* In fact, limestone and chalk deposits that occur throughout the world are the remnants of these tiny crusts.

Not all sarcodines are free living. Many occur, instead, as internal parasites of other animals. For instance, the small amoeba *Entamoeba histolytica* is a common intestinal parasite of humans. It is contracted by ingesting food or water contaminated with human feces that contain the parasite. The result is **amoebic dysentery,** in which an infected person experiences nausea, cramps, diarrhea, dehydration, and other flu-like symptoms. In severe cases, death can result.

A variety of sarcodines are shown below.

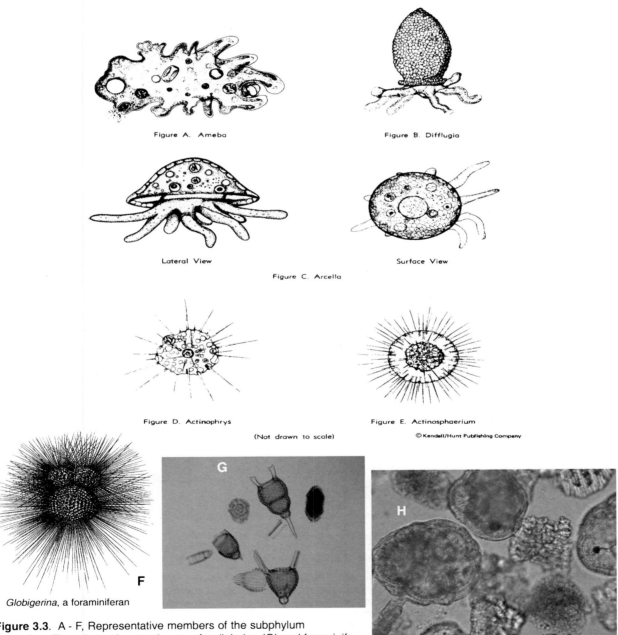

Figure A. Ameba

Figure B. Difflugia

Lateral View

Surface View

Figure C. Arcella

Figure D. Actinophrys

Figure E. Actinosphaerium

(Not drawn to scale)

© Kendall/Hunt Publishing Company

Globigerina, a foraminiferan

Figure 3.3. A - F, Representative members of the subphylum Sarcodina. The photomicrographs are of radiolarian (G) and foraminiferan (H) tests, which serve as protective shells around certain marine amoebas.

Subphylum Mastigophora
(the "flagellates")

Members of this group are quite unlike their sarcodine cousins. First of all, they lack pseudopodia and the consequent amoeboid movement. As a result, they tend to maintain a more consistent shape and their locomotion is accomplished, not by amoeboid flow, but instead by the thrashing action of one or more whiplike **flagella**.

Euglena is a common flagellate that exhibits both plant and animal features. For example, it moves animal-like with its flagellum, yet it is autotrophic. Scattered throughout its cytoplasm are chlorophyll-filled organelles called chloroplasts that carry out the complex processes of photosynthesis. Figure 3.4 below shows the structure of this rather small mastigophoran. Note the presence of the **stigma** (eyespot). This simple sensory device functions to detect the light necessary for photosynthesis.

A very different photosynthetic mastigophoran is *Volvox* (below). While most protistans spend their lives as unicellular, independent individuals, *Volvox* occurs as hollow, spherical clusters of up to 50,000 individuals. These clusters or colonies are thought by many zoologists to be the ancestor of multicellular animals. Indeed, some authorities actually view *Volvox* as a multicellular organism. Until more evidence is produced to the contrary, however, we will adopt the more traditional colonial theory for *Volvox*.

Each individual cell comprising the *Volvox* spheroid has its own nucleus, chloroplasts, and contractile vacuoles. In addition, each cell has two flagella extending outward. The entire colony is propelled in a slow, rotating fashion by the combined beating of thousands of flagella. The individual cells share cytoplasmic bridges to bind the entire colony into one spherical mass.

Reproduction in *Volvox* is a combination of asexual and sexual processes. As you observe various specimens, notice the smaller **daughter spheres** located within the large **mother colony**. The variable sizes of the daughter spheres represent different stages of asexual, mitotic development. The daughter spheres will eventually escape to become established as new independent mother colonies.

There are also certain specialized cells of the mother colony that can produce male and female **gametes** (sex cells) through meiosis. The sex cells, similar to spermatozoa and eggs, then undergo fertilization to produce a genetically distinct mother colony.

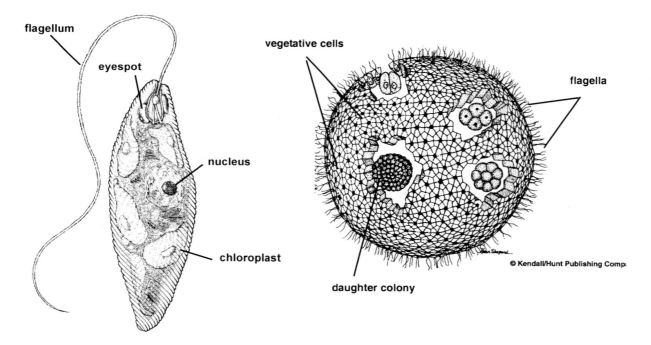

Figure 3.4. Diagrams of two mastigophorans. Plate **A** is of a common photosynthetic protistan, *Euglena*. Note the single flagellum, the presence of chloroplasts for photosynthesis, and the eyespot at the base of the flagellum. The diagram to the right is the colonial mastigophoran, *Volvox*. Each vegetative cell possesses two cilia and is attached to neighboring cells by way of a cytoplasmic bridge.

Imagine being bitten by a common fly in your neighborhood and dying from that single bite. This is what happens to over 10,000 Africans each year. But it's not really the bite of the tsetse fly (*Glossina*) that kills you, it is, instead, what is transported by that bite. One of the most ravaging of human diseases, African sleeping sickness, is caused by the tiny flagellate *Trypanosoma* (Fig.3.5). This parasite resides among red blood cells, where it can reach enormous numbers. If the parasite is transported to the brain, it may induce a coma, which is invariably fatal.

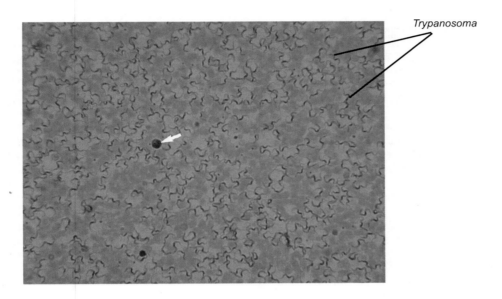

Trypanosoma

Figure 3.5. The photograph above shows numerous trypanosomes (the squiggles) clustered among red blood cells. The arrow points to a white blood cell (a monocyte). The diagram below depicts the life cycle of trypanosoma as it affects humans.

3. If the trypanosome enters the brain, a coma will eventually result, leading to certain death.

1. Tsetse fly picks up the *Trypanosoma* parasite while feeding on an infected person.

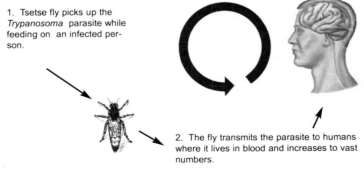

2. The fly transmits the parasite to humans where it lives in blood and increases to vast numbers.

Phylum Ciliophora

(the "ciliates")

Structure. The ciliates are among the largest and most complex cells known and are the most advanced members of the Protista kingdom (Figure 3.8, next page). All members of this large group possess a profusion of hairlike projections called **cilia,** which are used for a variety of purposes such as locomotion, feeding, and creating a current of water for respiration.

Ciliates are also unique among the protists in that they display **nuclear dimorphism**. Unlike other protozoans that possess a single nucleus, the ciliates may have two, three, four, or more nuclei. Some nuclei even have the rather odd appearance of beadlike strings as in the large ciliate, *Stentor* (Figure 3.7). Furthermore, these organisms are encased in a rather tough covering, the **pellicle** through which the cilia project. This protective covering helps to maintain a somewhat fixed shape. This covering, in some forms, is indented into an oral groove or **gullet,** which usually ends as a funnel-shaped **cytostome** (Figure 3.9). Food is brought into the gullet by the beating of the cilia.

Another unusual feature of ciliates is the presence of structures called **trichocysts,** which can be seen arranged in rows just beneath the pellicle. While there is some confusion over the function of these tiny capsules, we know that when disturbed, a ciliate will discharge its battery of trichocysts, dispersing long threads. Some trichocysts even seem to contain a toxin. Possible uses of such structures are for predation, protection (i.e., defensive), or as anchoring devices to attach the organism to a given substrate.

Like other protists, ciliates are endowed with **contractile vacuoles** used to expel surplus water. In some cases, the contractile vacuoles are fed by a series of radiating canals that serve as collecting tubes for water to enter the enlarged vacuole.

Locomotion. As the flagellates use their flagella to swim about, so, too, do the ciliates rely upon cilia for locomotion. Although cilia and flagella are quite similar to each other in structure, there are two major differences. First, cilia are usually much shorter and are considerably more abundant. They often form rows, rings, or striations along the circumference of the organism. These rings are especially prominent around feeding grooves. Some ciliates are covered with cilia giving them a "furry" looking coat.

A second difference is that cilia seem to beat together in unison. This allows the animal to move in a rapid, smooth, sometimes rotating, fashion. In fact, one of the most apparent differences between ciliates and other protists is this accelerated, smooth method of locomotion. It is very much unlike the plodding flow of an amoeba or the "herky jerky," spasmodic movement of the flagellates. Uniform ciliary beating also causes water to flow along their surface, a phenomenon called **wafting**. Wafting is used by many ciliates as a method of feeding and in respiration.

Although a few ciliates are parasitic on other organisms, the great majority are free living. Most are highly motile, swimming about in their incessant search for food. At times a ciliate may become temporarily attached to a given substrate, and some may even remain permanently attached (sessile) to a living or nonliving surface.

Nutrition. Ciliates use a variety of feeding methods. Some are predacious, even feeding on other ciliates. Others collect their food from the surrounding water from ciliary-induced currents. Still others are parasitic on invertebrate and vertebrate hosts, including humans.

A common feeding pattern is seen in *Paramecium*. A water current is established by cilia lining the gullet. From this current, food, such as bacteria, algae, or small protists, is isolated into food vacuoles where digestion occurs. A slight deviation of this pattern occurs with other ciliates such as *Vorticella* and *Stentor* (Figure 3.7). Here, the organism is attached by a stalk to the substrate with its oral groove extended upward. Cilia surrounding the oral surface beat to create a water vortex from which food particles are filtered. *Vorticella* is permanently attached to its substrate by a contractile, springlike stalk. When disturbed, the stalk quickly withdraws the organism out of harms way.

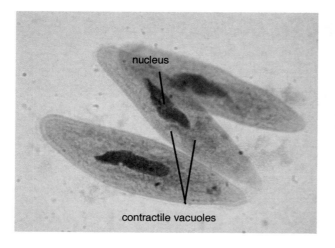

Figure 3.6. Three organisms of the species *Paramecium caudatum,* showing an elongated macronucleus and a number of bubble-like contractile vacuoles.

Figure A. Paramecium

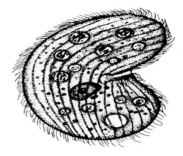

Figure B. Colpoda

Figure C. Stylonichia

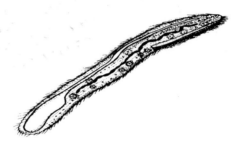

Figure D. Spirostomum

Figure E. Vorticella

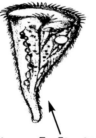

Figure F. Stentor

(Not drawn to scale)

Figure 3.7. Representative ciliates (above) and the large ciliate, *Stentor* (right).

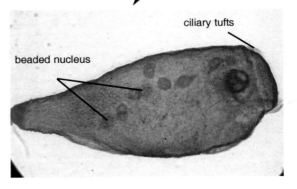

ciliary tufts

beaded nucleus

Reproduction. One of the most bewildering features of ciliates is their reproduction. Some reproduce entirely through **binary fission** (chapter 2), in which one cell simply undergoes mitosis to form two genetically identical organisms. However, many paramecia show a sexual-like pattern called **conjugation** (Figure 3.8). Ciliates are unique among organisms in possessing two different types of nuclei: a macronucleus and a micronucleus. The **macronucleus** is responsible for day-to-day activities of the cell and for *asexual* reproduction. The **micronucleus**, on the other hand, is responsible only for conjugation. In conjugation, sex cells (i.e., spermatozoa and ova) are *not* formed. Instead, two individuals of different mating types co-join and exchange genetic material by way of a **cytoplasmic bridge**. As a result, each conjugant receives DNA from its partner. *Once they separate, each undergoes transverse binary fission twice to produce four genetically unique paramecia.*

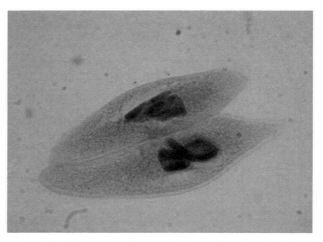

Figure 3.8. Two paramecia engaged in conjugation. Each will transfer genetic material to the other, separate, and then undergo fission twice to produce four genetically unique individuals.

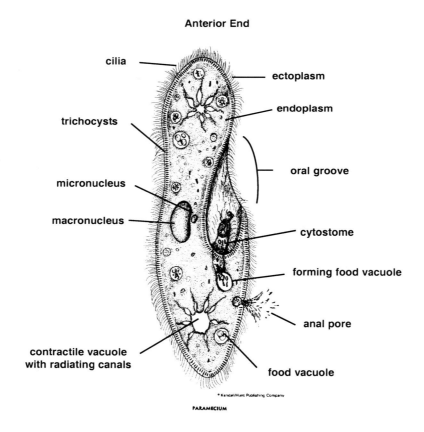

Anterior End

cilia — ectoplasm

— endoplasm

trichocysts

oral groove

micronucleus

macronucleus

cytostome

forming food vacuole

anal pore

contractile vacuole
with radiating canals

food vacuole

* Kendall/Hunt Publishing Company

PARAMECIUM

Figure 3.9. Structure of a typical ciliate, *Paramecium*.

Phylum Apicomplexa

All members of this phylum are internal parasites of animals. The name given to them refers to an "*apical complex*" of specialized organelles located at the tip of each organism. This complex apparently assists the parasites in invading their host. Otherwise, these protists are rather simplified. They lack locomotor structures, contractile vacuoles, and specific methods of feeding. Food merely diffuses through the general body surface.

Plasmodium spp., a member of this group, is a blood parasite responsible for malaria, one of the most devastating diseases of humans. The World Health Organization (WHO) estimates that almost half of the world's population (over 3.3 billion) live in areas at risk of malaria, and 190–311 million people suffer from the disease. There are 708,000–1,003,000 deaths annually. Most (90%) of the malaria cases occur in sub-Saharan Africa, but travelers can carry the disease anywhere, and 1500 cases are reported in the U.S., each year.

Malaria is transmitted from human to human through the bite of mosquitoes belonging to the genus *Anopheles*. Once in the mosquito, the parasite invades the gut where it undergoes **sporogony**, a type of asexual reproduction. The result is production of massive numbers of cells called **sporozoites**. These eventually enter the insect's salivary gland and are injected into another person during the next feeding episode.

Upon entering a person, the sporozoites invade the liver and undergo **schizogony**, another form of asexual reproduction. Once again, enormous numbers of parasites are produced, known as **merozoites**. Following a developmental period of about four weeks, these leave the liver to enter the bloodstream. Each merozoite invades a red blood cell where it divides repeatedly throughout a two-to-three day period. The blood cell eventually ruptures, releasing millions of new merozoites. These reinvade other erythrocytes (red blood cells), causing them to, likewise, burst as the cycle continues. Some of the parasites even undergo sexual reproduction, called **gametogony**. This course is repeated over and over until unimaginable numbers of invasive parasites are present in the blood. The end result is the destruction of blood cells and an invasive disease that causes headaches, intensive fevers, and chills. If the brain is under siege, disability and death often result.

During the middle of the 20th century, malaria was considered all but defeated. The disease was under attack on two fronts. DDT and other pesticides were used to control mosquito populations, while quinine drugs were used against the parasite. However, both the mosquito and the *Plasmodium* parasite have developed resistance to the respective chemicals. Consequently, malaria has made a terrifying comeback. Scientists and the medical community are now scrambling feverishly to establish new control over this horrendous disease.

GAMETOGONY

4. Merozoites leave liver and enter bloodstream. Here they invade red blood cells, reproduce asexually or sexually by gametogony, escape, reinvade other RBC's, and repeat this over and over causing a potentially lethal condition. Eventually the parasite is ingested by another mosquito and the cycle begins again.

1. Mosquito picks up parasite while feeding on an infected person.

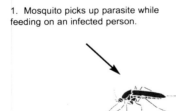

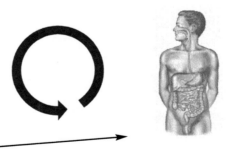

SPOROGONY

2. Parasite reproduces asexually in the mosquitoes' gut, producing sporozoites. These invade the salivary gland and await transfer to another human.

SCHIZOGONY

3. Sporozoites enter human liver and undergo asexual reproduction, producing merozoites.

Figure 3.10. Life cycle of the malarial parasite, *Plasmodium, spp.*

An Inside View 3.1

Some Major Protozoan Diseases

Malaria, the most virulent of protozoan diseases, is caused by the sporozoan *Plasmodium* and is transmitted to humans by a feeding *Anopheles* mosquito. Over 300 million people have the disease at any given time, and the annual death toll can exceed one million. There is little argument that malaria has been the most devastating human disease in history. Although there are more than 50 species of *Plasmodium*, only two regularly attack humans. Of the two, *Plasmodium falciparum* is the more serious and causes nearly all of the deaths. Symptoms include alternating chills, fever, and sweating, nausea and vomiting, muscle pain, anemia, jaundice, convulsion, and ultimately, coma. This is a blood parasite that kills by causing red blood cells to adhere to the walls of blood vessels, thus blocking blood flow to vital organs. In certain parts of the world, the mosquito has developed a resistance to insecticides while the parasite has similarly become resistant to antibiotics. Thus, previous high levels of control have diminished in recent years, resulting in an alarming resurgence of this disease.

African sleeping sickness. Another blood parasite is *Trypanosoma brucei*. This small flagellate is carried by a tsetse fly (*Glossina*) and injected as the fly feeds on a human host. Although African trypanosomiasis, commonly known as sleeping sickness, was virtually eliminated from Africa during the 1960s, it has recently made a resurgence throughout west and central Africa; in some states such as Angola, Uganda, and the Sudan, the disease is of epidemic proportions. Initial symptoms mimic influenza: fever, lethargy, malaise. The parasite may then enter a dormant stage, which may last for several years. Symptoms worsen as parasites invade the nervous system, causing extreme fatigue, malnourishment, alteration of mental state, sensory disorders, body wasting, somnolence, and irreversible coma. Two forms of the disease exist: *T. b. rhodesiense* and *T. b. gambiense*. The rhodesiense form is the more serious and can cause death in a matter of a few months, whereas the gambiense, or chronic, form may persist for several years. Both forms of the disease are fatal if left untreated. Control includes attack on the tsetse fly populations, surveillance of people in affected areas, and treatment with specific drugs during early stages of the disease. Once the parasite has crossed the blood-brain barrier, treatment is difficult and often ineffective usually resulting in death.

Chagas' disease is caused by another parasite related to the trypanosome that causes African sleeping sickness (above). This time *Trypanosoma cruzi* is the responsible agent, and 8 to 11 million people in Mexico, Central America, and South America are infected. It is carried from one human to another by a group of blood-sucking insects collectively known as "kissing bugs." This is a major health problem in South America. The infective stage of the parasite lives in the bug's gut and passes out with the feces. The insect (a hemipteran) feeds at night near the victim's nose or mouth. As it feeds, it invariably defecates, which provides the route to human infection. As the victim rubs the area where the insect fed, he or she inadvertently forces the parasite into the open flesh of the wound. Symptoms include swelling and reddening at the site of infection, swelling of the eye, swollen lymph nodes in the infected area, fever, malaise, heart arrhythmia, enlarged liver, difficulty in swallowing leadingto malnutrition, and cardiac

involvement. Death occurs in about 10% of the cases, usually due to cardiac arrest. There is no available treatment, although partial control has been achieved through insecticides and using door or window screens in areas where the insect vector is prevalent.

Amoebic dysentery. It is estimated that approximately 50 million people worldwide harbor the parasite *Entamoeba histolytica*, resulting in 40,000 to 100,000 deaths annually. There are 5,000 to 10,000 cases reported in the U.S. annually, mostly acquired by people while they were abroad. Amebiasis is present worldwide, but is most prevalent in tropical areas where crowded living conditions and poor sanitation exist. In the United States, institutionalized mentally retarded individuals and male homosexuals are considered high-risk groups. In most instances there are no symptoms, but at times the parasite may invade host tissue causing cramps, nausea, diarrhea, vomiting, pain on defecation, and overall malaise. The condition can be so severe as to cause death in children and other compromised individuals. The most common route of infection is through fecally contaminated food or water or direct contact with an infected person. Also, any insect (flies, cockroaches) that has a connection between human feces and human food can be a transmitter. In those parts of the world where human waste is used in the fertilization of crops, *E. histolytica* is a common human companion. Personal sanitation, especially when traveling in tropical countries, is an important avenue of prevention. The death rate from untreated amebiasis can be high, but the condition is easily controlled with early treatment.

Giardiasis. A person with a severe case of giardiasis experiences abdominal pain, nausea, loss of appetite, headache, intestinal gas, and "explosive" diarrhea, lasting as long as 14 days. *Giardia lamblia* is a common intestinal parasite throughout the world, including the United States. Nearly 5% of the people in developed countries, and 33% of the people in developing countries, have giardiasis. Many wild animals serve as reservoirs for this parasite and continually reinfect lakes, streams, and other waterways with their feces. A person then contracts the parasite by drinking contaminated water. The disease is contagious and can be passed from person-to-person, which has caused outbreaks in day-care centers, nursing homes, and among the male homosexual community. When camping or on outings, *Giardia* can be avoided by filtering drinking water or treating it with iodine or other appropriate chemicals. Drug therapy is available for chronic cases.

Trichomoniasis is a worldwide sexually transmitted disease caused by the protozoan *Trichomonas vaginalis*. In the United States, the highest incidence is in women between the ages of 16–35. Transmission is virtually always through some form of sexual contact, although in rare cases contact with contaminated surfaces (toilet seats, wet washcloths) may result in infection. The disease is asymptomatic in men and usually disappears spontaneously within a few weeks. At most, men may suffer a mild urethral itching or discharge. The infection steers a different course in women. A female with trichomoniasis may develop a frothy, foul-smelling, green-white voluminous vaginal discharge, vaginal inflammation, and severe itching in the genital area. In the U.S., an estimated 3.7 million people have Trichomoniasis, but only about 30% develop symptoms.. Abstinence or safe sexual practices are the best line of defense, while treatment with antibiotics is the curative.

Chapter 3 Summary

Key Points

1. All members of the kingdom Protista are unicellular organisms. As such, they carry out life processes, such as reproduction, respiration, digestion, excretion, and osmoregulation, within the confines of a single cell. However, certain protists, such as the large flagellate *Volvox,* have attained a colonial grade of organization. In such colonial forms, individual cells aggregate as a functional unit although each cell maintains the potential of independent life.

2. There are three basic forms of protists: amoeboids, flagellates, and ciliates.

3. The phylum Sarcomastigophora includes the amoeboid organisms (subphylum Sarcodina) and the flagellates (subphylum Mastigophora).

4. Ameboids move by cytoplasmic streaming of their pseudopodia, a process also known as *amoeboid movement.*

5. The flagellates locomote through the whiplike action of their elongated, threadlike flagella.

6. Many flagellates contain chloroplasts and are thus autotrophic organisms capable of manufacturing their own energy through the process of photosynthesis.

7. Most protists, including all sarcodinians, are heterotrophic and must rely upon external organic material for their energy source.

8. All organisms must be able to maintain an internal ionic balance with respect to their environment. This is especially true for salt (NaCl) and water. Maintaining this fluid balance is accomplished through the process of *osmoregulation* in which specialized organelles called *contractile vacuoles* are used.

9. Protists, as a group, are asexual organisms. The primary method of reproduction is *binary fission* in which a mother cell divides mitotically into two genetically identical daughter cells.

10. The large colonial *Volvox,* however, can reproduce asexually through fission as well as sexually through the production of separate gametes.

11. While the great majority of protists are freeliving, some are parasitic. For example, the sarcodine, *Entamoeba histolytica,* causes the intestinal condition known as amoebic dysentery, while the flagellate *Trypanosoma* is the agent for African sleeping sickness, a potentially fatal disease.

12. The phylum Apicomplexa contains *only* parasitic protists. Among these is the dangerous human blood parasite *Plasmodium,* the cause of malaria. *Plasmodium* exhibits a complex life cycle seen in many of these internal parasites that often alternates between sexual and asexual patterns.

13. All members of the phylum Ciliata are large and complex cells possessing cilia for locomotion and feeding.

14. Reproduction among ciliates varies from simple *binary fission* to the complex pattern of *conjugation.* During conjugation, two mating types undergo meiosis and then then exchange genetic material. They next separate and each completes two cycles of binary fission. Thus, for each conjugating organism, four genetically different individuals are produced.

15. Ciliates differ from other protists in that they possess more than one nucleus. In *Paramecium,* for instance, there is a large *macronucleus* used for primary functions and a smaller *micronucleus* involved in sexual reproduction. In the large ciliate *Stentor,* the nucleus is subdivided into a beadlike arrangement.

Key Terms

African sleeping sickness
Amoebic dysentery
amoeboid movement
asymmetrical
autotrophic
calcium carbonate tests
gametes
heterotrophic
wet mount
nucleus
osmoregulation
photosynthesis
plastids
pseudopodia
radiolarians
stigma
unicellular
cytostome
malaria
merozoites
nuclear dimorphism
oral groove
pellicle
radiating canal of contractile vacuole
trichocysts
wafting
gemmules
chloroplasts
colonial organization
contractile vacuole
cytoplasmic bridges
daughter spheroids
ectoplasm
endocytosis
endoplasm
euglenoid movement
flagellum
food vacuole
foraminiferans

Points to Ponder

1. What do you think is the advantage of being unicellular? Can you think of any disadvantages?

2. Contrast locomotion in an amoeba with that of *Euglena* and *Paramecium*.

3. Describe how an amoeba feeds.

4. What is the food source for *Euglena*? *Volvox*? *Paramecium*?

5. Using the concept of diffusion, explain how an amoeba exchanges oxygen and carbon dioxide.

6. Using diffusion as a basis, explain how an organism might maintain an osmotic balance. In other words, describe the process of osmoregulation.

7. Would you expect to find contractile vacuoles in freshwater or saltwater organisms? Explain.

8. Explain how *Paramecium* uses its contractile vacuole to osmoregulate.

9. Conjugation is considered to be a form of sexual reproduction. What arguments would you make in favor of that interpretation? What would you propose against that viewpoint?

10. What are trichocysts? How does *Paramecium* use them?

11. How is sporogony used by the malaria parasite?

Inside Readings

Brusca, R. C., and G. J. Brusca. 1990. Invertebrates. Sunderland, MA: Sinauer Associates.

Cheng, T. C. 1986. General Parasitology. New York: Academic Press.

Desowitz, R. S. 1991. The Malaria Capers. New York: W.W. Norton.

Ewald, P. W. 1994. Evolution of Infectious Diseases. New York: Oxford University Press.

Frenchel. T. 1987. Ecology of Protozoa: The Biology of Free-Living Phagotrophic Protists. New York: Springer-Verlag.

Jahn, T. L., E. C. Bovee, and F. F. Jahn. 1979. How to Know the Protozoa. Dubuque: Wm. C. Brown.

Kabnick, K. S., and D. A. Peattie. 1991. Giardia: A missing link between prokaryotes and eukaryotes. Am. Sci. 79:34-43.

Lee, J.J., et. al. (eds). 1985. The Illustrated Guide to the Protozoa. Lawrence, KS: Society of Protozoologists.

Noble, E. R,. and G. A. Noble. 1982. Parasitology: The Biology of Animal Parasites. Philadelphia: Lea & Febiger.

Pechenik, J. A. 1996. Biology of the Invertebrates. Dubuque, IA: Wm. C. Brown.

Sleigh, M. A. 1989. Protozoa and Other Protists. New York: Chapman and Hall.

4 Chapter 4

PHYLUM PORIFERA
THE SPONGES

Although simple in design and primitive in behavior, sponges are, nonetheless, the first true animals.

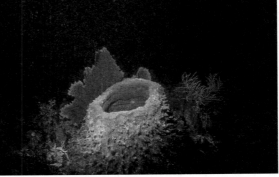

Chapter 4

PHYLUM PORIFERA:

Bit by bit, piece by piece the whole is constructed...
... until the whole becomes greater than the parts.

The Sponges

Introduction

Animals of the phylum Porifera, the sponges, are common fixtures in marine and, to a much lesser extent, in freshwater ecosystems throughout the world. While a rather small taxon numerically, with only about 5,000–10,000 species, in three clases, they are, nonetheless, relatively well known to most of us.

But that was not always the case. Sponges are strange creatures. So peculiar is their appearance that they were once considered not animals, but plants. After all, they look like plants...branched, asymmetrical, organless. They act like plants...attached and immobile. However, their anatomy, swimming larvae, and feeding habits give them away. Animals they are.

Sponge Anatomy

Sponges are composed of thousands of loosely arranged cells working together as a coherent unit. They are, thus, the first **multicellular** animals to appear on earth. However, they remain simple in design, never reaching the morphological sophistication of true tissues or organs as seen in higher animals.

The simplest sponges are nothing more than a hollow, asymmetrical tube (below). The outside of the tube is covered with flattened cells, **pinacocytes**, perforated by countless openings or **pores** through which water enters the animal. The pores, or **ostia**, are formed by donut-shaped cells called **porocytes**. The openings lead either directly to a central cavity, the **spongocoel**, or, in more complex sponges, through a **canal system** and then into the spongocoel. This central chamber is itself lined by special flagellated cells known as **choanocytes** (Figure 4.3, pg 88). These "collar cells" are important in pumping water through the sponge and in feeding. Situated between the outer and inner cell layer is a nonliving gelatinous matrix, the **mesohyl**. Imbedded within the matrix are mobile cells, **amebocytes** which have a variety of functions (see page 88). Sponges, therefore, consist of four different types of cells: pinacocytes, porocytes, choanocytes, and amebocytes.

Within the mesohyl layer are skeletal elements consisting of a lattice-like network of the protein **spongin**, a form of collagen unique to sponges. Also present in most sponges are interlaced, needlelike **spicules**. These are composed of mineralized calcium carbonate, or a glasslike silica. Whatever the composition, the skeleton is a supportive structure giving shape, form, and resiliency to the sponge. What is commonly known as a bath sponge is really nothing but the skeletal remains after all other living and nonliving material has been removed.

Finally, at the top of each tube is a rather large opening called the **osculum**. Water that has passed through the spongocoel exits by way of this vent. Waste materials, larvae, and sex cells also leave through the osculum. Refer to Figure 4.4 for a detailed sketch of sponge structure.

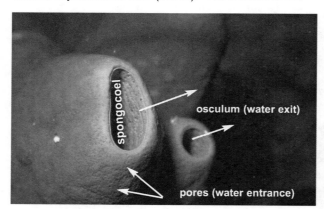

Figure 4.1. A cluster of sponges (above) showing their tubular structure and centrally located spongocoel. The large openings through which water exits the animals are termed oscula. Large numbers of pores perforate the external layers permitting water to enter. Both spongin and spicules (right) provide skeletal support.

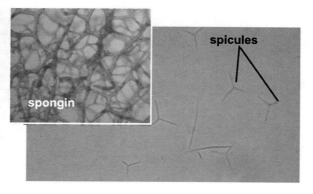

Defense

Sedentary animals must cope with some very problems. One of the most important is how to deal potential predators. Since they cannot escape they rely upon other defensive tactics that can be yed while standing still. Thus, sponges have ed such techniques as toxicity (having poisons or ms), being inedible (offensive taste), weaponry es), or escaping notice through camouflage. Many ges combine one or more of these tactics to survive t apparently works well since relatively few animals e fish and a turtle or two) are known to prey on ges.

Nutrition

Another distinct problem for a sessile animal is ing food. Do you wait in ambush until something comes by and then snatch it, or do you just take tage of the ample food supply floating around you? ill see in the next chapter that some attached als, such as the corals and sea anemones, use the sh, or bushwhack tactic. Sponges, on the other , are *filter feeders* and are able to simply extract directly from their watery environment. To do that, ver, requires a constant cascade of food. To assure teady supply, sponges have devised a rather simple on involving their many pores and the combined of flagellated cells, the *choanocytes*, lining the ocoel. Each choanocyte possesses a long central lum that can wave back and forth. A substantial current is created by the unified beating of these la, bringing food and valuable oxygen into the inner ber. The choanocytes then glean food from this n of moving water. Food items are then passed to *mebocytes,* which, in turn, pass it along to other

Respiration

The exchange of the respiratory gases, oxygen carbon dioxide, is essential to the existence of all als. Sponges accomplish this exchange by way of ion through their general body surface. They y extract oxygen from the surrounding water and carbon dioxide in the same manner. This type of nge is called **cutaneous respiration.**

Reproduction

One fascinating feature of sponges is their amazing power of **regeneration**. You can literally pass one through a screen and individual cells will regroup to form a collection of new sponges. This is asexual reproduction, **fragmentation**, to be exact, at its best. Other forms of asexual reproduction used by sponges include **budding**, and **gemmule formation**. Budding involves the formation of new cells by mitotic division and the subsequent growth of a nodule or "bud" into a new individual.

Gemmules (below) are specialized packets of amebocytes that are highly resistant to drying out (desiccation) and freezing. Thus, gemmules are used to escape unfavorable environmental states such as winters in temperate climates. When proper conditions return, the sac opens, individual cells escape, they grow into adults, and new sponges are established.

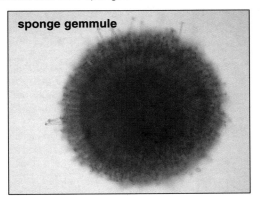

sponge gemmule

Reproductively, sponges are *monoecious;* they are hermaphroditic, each sponge being capable of producing both sperm and eggs. However, they rarely, if ever, self-fertilize. Instead, sperm are carried out the osculum of one sponge and channeled into the pores of another. There, sperm are trapped by the choanocytes, conveyed to amoebocytes, and finally transported to the waiting egg. Quite a trip for such unpretentious critters.

Once eggs have been fertilized they quickly develop into nonfeeding larva. These pass out of the osculum and then swim away from the parent sponge to a suitable spot where they settle down and develop into adult sponges. Thus, sponge populations expand into new territories because of the dispersal tendencies of individual larva.

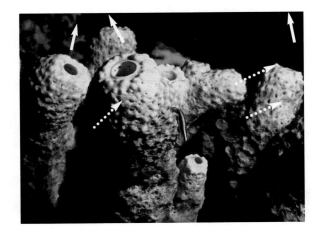

Figure 4.2. A cluster of sponges showing the arrangement of pores and oscula. Water enters the sponge via pores (dotted arrows) and exits through the oscula (solid arrows).

Division of Labor

Perhaps the greatest contribution of sponges to the animal kingdom is **cellular differentiation**. It's obvious, from the previous description, that sponges possess a variety of cell types, each performing different functions. No single cell can accomplish all of the various activities needed to maintain a living, healthy sponge. The chores, needs, and duties of each animal are, instead, divided among its various cells. As a result, the animal is more efficient and can accomplish things that individual cells cannot. Also, division of labor allows animals to reach sizes impossible for a single cell. Let's review the several cell types and their functions:

(1) **Pinacocytes**. These flattened cells make up the outer "skin" of the sponge. They give shape and provide protection.

(2) **Porocytes**. These "jelly-roll" cells form the channels for the pores (i.e., ostia) to permit water to flow into the inner chamber. Without pores sponges cannot survive.

(3) **Choanocytes** are special flagellated cells that line the spongocoel. These cells serve as a pump by creating a gentle current of water from which they trap food particles.

(4) **Amebocytes**. The most generalized cell type is the amoeboid-like cells patrolling the gelatinous, middle mesohyl layer. These cells have a number of functions:

- ◆ They form spicules and/or spongin.

- ◆ They form gametes (sperm and eggs).

- ◆ They transport food to other cells.

- ◆ They can differentiate into other cell types when needed. For example, if choanocytes are damaged they are replaced by amebocytes.

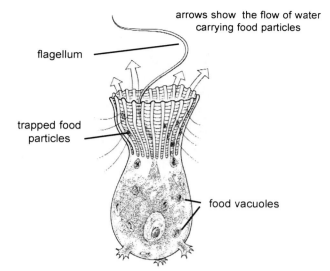

A Choanocyte ("collar cell")

Figure 4.3. Upper figure shows a series of choanocyte (collar cells) which are designed to filter food from the passing water. Lower photograph depicts several sponge groupings revealing their upright growth pattern.

IT'S A FACT

DID YOU KNOW THAT THERE IS ACTUALLY A CARNIVOROUS SPONGE? IT HAS RECENTLY BEEN DISCOVERED IN FRESHWATER HABITATS WHERE IT CAPTURES AND EATS OTHER SMALL ANIMALS!

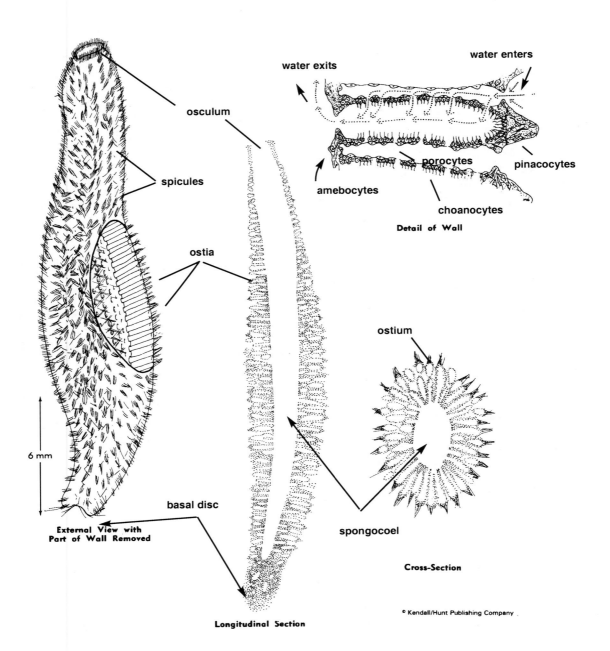

Figure 4.4. Sycon (*Grantia*) sponge structure.

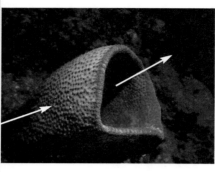

water exits via the osculum

water enters through pores

Figure 4.5. Representative sponges. Note the basal attachments and the route of water through the sponges as indicated by the arrows.

Water and Sponges

Without a consistent and adequate supply of water, sponges inevitably perish. Water that is pumped through the ostia (pores), into the spongocoel, and eventually out the osculum is a life-giving flow that meets the needs of poriferans in a variety of ways. It nourishes them by bringing food to the choanocytes. It also delivers oxygen needed by all cells. Even sperm from another sponge is carried into the spongocoel on this column of water; thus, sexual reproduction, itself, depends upon this gentle stream of water.

Water also carries things from the sponge. For instance, wastes such as carbon dioxide and general debris are flushed through the osculum. Sperm, too, are disseminated by the outflow as are tiny larvae.

Chapter 4 Summary

Key Points

1. Sponges are multicellular (Metazoan) animals.

2. They are aquatic; some are fresh-water animals but most live in marine habitats.

3. Sponges are sessile (attached), asymmetrical animals.

4. They possess multiple pores, canals, and a central cavity, the spongocoel.

5. They feed by a unique system of filtering food particles from water as it flows through the system of pores, canals, and the spongocoel.

6. The sponge body consists of a middle gel-like matrix (the mesohyl) sandwiched between two layers of cells.

7. The outer protective cell layer is composed of flattened pinacocyte cells and the inner layer consists of flagellated collar cells, the choanocytes. These cells pump water through the sponge and filter food from the flowing water.

8. These primitive animals contain an internal skeleton of calcium carbonate or silica spicules or a network of the collagenous protein, spongin.

9. Sponges reproduce by asexual means through budding, fragmentation, and gemmule formation, and sexually as monoecious individuals.

10. Sponges were the first animals to show cellular differentiation, which is an important process of cell specialization to enhance efficiency and to allow these animals to attain larger sizes.

Key Terms

amoebocytes
budding
cellular differentiation (division of labor)
choanocytes
fragmentation
filter feeding
gemmules
hermaphroditic
mesohyl
metazoa
monoecious
multicellular
osculum
ostia
pinacocytes
pores
porocytes
silica
spicules
spongin
spongocoel

Points to Ponder

1. Discuss the concept of cellular differentiation (division of labor) as it relates to the sponges.

2. Trace the flow of water through a sponge.

3. Describe how a sponge feeds.

4. What is the role of amoebocytes in a sponge?

5. How does a sponge reproduce?

6. Sponges are the first multicellular animals. What advantage does multicellularity convey to an animal?

7. What is the function of sponge spicules?

8. What is the advantage of cellular differentiation?

Inside Readings

Brusca, R. C., and G. J. Brusca. 1990. Invertebrates. Sunderland, MA: Sinauer Associates.

DeVos, L. e. al. 1991. Atlas of Sponge Morphology. Washington: Smithsonian Institution Press.

Pechenik, J. A. 1996. Biology of the Invertebrates. Dubuque, IA: Wm. C. Brown.

Rutzler, K. ed. 1990. New Perspectives in Sponge Biology. Washington: Smithsonian Institution Press.

Simpson, T. L. 1984. The Cell Biology of Sponges.
New York: Springer-Verlag.

Thorpe, J. H., and A. P. Covich. 1991. Ecology and classification of North American Freshwater Invertebrates. New York: Academic Press.

Wood, R. 1990. *Reef building sponges.* Am. Sci.78:224-235.

Chapter 4

Laboratory Exercise

The Sponges

Laboratory Objectives

✓ to learn the external and internal anatomy of a simple sponge,
✓ to identify representative examples of sponges, and
✓ to appreciate the economic and ecological importance of the phylum Porifera.

Supplies needed:

● Dissection microscope
● Compound microscope
● Prepared microscope slides
● Representative samples of sponges

Chapter 4
Worksheet

(also on page 589)

_____

(name)

Phylum Porifera

1. In the box to the right, sketch and label a simple sponge.
On the diagram include ostia (pores), an osculum, the spongocoel,
and the holdfast.
 ⌐Bottom on the
       surface

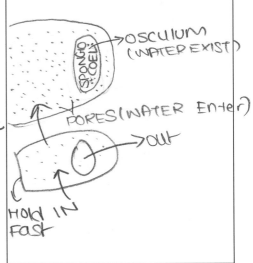

2. On your sketch use arrows to show the route of water through the
sponge. Then list (below) items that may be transported by the current
of water. Which cell type creates the water flow? choanocyte
Food, water, nutrients

3. Observe prepared slides of *Grantia* (=*Scypha*) and note the
presence of ostia perforating the outer wall and the centrally located
spongocoel. Refer to diagram on page 89.
 ⌐Small Dot ob pores

4. View a sponge skeleton (i.e., a bath sponge) using a dissection microscope. Note the abundance of pores and
the spongin that make up the specimen.
They look like web ob Fibers

5. Observe a prepared microscope slide of spicules. Then sketch several individual spicules in the space below as
well as the network of spicules that make up a sponge skeleton.

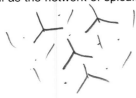

SKIP   6. If a live specimen is available, scrape a small area from its surface. Keep the tissue wet and then tease the
material apart with a needle probe. Sketch the spicules and compare them to the ones you saw above.

7. Observe a prepared slide of a gemmule and sketch one in the space below.

5 Chapter 5

THE CNIDARIANS

Whether it's in the bell-like form of a jellyfish or the upright stance of a sea anemone, with its tentacles waving wildly, the cnidarians were the first animals to develop muscles, nerves, coordinated movement, and cooperative behavior.

Phylum Cnidaria

The pools where the sea offers to our curiosity
The more delicate algae and the sea anemone
- Eliot

Introduction

Cnidarians are an unusual lot. With over 10,000 species, the group has some of the most conspicuous and, at the same time, some of the least recognizable forms in the animal world. For example, just about everyone is familiar with the pulsating clarity of jellyfish. And essentially all of us are aware of the richness and beauty of coral reefs. One of the most common animals in general biology lab's is the freshwater cnidarian, *Hydra*. Even sea anemones are relatively well-known residents of tide pools.

But the group also contains some of the most obscure animals on the planet. Very few of us, for instance, have encountered such forms as box jellies, sea pens, or sea pansies. And who but the most dedicated cnidophiles have witnessed dead mans' fingers, sea firs, gorgonians, or what about "by-the-wind-sailor"?

Cnidarians are, clearly, a widespread and varied assemblage of invertebrate animals. It is their striking diversity, however, that causes us to wonder about their commonalties. What exactly is the glue that holds the group together...just what features do they all share that allows us to classify them together into a single phylum?

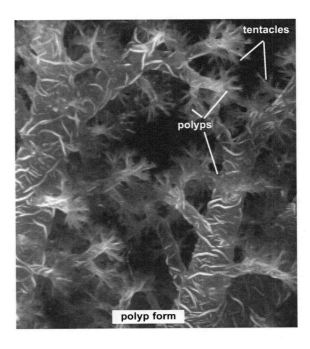

polyp form

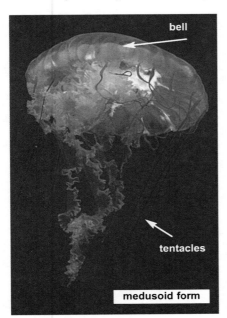

medusoid form

George Grall, © 1998, National Aquarium in Baltimore

Common Characteristics

Cnidarians display a medley of common features. None, however, is more pronounced than their **tentacles**, stinging tentacles to be exact. In fact, the very name of the phylum, Cnidaria, refers to the presence of these "stinging nettles." Located on their willowy tentacles, or scattered elsewhere on their bodies, are vast batteries of irritating cells called **cnidocysts**. Lurking within each of these cells is a harpoonlike device, the **nematocyst**, ready to dispatch an unwary prey or to repel an unwelcome enemy. The nematocyst injects a neurotoxin that induces paralysis. While most cnidarians are harmless to humans, some, such as box jellies and the Portuguese man-o-war can be dangerous, even lethal (see the *An Inside View 5.1,* on page 106).

Figure 5.1. The two basic body forms are the **polyp** (above) and the **medusa**, characterized by the jellyfish to the left. Both display the radial symmetry typical of the group. Note, also that the tentacles hang down in the medusa but project upward in the polyp forms.

Figure 5.2 below shows the common cnidarian *Hydra* with its armed and outward extending tentacles. The tentacles bear thousands of special stinging cells called **cnidocytes.** Each cnidocyte contains a **nematocyst** (the actual stinging part, see insert below). When touched, or when encountering potential food, the stinging cell explodes to impale its victim with a potent venom. Nematocysts are unique to the Cnidarians, and are possessed by all members of the group.

While stinging cells are the most common occupants of the cnidocytes, some harbor, not stinging cells, but adherent devices. These may be long threads to entangle prey or secretory cells that enable the *Hydra* to attach to a substrate.

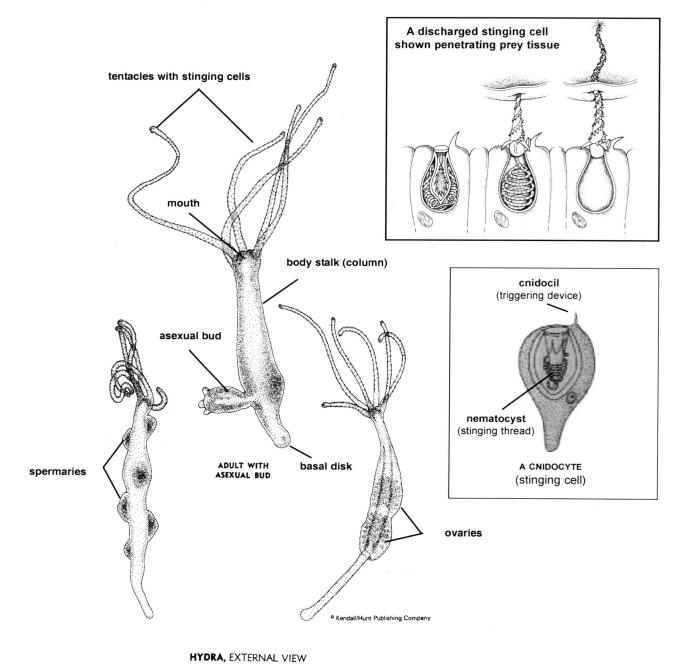

A discharged stinging cell shown penetrating prey tissue

tentacles with stinging cells

mouth

body stalk (column)

asexual bud

spermaries

ADULT WITH ASEXUAL BUD

basal disk

ovaries

cnidocil (triggering device)

nematocyst (stinging thread)

A CNIDOCYTE (stinging cell)

© Kendall/Hunt Publishing Company

HYDRA, EXTERNAL VIEW

Figure 5.2. The common cnidarian *Hydra* which exists only in the polyp form. The inserts show a stinging cell, the cindacyte with internal nematocyst. and it's action when discharged.

Structure. Another common feature among the cnidarians is their symmetry, they are all **radially symmetrical.** In fact, *they are the only animal group that is.* But why radial? Well, most zoologists think that radial symmetry is related to how and where these animals live. Slow-moving or attached aquatic animals can feed either by filtering food particles from a moving water current, as the sponges do, or by capturing food randomly *from any direction.* As a result, an animal has an advantage if it is arranged in a radial fashion. In other words, there is no distinct benefit to having anterior structures, such as a head, if your food, enemies, and companions are evenly distributed all around you.

As already stated, cnidarians have either an attached (**polyp**) or a free-floating (**medusoid**) body design. A typical polyp body plan is exhibited by *Hydra* and the sea anemones (below). The animal is attached to its substrate by a holdfast or **basal disc.** The body proper is termed a **stalk** (or column) and the mouth is located between the circle of **tentacles**. Thus, the animal consists of three body regions: (1) the oral zone (i.e., the region containing the mouth); (2) the aboral zone (away from the mouth); and (3) separating these two is the body stalk or column. The polyp body form is well designed to take advantage of a favorable local environment where an animal may remain attached. Although some polyps have a limited ability to move by crawling or looping, inch-worm fashion, they are, for the most part, sedentary animals.

Jellyfish, as in the photo below, demonstrate the medusoid (free-floating) design. Although medusae have some ability to "swim", it is greatly limited. They travel, instead, on ocean waves and currents trailing their tentacles to catch fish and other food items.

An advantage of an inactive lifestyle, as seen in the polyp forms, is that once an animal settles down within a favorable habitat it can remain there to exploit local resources. However, that also presents a problem. For instance, how does an otherwise immobile population of animals disperse into new locales if the current one becomes less desirable? Well, some cnidarians have solved this problem in a rather unusual way. They have adopted a two-tiered existence called **polymorphism**. They alternate between the attached polyp phase and a free-swimming medusa. During the first stage of its life cycle, the animal is in the form of a polyp, attached to its substrate and designed to exploit that sedentary lifestyle. During its alternate medusoid existence it is mobile and, thus, designed to swim about actively. This alternating, polymorphic pattern, also known as *alternation of generations,* is used by a wide variety of cnidarians (e.g., *Obelia,* page 100) and allows the same animal to take advantage of one local habitat for a period and then later relocate to another.

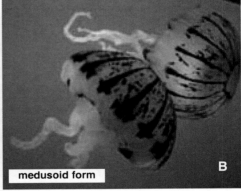

Figure 5.3. A cluster of large sea anemones (**A**) showing the typical polyp body plan. Note the ring of tentacles surrounding the centrally located mouth. Plate B is a photograph of two jellyfish, demonstrating the free-floating body design.

The medusa body shape is designed for a free, unattached existence. It is shaped like an inverted cup or bell with a fringe of tentacles hanging from the bell margins. The bell contains a gel-like **mesoglea** that assists in flotation.

Most medusae can slowly squeeze the bell, resulting in the typical pulsating motion of jellyfish. When the bell is constricted, water is forcibly expelled in a jet-like stream resulting in a backward propulsion. However, it is very likely that locomotion is not the only, or even the major, reason for the pulsating motion. The action actually causes a vortex of water to pass over the bell and into the trailing tentacles as shown in the figure below. It seems, then that the pulsation is designed primarily to propel food items toward the stinging nematocysts.

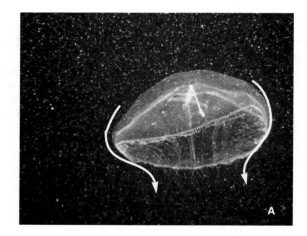

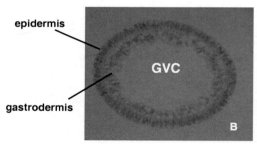

epidermis

GVC

gastrodermis

B

Figure 5.4. The bell-shaped body of a jelly fish (**A**) well illustrates the medusoid body plan. Note the thin tentacles suspended from the bell margin and the flow of water toward the tentacles. Also, note the "sea snow" adrift in the background. Plate **B** is a cross section of a cnidarian (*Hydra*) showing the arrangement of tissues (epidermis and gastrodermis) and the central gastrovascular cavity (GVC).

An inspection of the body wall of a cnidarian reveals yet a third feature of this group. Unlike sponges with their scattered cellular design, cnidarian cells are arranged into functional groupings or tissues. A tissue is an aggregation of similar cells designed to perform specific functions. Cnidarians have two tissue layers with a third nonliving zone, the **mesoglea,** sandwiched between. In this **diploblastic** arrangement, the outer layer, known as the **epidermis**, protects the outer surface of the animal. The inner tissue layer lines a central cavity where digestion occurs. This layer is the **gastrodermis** and is concerned primarily with nutrition and reproduction. Gastrodermis lines a central chamber, the **gastrovascular cavity** (GVC).

Even though cnidarians are primitive animals, they do have some rather sophisticated capabilities. For example, they have special chemoceptors and touch receptors that apparently enable them to detect prey and enemies in the water. They also possess photosensitive organs called **ocelli** that endow the animal with light-sensing abilities. They can thus avoid contact with objects in the water and maintain proper position relative to light conditions.

In addition, cnidarians, especially the medusoid forms, bear **statocysts** that monitor body posture. This is highly important in that a floating or swimming animal is extremely vulnerable if turned upside down. The working principle of the statocyst is remarkably simple (see chapter 2). A small capsule contains one or more small granules, **statoliths**, which when tilted stimulate hairlike sensors. These, in turn, send a signal to nerve cells that initiate muscle contractions to either maintain or correct the body's position. This is an elegant solution to a serious problem. In fact, most balancing systems in animals is based on this simple design. For instance, there is a rather surprising similarity between this system and that used by humans in our battle for balance and equilibrium!

Cnidarian have muscles that are arranged in two directions. **Circular muscles** wind around the animal and when contracted cause it to lengthen. Other muscles run up-and-down the length of the animal. When contracted, these **longitudinal muscles** can produce a variety of twisting and turning movements.

To coordinate their various activities the cnidarians have developed a *rudimentary nervous system*, the **nerve net**. This network of nerve cells is located throughout the epidermis and, to some degree, into the gastrodermis. As a result, the animal is able to respond to food, pressure, chemicals, vibrations, and temperatures in its environment. Also, the nerve net makes it possible for the animal to coordinate muscle pulsations during swimming or other movements.

However, to accomplish movement one final element is needed...a skeleton. In order for muscles to work, they must pull against some rigid structure. But, unlike sponges with their internal network of collagen and spicules, cnidarians have no such structures. Instead, they create a fluid skeleton. By engulfing water and exerting pressure they produce a rather rigid internal **hydrostatic skeleton** against which muscles can pull.

Feeding. With very few exceptions, cnidarians are carnivorous. Food captured by the tentacles is brought into the gastrovascular cavity for digestion. Cnidarians, as is true with many other invertebrates, can employ two basic methods to meet their nutritive needs. If the food particle is small enough, it will be engulfed by gastrodermal cells lining the GVC. Digestion then occurs within the individual cell by **phagocytosis** (i.e., intracellular digestion). However, if the item is too large for phagocytosis, it will be digested **extracellularly** through the release of digestive enzymes from secretory cells lining the GVC. These enzymes break down the proteins and carbohydrates so that food is ultimately available to the gastrodermal cells. The latter digestive process is similar to the method used by higher animals (e.g., humans).

Many cnidarians, especially corals, add to the above pattern through special interactions with other organisms. Living within many coral reef cnidarians, for instance, are green photosynthetic algae that provide some nourishment for the corals. Without the symbiotic arrangement between the corals and these algae hitchhikers, the reef-building cnidarians could not exist.

It is of particular interest that cnidarians have no specific outlet for undigested waste. Their digestive system is nothing but a blind tube. This, of course, means that all undigested material must exit as it entered, by way of the mouth. *Their mouth then also serves as an anus.* It doesn't take much to realize that this can present a real problem with mixing incoming food and outgoing waste...a problem known as **fouling**.

Respiration. Cnidarians have no special organ or structure devoted to the exchange of respiratory gases. Instead, oxygen enters through the general body surface by way of diffusion. This process, referred to as cutaneous exchange, is also responsible for the exit of carbon dioxide from the body. Regulation of ions (osmoregulation) also occurs cutaneously.

Reproduction. Budding, fragmentation, and binary fission are asexual reproductive means available to cnidarians. In addition, many reproduce sexually through both monoecious and dioecious pathways. To compensate for an attached lifestyle, many individuals combine both asexual and sexual reproduction. It is not unusual in some species of *Hydra*, for example, to observe a single animal with a well-formed testis, a mature ovary, and a developing bud at the same time!

Other species alternate between an attached, asexual form and a free-swimming, sexual one. *Obelia* is a small, common polymorphic cnidarian that exists as a sedentary, branched polyp colony (Figure 5.5 on the next page). Located on the branching network are two distinctly different polyps: feeding and reproductive.

Through budding and transverse fission, *Obelia's* reproductive polyps shed tiny **medusa** into the surrounding water. These, in turn, produce either sperm or eggs. Following fertilization, an embryonic, ciliated **planula** larva forms. The larva swims to a likely spot, attaches, and develops into a new polyp colony.

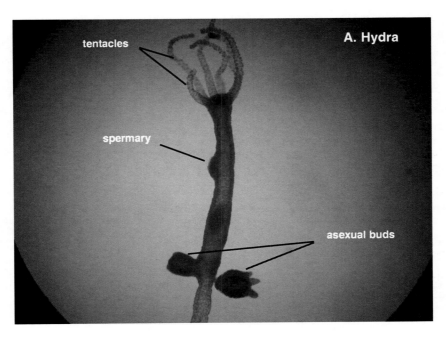

Figure 5.5. A, An adult *Hydra* with developing buds and a spermary containing sperm. On the next page, **(B)** The colonial *Obelia* showing feeding and reproductive polyps. **C,** An *Obelia* medusa and **D,** *Obelia* planula larvae stages are presented.

Obelia Life Cycle
(see text for explanation)

reproductive polyp

feeding polyp

feeding polyp

C. *Obelia*, medusa (sexual stage)

B. *Obelia,* colonial (tree-like) asexual stage

Figure 5.5. (continued from previous page) *Obelia* life cycle.

D. *Obelia*, planula larval stage

THE LIFE CYCLE OF *OBELIA* DEMONSTRATES THE FOLLOWING IMPORTANT CNIDARIAN FEATURES:

1. **polymorphism**; polyps and medusae in the same organism
2. **alternation of generations**; asexual and sexual reproductive phases
3. both sedentary *and* free-swimming life-history phases
4. a genetically unique, ciliated, **planula larva** able to actively relocate
5. **colonial arrangement** with two specific polyp types (reproductive and feeding)

Systematics

I. HYDROZOA - HYDRA, PORTUGUESE MAN-O-WAR, OBELIA
II. SCYPHOZOA - COMMON JELLY FISH
III. ANTHOZOA - SEA ANEMONES, CORALS

I. Class Hydrozoa.   Members of this class are typically polymorphic, possessing both polyp and medusoid body forms. Nematocysts are confined to the epidermis and the mesoglea lacks cells. Common examples are *Obelia, Hydra,* and *Physalia* (the Portuguese man-o-war) shown below.

While *Obelia* and *Physalia* are typical of the class in that each shows polymorphism, *Hydra* is not. *Hydra*, instead, occurs only as a polyp and never assumes a medusoid phase during its life cycle. However, its general tissue configuration, simple body plan, cell placement, mesogleal structure, and tentacle arrangement are reasons to place it within the Hydrozoa class.

Physalia, the Portuguese man-o-war (below) has a particularly confusing organization. It consists of a colony of polyps attached to a common flotation device, the **bladder** or **sail**. There are a number of specialized polyps that are suspended from the bladder with their tentacles hanging down: *reproductive polyps, feeding polyps, stinging polyps*, and a *bladder-producing polyp*. The remarkable thing about the man-o-war is that the entire colony began as a lone zygote. That single cell produced the bladder, and then all other polyps formed asexually by budding from the bladder! So, even though it appears to be one large individual, and is often mistakenly identified as such, it is, in reality, a colony of many individuals all clinging to the large, floating bladder.

II. Class Scyphozoa. The primary distinction of this group lies in its bell. Virtually all of its members exist as a medusoid only, or if a polyp occurs at all, it is small and an inconspicuous part of the life cycle. The bell, on the other hand, is large and filled with a thick and relatively firm layer of mesoglea. This thick mesoglea provides buoyancy and is the basis for the common name jellyfish as seen in the photo below.

Freedom from an attached existence permits these animals to reach relatively large sizes. The largest cnidarians, in fact, belong to this group. Some giants, such as the *Preya*, the Arctic's Lions Mane, can attain a bell of over two meters in diameter and tentacles that trail well over 120 feet below. It may well be the largest animal in existence!

The medusoid form has also led to an active, swimming locomotion. Water within the gastrovascular cavity is forcibly expelled through the mouth, resulting in a jetting, backward motion. Specialized muscle-like cells supply the needed contractile force.

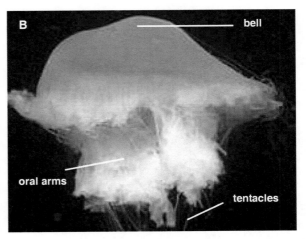

Figure 5.6. A. The venomous Portuguese man-o-war is a colony of polyps with an air inflated bladder (sail) and trailing tentacles consisting of various types of polyps. **B,** a typical jellyfish showing the mesoglea-filled bell, thin tentacles and the fleshy oral arms.

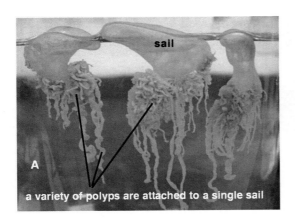

In the common moon jelly, *Aurelia aurita,* the medusoid (adult) animals are dioecious and fertilization results in a genetically distinct, swimming planula larva. The larva settles on a suitable substrate as a small polyp (scyphistoma) which eventually forms stacks of identical plate-like discs through an asexual process called strobilization. The individual discs soon break away from the strobila as a tiny medusoid ephyra. The ephyra then gradually matures into an adult *Aurelia.*

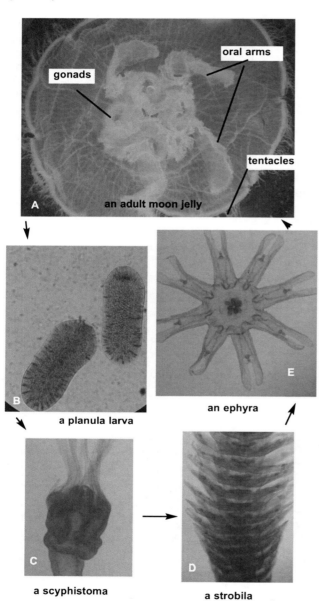

Figure 5.7. The life cycle of the moon jelly, *Aurelia.* Eggs and sperm from the adult moon jelly (**A**) produce a ciliated, swimming planula larva (**B**). The larva attaches to a rock or other substrate and grows into a tiny polyp called the scyphistoma (**C**). The scyphistoma transforms into a strobila which divides through transverse fission into a series of stacked discs (**D**). Each disc breaks away as a swimming medusa, the ephyra (**E**) which matures into an adult *Aurelia.*

Moon jellys are rare among cnidarians in that they practice internal fertilization. Males release sperm from their mouths as long strands which are collected by females. She pulls the strands into her mouth and passes them to the gastrovascular cavity where fertilization occurs. The life cycle of *Aurelia* is shown below.

Class Anthozoa. With over 6,100, Anthozoa contains more species than the other cnidarian classes, combined. Some of the most familiar and showy cnidarians belong to the class Anthozoa as illustrated by the photographs on the next page. The hauntingly beautiful solitary sea anemones and the spectacular colonial corals are included within this class. They are distributed throughout much of the world's marine environments being widespread throughout intertidal zones, occurring in tide pools, on the ocean's floor, and forming one of the earth's most diverse habitats...the massive tropical reefs.

The one morphological feature all anthozoans share is the absence of a medusoid body form. They only exist as polyps. As a result they are all attached, essentially sedentary animals. The common sea anemone, *Metridium,* Figure 5.13, demonstrates the polyp design. It, as with all polyps, has an oral-aboral axis. The oral surface bears the central mouth and an array of tentacles, sometimes numbering into the hundreds. The animal, in turn, is held in place by cement from the aboral pedal disc (holdfast), although some of the more primitive anemones can burrow into the soft ocean floor. Located between the oral and aboral surface is the column or scapus. The column is supplied with strong muscles that can cause the animal to extend, shrink, sway, and even swim clumsily. The muscles can also expel all water from the body cavity, allowing the animal to collapse into the familiar "bubblegum on a rock" so characteristic of harassed anemones.

The Anthozoan gastrovascular cavity (GVC) is unique among cnidarians. Unlike the single chamber found in *Hydra,* the GVC of anemones such as *Metridium* consists, instead, of a series of sub-chambers separated by a septum. Ciliated grooves (**siphonoglyphs**) and pores are present leading into the GVC to help maintain a constant supply of water. Resting on the floor of the central chamber are the beadlike gonads containing masses of long threads, the **acontia.** The acontia, bearing a defensive battery of stinging cells, can be extended through special pores for a considerable distance to ward off predators or even to engage in territorial combat with other anemones. The combat may, in turn, prevent overcrowding of local habitats.

Anthozoans are usually dioecious but may also reproduce asexually through various means. One such process, called pedal laceration, involves the breaking away of small pieces of their pedal disc. Each piece is then capable of growing into an adult polyp.

Figure 5.8. An array of anthozoans including a brain coral (top right) and a number of showy corals and anemones.

Anthozoans are involved in the formation of some of the world's most complex and rich ecosystems, the coral reefs. Of all of our planet's environments, the tropical rain forests and coral reefs comprise the most colorful and diverse assemblage of organisms. Reefs occur in three geologic patterns:

(1) **fringing** reefs are found close to shores along rocky coastlines, and follow the contour of the shoreline;

(2) **barrier** reefs are separated from land by variable-sized lagoons or channels; and,

(3) **atolls** are ringlike reefs encircling receding volcanoes.

All corals exist as polyp colonies of two types: reef making (**hermatypic**) and nonreef making, or **ahermatypic**. The reef-making forms, in turn, are all "stony" corals. These produce an external "skeleton" of calcium carbonate that makes up the hard portion of the reefs. Individual polyps, by the millions, secrete a thin layer of $CaCO_3$ creating tiny calcareous cups into which they can retreat if disturbed. When a polyp dies it is soon replaced by another, which deposits its own cup. Thus, the entire reef grows ever so slowly, on the order of only 30–50 mm per year.

The nonreef builders, on the other hand, can be either hard or "soft" corals. Those designated as hard form an internal rod or lattice as a support. Polyps are then draped in a colonial fashion over the support. Soft corals (e.g., *Alcyonium*) occur as fleshy, sometimes erect, aggregations on various substrates.

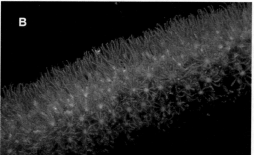

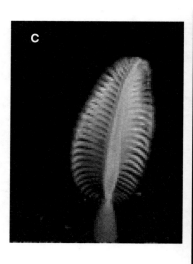

Figure 5.9. Examples of nonreef building corals. Sea fans (**A** & **D**) exhibit a spreading growth habit whereas the sea pen (C) is more compact. Branches of sea fans are adorned with thousands of miniature polyps as shown in plate **B**.

The reef corals are so productive only because they have developed a symbiotic partnership with a group of dinoflagellate protists, the **zooxanthallae**. These algae-like organisms live within the coral polyps for a number of reasons: protection, to obtain CO_2 for photosynthesis, and to use nitrogenous wastes for growth. Corals, too, benefit greatly from the relationship by receiving a vital source of nourishment. In fact, without the fatty acids, amino acids, and glucose supplied by the algae, coral polyps, and by extension, the reefs, could not exist. Some studies have shown that as much as 95% of the photosynthetic energy fixed by the zooxanthellae is passed along to their coral hosts!

Corals provide a tremendous opportunity for people to appreciate and understand ecosystem diversity. The whole concept of eco-tourism has developed, in large part, due to the accessibility of coral reefs and rain forests. But corals have other practical uses as well. Because of their calcareous structure, some corals can be incorporated into bone grafts to repair facial blowout fractures. The coral skeleton is implanted in the region of damage and forms a base upon which new bone can grow. Still other corals, again because of their lattice-like growth patterns, have been used as a support for artificial eyes. In fact, a coral implant at the back of the eye socket allows infiltration of nerves, muscles, and blood vessels. As a result the prosthetic eye can achieve near normal cosmetic movement!

Figure 5.10. Coral reefs are exceedingly rich in their animal and plant diversity.

An Inside View 5.1

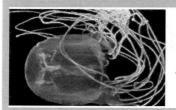

Chironex - the Killing Box

AS A RULE CNIDARIANS POSSESS STINGING CELLS THAT POSE LITTLE THREAT TO HUMANS...
BUT THEN THERE ARE EXCEPTIONS.

Swimmers from Australia to Africa to Texas face a common, virtually invisible, threat. *Chironex fleckeri*, the Indo-Pacific box jelly (AKA the "fire medusa" or "sea wasp"), is an unwilling but deadly menace. It carries enough poison in its transparent tentacles to kill more than 20 grown men. It kills more quickly than a pit viper. It's venom is more deadly, by far, than its cousin, the Portuguese man-o-war. It is as dangerous as the infamous blue-ringed octopus and it rivals the tiny, poison dart frogs in its lethal discharge. This gelatinous ghost is truly an ominous shadow in marine ecosystems. Many experts consider it the most venomous marine animal alive.

The box jelly is a typical, although large, medusoid. It has a soccer-sized bell with up to 60 trailing tentacles, some over three meters in length. The tentacles bear billions of stinging cells, each housing a poison-filled capsule, the nematocyst. The massive amount of venom carried by this jellyfish is, in fact, somewhat puzzling. Why possess so much more toxin than needed to snatch an assorted buffet of small fish and crustaceans? Why the need for overkill? The answer most likely lies in the jelly's delicate construction. A struggling fish or prawn could easily rip the flimsy tentacles to shreds. But by injecting an extremely potent toxin the prey become immobilized within a few seconds, thus minimizing the risk of damaged tissue. In this case too much of a good thing is fine for the jellyfish but not so fine for their prey, and surely not for an unsuspecting swimmer.

Chironex typically spends the daylight hours at considerable depths, traveling toward the surface at night to feed. At times, however, tidal surges can transport them ashore in relatively large numbers. It is then when humans and box jellies come into fateful contact. Their presence is unpredictable and almost undetectable until an unfortunate victim collides with one. And this doesn't always have to be in the water as even beached box jellies can deliver their chilling kiss.

But they do not stalk or seek out large animals, and that certainly includes humans. Envenomation, instead, occurs when a person accidentally brushes against the fleshy tentacles. And that meeting can be gruesome. Someone who has received a large dose has little chance of survival. One authority indicated that if a swimmer becomes entangled in the weblike tendrils, his chance of living or even reaching shore are virtually zero. Those who have survived their encounter with the box jelly all express the same thing, numbing, excruciating pain. One woman who weathered a face-lashing declared that the pain could only be described as "...if a hot poker was thrust through my cheek." Another victim compared the pain to "being immersed in a vat of fire."

The venom is a complex neurotoxin consisting of a mixture of proteins, enzymes, polypeptides, and tetramine. Exposure causes immediate red welts on the skin. Fever, nausea, vomiting, and severe cramps soon follow. The venom then induces paralysis of the central nervous system leading to respiratory failure, and finally to cardiac arrhythmia and collapse. Death can occur within two to three minutes; children being especially vulnerable.

Immediate treatment involves removing any remnants of the tentacles and then rinsing with sea water or vinegar (~4% acetic acid). Although anecdotal cures often mention human urine as an antidote, that home remedy has yet to be proven scientifically. The best chance of survival depends upon rapid medical intervention. Hospital treatment involves injections of antivenom, epinephrine, Decadron, and antihistamine.

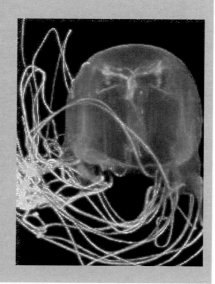

Chapter 5 Summary

Key Points

1. Cnidarians are the only radially symmetrical animals.

2. They are at the tissue grade of organization. As such they are diploblastic with two tissue layers (epidermis and gastrodermis) derived from the embryonic ectoderm and endoderm, respectively.

3. Members of this phylum are polymorphic. They exist as two distinct adult body forms, an attached polyp and a mobile medusa.

4. They show alternating patterns in which one generation, the polyp, reproduces asexually and the other, the medusa, are sexual. At times, one of the forms, usually the polyp, is the only morph, and may then reproduce both asexually and sexually.

5. All cnidarians possess tentacles bearing toxic stinging devices, the nematocysts.

6. They have no distinct circulatory, respiratory, osmoregulatory, or excretory systems. Exchange of respiratory gases and osmoregulation occurs cutaneously through the body surface.

7. There are a number of primitive organs present: chemo-receptor, touch receptors, organs of balance (statocysts), light-sensing organs (ocelli), and an early nervous network, the nerve net.

8. Reproductively, they show considerable flexibility. Asexually, they use fragmentation, budding, and fission. Sexually, they show both the monoecious and dioecious patterns. Also, they use a genetically unique planula larva.

9. Finally, cnidarians have fully exploited the colonial life-form, occurring in sheets, attached to bladders, and even molding massive calcareous reef ecosystems.

Key Terms

acontia

ahermatypic

alternation of generations

basal disc

calcareous cups

chemo-receptor

cnidocyte

column

diploblastic

ectoderm

epidermis

epithelio-muscular cell

extracellular digestion

gastrodermis

gastrovascular cavity (GVC)

hermatypic

holdfast

intracellular digestion

medusa, medusoid

nematocyst

nerve net

ocelli

oral-aboral axis

planula larva

polymorphism

polyp

radial symmetry

septum

siphonoglyph

statocyst

tissue grade of organization

touch receptors

zooxanthallae

Points to Ponder

1. What is the advantage of radial symmetry?

2. What is diploblasty? Why is it important?

3. What special technique do cnidarians use to capture their food?

4. Compare intra- and extracellular digestion.

5. What makes a box jelly so dangerous?

6. What is alternation of generations as demonstrated by the cnidarians? What are its advantages.

7. How is polymorphism displayed by this group?

8. What is a planula larva and what is its function?

9. How is respiration and osmoregulation accomplished in cnidarians?

10. Describe the nervous system of a *Hydra*.

11. What functions do ocelli and statocysts perform?

12. Describe how a statocyst functions.

13. Describe how a hydrostatic skeleton works.

14. Describe the three types of coral reefs.

15. What is the advantage of a colonial existence? Name two types of cnidarians that are colonial.

16. What are zooxanthallae? What role do they play in the formation of coral reefs.

Inside Readings

Brusca, R. C., and G. J. Brusca. 1990. Invertebrates. Sunderland, MA: Sinauer Associates.

Derr, M. 1992. Raiders of the reef. Audubon. March/April pp. 48-56.

Faulkner, D., and R. Chesher. 1979. Living Corals. New York: Clarkson N. Potter.

Hessinger, D. A., and H. M. Lenhof, eds. 1988. The Biology of Nematocysts. San Diego: Academic Press.

Murata, M., et al. 1986. Characterization of compounds that induce symbiosis between sea anemone and anemone fish. Science. 234:585-587.

Muscatine, L., and H. M. Lenhof, eds. 1974. Coelenterate Biology. New York: Academic Press.

Pechenik, J. A. 1996. Biology of the Invertebrates. Dubuque, IA: Wm. C. Brown.

Sebens, K. P. 1994. Biodiversity of coral reefs: What are we losing and why? Am. Sci. 34:115-133.

Shick, J. M. 1991. A Functional Biology of Sea Anemones. New York: Chapman and Hall.

Thorpe, J. H., and A. P. Covich. 1991. Ecology and Classification of North American Freshwater Invertebrates. New York: Academic Press.

Chapter 5

Laboratory Exercise

Phylum Cnidaria

Laboratory Objectives

✓ reproduce the classification system of the specimens studied,

✓ identify both live and prepared specimens of cnidarians,

✓ identify anatomical features of various specimens of cnidarians, and

✓ identify habitats, locomotor capabilities, and behaviors of the specimens studied.

Supplies needed

● Dissection microscope

● Compound microscope

● Prepared microscope slides

● Live *Hydra*

● Display specimens of the phylum Cnidaria

Chapter 5
Worksheet

Phylum Cnidaria
l. Class Hydrozoa

~~*Hydra*~~ Anemone

1. Obtain a living specimen of *Hydra* and observe the following features:

    ✓ basal disc (aboral surface)
    ✓ mouth
    ✓ body (column)
    ✓ look for buds (may or may not be present)
    ✓ tentacles
    ✓ look for ovaries and/or testes (may not be present)
    ✓ hypostome (mound surrounding mouth)

2. Observe feeding behavior of *Hydra* by giving them the small crustacea, *Daphnia*.
   Explain how *Hydra* uses its tentacles in feeding.

   It intengales its food and sucks it.

   Anemone
3. ~~*Hydra*~~ doesn't show polymorphism. Is this typical of other members of the class Hydrozoa?
   Explain.

   No because its polyp

4. Using a whole-mount (w.m.) slide of *Hydra* and referring to the figure on the
   next page, locate the following features:

    ✓ basal disc or holdfast (aboral surface)
    ✓ mouth
    ✓ body (column)
    ✓ look for buds (these may or may not be present)
    ✓ tentacles with nematocysts in cnidocytes (cnidoblasts)
    ✓ look for ovaries and/or testes (may not be present)
    ✓ hypostome (mound surrounding mouth)
    ✓ epidermis
    ✓ gastrovascular cavity (GVC)
    ✓ gastrodermis

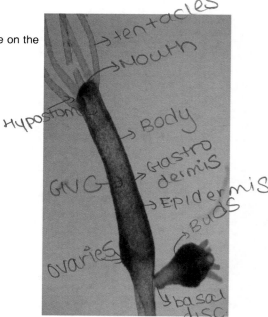

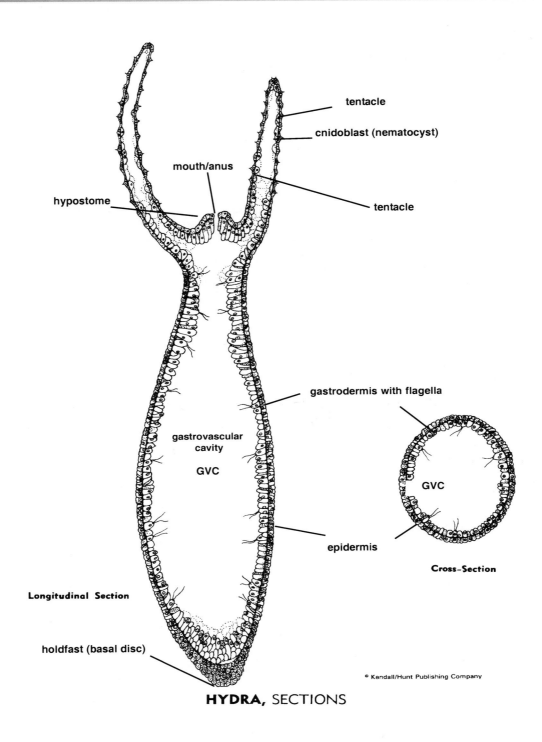

tentacle

cnidoblast (nematocyst)

mouth/anus

hypostome

tentacle

gastrodermis with flagella

gastrovascular cavity

GVC

GVC

epidermis

Cross-Section

Longitudinal Section

holdfast (basal disc)

© Kendall/Hunt Publishing Company

HYDRA, SECTIONS

Figure 5.11. Internal anatomy of a *Hydra*.

5. This phylum is described as diploblastic. Describe how cnidarians demonstrate this condition.

   Because they have two dirved tissue layer endodermis and gastrodermis

6. Observe a whole-mount slide of a *Hydra* containing a bud. Are buds used for <u>sexual</u> or <u>asexual</u> reproduction?

   Asexual

7. Now obtain two cross-section (c.s.) slides of *Hydra*; one showing testis (male) the other containing an ovary (female).

   a. On the photo to the right, label the epidermis, gastrodermis, mesoglea, GVC, and developing sperm.

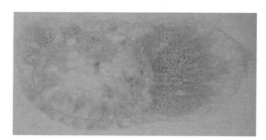

   b. Why are the sperm cells smaller and more numerous toward the tip of the spermary?

   c. On the photo to the right, label the epidermis, gastrodermis, mesoglea, GVC, and developing ovum (egg).

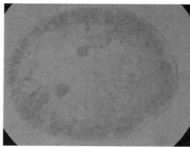

   d. Give two reasons that explains why eggs are so much larger than sperm.

Obelia

8. Using a whole-mount (w.m.) slide of *Obelia*, hydroid (polyp) and the figure below, label the photograph to the right as to

      ✓ reproductive and feeding polyps
      ✓ gastrovascular cavity
      ✓ tentacles
      ✓ branching stalk

Is this the <u>sexual</u> or <u>asexual</u> stage in the life cycle?

      Asexual

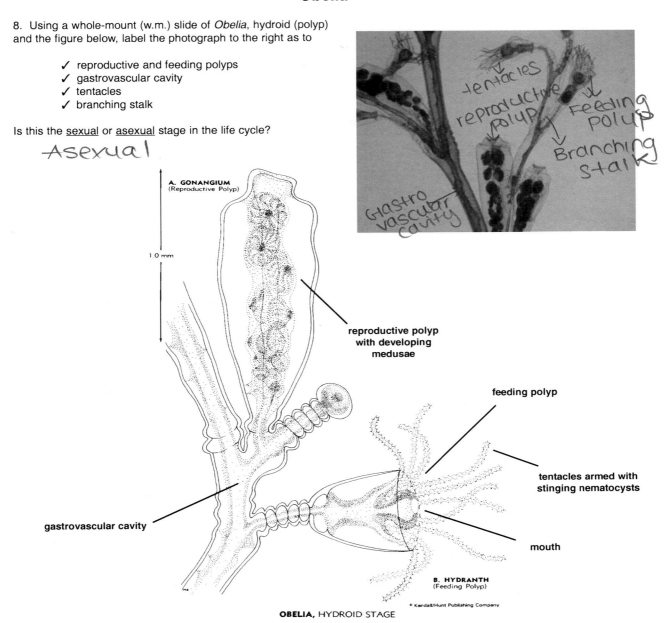

Figure 5.12. The anatomy of *Obelia*.

9. View a slide of *Obelia* medusa and refer to the photograph to the right.

    Is this the <u>sexual</u> or <u>asexual</u> stage in the life cycle?

      Sexual Stage

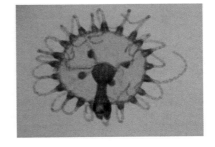

Physalia

10. Examine a preserved specimen of *Physalia*.

11. What is *Physalia's* common name? <u>Portugese</u>

12. Is this an individual or a colony? Explain.
colonies ob individual·

13. Why should you avoid *Physalia* if you are swimming in the ocean?
Stinging tentacles

II. Class Scyphozoa

Aurelia

14. Observe a preserved specimen of the moon jelly, *Aurelia* (see photograph below).

15. Note the tentacles and inflated bell-shaped structure.
What substance located within the bell helps to keep it
and other jellyfish afloat?
Mesobia

16. The immature Aurelia is known as an ephyra. Observe a
slide of an ephyra and note its radial shape and similarity to
an adult jellyfish.

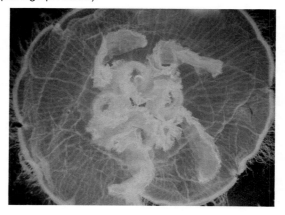

They have round shape
rather then the star
chase bor ephura.
They also have formed
oral arms.

III. Class Anthozoa

Metridium

17. View preserved specimens of the sea anemone *Metridium* and refer to the figure on the the next page.

   a. How is it similar to Hydra?
    Both have dipolyps

   b. How does it differ from Hydra?
    Hydra are asexual, where corals are both.

   c. How does it differ from corals?
    They are reeb building

18. Examine the dissected *Metridium* on display. Locate:
    ✓ pedal disc (holdfast)
    ✓ column (scapus)
    ✓ tentacles
    ✓ mouth
    ✓ acontia
    ✓ gastrovascular cavity (GVC)

19. Survey the various examples of coral "skeletons" on display. Locate the calcareous cups. What is their function?
Calcareous cups it is draws escape brom
the skelton

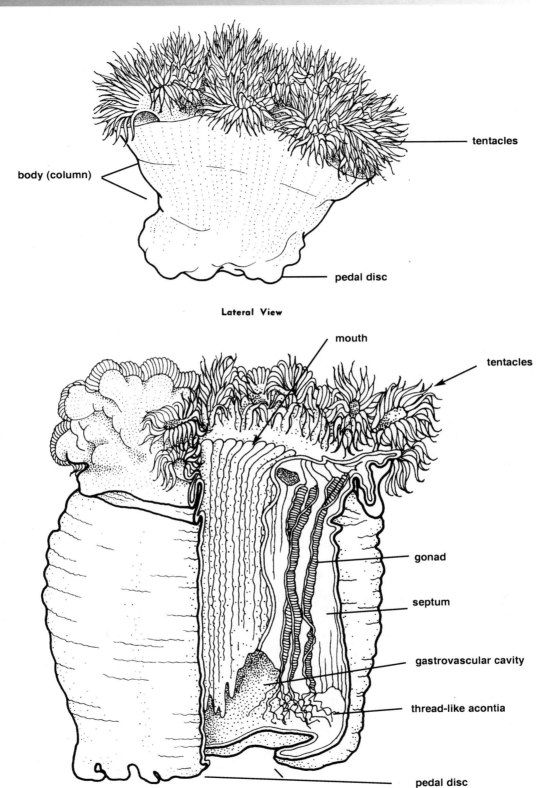

tentacles

body (column)

pedal disc

Lateral View

mouth

tentacles

gonad

septum

gastrovascular cavity

thread-like acontia

pedal disc

Figure 5.13. *Metridium,* the common sea anemone

6 Chapter 6

PHYLUM PLATYHELMINTHES

THE FLATWORMS

Chapter 6

Phylum Platyhelminthes

...a worm's-eye point of view.
- Lyle

Introduction

*P*latyhelminthes are flatworms. With a name like that, it doesn't take much to imagine what they look like: long and thin, rather leaflike in their overall design. However, as descriptive as it might be, being **dorso-ventrally flattened** represents only a small fraction of the platyhelminth story. We are about to discover, for example, that this primitive group has introduced three of the most significant biological advances ever witnessed in the long history of animals. These three developments are so important that they have earned flatworms a very special place in the rank-and-file of animals: that of a *transitional group* between lower and higher forms. There are over 22,000 known flatworm species.

(1) **Triploblasty**. You will recall that the previous animal group, the cnidarians, were organized into two germ (tissue) layers, the *ectoderm,* forming the animal's outer layer, and the *endoderm,* contributing to internal structures. Consequently, jellyfish and their cousins are *diploblastic* organisms. Our current group, the flatworms, however, have carried animal body design one step further. They have added a third layer; they are triploblastic. This third layer, the **mesoderm**, is wedged between the other two and allows even greater tissue and organ development, and thus, greater overall efficiency. As mentioned above, the ectoderm gives rise to the outermost layers of the animal, the *epidermis,* and also to the nervous system. The endoderm, on the other hand, produces the inner lining of the digestive system, the *gastrodermis.* The new middle layer, mesoderm, then gives rise to almost everything else: muscles, skeleton, the circulatory system, much of the excretory system, and portions of the reproductive organs. Triploblasty is such an important feature that all animals from this point on have the same three layers.

Fate of the Three Primary Layers

ectoderm → epidermis (skin) and nervous system
mesoderm → muscles, skeleton, circulatory system, etc.
endoderm → digestive system & related structures

(2) **Bilateral Symmetry.** Even though there appears to be an endless array of shapes and designs, in reality, there are only three body forms an animal can assume: (a) *asymmetry*, such as the sponges in which there is no ordered or discernible pattern; (b) *radial symmetry* displayed by the cnidarians, whereby any plane passing at the same time through both the center and the long axis will produce near-identical halves; and (c) *bilateral symmetry* (all remaining animals). Basically, a bilateral animal is one that has a right and left side, as well as both an anterior-posterior and a ventral-dorsal axis. An animal with such a design can only be split into two equal pieces if the cleavage plane cuts through the longitudinal midpoint. This is what is known as a sagittal section. The result is two halves, each a mirror image of the other. The appearance of bilateral symmetry is one of the pivotal developments in the evolution of animals. Its significance is more fully discussed later in this chapter.

(3) **Cephalization**. The great majority of animals possess a concentration of sensory and nervous structures anteriorly. We, of course, call this anterior concentration a **head,** and the process of forming such a structure is cephalization. Without cephalization it is doubtful that higher animals, as we know them, would ever have appeared.

Figure 6.1. Cephalization, the concentration of nervous and sensory structures anteriorly, was one of the crowning achievements in animal evolution.

Integration:

A Case of Forward Locomotion

As you are now discovering, in evolution's arena events rarely occur in isolation. Developments, regardless of their importance, are almost always **interrelated**. This is certainly the case with the three major advances just described: triploblasty, bilateral symmetry, and cephalization. These are, indeed, indelibly intertwined. Each is dependent, in some fashion, upon the other. Each expresses itself more fully in accord with the others. And it all has to do with locomotion, **directed locomotion**, that is.

Animals that are either sedentary or just float freely in the water meet basic needs in quite a different way than do active animals. For one thing, attached or floating animals, such as sponges, sea anemones, and jellyfish, receive stimuli from any direction. Likewise, their food may also arrive from any direction. As a result, asymmetry or radial symmetry are efficient body designs. Having a *circle* of tentacles and sensory organs (e.g., nerve net, ocelli, chemo-receptor, statocysts, etc.) makes perfect sense for attached or even for floating organisms; it allows them to take full advantage of multidirectional stimuli.

But sessile animals do face some problems. First, they must be patient. They must sit and await the rather random arrival of food, which they then capture or filter from the water stream. Second, they are highly vulnerable to predation. A sessile animal, with tentacles waving about is highly noticeable. To counter this vulnerability they have developed a number of safety devices: protective spines, stinging cells, chemical defenses, and so on. Attached animals are also extremely vulnerable to a changing or capricious environment. If their habitat is altered significantly, it may be difficult or even impossible for them to relocate. They are threatened with the very real possibility of local extermination.

Active animals, on the other hand, can enjoy survival strategies beyond the wildest expectations of their sedentary cousins. To do so, however, requires locomotion. Not just any locomotion; no, it requires efficient, directional movement...active maneuvering toward or away from objects. Directional locomotion allows an animal to actively explore and to exploit its environment. To maximize forward locomotion, a system of powerful muscles are needed. Muscles that can be contracted deliberately and with sufficient control to permit active locomotion. The problem is, we have yet to encounter muscles of this sort in any animal. But that is about to change. Flatworms have solved the dilemma through triploblasty and the resultant arrival of *mesoderm*. From this middle layer, strong locomotor muscles and, later, an entire host of advanced organs, are derived.

But how does forward locomotion relate to having a head? Let me suggest the following. When an animal moves forward, stimuli no longer come equally from all directions; instead, they arrive in a predictable sequence: anterior first, sides next, and then finally the posterior. It is only logical that the part of the animal first entering a new environment receives information first. In point of fact, when an animal moves forward the anterior surface is literally bombarded by new stimuli. Next, the signals sweep, rather uniformly, over both sides of the animal. The wave is then recorded ultimately by the rear of the organism. As long as an animal is moving forward, this is the expected progression of information.

Radial symmetry then, as you might conclude, doesn't work well for an active, forward-moving animal. Indeed, when environmental stimuli occur in a sequence, as described above, the best body design happens to be bilateral symmetry. It allows the animal to move forward with the least resistance. But more importantly, it permits the animal to process similar stimuli from both sides...in stereo... as it travels. Consequently, paired, lateral sensory structures become a priority. Just think of the eyes, ears, lateral lines, whiskers, and antennae of higher animals. They all occur in pairs, one on each side.

Finally, concentrating sensory and nervous structures anteriorly (where signals first arrive) is the most efficient placement. A head is, then, no mere morphological accident. In fact, a good case can be made that cephalization is a *prerequisite for any active, forward-moving animal*. It permits the immediate processing of vital information so that any new environment can be quickly evaluated: food more easily obtained; mates located; danger avoided. A true hunter has now been designed. To appreciate the importance of cephalization, just envision what might befall an animal that moved in the wrong direction...tail first!

Figure 6.2. A typical bilateral animal. Note the emphasis on cephalization, bilateral symmetry, and forward locomotion.

But Why Flat?

We have just learned that triploblasty provided the muscle power for forward motion; bilateral symmetry added equilateral sensory input; and cephalization packed sensory and processing centers where it's most needed...into a head. But why flat? What advantage is there to an animal being flattened like squashed road-kill? The answer lies in the physical relationship between surface and volume. The larger an animal becomes, the more difficult it is for respiratory gases, fluids, nitrogenous wastes, and other materials to diffuse throughout its tissues. As long as an organism is small, its surface-to-volume ratio is adequate for simple diffusion to occur. But as soon as it gains some size, especially bulk, diffusion becomes less and less efficient as a means of transport. Internal cells, those buried deep within an animal, are then at a distinct disadvantage, and the bulkier the animal becomes, the greater the disadvantage. Since flatworms lack both a circulatory and a respiratory system, a bulky body would not permit innermost cells to receive vital materials by diffusion alone.

A flat body design, on the other hand, readily compensates for the missing systems (especially the circulatory system) and thereby solves the problem of diffusion distance. In other words, each cell is close enough to the exterior surface to allow efficient exchange of oxygen and carbon dioxide by way of diffusion.

The same applies for osmoregulation and excretion, to a point. Flatworms are acoelomate, that is, they lack an internal body cavity. Instead, the space between the outer wall and the inner endoderm is filled with a mesodermal material called **parenchyma**. This material is under hydrostatic pressure and thus serves as an internal skeleton. However, the presence of parenchyma tends to impede the free flow of diffused material, especially nitrogenous wastes. The flat design does permit a portion to diffuse to the outside but the remainder must be removed in another way. To do this the flatworms introduce the first excretory system, a "kidney-like" structure called the **protonephridium**. This organ consists of a small sac bearing an internal ciliary tuft, looking very much like the end of a small paintbrush. As the tuft, also called a flame cell, flutters, it draws fluid into the sac from the surrounding tissue and transports it to through a short duct to the outside. As a result a proper ionic and fluid balance is maintained within the organism.

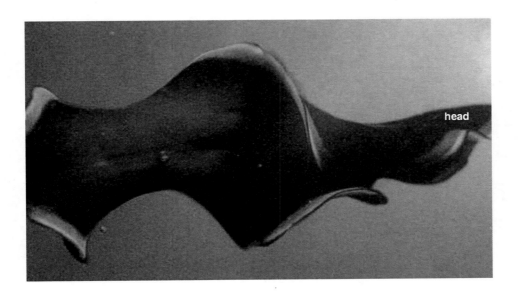

Figure 6.3. A purple marine flatworm. Note the head and bilateral symmetry. Its bold pink and purple color pattern indicates its rather unpleasant taste.

Systematics

~ Kinds of Flatworms ~

*T*here are three major taxonomic classes of flatworms. Of these, two are entirely parasitic and the other is free living. Most flatworms (> 80%) are parasitic and these have undergone extensive alterations in accord with that lifestyle. The free-living forms, on the other hand, have not experienced such modification and, thus, are more representative of the basic plan of the phylum. Since features such as locomotion, nutrition, reproduction, and basic anatomy vary so much among the three classes, they are discussed along with each group.

Class Turbellaria

All of the over 3,000 species of free-living flatworms belong to this class. While the great majority inhabit marine ecosystems (see photo on previous page and those below), there are several freshwater species and even a few terrestrial forms that survive in moist habitats. Some reach a relatively large size (> 50 cm), but most are less than 1 cm in length.

The common planarian, *Dugesia,* is typical of the class (Figure 6.5 on the next page). They are residents of ponds and other freshwater habitats gliding easily among aquatic vegetation, patrolling muddy substrates or seeking food among rocks or other debris.

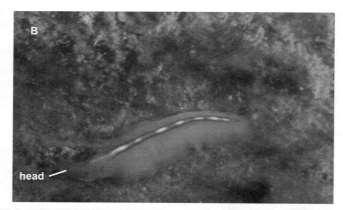

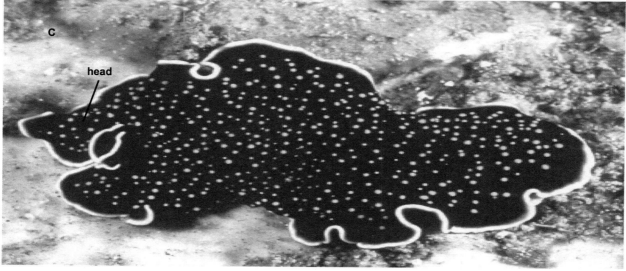

Figure 6.4. Three marine turbellarian flatworms belonging to the genus *Pseudoceros* are indicated in the photographs above. These colorful animals are not well known and many have yet to be fully described or named.

Sensory Structures. *Dugesia* is shaped rather like an elongated arrowhead and thus it is easy to detect the cephalic nature of this animal. The head bears two major sensory structures: a pair of eyes along the mid-dorsal line and a pair of "ear-like" lobes, the **auricles**, laterally. The eyes, although unable to form distinct images, do allow the animal to detect variations in light intensity and light direction. Planarians are repulsed by bright light, a condition known as **negative phototrophism**.

Even though the auricles look like ears, and the term auricle actually means "ears," they have nothing to do with hearing. Instead, they bear tiny clusters of differentiated cells to monitor the environment: there are *chemo-receptor* to detect chemicals in the water; *touch receptors* to discern mechanical stimuli; and cells to detect water vibrations (*pressure receptors*). Since these receptors are distributed in stereo fashion on each side of the animal, they provide information equally during forward locomotion.

Flatworms have a primitive brain, the **anterior ganglion**, nestled just beneath the eyes. The ganglion receives nervous input from the various sensory receptors in the head. In addition, the animals possess a pair of **longitudinal nerve cords** that interconnect by way of commissures. The nerve cords transmit sensory impulses to the brain and carry motor responses to the muscles. So, while primitive, the flatworm nervous system is, nonetheless, able to achieve the basic activity of any nervous system...sensory input, integration, and motor output. The entire nervous apparatus has the appearance of a ladder, indeed it has, at times, been called a "ladder system" (Figure 6.13).

Locomotion.   Planarians move by a combination of muscular contractions and ciliary action. They possess unique epidermal glands, **rhabdites**, that secrete a thick mucus covering. The slimy mucus undoubtedly helps to prevent desiccation (drying out) but also assists in locomotion. Ventral cilia beat in unison within the mucus envelope to produce a smooth gliding motion. Muscle action, in turn, gives the animal the additional ability to twist and turn as it glides along.

Free-swimming flatworms actually have three layers of muscle lying beneath the epidermis. Fibers of each layer overlap so that complex movements can be achieved. The result is an animal that can maneuver easily over objects within its environment. Some turbellarians can also loop the front of their body off the surface and move in a rather inchworm-like manner.

Nutrition. Planarians are carnivorous, feeding on small organisms such as crustacea, worms, and protozoans. They are unable to actively catch prey but tend, instead, to entrap it in a mucus wad. They then wrap their body around the prey, and slowly draw it into the everted pharynx. Planarians also serve as scavengers, feeding on dead animals.

One of the real oddities of planarians is the arrangement of the digestive system. For example, the mouth, rather than being located in or near the head, where you might expect, is instead situated midway down the animal. Ingested food is transported from this mid-ventral mouth into a tubular, protrusible **pharynx**, a muscular device much like a soda straw that actively sucks up either whole prey or food fragments.

The rest of the digestive system is shown in the photograph below and in Figure 6.13. Note that the pharynx leads to a **tripartite**, or three-part, intestine. The gut passes forward as an anterior intestine and sends two branches laterally and posteriorly. The intestines, in turn, give off tiny branches, the **diverticula**. The resulting digestive system is a branching network of tubes delivering food to within millimeters of every part of the animal. This spreading apparatus takes the place of a circulatory system, which is lacking in flatworms. Finally, like cnidarians, flatworms have no separate exit for undigested food, so its mouth also serves as its anus.

Reproduction. Flatworms can reproduce both asexually and sexually. In fact, their regenerative powers are remarkable. If one is cut in two, or even into several pieces, each piece can replace the missing parts. In fact, in nature, they are known to literally tear themselves in half and then grow into two new adults. They have also been observed pinching themselves into a number of sections, much like a string of beads, then breaking into individual wedges and each, again, regenerating the missing parts.

Although flatworms are *monoecious,* each animal capable of producing both sperm and egg, self-fertilization is uncommon. Instead, each exchanges sperm with a partner through mutual copulation. The eggs are then fertilized internally, deposited in the surrounding water, and new individuals hatch within a few days. However, if reproduction occurs in late autumn the eggs may lie dormant until temperatures rise in the spring when the new worms hatch.

A bizarre form of mutual copulation is seen in certain flatworms in what is called **hypodermic impregnation**. It seems they have developed penises but nothing analogous to a vagina. Thus, to insert sperm they stab each other with razor sharp, harpoon-like, penises in prolonged bouts of "penis fencing".

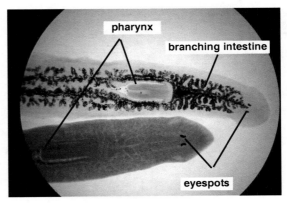

Figure 6.5. The common planarian, *Dugesia*. A head is well developed, possessing eyes and lateral bulges (auricles) that serve as chemo-sensors. Note also the tripartite intestine that branches throughout the animal.

Class Trematoda

There are 18,000 to 24,000 trematode species, all of which are parasitic and are commonly known as **flukes**. Some are ectoparasites, occurring only on the outside of their hosts, such as on the gills of fish. Most, however, are endoparasites of the liver, lungs, intestines, mouth, bladder, or blood vessels. We will confine our discussion to these internal types.

Physically, flukes look very much like planarians, but it's only a superficial similarity. Their behavior, food habits, habitats, sensory structures, and especially, their life cycles, are very different from their planarian cousins.

One of the most apparent structural differences lies in their body coverings. Unlike the cellular, ciliated epidermis of planarians, flukes possess a **non-ciliated syncytial integument.** The integument forms a jacket that enwraps the animal in a large *unicellular* mass. It serves as a protective coating, helping to defend against acidic, erosive secretions of the host. Furthermore, the integument is ridged, much like fingerprints, enhancing not only the passage of respiratory gases (cutaneous respiration), but also permitting the absorption of some nutrients directly from the surrounding environment.

The bulk of their food, however, is taken in through the *anterior mouth* and passed to a V-shaped, blind intestine. There, extracellular and intracellular digestion occurs. Food consists of tissue fluids, blood, mucus, and other cellular debris. Surrounding the mouth is a muscular oral sucker that assists in attaching to the host tissue. Some forms also have a ventral sucker for additional adhesion. As you would expect with an internal parasite, sensory structures are poorly developed. They lack ocelli, and auricles are, likewise, absent. Overall, their nervous systems are less evolved than the free-living flatworms.

Like planarians, trematodes are monoecious. But their reproductive potential is much, much higher; a fluke may produce 100,000 eggs for every single egg produced by a free-living flatworm. Following *mutual copulation*, sperm, from posterior, **branching testes**, are transported to a sac, the **seminal receptacle** in each partner. There they are stored to be retrieved when needed. Eggs are produced in an *ovary* located just anterior to the receptacle. As eggs are formed, they are fertilized by a spermatozoan, one by one, much like an assembly line. After fertilization, each egg (actually an embryo) is enveloped by a yolky mass from the large **yolk** (vitelline) **glands**. The embryo, now equipped with a source of nourishment, is stored with thousands of others in a bulging **uterus.**

Stored eggs are released in huge quantities into the host's passages (e.g., gastrointestinal or urinary system) to be deposited along with the host's bodily discharges. Should the eggs reach an appropriate substrate, usually a wet pasture, pond, or other moist area, they undergo a series of larval transformations before locating another final host. These larval changes are described in more detail in the next section.

A Word to the Wise
-or-
Can This Course Save My Life?

● Members of the next two groups are all parasites of both invertebrate and vertebrate animals, including humans. Some cause debilitating, crippling diseases, while others are lethal. Indeed, the second most deadly human parasite is a flatworm. In some parts of the world every person carries at least one.

● So what does that have to do with you? Well, one day you may just find yourself traveling to the four winds: South America, Africa, Asia, the Pacific Islands...who knows? And how will you safeguard against tiny unwanted hitchhikers...the internal ones?

● Perhaps an understanding of these parasites, their life cycles, pathology, and prevention might just help you avoid some personal distress. Conceivably, it could help avert a little intestinal discomfort or maybe ward off a dose of psychological anguish. Possibly your knowledge of parasites will prevent a wrinkle in your travel plans. *Then again, it may even save your life.*

Three Important Trematode Parasites of Humans

*I*t is in their life cycles that flukes are most interesting, biologically. Most show complex cycles involving one or more intermediate hosts. These are known as **digenetic** flukes and are the ones most responsible for human infection. Although there are several that invade human tissue, only three are widespread enough to deserve treatment here.

(1) ***Fasciola hepatica,*** the sheep liver fluke. Although sheep are the most common host of *Fasciola,* close to as many humans have the sheep liver fluke as the human liver fluke, with anywhere between 2.8 million and 17 million cases annually. Adult parasites reproduce sexually in bile passage near the sheep liver. Large numbers of eggs (>300,000 per day) are deposited in wet pastures, drainage ditches, ponds, or other aquatic areas along with the sheep's fecal material. The eggs soon hatch, liberating a **miracidium** larva. The miracidium actively seeks a specific snail into which it bores. Upon reaching the snail's gut, the larva reproduce asexually producing vast numbers of another larval form, the **redia**. These escape the snail after a developmental period of about eight weeks and then burst open to release numerous fork-tailed **cercaria** larva. These swim toward, and attach to, emerging vegetation.

Once secured on the vegetation they are again transformed, but this time into dormant cysts, the **metacercaria**. The cysts develop a protective covering and thus become resistant to desiccation. They can remain inactive for up to several months until eaten by an appropriate final host. This is usually a grazing sheep. However, humans can become infected by ingesting eggs through accidental contact or by eating wild plants.

The fluke then penetrates the human's gastrointestinal wall and migrates to the liver. Once there it bores through liver tissue and after about an eight-week developmental period it takes up residence in the nearby bile duct as an adult fluke.   It is during the migration period that *Fasciola* can be most damaging, causing liver inflammation. The adult fluke can reach an excess of 3 cm and thus can cause blockage of the bile duct or other damage due to its large size.

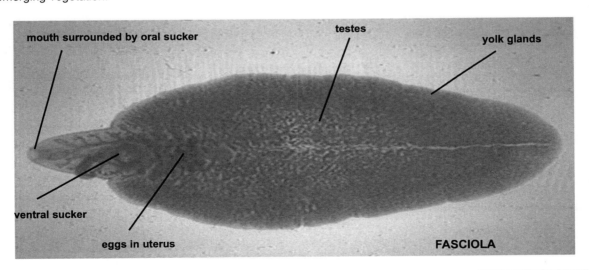

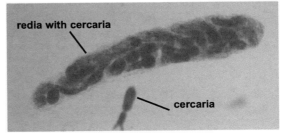

Figure 6.6. An adult sheep liver fluke (top diagram) showing overall flat shape and a number of structures. This is a large fluke that normally parasitizes sheep, but is also a common parasite of humans. The photo to the right shows two larval stages in the development of this fluke. The redia contains numerous cercaria, one of which is also shown. The cercaria locate vegetation and become dormant awaiting a feeding sheep.

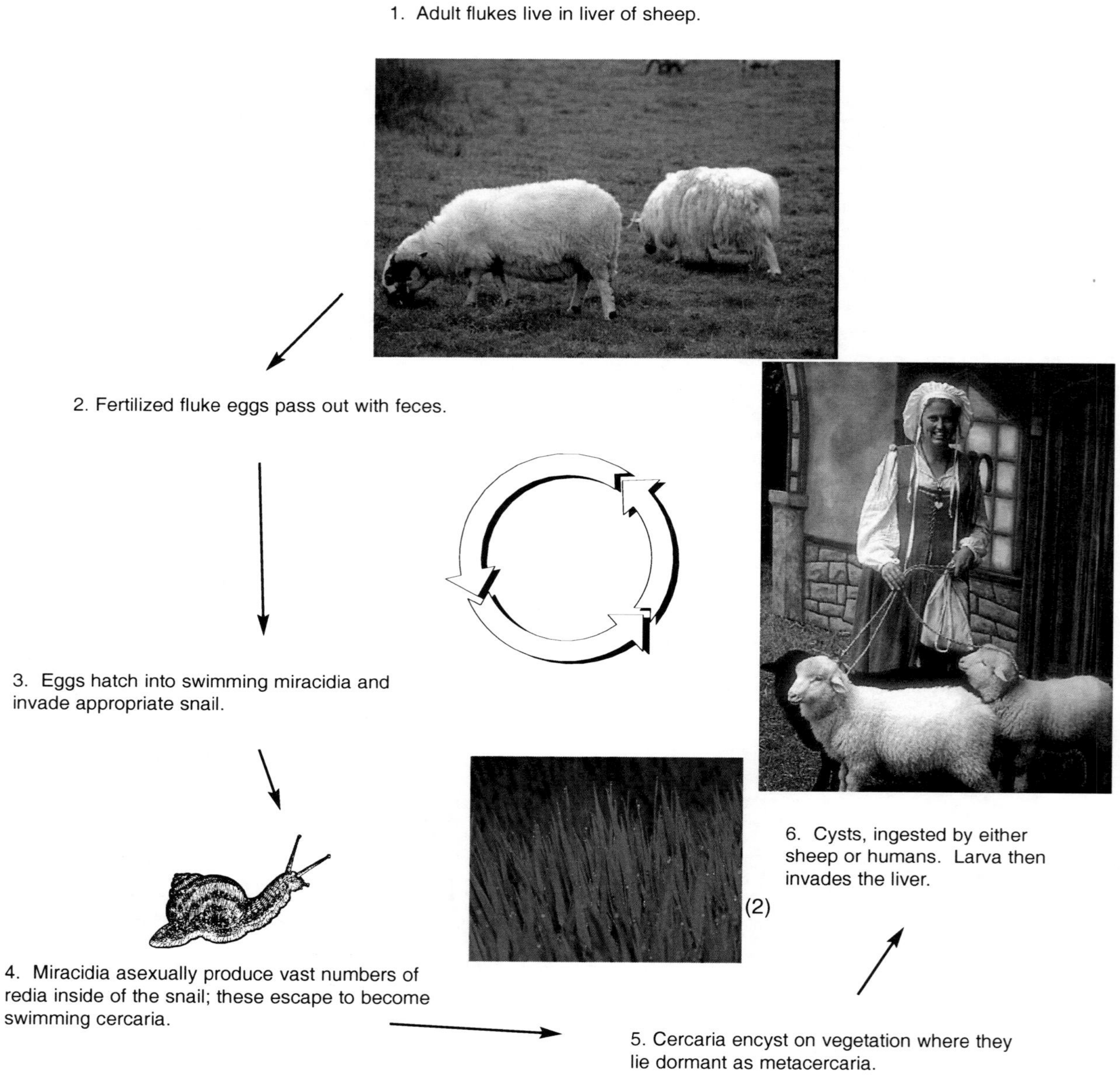

1. Adult flukes live in liver of sheep.

2. Fertilized fluke eggs pass out with feces.

3. Eggs hatch into swimming miracidia and invade appropriate snail.

4. Miracidia asexually produce vast numbers of redia inside of the snail; these escape to become swimming cercaria.

5. Cercaria encyst on vegetation where they lie dormant as metacercaria.

(2)

6. Cysts, ingested by either sheep or humans. Larva then invades the liver.

Figure 6.7. Life cycle of *Fasciola hepatica*, the sheep liver fluke.

Clonorchis sinensis is the human liver fluke (Figures 6.8 and 6.14 on the following pages).   Although the life cycle of the human fluke, *Clonorchis,* is, in general, the same as that of *Fasciola*, there are some notable differences. The most prominent one is that *the human, not the sheep, is the primary host.* The adult fluke lives in the human liver where mutual copulation produces huge numbers of fertilized eggs.

Following fertilization, eggs are deposited and released along with human feces. If defecation occurs near water, or if fecal material is dumped in waterways, then the eggs erupt as *miracidia* larva. These invade a *snail*, and undergo asexual reproduction to produce vast numbers of *cercaria* larva. It is estimated that a single miracidium can produce on the order of 250,000 individual cercaria!

These swimming cercaria larva escape from the snail, but unlike the sheep liver fluke, *Clonorchis* cercaria do not become dormant on vegetation; *they instead must locate another host...a suitable fish.* They burrow into the fish flesh and there become *metacercaria* cysts to await ingestion by a human. A person becomes infected by eating uncooked, or undercooked, fish containing the encysted stage.

Upon reaching the human stomach the cyst ruptures, releasing immature flukes that make their way to the liver to repeat the cycle. Adult flukes can survive for years in the liver. They are quite small and several may coexist in the same liver without causing any symptoms. However, should a person continue to be reinfected, large numbers of flukes may accumulate causing severe liver damage. Over 20 million people, mostly in east, and southeast, Asia, are infected with the human liver fluke.

(3) Schistosomiasis is a dangerous disease caused by a number of flukes of the genus *Schistosoma*. Over 200 million people are thought to be infected by this human blood fluke. Indeed, in areas of Africa, the West Indies, Asia, and South America, this is a serious, widespread disease.

These trematodes differ in several ways from either *Fasciola* or *Clonorchis*. The adults do not reside in the liver, but instead, become lodged within blood vessels of the intestines or the urinary bladder where they reproduce sexually. Unlike the liver flukes discussed earlier, these flukes are dioecious. The male is larger than the female and wraps her within an elongated groove, much like a hot dog in a bun. Trapped there, she remains to continually release fertilized eggs into the small blood vessels. The eggs contain a spiny projection used to burrow from the blood vessels through intestinal tissue to reach the gut interior. They are then passed to the outside along with feces.

Miracidia larva hatch and burrow into a snail where, again, large numbers of cercaria are produced. These, in turn, escape as fork-tailed swimming larva. But they do not seek snails, vegetation, or a vulnerable fish. Instead, *these larva are able to penetrate directly into human skin.* And they just need a few seconds to do so. This causes an irritating response known as "swimmer's itch." If a waterway is heavily infested with cercaria, a human bather or swimmer can become infected with hundreds or thousands within less time than it takes to read this sentence!

Pathology occurs at two levels. First, when eggs burrow through the intestinal wall some bleeding and pain may result, although this is rarely a serious complication. The most serious symptoms occur when eggs are swept to the liver where their burrowing may cause severe liver damage. Large numbers of eggs, in fact, may cause death due to progressive liver failure.

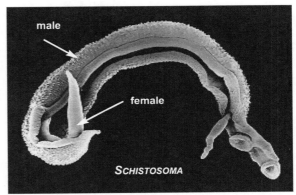

SCHISTOSOMA

Trematode Parasites of Humans

Parasite	Location in Humans	Intermediate Host	Pathology
1. *Fasciola hepatica* (sheep liver fluke)	• in the bile duct of the liver	• snail	• none unless there's a heavy infestation
2. *Clonorchis sinensis* (human liver fluke)	• in the bile duct of the liver	• snails and fish	• none unless there's a heavy infestation
3. *Schistosoma spp.* (human blood fluke)	• in the blood vessels of the intestine	• snails	• can cause serious liver damage and even death

1. **Adult Fluke** in human liver
2. **Egg** leaves in human feces
3. **Miracidium** in snail
4. **Sporocyst** produces redia
5. **Redia** produces cercaria
6. **Cercaria** escapes from the snail
7. Cercaria penetrate fish
8. **Metacercaria** ingested by human
9. **Immature fluke** in human migrates to liver where it matures.

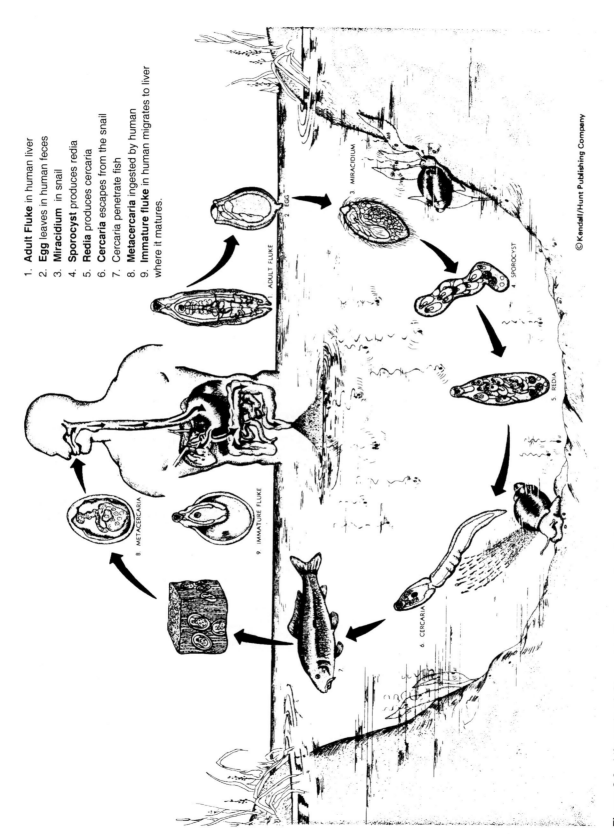

© Kendall/Hunt Publishing Company

Figure 6.8. Life cycle of *Clonorchis*, the human liver fluke.

Class Cestoda

Commonly known as tapeworms, the cestodes constitute the other parasitic flatworm group. There are over a thousand known tapeworm species. They are all elongated, thin, ribbon-like parasites containing three body regions:

> (1) the **head** (scolex)
> (2) the **neck**
> (3) the **body** (strobila)

The head's primary function is for attachment. It often bears hooks, suckers, or spines to anchor itself to the intestinal wall of the host. The head typically buries within the soft intestinal lining, allowing the rest of the animal to trail within the intestinal lumen much like a ribbon inside of a tube. Although the head may bear suckers, it does not contain a mouth.

The neck is a short, but highly important, area of a tapeworm. Known as the "growth zone", it is here that the animal elongates producing a continual stream of repeating units called **proglottids.** Each proglottid, in reality is considered a separate individual. In fact, the remainder of the animals' body, or **strobila**, consists of nothing but these repeating units which are formed from the neck in a process called **strobilization**, not unlike budding. As new proglottids are formed, older, mature ones are pushed posteriorly—the oldest proglottids being those at the posterior end of the animal. These are filled with fertilized eggs and are often called ripe or **gravid** segments. The number of proglottids in a given tapeworm depends upon the species and the age of the animal. Some consist of but two or three short proglottids while others contain thousands and can be up to 20 meters in length. In fact, the longest human tapeworm on record is approximately 25 meters (that's over 75 feet)!

Tapeworms are among the most highly specialized and widespread parasites in the animal kingdom. Virtually every vertebrate animal has its own species of tapeworm, humans included. It is estimated that over 135 million people (that's one in every 50) harbor one or more tapeworms.

Basically, tapeworms are little more than reproductive machines. They have sacrificed most of their internal organs to improve their reproductive capability. For instance, they have no digestive system...no mouth, no pharynx, and no intestines. To compensate for the lack of a digestive system, their outer integument is thrown into a series of microfolds, looking very much like corduroy material. These folds increase the overall surface area of the tapeworm tremendously. As a result, they can absorb nutrients directly from their host's intestines. Indeed, this is the only source of food for cestodes.

As you might expect, their sensory and nervous systems are also quite limited. They possess no statocysts or ocelli. Furthermore, they have only minimal locomotor capability; a weak wriggling action is all that tapeworms can accomplish.

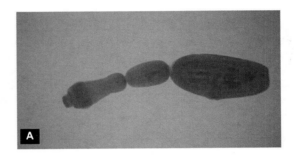

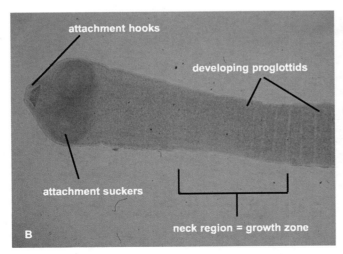

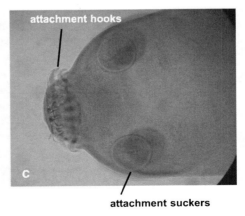

Figure 6.9. Tapeworm anatomy. A small but potentially dangerous tapeworm, *Echinococcus granulosus,* is shown in the upper plate. Plate **B & C** is of *Dipylidium caninum*, the dog tapeworm. Note the attachment hooks and suckers and the growth zone where new proglottids are added.

But their reproductive apparatus is another story altogether, they are reproductive machines. Each proglottid contains a complete male and female reproductive system. Now, just imagine...an average tapeworm may have between 2,000–4,000 segments. That means they have that same number of male structures and, of course, an equal supply of female parts. To further enhance their reproductive prowess, each proglottid can copulate with other segments or with any segment from a neighboring tapeworm! And if that's not enough, each proglottid can also fertilize itself!

There is perhaps no other animal on earth so devoted to reproduction. As you might expect, their reproductive output reflects this specialization. As new proglottids are being produced anteriorly, the most posterior ones are resorbing their reproductive structures and become nothing but egg-filled sacs. A single gravid proglottid may contain 50,000 ripened eggs. Usually the whole apparatus, proglottid and all, is then shed along with the host's feces to the outside. Typically, one to several proglottids per day are passed in this manner as the host's feces are voided.

The proglottids can squirm adequately enough to untangle themselves from the fecal mass. They crawl away to a suitable habitat where the eggs are liberated. The eggs are transformed into a resistant **onchocercus** larva, which awaits an intermediate host. If eaten by the appropriate secondary host, the onchocercus burrows into host tissue and becomes encysted as a **bladderworm** (cysticercus). Bladderworms remain in the intermediate host tissue until eaten, in turn, by either another intermediate host or by the final host, usually a carnivore. Tapeworms do not reproduce within the intermediate host.

Upon reaching the final host's intestine, the bladderworm evaginates, freeing the small tapeworm inside. It immediately becomes attached to the intestinal wall and begins the process of proglottid formation. Tapeworms can remain for many years in their vertebrate host releasing five or six proglottids each day.

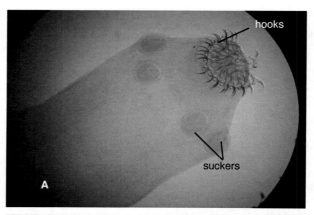

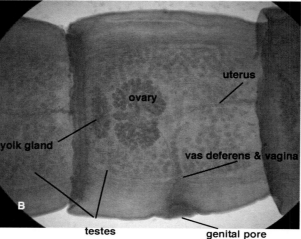

Figure 6.10. The tapeworm *Taenia pisiformis*. The head (scolex) contains attachment devices in the form of hooks and suckers (**A**). Immediately behind the head, the neck is the growth zone of the animal. Plate **B** shows a mature proglottid, which contains both male and female reproductive structures. The figure below (C) is a depiction of a complete tapeworm.

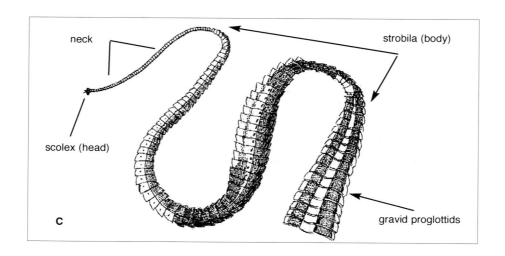

Some Common Tapeworms

There are a number of tapeworms of the genus *Taenia* that infect wild animals as well as humans. For example, *Taenia pisiformis* is a ubiquitous tapeworm of wild and domestic carnivores: wolves, coyotes, foxes, and pet dogs. Carnivores obtain the parasite when they eat rabbits contaminated with the bladderworm stage. Adult tapeworms reside in the intestine of the final host. Eggs pass in the feces and become attached to vegetation where they are eventually eaten by rabbits.

Tapeworms of this same genus are important parasites of humans. Of these, the two most important are the beef tapeworm (*Taenia saginata)* and the pork tapeworm (*Taenia solium*). As the life cycles of both are very similar, the pattern seen in the beef tapeworm will be discussed here as typical for the group. The adult beef tapeworm inhabits the human small intestine where reproduction occurs. Egg-filled proglottids are released along with fecal material to reach a suitable substrate where cattle may graze. The eggs enter the cow's intestine, burrow into blood vessels, and encyst within muscle tissue. The dormant **cysticercus** (bladderworm) remains in muscle until consumed by a human. Once inside the human's intestine, the bladderworm turns inside out, the tiny scolex affixes to the intestinal wall, and the cycle begins once more.

The presence of a beef, or pork, tapeworm in the intestine is basically *asymptomatic;* unless a person becomes repeatedly reinfected the interloper would most likely go unnoticed. Therefore, although slightly under 1,000 tapeworm infections area reported annually in the U.S., many more people may be hosting tapeworms. Thorough cooking of beef and pork is an adequate measure to prevent tapeworm infection.

The real threat with tapeworms is not so much the adult worms, but rather the complications that may occur as a result of swallowing their eggs. This condition, called *cysticercosis*, is potentially fatal A person who comes into contact with human feces, or eggs that have been transferred to objects, may inadvertently ingest eggs in large numbers. Continued exposure can come from pets, especially dogs, who may pick up eggs when cleaning themselves. Think about that the next time you let a dog or cat lick your face!

When a person ingests tapeworm eggs, the tapeworm reacts as though it was within its normal intermediate host (cow, pig, rabbit, fish, flea, etc.). It makes its way to various tissues and encysts as a bladderworm. There it awaits until a final host feeds on the tissue. However, since the cyst is actually in a human and not in its normal intermediate host, that will not happen. Instead, cysts may remain for years in muscle, bone, heart, liver, brain, and elsewhere. Repeated reinfection may result in muscle fatigue, soreness, toxic reactions, epileptic symptoms, and even heart failure.

Now, at this point in our discussion you may very well have some concerns about human tapeworms. In fact, you may have questions and want some answers. The *Inside View 6.1,* on the next page, addresses some of the more common questions regarding humans and tapeworms.

3. Humans eat uncooked or undercooked beef contaminated with the bladderworms. Tapeworm makes its way to the intestine where it attaches and begins to grow.

1. Tapeworm lives in human small intestine where it undergoes sexual reproduction. Eggs are then passed with feces.

2. Cattle ingest eggs while grazing. Larvae invade muscle tissue and become encysted as a bladderworm (cysticercus).

Figure 6.11. Life cycle of *Taenia saginata,* the beef tapeworm that infects humans.

An Inside View 6.1

The Four Most Common Questions about Tapeworms

1. Are they harmful?

✓ Pathology can occur on two fronts. First of all, an adult tapeworm within a human intestine produces very few, if any, symptoms. Unless a person has a toxic reaction or the number of tapeworms becomes inordinately high, there will be few, if any, manifestations. On the other hand, if an individual swallows tapeworm eggs, a condition called **cysticercosis**, serious complications, including chronic disability and even death can result. This is especially true if a person swallows eggs over and over.

2. How do I know if I have one?

✓ Most often an infected person is unaware of a tapeworm's presence. This is especially so if the infestation is light. However, vigilance is important...be on the lookout. The presence of tiny proglottids in human stool is a sure sign of a tapeworm freeloader. If you suspect a tapeworm is present, consult your physician, who may request a fecal sample from which eggs can be identified.

3. How do I get one?

✓ People normally become infected with tapeworms by eating contaminated meat. The most usual exposure comes from eating undercooked or uncooked beef, pork, or fish containing the cysticercus (bladderworm) form of the parasite. Once the cyst enters the intestine it evaginates, thus exposing the scolex to the intestinal wall. There it attaches and begins to grow. It can remain there for many years.

✓ Swallowing eggs is potentially very serious. This happens when a person is exposed to human feces or objects to which tapeworm eggs have become attached. The practice of using human wastes as fertilizer is still prevalent in certain parts of the world. In this case, eggs could easily find their way to edible vegetation. Another pathway is through contact with animals, especially pets, that can transfer eggs to a person. Dogs and cats use tongue-licking as a way to clean their rectal area. Eggs can certainly adhere to the tongue and be transferred to an owner, child, or anyone in close contact.

4. How do I get rid of it?

✓ Tapeworms can coexist with humans for many years, although usually without symptoms. If a tapeworm is present it can be removed. There are purging medications which are successful in extracting the parasite. Keep in mind that unless the scolex (head) is removed, the tapeworm will regrow. So, beware of any home remedies claiming to remove tapeworms. Again, if you suspect that one is present, consult your physician.

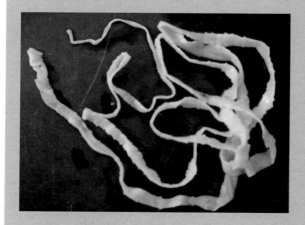

Chapter 6 Summary

Key Points

1. Platyhelminthes are flatworms; they are dorso-ventrally flattened.

2. As a result of their active lifestyle, they became the first group to show cephalization, bilateral symmetry, and three germ layers (triploblasty). They are the first hunters.

3. Well-coordinated muscles have developed to permit active locomotion.

4. They are acoelomate in that they lack an internal body cavity. A mesodermal material, parenchyma, fills the internal space between the body wall and the digestive system.

5. They use cutaneous gas exchange (respiration) and osmoregulation, thus, they are found either in aquatic environments or in moist terrestrial ones.

6. Some forms use a protonephridial (flame cell) system to supplement cutaneous osmoregulation.

7. Flatworms lack respiratory and circulatory systems and have an incomplete digestive system. The tapeworms lack a digestive system altogether.

8. Flatworms have complex reproductive systems and, at least in the parasitic forms, very complex life cycles.

9. As a group, they can be monoecious or dioecious. Some also use internal fertilization.

10. Hypodermic impregnation is practiced by some flatworms to deliver sperm to a monoecious partner.

11. Many flatworms have adopted a parasitic existence and, as such, show considerable behavioral and anatomical modification.

Key Terms

acoelomate

anterior ganglion

auricles

bilateral symmetry

bladderworm

cephalization

chemoreceptor

cysticercosis

digenetic life cycle

dorso-ventrally flattened

ectoparasite

endoparasite

final host

flame cell

gravid

intermediate host

longitudinal nerve cord

mesoderm

metacercarial larva

miracidium

negative phototrophism

oral & ventral sucker

parenchyma

pharynx

protonephridia

rhabdites

scolex

seminal receptacle

strobila

syncytial integument

tripartite intestine

triploblasty

yolk (vitelline) glands

Points to Ponder

1. What is triploblasty? Why is it important in the evolution of higher animals?

2. Define bilateral symmetry. What advantages does it provide?

3. Define cephalization. Why is it important to an animal?

4. Relate triploblasty, bilateral symmetry, and cephalization to directed locomotion.

5. Why are flatworms flat?

6. Why does a planarian have a tripartite intestine?

7. Compare the life cycle of *Fasciola hepatica* to *Clonorchis sinensis.*

8. How does *Schistosoma* differ from the above two parasites? Why is it so dangerous to humans?

9. Give the three parts of a tapeworm and indicate the functions of each.

10. What is a bladderworm?

11. What is cysticercosis? Why is it so dangerous to humans?

12. What is the normal way that people get tapeworms?

13. How might you know if you are infected?

14. How does a tapeworm feed?

Inside Readings

Brusca, R. C., and G. J. Brusca. 1990. Invertebrates. Sunderland, MA: Sinauer Associates.

Desowitz, R. S. 1981. New Guinea Tapeworms & Jewish Grandmothers: Tales of Parasites and People. New York: W. W. Norton.

Dineen, J. K. 1963. Immunological Aspects of Parasitism. Nature. 197: 268.

Morris, S. C., et al. 1985. Origins and Relationships of Lower invertebrates. Oxford: Claredon Press.

Pechenik, J. A. 1996. Biology of the Invertebrates. Dubuque, IA: Wm. C. Brown.

Chapter 6

Laboratory Exercise

Platyhelminthes

The Flatworms

Laboratory Objectives

✓ to reproduce the classification of the specimens studied,
✓ to identify both live and prepared specimens of flatworms,
✓ to identify anatomical features of various specimens of flatworms,
✓ to describe the life cycle of a fluke and a tapeworm.

Supplies needed:

● dissection microscope
● compound microscope
● prepared microscope slides of planarians, flukes, and tapeworms
● live *Planaria*
● preserved flatworm specimens

Chapter 6
Worksheet

Phylum Platyhelminthes
Class Turbellaria
Planaria

1. Select a whole-mount slide of the turbellarian, Planaria. Using Figs 6.12 and 6.13 as guides, locate the following features:

      ✔ **branching intestine** - note the anterior and two lateral branches.

      ✔ **auricles** - what is their function? *Monitor environment*

      ✔ **pharynx** - this is protrusible. What does that mean? *Sucks up the food*

      ✔ **eyes** - what do you think a planarian sees? → *Detect light & intensity of light*

2. On a Planaria cross-section slide (Fig. 6.12), locate the following:

      ✔ **parenchyma** - what is its function? *Support the structure*

      ✔ **intestinal branches** - why are there so many branches and sub-branches? *Distribution of nutrients*

      ✔ **gastrodermis** - does this animal use extra- or intracellular digestion, or both? *Both*

      ✔ **ventral nerve cords** - to what do they connect anteriorly? ~~pharynx~~ *Gangila*

3. Living Planaria.

    1. Obtain one or two living Planaria from the dish, being careful not to stir the water.

    2. Use a dissecting scope to observe your specimens anatomy, locomotion, and behavior.
      Then answer the following based upon your observations:

      ✔ Is Planaria a free-living or parasitic animal?
          Free living

      ✔ Describe planarian locomotion.
          cilia movement and muscles

      ✔ This animal is positively thigmotactic (drawn to surfaces) on its ventral surface but negatively thigmotactic on its dorsal surface. How would you explain this?

      ✔ Place a small amount of food (liver or egg yolk) near your Planaria. Observe feeding for about 10 – 15 minutes and describe this behavior.

      ✔ This animal is said to be negatively phototrophic (i.e., repelled by bright light). See if you can test this. Place an object so that it causes one area of the container to be dark and the other brightly illuminated. How does your animal respond?
          Go towards the dark

      ✔ Now perform an osmoregulation experiment. Place a pinch of salt near the Planaria and describe what happens. How do you explain this *at the <u>cellular</u> <u>level?</u>*
          water flowing out become Hypertonic. As we made the water now saltly. so, they become irritated with disturb condition.

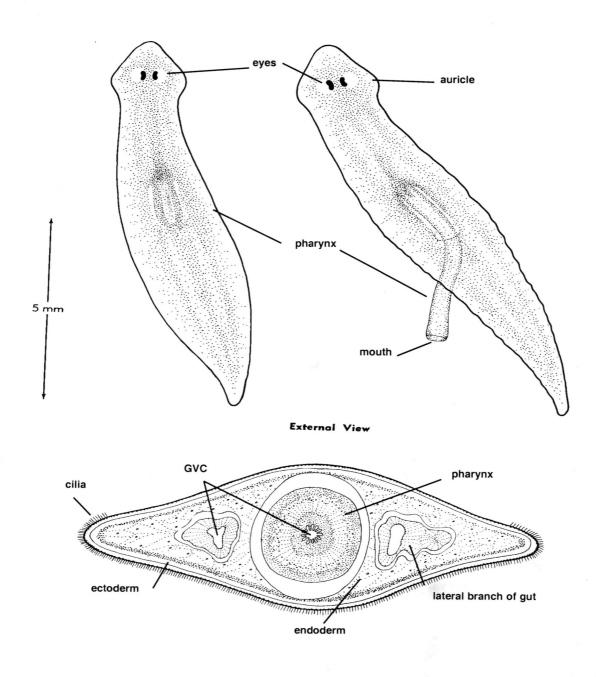

External View

Figure 6.12. The common turbellarian, Planaria.

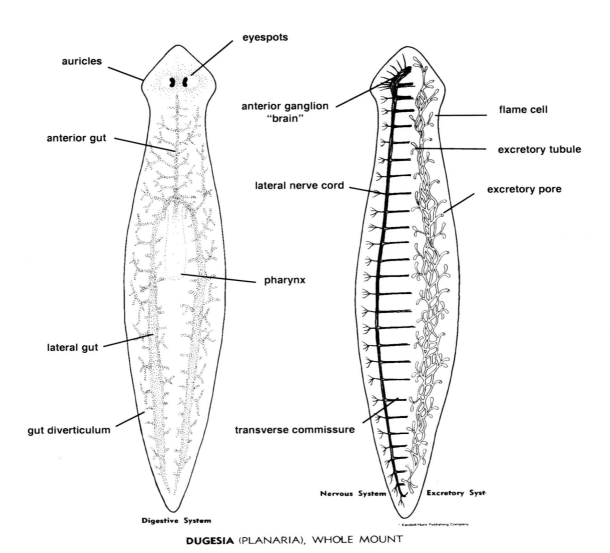

eyespots

auricles

anterior ganglion
"brain"

flame cell

anterior gut

excretory tubule

lateral nerve cord

excretory pore

pharynx

lateral gut

gut diverticulum

transverse commissure

Nervous System   Excretory Syst

Digestive System

Kendall/Hunt Publishing Company

DUGESIA (PLANARIA), WHOLE MOUNT

Figure 6.13. Internal anatomy of a planarian.

Class Trematoda

Clonorchis

1. On a whole mount-slide of the human liver fluke *Clonorchis*, locate the following features.

       ✔ **uterus** - is it filled with eggs? Yes
       ✔ **yolk gland** and **duct** (vitellaria) - what role do they play in reproduction? Source of fertilized eggs
       ✔ **ovary** - what is its function? Eggs are produced
       ✔ **testes** - are the sperm produced here used by this fluke? NO, exchange sperm
       ✔ **oral** and **ventral suckers** - what is their function?
       ✔ **mouth** - does the mouth also serve as an anus? Yes
       ✔ **pharynx** and **intestines** - what is this animal's food? Food, mucus, cellular division

2l. Life cycle (refer to page 126 in the text)

  a. Where exactly would you find *Clonorchis* as an adult? Human liver
    ...as a larva? Snail

  b. View microscope slides of *Clonorchis* cercaria & rediae.

  c. What is the host for the redia stage? Snail   the cercaria stage? Fish

  d. How does a human become infected with *Clonorchis*?
    People eat undercooked food -

Clonorchis sinensis
(human liver fluke)

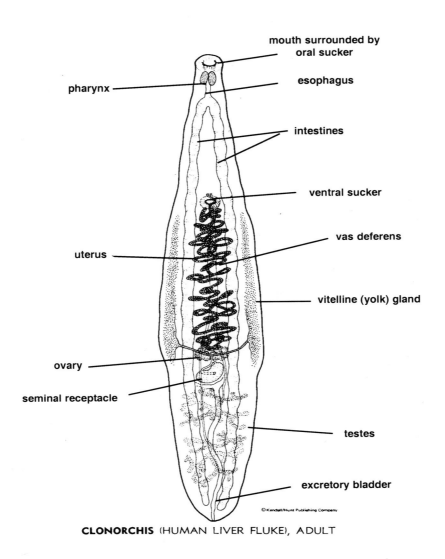

CLONORCHIS (HUMAN LIVER FLUKE), ADULT

Figure 6.14. Anatomy of *Clonorchis*, the human liver fluke.

Class Cestoda

The Tapeworms

1. Locate the following features on a whole-mount slide of the dog tapeworm *Taenia pisiformis*:

    ✔ **scolex** (head) - what is its function? *Attatchment*

    ✔ **sucker** - how many are there? *24*

    ✔ **proglottids** (mature and ripe) - note the abundance of eggs in the ripe proglottids

    ✔ **strobila** (body) - where do individual proglottids come from? *It comes from the neck which is the growth zone*

    a. What happens to the posterior-most proglottids? *So they can form new ones*

    b. What is the final host for *Taenia pisiformis*? *Some common* (page 129 & 168) *tapeworms, T·solium, canium*

    c. What is the intermediate host for this parasite? *Rabbits*

    d. Describe a situation in which a human can become infected with *Taenia pisiformis*. *come in contact with the pet that they*

    e. Describe how a tapeworm gets its nourishment (i.e., food). *It gets its nourishment from its intermediate host, which can cause the host nutritional deficiencies.*

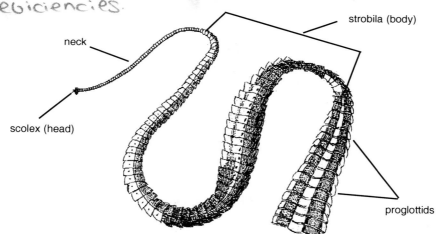

Preserved Specimens

There are a number of specimens of the phylum Platyhelminthes on display that represent the different classes of flatworms. For each specimen you should note its name, taxonomic class, and distinguishing characteristics.

❖ Making labeled sketches can also help you remember the various animals.

7 Chapter 7

PHYLUM NEMATODA

THE ROUNDWORMS

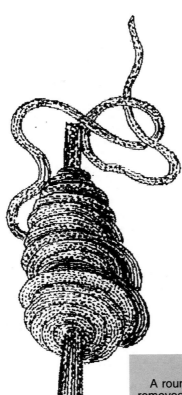

A Simple Solution

A roundworm parasite, the Guinea worm, can be removed from beneath the skin of its human host by a simple procedure. The end of the long worm is wrapped around a stick and then, over a day or so, slowly extracted from an arm or a leg.

Phylum Nematoda
The Roundworms

We humans are a minority of giants stumbling around in a world of little things.

- Hubbell

Introduction

*T*here is a rather odd assemblage of animals so different from each other that establishing phylogenetic relationships is nearly impossible. To look at them you would not suspect that they had anything at all in common. Yet, in one respect, they are, indeed, closely related; they do share one important feature. Each is endowed with a central body cavity, a **pseudocoel** (more on that later). And it is for that reason they are grouped together into an unorthodox conglomeration called the **pseudocoelomates**. While this term has no real taxonomic validity, it does underscore the common connection between the various phyla.

Many of the animals belonging to this assemblage are relatively unknown. Others, such as the rotifers and, especially, the nematodes, are more commonly recognized. It makes little sense, then, for us to give a detailed coverage to each of the minor groups, and we won't; a brief outline will suffice for them. That will allow us to concentrate our efforts on the much more extensive and important group, the nematodes. But first, a short review of some of the lesser phyla.

Phylum Acanthocephala. Members of this taxon are small (< 2 cm), elongated, thread-like parasites of vertebrate digestive systems. They are especially prominent in freshwater fishes but are also found in birds, amphibians, reptiles, and mammals. They are commonly called the "spiny headed" worms due to a thorny anterior proboscis, which is used to attach to the host. Behind the proboscis is a short neck and a segmented body. This parasite, like tapeworms, lacks a digestive system. But unlike the cestodes, the sexes are separate. Eggs are deposited along with the host's feces and larval stages develop within an intermediate host, usually an insect. The final host is infected upon eating the insect. Humans rarely harbor this parasite.

Phylum Nematomorpha. Another thread-like group, the "horsehair worms" are much longer than their "spiny headed" kin; some reach over a meter in length. The adults are all free living, whereas the juveniles are internal parasites of insects and other arthropods. Within an insect host, the juvenile develops into the adult worm and then literally bursts out of the host, usually killing it in the process. The dioecious adults lack a functional digestive system and, in fact, do not feed. Instead,

they reproduce. A single female may lay in excess of 1,000,000 eggs in her short lifetime.

Phylum Rotifera. These small (< 1 mm) omnivores are widely dispersed in aquatic environments. The great majority (> 95%) occur in freshwater and are free living. Only a few parasitic forms have been identified. As a group, the free-living forms feed on protozoa, algae, and even other rotifers. They are characterized by a crown of cilia, which, when beating, gives the appearance of rotating wheels. Thus, the name rotifers. It is with the "wheels" that feeding is accomplished, much like some of the ciliated protozoa. Although they are dioecious, many forms reproduce **parthenogenically**, no sperm being necessary for embryonic development. Indeed, in many rotifer communities males are a rare commodity!

Phylum Gastrotricha. Like the rotifers, members of this taxon are small aquatic animals. But they lack the ciliated crown, possessing ventral cilia instead. As a matter of fact, they seem more closely allied with free-living flatworms (e.g., planaria) who also have ventral cilia. However, since they have a pseudocoel, they are usually placed among those groups we are now discussing. They are monoecious and, as with rotifers, may use parthenogenesis as a mode of reproduction. One huge egg is produced, which, like the planarians, may be able to over-winter to escape an unfavorable environment. Gastrotrichs live in bottom sediments feeding on detritus, bacteria, or protozoans. They are a small, puzzling group of animals that seem to lie somewhere between the flatworms and the roundworms.

Phylum Loricifera. Just when you think modern science has discovered all of the major groups of animals, a new one appears. Less than 20 years ago these animals were unknown. Now they're a *bone fide* phylum. Albeit, a small one. Only 100 or so species have been described so far. They are all marine and tiny — very tiny, less than 0.5 mm — so small that each one clings desperately to one or two grains of sand. They are shaped like a tiny lemon with an anterior pointed snout surrounded by spines. At the posterior end are two leafy swimming appendages. These are, in turn, encircled by a series of sensory hairs. Little is known of their behavior, feeding habits, or reproduction. Loriciferans are, indeed, bizarre little members of the animal kingdom.

The Roundworms: Phylum Nematoda

*S*ome scholars of nematode biology suggest that roundworms may very well be the most abundant animals on earth. While "only" a little over 28,000 species have been described, literally millions of roundworm species may exist! They seem to rival the insects in diversity, distribution, and sheer numbers. Roundworms have invaded essentially every known habitat: benthic, littoral, pelagic, brackish, freshwater, terrestrial, and subterranean; some even have airborne stages! They have also adopted just about every known lifestyle: carnivorous, herbivorous, detritus feeders, plankton feeders, ectoparasites of plants and animals, endoparasites of plants and animals, free living, and commensalistic.

But they are especially abundant as parasites. Indeed, virtually every plant and animal species harbors its own nematode parasite. If there are over 40 million species of plants and animals on earth, there are, most likely, close to that many parasitic nematodes. And keep in mind that figure doesn't include the army of free-living nematodes. Some authorities submit that the number of free-living nematodes might equal their parasitic cohorts. There's no doubt that their diversity is impressive.

However, the vastness of nematode diversity doesn't compare at all with the immense number of individuals that exist. They occur everywhere, and in massive aggregations. They probably exceed thousands in each shovel of garden soil. Tidal silt or pond mud holds many times that number, perhaps as high as 4 million individuals per square yard. And it is estimated that a single rotting apple bears in the neighborhood of 100,000 individuals. Nematodes are, indeed, here, there, and everywhere.

Nematode Structure

The average nematode is a rather small, thread-like, cylindrical animal. In other words, a tiny roundworm. Most, in fact, are under 5 mm in length and, with their secretive habits, usually go unnoticed. But that isn't always the case. Indeed, roundworms can reach formidable size. The one that invades the human small intestine, for instance, can attain a body length of well over a foot (~ 40 cm), and the record holder is a whopping 28-foot titan from the placenta of a sperm whale!

External Anatomy. While roundworms may differ considerably in size, they all, nonetheless, share the same general body features. Although Figure 7.1 shows the anatomy of the human parasite *Ascaris lumbricoides*, it, in reality, represents the basic architecture of all roundworms.

Nematodes are vermiform (worm-shaped), whitish, weakly cephalic animals. They are elongated and tapered at each end and are covered by a sheath, the **cuticle**. The sheath is formed by cells in the epidermis, but is, itself noncellular. It is an extremely important feature of nematode biology as this list will attest:

(1)   The cuticle is inelastic yet flexible; it can bend but cannot stretch.

(2)   The cuticle is permeable to respiratory gases, water, and nitrogenous wastes. Thus, respiration and much of the process of osmoregulation is cutaneous.

(3)   Since water can diffuse into a nematode, it can just as easily diffuse out. As a result, nematodes are subject to dehydration and must be aquatic or live in an otherwise moist environment.

(4)   In parasitic nematodes, the cuticle is thought to provide some protection from corrosive secretions of the host.

(5)   The cuticle assists in the establishment of an internal hydrostatic skeleton. As long as the animal remains hypertonic, water will diffuse into the pseudocoel. This creates pressure against the cuticle (which cannot expand). Thus, the animal becomes turgid. This is an important factor in locomotion as described later.

Internal Anatomy. One of the universal realities of animal evolution is the march toward complexity. As we have already witnessed, animals have undergone a tremendous change from the simple unicellular protozoan, on through the cnidarian diploblastic tissue plan, to the three-layered, organ pattern of flatworms. And as we shall continue to see, animals will consistently mount an ongoing campaign of advancement.

The Body Cavity. One of the results of this campaign is organ development. As internal organs, such as the gut, protonephridia, and the reproductive machinery appeared, an internal space was needed to accommodate them. That is because organs can be most efficient if they have a space in which they can flex, expand, and contract.

The acoelomate flatworms tried, but didn't get it quite right; a mesodermal-filled cavity is all that resulted. Now it's the roundworm's turn. They introduce the **pseudocoel**, a cavity derived, not from mesoderm, but from an embryonic cavity called the blastocoel. It works, but it, too, is not quite right. While they have certainly improved upon the flatworm design, they also come up wanting, albeit by a small margin. From this point on, however, all animals, from snails to humans, will have a body cavity derived from mesoderm and lined by a membrane, the **peritoneum**. The true coelomic cavity will have dawned and, as we shall see, it becomes one of the hallmarks of higher animals.

The Digestive System. Roundworms are simple in their internal design, demonstrating a "tube-within-a-tube" arrangement. The outer tube is the body wall, while the inner tube consists of an elongated intestine. Suspended within the pseudocoel is a **complete digestive system**: an anterior mouth leading to a tube and ending in an anus. The mouth is surrounded by three, and sometimes six, elevated lips helping to manipulate food into the mouth. *For the very first time we see animals that do not regurgitate undigested food.* An anus seems like such a simple solution to the fouling problem, but a solution that animals had some difficulty discovering. Nonetheless, once introduced, the complete digestive system is quickly and uniformly adopted by all animals.

Although the digestive system seems simple in its structure, it suffers somewhat in its function. Because of the hydrostatic pressure within the pseudocoel, the intestine is a compressed and flattened tube. So, in order to force food through this tube, a muscular, anterior pharynx is required. The pharynx not only helps to pull food into the digestive system, it also serves as a pump to propel food toward the anus where undigested food can be voided.

Roundworms gain nourishment in a variety of ways. The free-living forms may be carnivorous on any animal smaller than themselves, including other nematodes. They also feed on bacteria, fungi, algae, yeast, and detritus, and even scavenge dead remains. The parasitic nematodes infect both plants and animals. They attack all parts of plants: roots, stems, leaves, buds, flowers, and fruit. Similarly, they are found throughout animals: digestive system, kidney, lungs, heart, muscle, eyes, brain, and elsewhere.

However, because of the cuticle, roundworms cannot absorb food directly through the body surface. Thus, even the parasitic forms must ingest their food, such as cell debris, tissue fluid, dead blood cells, mucus, and predigested chyme, through the mouth.

Nematode Reproduction. Roundworms are virtually all dioecious, and fertilization is internal. With internal fertilization it is necessary for the two sexes to get together on the copulatory battlefield. While this is usually not a problem for parasitic roundworms, squeezed together into rather cramped quarters, it is a distinct challenge for some of the free-living forms. Thus, certain female nematodes actually attract the male by emitting sex signals in the form of chemical pheromones!

The female is, for the most part, larger than the male to accommodate the huge number of eggs she develops. For example, female *Ascaris lumbricoides*, release about 200,000 eggs per day, and most parasitic forms lay well into the tens of thousands each day! By the way, she usually outlives the male; he, often, dies shortly after mating is accomplished. Figure 7.1 and 7.2 shows the components of the female reproductive system and Figure 7.3 shows the same, but in cross section. The system consists of thread-like **ovaries** and **oviducts** leading into a larger Y-shaped **uterus**. Eggs formed in the ovary travel up the oviduct to the uterus, where fertilization occurs. Fertilized eggs (i.e., embryos) then exit through the **genital pore**. Humans then contract the parasite by directly ingesting the eggs through food, water, or other contact.

The male is characterized by a posterior curved "hook" and sharp **copulatory spicules** (Figure 7.1). The spicules are inserted into the female genital pore to anchor the partners and to open the vulva to permit the passage of sperm during mating. The **testes** are usually paired and, like the ovaries, are long and thread-like. Sperm mature as they move down the testes and are stored in a saclike **seminal vesicle** awaiting ejaculation into the female. The sperm of roundworms is rather peculiar. Unlike the streamlined, flagellated spermatozoa of most animals, roundworm sperm are **amoeboid**. Thus, they cannot swim rapidly toward eggs as in other animals but must slowly pursue the ova within the uterus.

Certain nematodes have a special stage called the **dauer larva** that becomes inactive in response to adverse environmental conditions. During dormancy the animal slows its metabolic processes, stops feeding, and resists desiccation. When favorable circumstances return, the larva emerges and assumes its free-living lifestyle. This is a classic example of **cryptobiosis**, the use of a dormant stage to avoid unfavorable conditions.

Other patterns of nematode development, especially among the parasitic roundworms, varies considerably. Many, for example, show what is called **eutelic growth**. As juvenile nematodes grow mitotically they shed the cuticle four times. *Once adulthood is reached mitosis ceases.* From that point on, growth is eutelic, that is, cells increase in size but not in number.

Musculo-Nervous Interaction. Nematodes are cephalic, but only marginally so. They do have an anterior ganglion that serves as a primitive brain and two elongated, longitudinal **nerve cords,** a dorsal one and a ventral one (Figure 7.3). Their sensory apparatus is not well developed although they do have a few anterior touch and chemosensors. It is thought that they may have primitive photosensors but lack any semblance of eyes, even simple ocelli.

What is unusual about the nervous system is the way it interacts with the longitudinal muscle fibers. In virtually all animals, the nervous system sends out extensions to contact muscles. Each muscle cell is served by a thread-like appendage extending from a nerve cell. This, however, is not the case with nematodes, in fact, the opposite is true. Instead of muscle cells receiving cytoplasmic projections from the nerves, they send their own spurs to contact the longitudinal nerve cords. This odd arrangement can be seen in the figure to the right and in Figure 7.3.

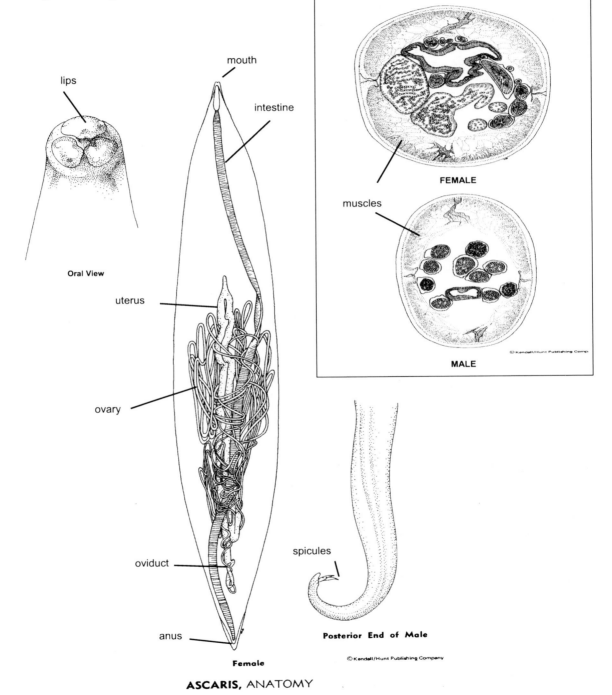

ASCARIS, ANATOMY

Figure 7.1. The anatomy of *Ascaris*, the human roundworm. The insert above shows cross sections of *Ascaris* male and female.

Nematode Locomotion

Strangely, nematode locomotion is much less developed than that seen in the earlier flatworms. Instead of the smooth gliding motion of planaria, for instance, roundworms can do little more than thrash around in a wild, whipping action. This S-shaped, or **sinusoidal motion** as it's called, works tolerably well for terrestrial roundworms as they can push against soil particles to thrust forward. The locomotion pattern is not much of a problem for parasitic roundworms as locomotion is not very important in their scheme of things. After all, they are fairly content to stay attached in their comfortable lairs.

However, sinusoidal motion does present a serious problem for swimmers. As a matter of fact, for a roundworm, swimming is virtually impossible. There are two reasons for this. First of all, the presence of a cuticle precludes the ventral cilia of flatworms. Without cilia upon which to gracefully glide, roundworms are left to rely solely upon muscular action. But the muscle arrangement of roundworms is insufficient to provide smooth and directed locomotion. Roundworms lack the layered muscular arrangement of their flatworm predecessors. In fact, they possess only longitudinal bundles; the circular and oblique layers of flatworms are missing. As a result, when nematodes constrict a lateral bundle against the hydrostatic skeleton, the body contorts toward that side, followed by a similar, but opposite, motion of the other side. Thus, a whiplike action is all that occurs and forward motion is severely restricted.

The Thrashing of Vinegar Eels

Vinegar eels (*Turbatrix spp.*) thrive in the sediment of unpasteurized (homemade) vinegar where they feed on yeast and bacteria. In this acidic bath they seem determined to swim, but their efforts are all but wasted. It is rather comical to watch these roundworms trying to swim. They flail about feverishly, expending a tremendous amount of energy, only to go nowhere. As in a dream, being chased by an invisible shadow, they get nowhere. It is something to watch several dozen vinegar eels in a desperate race to gain an advantage on their brothers only to realize, after viewing for some seconds, that they haven't even traveled the mere distance of the microscope field! But this is easy to remedy. Just add a small amount of sand to their vinegar bath and observe the transformation. They immediately become wormy marathoners, moving forward easily and swiftly.

This should help to explain why free-living nematodes have preferred a terrestrial rather than an aquatic existence. If an animal lives within the interstitial spaces among soil or sand particles, then it can use these particles as leverage "pegs" to push against. So, a terrestrial nematode can achieve some credible forward locomotion while its aquatic counterparts struggle in vain. In water, where there aren't any leverage pegs, swimming becomes an aerobic exercise — running, or in this case, swimming, in place. Thus, open water is one of the very few habitats that roundworms have not conquered. Aquatic forms are found, instead, threading their way through the bottom sediments and in contact with plants or other debris; and, of course, in homemade vinegar.

Chapter 7 Summary

Key Points

1. Nematodes are vermiform, cylindrical pseudocoelomate animals covered by a syncytial cuticle. The cuticular covering plays an important role in protection, locomotion, food transport, cutaneous respiration, and in forming the internal hydrostatic skeleton.

2. Organs are suspended within a pseudocoel under hydrostatic pressure.

3. Respiration and osmoregulation, in part, are cutaneous.

4. Nematodes are weakly cephalic, lacking the well-pronounced head of planarians. Their sensory structures are, likewise, not prominent.

5. Roundworms are the first group of animals to develop a complete digestive system consisting of a mouth, intestine, and an anus.

6. Roundworms display an unusual growth pattern in which cells stop dividing and the animal grows thereafter by increase in cell size only. This is termed eutelic growth.

7. Nematodes are endowed with longitudinal muscles only. As a result, movement is restricted to a whiplike, sinusoidal action.

8. For the most part, nematodes have separate sexes and show some sexual dimorphism. They engage in copulation, fertilization being internal. The number of eggs produced can be vast.

9. They are cosmopolitan in distribution: freshwater, marine, moist terrestrial habitats, and as important parasites on both plants and animals.

Key Terms

amoeboid sperm
blastocoel
complete digestive system
cryptobiosis
cuticle
dauer larva
eutelic growth
hydrostatic skeleton
parthenogenesis
peritoneum
pheromones
pseudocoel
sinusoidal movement
syncytium

Points to Ponder

1. What is so distinctive about the roundworm digestive system?

2. What is a syncytium?

3. What is the role of the roundworm cuticle?

4. How does the cuticle contribute to roundworm locomotion?

5. How can you differentiate between an acoelomate, pseudocoelomate, and a coelomate animal?

6. How does a roundworm exchange respiratory gases?

7. What is cryptobiosis? What form of roundworms practice this phenomenon?

8. Describe roundworm locomotion.

9. What is meant by the tube-within-a-tube body arrangement?

Inside Readings

Bird, A. F., and J. Bird. 1991. The Structure of Nematodes. New York: Academic Press.

Brusca, R. C., and G. J. Brusca. 1990. Invertebrates. Sunderland, MA: Sinauer Associates.

Crowe, J. H., A. F. Cooper, Jr. 1971. Cryptobiosis. Sci. Am. 225(6): 30-36.

Hope, W. D. ed. 1994. Nematodes: Structure, Development, Classification, and Phylogeny. Washington, DC: Smithsonian Institution Press.

Levine, N. D. 1980. Nematode Parasites of Domestic Animals and of Man. Minneapolis:Burgess.

Maggenti, A. 1981. General Nematology. New York: Springer-Verlag.

Malakhov, V. V. 1994. Nematodes. Washington: Smithsonian Institution Press.

Moore, J. 1984. Parasites that change the behavior of their hosts. Sci. Am. 250(5):108-115.

Nicholas, W. L. 1984. The Biology of Free-living Nematodes. New York: Oxford University Press.

Pechenik, J. A. 1996. Biology of the Invertebrates. Dubuque, IA: Wm. C. Brown.

Poinar, G. O. 1983. The Natural History of Nematodes. Englewood Cliffs NJ: Prentice-Hall.

Wharton, D. A. 1986. A Functional Biology of Nematodes. Baltimore: Johns Hopkins University Press.

Chapter 7

Laboratory Exercise

The Roundworms

Nematoda

Laboratory Objectives

✓ to reproduce the classification of the specimens studied,
✓ to identify both live and prepared specimens of roundworms,
✓ to identify anatomical features of various specimens of nematodes and pseudocoelomates, in general,
✓ to describe to the life cycle of *Ascaris lumbricoides*, and
✓ to identify habitat, parasitic cycles, and diseases associated with each of the specimens studied.

Supplies needed:

✓ Dissection microscope
✓ Compound microscope
✓ Prepared microscope slides of *Ascaris* and selected nematodes
✓ Preserved *Ascaris* for dissection

Chapter 7
Worksheet

(also on page 591)

Phylum Nematoda
Ascaris lumbricoides External Anatomy

A. Observe the external anatomy on specimens of both a male and female *Ascaris*. Use a dissecting microscope to find the mouth of each. This is done by locating the three anterior lips surrounding the mouth (Figure 7.1 & 7.2). Also refer to the figures to help locate the copulatory spurs (spicules) of the male. Finally, find the female genital pore, which is where sperm enters the female during copulation. This is a small ventral slit located about one-third of the way from the mouth.

Give three *external* ways that you can distinguish males from females.

1. _Males have spicules, females don't have_
2. _Males are tiny, compared to female_
3. _Males are thinner, compared to female_

Ascaris Dissection
(be certain to read the cautionary note below)

B. After the external examination, you will need to dissect a female *Ascaris*. First note the superficial, thin, almost transparent cuticle covering the animal and its underlying epidermis. Using a sharp scalpel or a needle probe, cut through both the cuticle and the epidermis along the entire length of the animal. Do not go too deep or you risk disturbing the underlying organs. Now pin open the worm so that the internal structures are exposed. Add a little water to keep the specimen moist. Note that the organs are freely suspended in an open space inside of the animal.

What is this space called? _Pseudocoelom_

C. Now use Figure 7.1 to locate the following features on the female *Ascaris*.
- ✔ mouth
- ✔ anus
- ✔ intestine
- ✔ ovary/ oviduct (you do not need to differentiate between these two)
- ✔ uterus*
- ✔ vagina
- ✔ genital pore

*Note: The uterus forms an inverted Y. The tail of the Y points toward the mouth.

D. What is the normal food of an *Ascaris*? _Nutrients of your intestine_

(you may also wish to refer to page 159 in chapter 8 for a further discussion of *Ascaris*).

Caution

Ascaris eggs can withstand adverse conditions: high temperatures, low temperatures, desiccation, and very likely, formaldehyde preservatives. Therefore, as a precaution against the possibility of infection by eggs that might have survived the preservation process, be sure to observe the following:

1. Do not put your hands near your mouth during lab.
2. There is to be no food or beverages during today's lab.
3. Gloves are to be worn during your dissections.
4. Thoroughly wash your hands with soap and water after today's lab.

Ascaris (Microscopic Anatomy)

E. Examine a c.s. microscope slide of an *Ascaris* female (Figure 7.1 & 7.3). Locate the following:

- ✔ pseudocoel
- ✔ uterus and eggs
- ✔ ovary/oviduct (you do not need to tell them apart)
- ✔ intestine
- ✔ nerve cords (ventral and dorsal)
- ✔ longitudinal muscle
- ✔ cuticle and underlying epidermis

1. Approximately how many eggs/day can a female *Ascaris* produce? *200,000*

2. How does a human become infected with *Ascaris*? *Directly ingesting egg in food*

3. On the cross-section slide of a female *Ascaris*, locate the uterus. Then increase magnification so that the eggs within the uterus are in view. Note their scalloped edges.   Sketch several eggs in the space below:

→uterus
→eggs

F. Examine an *Ascaris* male microscope slide — cross section (Figure 7.3). Locate the following:

- ✔ pseudocoel
- ✔ testis
- ✔ intestine
- ✔ nerve cords (ventral and dorsal)
- ✔ longitudinal muscle
- ✔ cuticle and underlying epidermis

4. Where in the female *Ascaris* does fertilization occur? *uterus*

5. What is the function of the testis? *Sperm maturation*

6. What is the function of the male copulatory spicules? *Open up the pore st*

7. How does the sperm of *Ascaris* differ from other animals? *They don't have Flagellum in them. More peculiar*

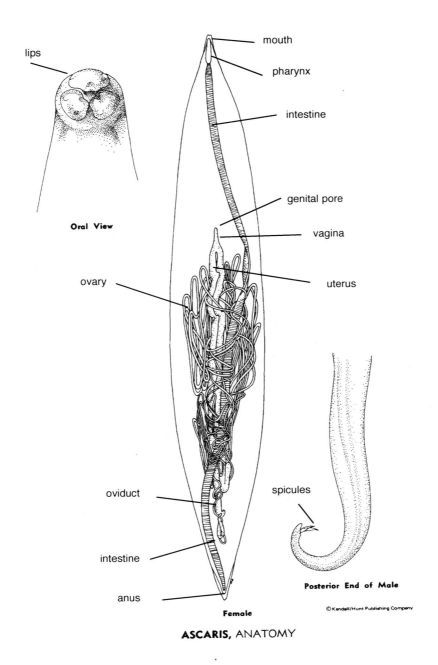

ASCARIS, ANATOMY

Figure 7.2. External and internal anatomy of *Ascaris*.

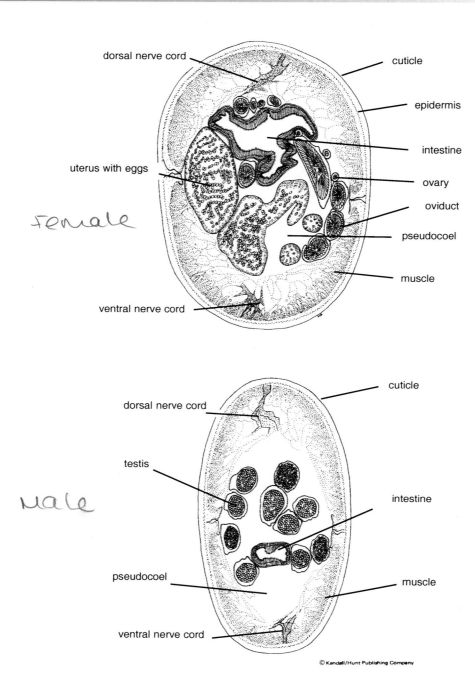

Figure 7.3. Cross sections of a male and a female *Ascaris*.

8 Chapter 8

ANIMAL INTERACTIONS

SYMBIOSIS

Chapter 8

ANIMAL INTERACTIONS:
Symbiosis

Big fleas have little fleas upon their backs that bite 'em,
and little fleas have lesser fleas,
and so on ad infinitum.
- Anonymous

Introduction

*I*n nature, animals never live in total isolation; there are no hermits in the world of animals. A single, unattached individual just doesn't exist. Instead, animals show a vast facility to interrelate with each other. In many ways, and on many levels, they associate together. Often the interactions are beyond imagination.

Sometimes the relationship is required, or **obligatory**, in that one or another participant cannot survive without it. It is necessary for survival. An example would be a tapeworm and its host. The tapeworm is incapable of surviving without that special relationship. At other times the connection is not essential but is exploited if, and when, the opportunity presents itself. This type of relationship is known as a **facultative interaction**. Some leeches, for example, will assume a parasitic lifestyle and feed on blood if given the chance. But they can live very well on their own as carnivores or scavengers.

Another way of viewing animal interactions is to consider the kinship of the participants. If an interaction occurs between members of the same species, it is then an **intraspecific relationship** (e.g., male-female contacts; parent-offspring relationships; play behavior; group foraging efforts, social grooming, and so on). On the other hand, the relationship may be between members of different species. This latter interaction, an **interspecific relationship**, is what concerns us here, for, in biology, the term **symbiosis** is used to identify *any long-lasting, intimate association between members of two or more species.* In other words, symbiosis is a close, enduring interspecific relationship.

Figure 8.1. Intraspecific relationships occur between members of the same species such as the strong bond between mother and infant (**A**) or the interaction that occurs during mating as in the snakes above (**B**). These are *not* examples of symbiosis, which, instead, refers to interspecific interactions between members of *different* species...the subject of this chapter.

Such interspecific relationships are widespread. In fact, virtually all animals serve as a natural habitat (**host**) for hundreds, if not thousands, of other organisms (**symbionts**). The interactions can also be complex and, at times, bizarre. Nonetheless, although wide-ranging, symbiosis, seems to follow only four patterns.

(1) **Mutualism**: There is a special relationship that exists between termites and their intestinal protozoa. These gut parasites break down cellulose as a food source and in doing so release nutrients for their termite host. Without their microscopic hitchhikers termites could not survive on their wooden diet— and, of course, neither could the protozoa. In this case, and countless others, a *mutualistic* relationship has been formed and both the symbiont and the host benefit. These associations provide advantages such as housing, protection, food, or some other mutually beneficial resource. Mutualism is a common type of symbiosis and is often the result of complex adaptive interdependence. That is, both parties adapt to each other throughout the association and, in fact, **co-evolve** in response to the relationship. As one party becomes better adapted to the interaction so, likewise, does the other.

Examples of mutualism are many; here are but a few: photosynthetic algae (zooxanthellae) provide an energy source for corals while benefiting from a protective habitat; sea anemones hitch a ride on the back of certain crustaceans, thus allowing the otherwise sedentary anemone a chance to relocate, while the anemone's stinging tentacles provide the crustacean with some protection from its enemies; gleaning fish and shrimp remove ectoparasites from a variety of marine fish; oxpeckers, likewise, remove external parasites from many large mammals; and a species of ant serves as bodyguard to a caterpillar who, in turn, awards drops of sweet nectar to the waiting attendants.

Figure 8.2. Mutualism is a common interaction between members of different species in which both parties derive a benefit. Some symbionts are adapted to pick ectoparasites from their hosts as in the cleaner wrasse (**A**) and oxpeckers (**B & C**).

(2) **Commensalism**: This is a somewhat less commonly known association in which the symbiont is benefited while the host remains unaffected. Remoras are small fish that attach to sharks and other larger fish to forage on leftover scraps of food...they profit while the shark is not affected. Certain fish and crab species receive protection as they hide among the sharp, toxic spines of sea urchins; the urchin is indifferent. Many fish cruise along with large predators receiving food scraps along the way.

There is one special example of apparent commensalism that is somewhat confusing. An uncertain relationship exists between several species of clown fish and venomous sea anemones. The fish live among the anemone's stinging tentacles, but it is unknown if the anemone gains from the relationship. Perhaps the clown fish remove debris from the tentacles; if so, then the relationship would be more likely one of mutualism.

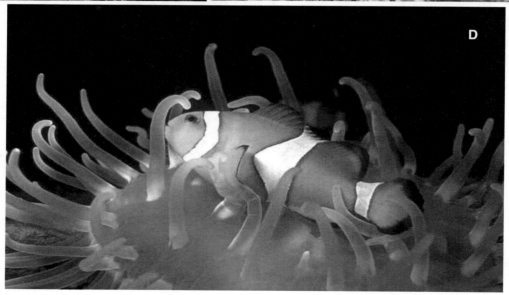

Figure 8.3. Remoras (**A**) have a modified fin to attach to large fish. They most likely receive scraps of food as the larger fish feeds, as do the small fish in plate (**B**) shadowing the hammerhead shark. Some fish can use the spines of sea urchins to protect them from attack (**C**). Several species of clown fish (**D**) live within the stinging tentacles of sea anemones. It is suspected that a mucous covering keeps the fish from being stung. It is unlikely that the anemone derives any benefit.

(3) **Phoresis**:   In this simplest of symbiotic encounters the symbiont uses the host only for transport and nothing else.   A wading bird may inadvertently pick up fish, insect eggs, or larvae and transport these to another locale.   Barnacles affix to the surface of whales or other large marine animal...they get a ride and the host is untouched.   In simple terms, phoresis involves one organism using another to "hitch a ride"; thus, the symbiont is dispersed into new surroundings. Although the phoront derives some benefit from the relationship, its survival does not depend upon it.   And, of course, the host is not affected.   In this regard, phoresis is similar to commensalism.

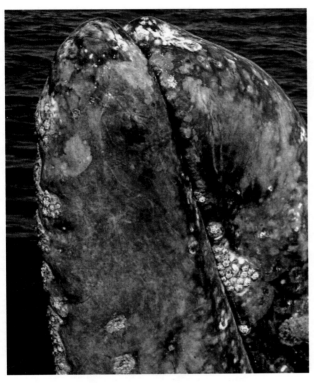

Figure 8.4. Barnacles demonstrate the phenomenon of phoresis by hitching a ride on the surface of whales.

(4) **Parasitism**: There's little doubt that the most common symbiotic relationship among animals is parasitism.   Here the symbiont (parasite) *benefits at the expense of the host,* usually by gaining some metabolic advantage. Sometimes, in fact, this relationship is referred to as "destructive symbiosis."   In one way or another the parasite exploits the host to the latter's detriment.

However, while the host is adversely affected it is usually not critically harmed.   It is to the parasite's advantage not to seriously injure or kill its host, for to do so would result in the parasite's demise as well.   In parasitism, it seems that the longer the relationship, the more benign it becomes.   **Co-evolution** apparently works its genetic magic to allow the two to live in relative harmony.   But there are many exceptions, and trying to understand exceptions can be troublesome.

Generally, in pathogenic cases wherein the parasite causes severe disability or death, the relationship is considered to be of recent origin.   An example can be cited between large African mammals and a species of *Trypanosoma*, a blood protozoan.   In this instance, wild hosts are not seriously affected by the parasite but when domestic cattle were introduced into the region, they became infected and, as a result, large numbers of the cattle died.   Those that survived will pass on the survival trait, and, over time, a state of relative accord should develop...or so the explanation goes.

Parasites wear many costumes.   Those that occur on the host's exterior (*ectoparasites*) are modified in different ways than their internal cohorts (*endoparasites*).   Whether occurring internally or externally, the parasite is specifically adapted to gain safe refuge, food, reproductive opportunities, and other advantages. In fact, the parasite is often so intricately adapted that it can reside only in certain specific locations within or on the host, and nowhere else.   Examples of **host** and **tissue specificity** of this sort are many: *Fasciola hepatica* is a fluke confined to the liver of a sheep; *Clonorchis sinensis,* another fluke, is only found in human livers; tapeworms, likewise, are limited to the intestine of specific hosts; heartworms to the heart cavities of dogs; lungworms reside only within the lungs of frogs; and so on.   There are even larval mollusks, baby clams, that parasitize only the gills of certain fish.

Whatever you might say about parasitism, one thing is certain...the parasitic lifestyle is extremely popular.   Of the 34 animal phyla, 14 contain parasitic members, and 5 are entirely parasitic.   From protozoa to insects, from fish to bats, the parasitic stage has seductively lured its players, and the results can be hideous: a catfish that prefers the inside of the human urethra; roundworms that cause the ghastly swellings of elephantiasis; tapeworms that cause blindness; mites living in the ear of a moth; a protozoan that slowly dulls the mind in African sleeping sickness. There are even worms that romp in the sinuses of narwhals, driving their host nearly insane with their endless cavorting.

Every known vertebrate animal (and most invertebrates, as well) supports an entire congregation of parasites.   Flukes, ticks, mites, protozoa, fleas, lice, tapeworms, bedbugs, leeches, biting flies, roundworms, lampreys, and even vampire bats are members of this exclusive fraternity— each with its own special needs; each with its own special contraptions, modifications, and adaptations.

One of the things that makes the study of parasites so fascinating to the zoologist, and anyone else who looks closely at the parasitic relationship, is that the "creativity" of parasites is simply astonishing. Yes, they can be repulsive vagabonds from our viewpoint as potential victims, but they are also, without question, awesome creatures. Their specialized adaptations are simply anatomical and physiological wonders. The degree to which parasites have exploited evolution's promise is nothing short of spectacular. They are grotesquely beautiful examples of evolution's power to invade the most unimaginable niches.

Figure 8.5. Virtually every animal has parasites. Here one baboon removes ectoparasites from another in an act of social grooming. In addition to the external parasites, there's little doubt that both animals also harbor a number of internal parasites.

Parasitism: What's the Attraction?

The question then becomes, why so many? Or stated another way: What is the lure that draws such an ungainly gang to this strange existence? Well, the answer is not too difficult to find. Even a casual examination of parasitism reveals three possible forces that might cause a free-living animal to become fettered to this lifestyle. But three powerful, magnetic forces they are: *food, safety, and sex.*

There would be little argument in the scientific community that the primary motive of parasitism is a readily available food supply. Whether it's a blood meal for a mosquito, flea, tsetse fly, leech, or vampire bat, the attraction is powerful. Likewise, the cell debris, tissue fluid, and mucus meals of flukes and roundworms is equally tempting as is the predigested paste of the tapeworm terrain. Lice and mites apparently get excited over scalp flakes, while tiny catfish go ballistic over urine sediments. Food is, indeed, a compelling excuse for the parasitic way.

And to this just add the fact that the banquet can be served in a safe, secure hideaway, free from predators, free from extremes of climate, free from excessive competition. To have a place where food is plentiful, easy to obtain, and where life is secure must have been a truly powerful attraction.

Then, as if plenty of food and a safe shelter weren't enough, along comes one more temptation: ready and willing reproductive partners. In the everyday life of many parasites, there is an ample supply of potential mates, indeed, you may be surrounded by them. Parasites don't need to spend time and energy seeking out partners; their kitchen and bedroom are one in the same. Surely, the availability of reproductive partners further spurred the rush to enter the parasitic lifestyle.

These three, ready food, a safe harbor, and accessible partners must have combined to present an overwhelmingly tempting new life. So attractive are these three forces that they served as an invitation to thousands of free-living animals to forsake an independent existence in favor of willing bondage. So, just maybe we're asking the wrong question. Instead of wondering why there are so many, many parasites, perhaps we should marvel at why there aren't more!

Is It Really That Easy?

The answer is no. Attractive, maybe; easy, no. Becoming a successful parasite isn't without its share of difficulties. In fact, there are a number of very real barriers that parasites must overcome in order to take advantage of the Utopia that awaits them.

(1) **Finding the host.** Just locating the proper host can be a major hurdle for a parasite. But we have already seen that they can be extremely inventive in circumventing this obstacle. For instance, parasites have discovered that by linking to the host's food they stand a better chance of ending in or on that host. So we find tapeworms ensconced in beef, pork, rabbit. We find flukes in fish, in a snail tentacle, or on edible vegetation, and we see roundworms buried in the flesh of a pig. Furthermore, parasites have recruited a number of insects to join their search team: mosquitoes, biting flies, fleas, lice, ticks, and many others all serve as active transporters (vectors) of parasites.

(2) **Attachment in the right place.** However, finding the proper host is only half the battle. Parasites must then pinpoint their favorite, specific tissue and once there they must stay put. Parasites have, therefore, invested much of their evolutionary effort toward attachment devices. They are adorned with hooks, suckers, spines, hairs, glue, claws, teeth, and even lips to vigorously cling to their secret places.

(3) **Overcoming host defenses.** So far the attack seems to be one-sided, all in the favor of the parasite. However, that is far from the case. Hosts are not passive recipients in the game of parasitism. Indeed, they have an entire battery, a virtual armory of defenses at their fingertips. Acidic intestinal fluids, an anaerobic milieu, high temperatures, and explosive muscular contractions are all forces opposed to parasitic survival. To be a successful parasite, these defenses, too, must be conquered.

The inside of an intestine is pretty inhospitable. It is a dark, hot, acidic, anaerobic, and pulsing environment. But as stark as the intestine might be, it's not the front line of defense against parasitic attack. No, that honor goes to the immune system. Imagine, an entire body system designed just to remove foreign material, just to repel tiny invaders. If the immune system recognizes a parasite as an alien intruder, the parasite is, in most cases, doomed. Thus, the parasite must disguise itself as friend rather than foe, or else face inevitable eviction. There are many parasites that are quite adept at doing just that. Through a process called **molecular mimicry** some parasites can imitate secretions or other chemical features of host tissue. By doing this they con the immune cells into accepting them as a natural part of the host rather than as invaders.

(4) **Surviving outside of the host.** For a parasitic animal to be really successful, it must spread from one host to another. That means that it must survive, for a period of time, outside of the final host. The adult parasite, or one of its larval stages, must find a way to leave one host and then survive long enough to re-enter another one. Parasites have, for the most part, chosen to do that in immature states: eggs, larva, or juveniles.

The methods used are numerous and often far-fetched, yet they fall under only three general strategies: (1) massive egg production; (2) endurance; and (3) the use of intermediate hosts. Parasites are legendary for their prolific egg production. A single female *Ascaris* can issue 200,000 eggs (actually, embryos) per day, during her 1-year lifetime. Not to be outdone, tapeworms can release one or more gravid proglottids daily, each containing upwards of 100,000 eggs. It seems that this strategy recognizes the nearly impossible survival odds and responds by flooding the market with tiny, host-seeking vagrants.

The second survival strategy involves an almost unbelievable capacity for tolerance. Both eggs and larvae can withstand the harshest of conditions. They can become drought or temperature resistant, can survive without food, or even resist abrupt changes in pH, and often they will become dormant. In a state of dormancy, the parasite reduces its metabolic activities, shrivels up, and just waits. Their patience is astounding, since the wait can be a long one. Parasites have been known to re-emerge from a cryptobiotic state decades after entering, none the worse for wear. There is even considerable anecdotal evidence of much, much longer dormant periods.

But it's the last strategy that truly stretches our credulity. How are we to believe that a parasite located in our innermost recesses, say our blood vessels, can enlist the aid of a tsetse fly to transport it to another unknowing victim? Consider these lengths to which some parasites reach to use insect bloodsuckers in their quest to locate another host. There's a nematode, *Wuchereria bancrofti,* that lives within human lymph nodes causing the hideous condition of elephantiasis. It produces a larva that lives, during the daytime, within the deep blood vessels of the body, but which travels each evening toward the skin where mosquitoes are feeding. The parasite actually synchronizes its behavior to that of a mosquito it has never met and cannot see. This synchronization of parasite and vector behavior is called **periodicity**.

It's beyond belief that a parasite can even change the behavior of another animal to improve its own chances of survival. But there are, in fact, many examples of just that. There's a species of snail that becomes so confused by the presence of parasites that it exposes itself to possible bird predation. And then there's a fish that becomes so disoriented by the presence of internal parasites that it fails to flee oncoming predators— predators that will then become another host for that parasite.

Then there's the little fluke, *Dicrocoelium dendriticum,* whose larvae invade ants. The infestation causes the ants to forego normal secretive behavior to, instead, plunge themselves upward to clutch blades of grass with their powerful jaws. There they hang like miniature Christmas tree ornaments until along comes a herbivore, leisurely grazing and unwittingly ingesting a little parasitic surprise along with its grassy feast. Truly, parasites have left no leaf unturned in their greedy pursuit of one host after another.

Problems faced by Parasites

1. finding the host
2. finding the right place in the host
3. overcoming defenses of the host
4. surviving outside of the host

Some Human Infestations

*I*t is unfortunately true that *we* are not exempt from this parasitic battlefield. As a matter of fact, humans are equal opportunity employers when it comes to engaging parasites. We have our share of critters crawling over our skin, seeking special treats at the base of our hair follicles, looking for private sanctuaries from which to withdraw a sip of blood, creeping over our surface devouring fleshy flakes, gushing within our blood vessels, hanging about our intestines, lurking within our livers, or seeking refuge in our muscles, skin, eyes, lungs, urinary tract, and so on. Yes, we have our share, and while most are relatively harmless, some do exact a punishing toll. See *Inside Views* 10 and 11 at the end of this chapter for a detailed discussion of an important human parasite and also some suggestions on how you can protect yourself from parasitic infestations.

It's quite obvious that a detailed presentation of every human parasite is not feasible here. Nonetheless, a review of some of the more common ones could help us better understand the nature of this special symbiotic relationship. But before we do, you may want to review those basic parasitic terms listed below.

A Review of Basic Parasitic Terminology

ectoparasite - an external parasite; ticks, fleas, lice.

endoparasite- an internal parasite; tapeworms, nematodes, flukes.

permanent parasite - one that has no free-living stage; tapeworms, nematodes.

periodic parasite- one that visits the host occasionally; mosquitoes, blood-sucking flies; leeches.

erratic parasite- a parasite within the proper host but in the wrong tissue; *Ascaris*, some tapeworms.

incidental parasite- one in the proper place, but in a non-preferred host; dog tapeworm in humans.

parasitoid- a parasite that uses the host as a place to lay eggs; tarantula killer wasp; many spiders.

pathogenic parasite- one that causes an illness, disability, or death of its hosts; *Trypanosoma*, *Plasmodium*.

definitive (final) **host**- where parasites reach adulthood, generally reproduce sexually here; most animals.

intermediate host- where parasites spend immature stages, often reproduce asexually; snails, insects, etc.

vector- a host, often an insect, that actively transmits the parasite from one final host to another; mosquitoes.

reservoir host- where the parasite builds large numbers, often used in conjunction with human infestation; malarial mosquito.

indirect life cycle- a pattern in which a parasite uses one or more intermediate hosts; flukes, nematodes, tapeworms.

direct life cycle- a pattern lacking intermediate hosts; many nematodes.

superparasitism - when a host harbors more than one kind of parasite; a human with lice, mites, tapeworms.

autoinfection - a pattern in which a final host becomes reinfected with the same parasite; a child touching its anal area, picking up eggs, transferring them to his or her mouth.

hyperparasitism- when parasites have parasites; fleas harboring mites.

Ascaris lumbricoides
~ the human roundworm ~

*T*he most common human roundworm the world over is *Ascaris lumbricoides*. And at over 18 inches (>40 cm), it is also the largest. It is estimated that almost 807–1,221 million people are host to this parasite. That means that close to one out of every seven persons carries this huge worm!

Adult *Ascaris*, both male and female, reside in the human small intestine feeding on the semifluid mass of partly digested food. There, they go virtually unnoticed. Unnoticed, that is, unless they gather in large aggregations blocking intestinal flow. That's painful... maybe even fatal

This parasite has the nasty habit, especially when immature, of wandering about the body like little vagabonds. They get into all kinds of unwanted nooks and crannies where they can wreck havoc. The migrants, for instance, can cause hemorrhage of blood vessels, appendicitis, infection of the peritoneum (i.e., peritonitis), or even fatal pancreatic infection. When straying about the lungs, they may cause pneumonia-like symptoms, or they may roam into the heart, inducing cardiac failure. *Ascaris* worms are even known to illicit seizures by invading the brain. These roundworms can also exact considerable psychological distress in their meandering if, for example, they exit a nostril, the mouth, the anus, or even a tear duct! *Ascaris* can, indeed, be an insidiously malicious companion.

This parasite has a direct life cycle, requiring no intermediate host. Humans contract it by ingesting infective eggs, usually on contaminated food, in water, on soiled utensils, or from hand-to-mouth. The larvae emerge within the small intestine, bore into neighboring blood vessels, travel throughout the circulatory system, and eventually reach the lungs where they continue to develop. After a few days in the lungs they crawl up the trachea, reach the back of the mouth, and are again swallowed. Upon reaching the small intestine for the second time, the worms mature and may remain there for up to a year. These roundworms are unusual in that they have no special attachment devices. Since they are relatively large, they are apparently able to stem the tide of intestinal contractions and the constant flow of semidigested food to remain in place.

As with all roundworms, fertilization in *Ascaris* is internal. The worms copulate while inside of the human intestine and produce fertile eggs in enormous amounts. Some estimates place the number of fertilized eggs stored in the uterus of a single female at around 27 million. These are released along with the host's bowel movements at a rate of around 200,000 per day. If eggs reach surface soil, they become infective and can remain dormant there for two years or more. Once swallowed they resume the life cycle just described.

Fertilized eggs of *Ascaris* have a characteristic wavy or scalloped appearance. Therefore, adult worms can be rather easily detected through examination of fecal smears. If present, the adults can be removed with oral medications.

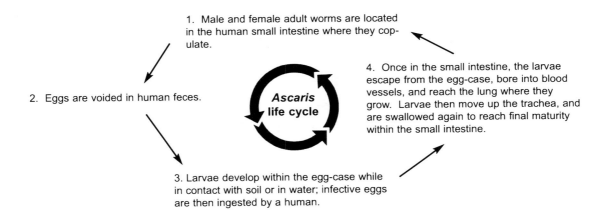

1. Male and female adult worms are located in the human small intestine where they copulate.

2. Eggs are voided in human feces.

Ascaris life cycle

4. Once in the small intestine, the larvae escape from the egg-case, bore into blood vessels, and reach the lung where they grow. Larvae then move up the trachea, and are swallowed again to reach final maturity within the small intestine.

3. Larvae develop within the egg-case while in contact with soil or in water; infective eggs are then ingested by a human.

Figure 8.6. Life cycle of *Ascaris lumbricoides*, the human roundworm.

Trichuris trichiura
~ the human whipworm ~

*S*haped very much like Zorro's famous bullwhip, adult whipworms embed their tiny threadlike heads into the epithelial lining of the human colon (i.e., large intestine). Although infection with this roundworm is usually asymptomatic, at times it can cause problems ranging from bloody stools, nausea, abdominal pain, colitis, to weight loss. And if the burden becomes excessive, this nematode can induce a prolapsed rectum...a condition wherein several inches of the rectum, covered by writhing worms, protrudes from the anus.

The adult worms reproduce sexually in the colon and the female releases up to 10,000 fertilized eggs per day. The eggs, passing along with human excrement, reach soil and become infective within about three weeks. Humans then pick up the parasite through contact with infected soil or by ingesting contaminated food or water. Ingested worms reach the colon and mature within 30–90 days. They may remain there for as long as eight years.

The whipworm is primarily a parasite of warm, tropical climates and does occur commonly in the southern United States. Roughly 604–795 million people are infected worldwide, making the human whipworm the third most common human nematode. In the U.S., it is the second ranking nematode parasite of humans, behind the pinworm.

Infections are most readily diagnosed by identification of eggs in the stool, and oral medications are effective in removing this roundworm.

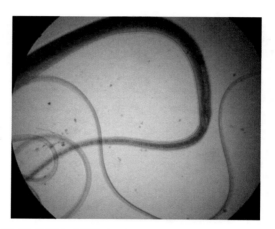

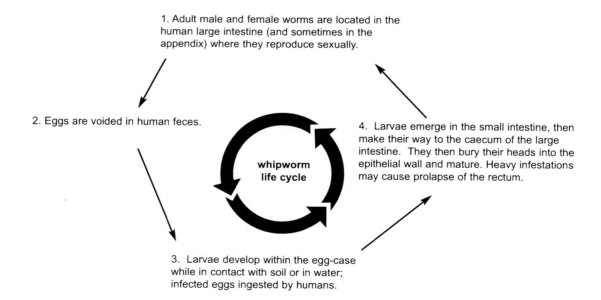

1. Adult male and female worms are located in the human large intestine (and sometimes in the appendix) where they reproduce sexually.

2. Eggs are voided in human feces.

whipworm life cycle

4. Larvae emerge in the small intestine, then make their way to the caecum of the large intestine. They then bury their heads into the epithelial wall and mature. Heavy infestations may cause prolapse of the rectum.

3. Larvae develop within the egg-case while in contact with soil or in water; infected eggs ingested by humans.

Figure 8.7. Life cycle of *Trichuris trichiura*, the human whipworm. The photograph above is of a mature whipworm; note the long threadlike anterior.

Taenia solium
~ the pork tapeworm ~

*T*here are two routes that this tapeworm can take in the human. First, the normal pattern requires consumption of undercooked infested pork. If the meat contains viable **bladderworms** (the cysticercus larval stage), then once swallowed, the larvae will evaginate and attach to the intestinal wall. There they may grow to a length in excess of 25 feet. You will recall from our earlier discussion of tapeworms that they grow from the neck region, thus the egg-filled proglottids are always located distally (away from the head).

Upon maturation *T. solium* reproduces either by cross- or self-fertilization producing voluminous numbers of eggs; each proglottid may, in fact, contain up to 100,000 fertilized eggs. Five or six gravid proglottids are dispersed daily from the human during a bowel movement. If infected eggs reach garbage or are otherwise ingested by a pig, the **onchospere** larvae travel to various organs where they become encysted as **bladderworms**. Humans then become infected by consuming tainted, inadequately cooked pork. Intestinal tapeworms cause very little, if any, damage. In fact, unless an infected person detects the whitish proglottids in the stool, he or she is usually oblivious to their presence.

Approximately 250,000 people in the United States are thought to harbor the pork tapeworm. This relatively low level of infection is due, in part, to religious practices and the tendency of most people to eat only well-cooked pork. Oral medication is effective in ridding the host of adult tapeworms.

The second, or alternate, route by which humans may contract pork tapeworm is through swallowing infective eggs. In this case, the larval cysticercus follows its normal pattern *for pig infestation*. In other words, the tapeworm "behaves" as though it were inside of a pig instead of a human. It bores through the intestinal wall to assume residence in essentially any organ, especially muscle. The bladderworm encysts and remains as permanent residents of the unlucky host, resulting in the condition termed **cysticercosis**.

This condition may elicit severe pathologies, making the pork tapeworm one of the most dangerous human intestinal parasites. Neurological disorders, cardiac failure, muscle spasms, local paralysis, and blindness can all result from human infestation with the cysticercus of this tapeworm. Unlike adult tapeworms, which can be easily extracted, removing bladderworms, especially heavy infestations, is virtually impossible.

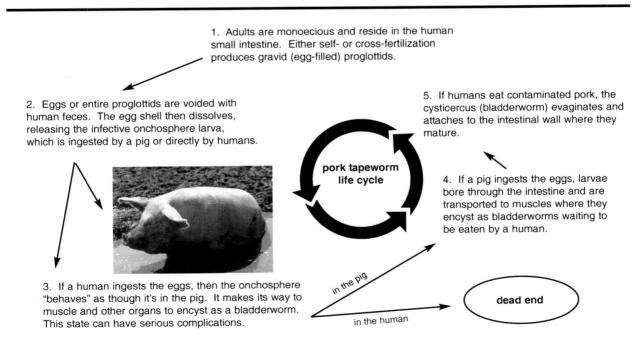

1. Adults are monoecious and reside in the human small intestine. Either self- or cross-fertilization produces gravid (egg-filled) proglottids.

2. Eggs or entire proglottids are voided with human feces. The egg shell then dissolves, releasing the infective onchosphere larva, which is ingested by a pig or directly by humans.

pork tapeworm life cycle

5. If humans eat contaminated pork, the cysticercus (bladderworm) evaginates and attaches to the intestinal wall where they mature.

4. If a pig ingests the eggs, larvae bore through the intestine and are transported to muscles where they encyst as bladderworms waiting to be eaten by a human.

3. If a human ingests the eggs, then the onchosphere "behaves" as though it's in the pig. It makes its way to muscle and other organs to encyst as a bladderworm. This state can have serious complications.

in the pig

in the human

dead end

Figure 8.8. Life cycle of *Taenia solium*, the pork tapeworm.

Enterobius vermicularis
~ the human pinworm ~

*T*he most common human roundworm in the United States is *Enterobius*, the pinworm. Estimates of infestation are as high as 33% among children and about one-half of that for the adult population. Incidence is particularly high in cooler, temperate zones. The high incidence of infection is related to the parasite's small size and reproductive characteristics, as described below.

Adult worms within the large intestine adhere to the epithelial wall where they feed on tissue debris and bacteria. There they mate and the male soon dies. During the night, the lowered body temperature of the host causes the female, containing up to 16,000 eggs, to crawl out of the host's anus where she releases her ripe burden. She, too, then dies. The eggs have a sticky gelatinous coat that allows them to adhere to the anal area, to bed clothes, to linen, and any other contacted surface. The movement of the worms and the release of eggs causes an intense rectal itching. The host, especially a child, may scratch the area and thereby pick up a mass of eggs, which may be subsequently swallowed as fingers are placed in or near the mouth (**autoinfection**).

Within a few hours the larva hatch. Those in the anal region may crawl back into the rectum and thus reinfect the host (**retroinfection**). Eggs may also dry out, become airborne, and remain infective for two to three days. *That, of course, means that you can contract this parasite just by breathing contaminated air.* Eggs can be found in household dust or in public places such as theaters, schools, restaurants, or churches. An unsuspecting person may then become infected unknowingly and without much recourse. As a result, infection with *Enterobius* is highly contagious. Infestation of entire families, friends, and associates is common.

Pinworm contamination is usually asymptomatic. Heavy infestations, however, may cause sleeplessness, abdominal pain, and vomiting. Also, migrating worms may crawl up a female's reproductive tract causing inflammation of the vagina, uterus, or oviducts. The presence of pinworms is easily detected. Night restlessness and anal irritation are signs of pinworm infestation. Inspection of the anal area with a flashlight may reveal the tiny female worms as they migrate out of the rectum. Also, eggs will adhere to a piece of scotch tape placed over the anal region. A physician can then determine if this or any other intestinal parasite is present. Oral medications are effective against this nematode and, due to its high level of contagion, all members of the family and even friends and relatives may require treatment.

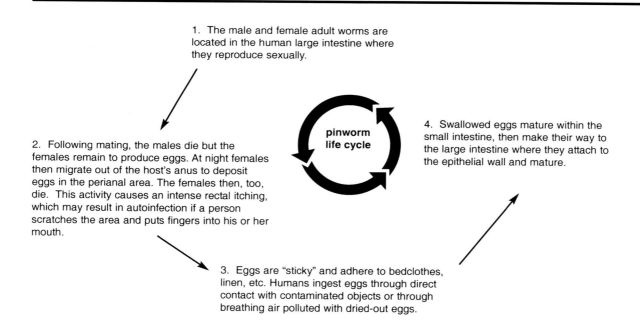

1. The male and female adult worms are located in the human large intestine where they reproduce sexually.

pinworm life cycle

2. Following mating, the males die but the females remain to produce eggs. At night females then migrate out of the host's anus to deposit eggs in the perianal area. The females then, too, die. This activity causes an intense rectal itching, which may result in autoinfection if a person scratches the area and puts fingers into his or her mouth.

3. Eggs are "sticky" and adhere to bedclothes, linen, etc. Humans ingest eggs through direct contact with contaminated objects or through breathing air polluted with dried-out eggs.

4. Swallowed eggs mature within the small intestine, then make their way to the large intestine where they attach to the epithelial wall and mature.

Figure 8.9. Life cycle of *Enterobius vermicularis*, the human pinworm.

Necator americanus
~ the human hookworm ~

Any skin contact with moist soil containing the larval form of this nematode may very well result in infection. The tiny, microscopic larvae can quickly penetrate human skin, and can do so in large numbers. They have been so successful that 740 million people have hookworms (*Necator americanus* or *Ancyclostoma duodenale*), worldwide. Within the skin, they travel by way of the circulatory system to reach the lungs. Once there, they bore into the respiratory spaces (alveoli) and migrate up the bronchi, to the trachea, and then to the back of the throat. The larva are eventually swallowed and finally reach the small intestine where they mature.

Adult worms attach firmly to the intestinal wall with a series of buccal plates and teeth. They break through the mucosal lining, inject an anticoagulant and begin to vigorously feed on blood. The resulting activity of several of these worms can induce anemia, abdominal pain, stunted growth, swelling of the abdomen, mental lethargy, and even death. Strangely, and for some unknown reason, infection with this parasite may cause a person to have a strong desire to eat soil, a condition termed **geophagy**!

The adult male has an enlarged posterior bursa to enwrap the female during copulation (see the photo to the right). She may lay up to 5,000 eggs per day over a 15-year period. Eggs are liberated along with human excrement and upon contact with moist soil a small larva escapes. This stage is free living and can survive for up to six weeks on its own, feeding on bacteria and other organic debris. The larva invades human tissue through skin pores, sores, or between the toes where the skin is especially thin. The invasion is usually accompanied by intense irritation called "ground itch"

It is estimated that upwards of one billion people worldwide are infected with this parasite. Within the United States, however, it is pretty much confined to the southern tier of states where several million people are infected.

The worm is best diagnosed by identification of eggs in a fecal smear. While a high rate of cure can be reached through administration of oral medications, prevention is a more realistic control approach. Avoiding contact with contaminated soil or with human sewage is the primary safeguard.

One more point about hookworms is important. In addition to humans, *Necator* also commonly infects dogs. Keeping people, especially children, from contact with dog droppings is, therefore, important. As with certain tapeworms, a licking dog may transfer hookworms to the hands or face of an adoring pet owner.

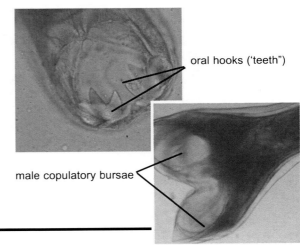

oral hooks ('teeth")

male copulatory bursae

1. The male and female worms reside in the human small intestine where they reproduce sexually.

2. The eggs are then voided in feces.

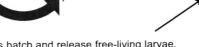

hookworm life cycle

4. Infective larvae penetrate human skin upon contact, and travel through the circulatory system until they reach the lungs, where they bore into alveoli, travel to the trachea, and are then swallowed to finally reach the small intestine where they mature.

3. Eggs hatch and release free-living larvae, which feed for a short time and are later transformed into infective filariaform larvae.

Figure 8.10. Life cycle of *Necator americanus,* the human hookworm. The top photo above shows the "teeth" located within the mouth cavity of a hookworm, and the lower photograph shows the copulatory bursae of a male hookworm.

Onchocerca volvulus
~ the blinding worm ~

*U*nlike many other nematodes, *Onchocerca* does not invade the intestines. Instead, this parasite matures within fibrous nodules just beneath the skin, usually one male and one female per nodule (see photo to the right). These tumor-like growths may occur on the head, pelvic area, chest, lower arms, and legs. Within each nodule, adults reproduce and emit extremely small larvae known as microfilariae. These tiny larvae patrol just beneath the skin surface or within the anterior region of the eyes. A feeding blackfly, *Simulium* spp., picks up the microfilariae, which reach the infective stage in about two weeks. *O. volvulus* is then transferred to another human when the fly feeds again.

This is a particularly virulent parasite. Among the complications produced by *Onchocerca* is a severe form of dermatitis caused by migrating filariae. In fact, the itching caused by this parasite may become so intense that suicide is a not uncommon reaction to this infestation. Furthermore, as the larvae invade structures within the eye they invariably induce blindness. In some areas of its range, the rate of infection is so high that virtually everyone over the age of 50 is blind!

Worldwide, 37 million people are infected with the worm. The parasite is limited primarily to streamsides in tropical areas of Africa, South America, and Mexico. In essence, the disease occurs wherever blackflies are

found. Diagnosis is, of course, made from evidence of nodules and complications within the eye. Chemotherapy is effective against the microfilariae and surgical intervention can remove adults from within the nodules. Control of blackfly populations is another line of defense. The use of netting and window or door screens to prevent household invasion by the fly has been effective. The use of insecticides along streams has also yielded positive results against fly populations.

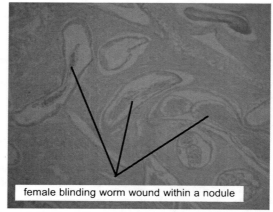

female blinding worm wound within a nodule

Figure 8.11. A section through a subcutaneous nodule containing a female *Onchocercus*. This parasite is also known as the blinding worm because of its inclination to invade the eye.

1. In humans, the adult worms reside just beneath the skin (subcutaneously), forming worm-filled nodules. Microfilarial larvae migrate in blood vessels of the skin where they are picked up by feeding blackflies.

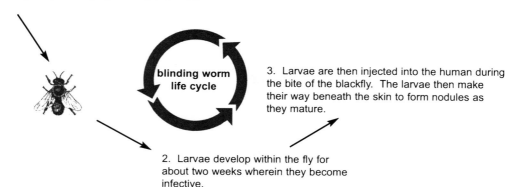

blinding worm life cycle

3. Larvae are then injected into the human during the bite of the blackfly. The larvae then make their way beneath the skin to form nodules as they mature.

2. Larvae develop within the fly for about two weeks wherein they become infective.

Figure 8.12. Life cycle of *Onchocerca volvulus*, the blinding worm.

Trichinella spiralis
~ the trichina worm ~

*T*he study of *Trichinella spiralis*, the trichina worm, is complicated by the fact that it has one of the most complex life cycles of any parasitic roundworm. As we shall see, both humans and pigs may serve as final hosts; the pig is the normal depot while man is a dead-end site.

Humans contract *Trichinella* by eating improperly prepared, contaminated pork. Within the human, the tiny adult worms live in the small intestine. Mature *Trichinella* mate and the male dies shortly afterward. The female remains to issue larvae over the next few months after which she, too, then dies. The larvae penetrate the intestinal wall, are carried to the heart, and enter the general circulation. They eventually invade muscle tissue where they encyst in characteristic spirals (thus, their name). In an attempt to combat the cysts, the host reacts by calcifying the capsule. However, the larvae become dormant within this capsule and may remain infective indefinitely, a condition called **trichinosis**. Since human tissue is not eaten, the cysts remain for the life of the individual, thus, the designation as a dead-end host.

Symptoms, if they occur, are usually mild, resulting in nausea, muscle cramps, edema, and diarrhea. However, in some cases severe reactions may appear. The harsher symptoms include muscular pain, respiratory distress, heart damage, and neural complications. Death can result from respiratory failure, cardiac arrest, neurological dysfunction, and peritonitis (severe infection of the peritoneum). The disease is found primarily in Europe and the United States.

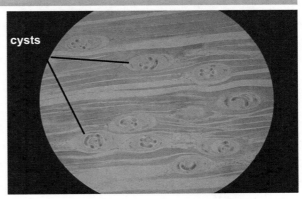

cysts

Figure 8.13. *Trichinella spiralis* encysted within skeletal muscle.

An average of 12 cases are reported each year in the U.S., and roughly 10,000 occur annually, worldwide. Likewise, the number of hogs carrying the disease has fallen dramatically in the U.S., whereas the incidence in some herds in other parts of the world may be as high as 100%! There is no affective treatment for this disease although careful preparation of pork and keeping pigs from contact with contaminated wastes are effective preventative steps.

Keep in mind, however, that *Trichinella* is really a parasite of pigs. That is, swine are the usual final host for this nematode. The life cycle in pigs is essentially the same as that just described for humans. The only major exception is that pork tissue is eaten and human tissue is not. Pigs contract the parasite by eating garbage that contains infested pork wastes or by eating rats who have become infected. Rats, in turn, become infected by eating either pork scraps or by cannibalizing other rats who are already infected.

Figure 8.14. Life cycle of *Trichinella spiralis*, the trichina worm.

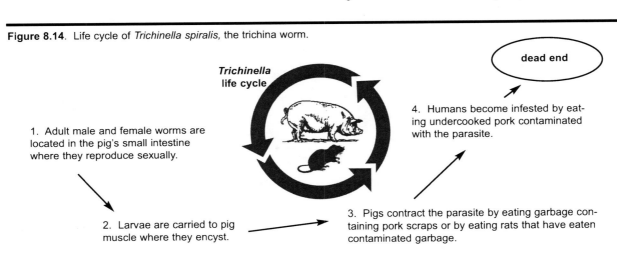

Trichinella life cycle

dead end

1. Adult male and female worms are located in the pig's small intestine where they reproduce sexually.

2. Larvae are carried to pig muscle where they encyst.

3. Pigs contract the parasite by eating garbage containing pork scraps or by eating rats that have eaten contaminated garbage.

4. Humans become infested by eating undercooked pork contaminated with the parasite.

Wuchereria bancrofti
~ human elephantiasis ~

*P*erhaps the most ghastly parasitic disease in humans is caused by the filarial worm, *Wuchereria bancrofti*. This nematode infiltrates the lymphatic system (see box below) where accumulated living, dying, and decomposing worms block the free flow of lymphatic fluid. As a result, fluid from the extremities cannot be returned to the heart and edema (swelling) results.

The incubation period for this parasite in the human is several years; thus, severe stages of edema are rarely seen in individuals younger than 40 years. But the resulting enlargements can reach mammoth proportions. Lower legs, arms, scrotum, and breasts are commonly affected. Feet and calves can attain the size of basketballs, or larger. Breast and scrotal involvement can be absolutely gigantic. In one case, the scrotum of a male weighed over 18 kg. That's almost 40 pounds; or put another way...equivalent to almost 5 gallons of water!

The adults release miniature microfilarial larvae that migrate to the circulatory system. During the day they remain deep within the body's vessels, but at night they migrate to the peripheral circulation where they are ingested by a mosquito as it feeds. Amazingly, the parasite has synchronized its behavior to that of the vector, in this case a mosquito. This incredible migration between internal and peripheral blood vessels is called **periodicity**.

Elephantiasis (so-called because of the elephantine appearance of the affected body part) is one of the fastest growing human diseases worldwide. Although originally confined primarily to the tropics, it is spreading throughout the middle latitudes. Current estimates place the number of people inflicted worldwide with this parasite at about 546 million. In the past it has occurred in the southern U.S., thought to have arrived with slave traffic.

Treatment depends upon the stage of infection. Oral medications are successful against the microfilariae and, at times, against the adults. However, there are few successful regimens against advanced stages of the disease; surgery is rarely an option.

The Human Lymphatic System

Our lymphatic systems perform two basic functions: (1) the return of tissue fluid (lymph) to the circulatory system, and (2) the filtering of that fluid. Tissue fluid that accumulates around individual cells can seriously disrupt cell activity; thus, it must be removed constantly. It is the role of the lymphatic system to collect the fluid by way of a series of tubules and carry it to the heart where it is unloaded into the right atrium. On its way to the heart, the fluid passes through at least one lymph node. There it is filtered and cleansed by specialized phagocytic cells. Bacteria, cell debris, and other foreign items are thereby removed.

---

1. Adult worms live in human lymph nodes (see above) where they may accumulate causing lymphatic blockage.

2. Microfilariae result from the mating of adult worms. The larvae migrate to peripheral blood and are ingested by a feeding mosquito.

elephantiasis life cycle

3. Larvae develop within the mosquito.

4. Humans become infected when bitten by a mosquito vector. Larvae enter the lymphatic system and become lodged within human lymph nodes where the parasite matures and causes blockage and ultimately grotesque enlargement of extremities..

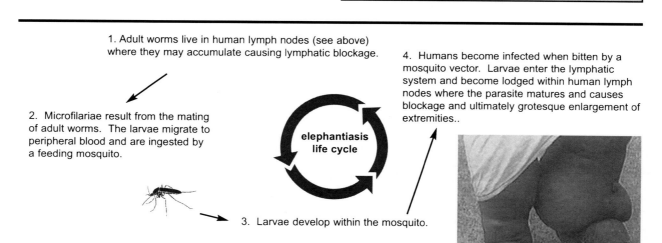

Figure 8.15. Life cycle of *Wuchereria bancrofti*, the parasite causing human elephantiasis.

Man's Best Friend?

If you love dogs, cats, and good health, as I do, be sure to read the next few paragraphs. Your "best friend" may be giving you more than companionship, obedience, and a warm, fuzzy feeling. Our pets carry a number of **zoonotic** parasites that can be transferred from animals to humans. For example: reptiles can transmit salmonella (a potentially serious bacterium); birds may spread psittacosis (a virus attacking the human respiratory system); cats can transfer *Toxoplasma*, a nasty protozoan (as well as a number of other diseases); and fido, our closest ally, can be pretty generous in his contributions. Domestic dogs can transmit *Cryptosporidium*, *Giardia*, hookworm, flukes, roundworms, and tapeworms, to name just a few. Here are details of four of the more common parasites we can contract from our pets.

(1) ***Taenia pisiformis.*** This is a tapeworm of wild canines and domestic dogs (as well as cats). Adult worms, as large as 200 cm (~7 feet), reside in the small intestine of dogs. Egg-rich proglottids are shed along with feces whereupon they reach soil and adjacent vegetation. A rabbit, the intermediate host, ingests the eggs, which then hatch in its small intestine. Larval tapeworms burrow through the wall of the rabbit's intestine, invade the liver, and eventually gain entrance into the body cavity. There they are transformed into bladderworms, which attach to various organs, especially the liver. Dogs become infested with the tapeworm when eating contaminated rabbits. Hunters who give the viscera of a killed rabbit to their dogs as a reward or as a special treat don't realize that virtually all rabbits carry *T. pisiformis* larva.

Humans can contract cysticercosis by ingesting eggs that accompany dog excrement. If an infected dog tongue-cleans itself then licks your hands or face he can inadvertently transfer the eggs to you!

(2) ***Dipylidium caninum,*** the dog tapeworm. This one-foot-long tapeworm produces large proglottids that are evicted from the anus. These proglottids are capable of crawling about on their own and, thus, can be easily identified. Eggs are ingested either by fleas or lice who serve as intermediate hosts. Dogs then pick up the parasite while accidentally ingesting either of these blood-sucking pests. Humans, particularly children, can contract either the eggs or larva of this parasite directly from the dog through petting, fondling, or allowing the pet to engage in face licking. Infestation is usually asymptomatic.

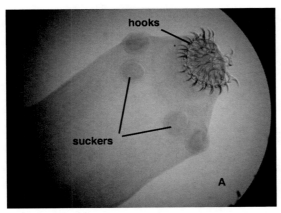

Figure 8. 16. Three photos of *Taenia pisiformis* showing **A**, the scolex (head) with suckers and hooks; **B**, a mature proglottid; and **C**, a gravid proglottid filled with thousands of eggs.

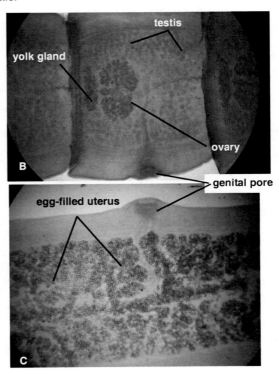

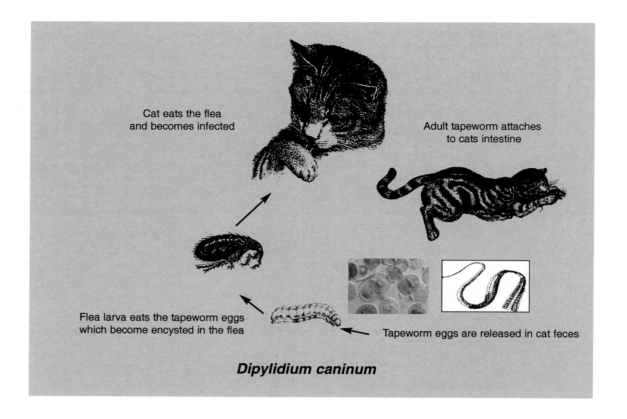

Dipylidium caninum

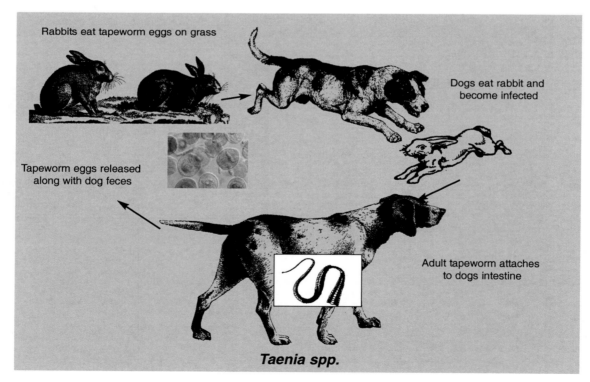

Taenia spp.

Figure 8. 17. The life cycles of two important tapeworms infesting household pets. The upper diagram depicts the infestation of cat or dogs by *Dipylidium caninum* a tapeworm carried by fleas. The bottom plate shows the life cycle of a member of the genus *Taenia pisiformis* that is transmitted to dogs and other canines by eating wild mammals, especially rabbits.

(3) **Toxocara canis,** the dog roundworm. If you ever owned a puppy you are most likely familiar with this common parasite. And common is the key word. Almost every puppy carries this roundworm; some estimates place the rate of infection as high as 98%. This is the parasite removed by "worming" a puppy. The three-inch-long adult worms live in the intestines of puppies while the larval stages reside in adult females. During birth, the larva migrates across the placental barrier from mother to puppy, explaining the high rate of infection.

Humans usually contract the worm by swallowing eggs or even the larva, and almost 14% of the U.S. population has been infected with *Toxocara*. Children, because of their close proximity to puppies and their inattention to personal hygiene, are at greatest risk. Eggs hatch in the human intestine. The larvae then bore into the gut lining and migrate through the body reaching the liver, eyes, brain, and other locations. As they wander about they may cause inflammation, lesions, and, at times, more serious complications, particularly in the eye. Personal hygiene and worming puppies are the best measures against infection.

(4) **Echinococcus granulosus.** This is another common dog tapeworm. Although among the smallest adult cestodes, averaging only 2–8 mm, it is by far the most dangerous to humans. Eggs that have been shed from a dog are ingested by the intermediate host, usually a sheep, pig, or rat. Domestic dogs generally contract the parasite by feeding on infested meat or by eating an infected rat. The eggs, however, are very small and easily cling to dog hairs, as well as to anything coming into contact with dog fecal material. Humans may become infected with eggs by simply petting a dog or by allowing a dog who has just groomed to lick the face or hands. Once an egg is swallowed, it burrows into the body cavity and attaches to any of a number of organs, including the brain. The larva then turns into a bladderworm, but not just any ordinary bladderworm. This one is different. This one begins to grow into a **hydatid cyst,** which buds additional tapeworms inside the ever-expanding bladder. A cyst may become watermelon size and contain thousands of tiny internal larvae called hydatid sand. Each "sand granule" is a miniature tapeworm capable of reinfecting the host.

The enlarging cyst is similar to a slow-growing tumor. In fact, symptoms of the two are similar: internal pressure, seizures, kidney failure, liver dysfunction , and so on, depending upon the organ infected. If the cyst ruptures, either on its own or during attempts at surgical removal, the released "sand" dooms the patient. Careful surgical intervention may be successful, especially for those cysts lodged in the body cavity. However, for cysts of the lungs, bony tissue, or brain, surgery may be impossible. Death, in these cases, is imminent. *Echinococcus* infects dogs around the globe, particularly in the northern hemisphere, but fortunately, cases in humans are rare.

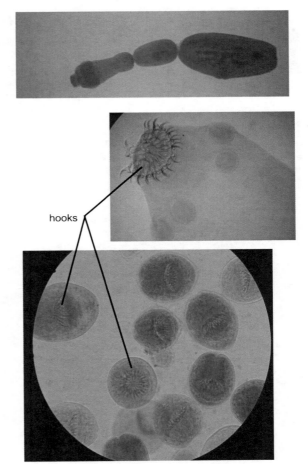

hooks

Figure 8.18. *Echinococcus granulosus,* a small tapeworm that usually infects dogs but can be extremely dangerous should a person swallow the eggs and consequently have a cyst break open within the body cavity. The lower figure shows the tiny secondary cysts known as "hydatid sand" which are so potentially harmful. Note the hooks on the adult head and within each minute cyst.

An Inside View 8.1

The Shameless Tormentor
Pediculus humanus

*H*aving another lousy day? Are you surrounded by general lousiness? Do you louse up from time to time? Was your professor just too nit-picky on your last zoology exam? Truly, lice are ingrained in our human experience. They have been inseparable partners with us for a long, long time. Probably as soon as we gathered into bands of Neolithic foragers, we became home for these six-legged tormentors. And an encounter with *Pediculus* is just one more reason to feel lousy. In fact, fighting off a louse infestation can be an unpleasant, embarrassing, and a rather lopsided affair; all tactics seem to favor the louse.

In his marvelous little book *Rats, Lice and History* Hans Zinsser places lice-borne typhus among the true scourges of mankind, right up there with bubonic plague, malaria, leprosy, small pox, and war-time casualties. He recounts the many creative ways we have tried, for thousands of years, to rid ourselves of this pest. So you would think that such a nuisance would by now be pretty much eradicated. Right? Of course not. Indeed, it seems that an honest-to-goodness present day outbreak is among us. Estimates place the number of Americans infested at around 6–12 million. Most of these are children under the age of ten. Public health records indicate that the infestation is growing with three reasons generally cited:

(1) a sharp increase in the number of preteen children,

(2) overcrowded schools, and

(3) the emergence of a "super louse," resistant to modern-age chemicals.

That sounds just great...a super louse. The plain old louse of our ancestors wasn't enough; now we have our very own creation, and it's bionic!

One of the factors that make lice so hated is the ease with which we contract them. Just casual contact with an infected person, his clothes, or personal belongings, can transfer eggs or adult lice to me or to you. And while infestations are often correlated with crowded, filthy, impoverished conditions, lice are really not all that particular. They infest the wealthy along with the poor, the clean and the unkempt, the socialite and the homeless, the urbanite and the suburbanite.

To make matters worse, as easy as it is to acquire lice, it's even easier to keep them. They have a pure affection for our hairy domes and once ensconced among blondes and brunettes, redheads and silverheads, they prefer to stay. They are efficient, enterprising, and, yes, loathsome creatures that can do four things very well: suck blood, stay put, glue eggs (called nits) to hair shafts, and cause general humiliation. No parasite elicits the sense of repulsion like that of a head full of lice. If you can get anyone to admit to an infestation, you will be told of the shame, the stigma, the embarrassment, and the sheer humiliation. So, it's not bad enough that they plunder our scalps, use our blood as an unending reservoir of food, and incubate baby nits among our curls, they add a healthy dose of insult as well.

But humans are a resourceful and persistent lot. We fight back. In the U.S. alone, there is a multi-million dollar industry built up around delousing. In 1996 Warner Lambert sold $52 million of Nix, an over-the- counter treatment; stainless steel combs designed just to remove the nits are available at any drugstore, or any decent web site. Great Britain is so infested that the country has a semiannual "Nitpicking" day, everyone's included.

The truth is, lice are difficult...no, very, very difficult to control. There are several home remedies that are virtually worthless (Vaseline, kerosene, alcohol, baking soda, gasoline). The only certain method is the tedious removal of the rice-sized adults or the even smaller nits with a fine-toothed comb. Delousing of this type is a four to six hour operation. And all the while, shame reins supreme.

Lice can transmit three diseases to humans:

✓ **relapsing fever**
✓ **trench fever**
✓ **epidemic typhus**

Of the three, only typhus has presented a gruesome past. At one time, it was one of the world's most feared and devastating blights, causing widespread pestilence and, as late as the 1920s, was responsible for over three million deaths in post-war Europe. The causative agent, a spirochete (similar to the one that causes syphilis), is passed to the louse while feeding on an infected human. Then, in turn, it is passed in the feces of the louse to enter humans by way of the louses' excrement. We inhale it, or rub it into sores or into the fluid around the eyes and mouth. But we're not the only ones to suffer. Typhus also kills the louse. So, from the view of the louse, it's the human who is the curse— the shameless tormentor. Typhus, by the way, is no longer a major threat to human populations. But to paraphrase Dr. Zinsser: As long as lice prefer the taste of our blood, the threat of typhus remains.

An Inside View 8.2

But What Can I Do?

Here are 15 simple and practical precautions that you can take to avoid parasitic infection.

Personal:

1. Practice common-sense personal hygiene; wash your hands before you eat, after playing with or handling pets, after handling pet refuse, or after grooming your pet.

2. Don't let pets lick your face.

3. Perform a "tick check" on yourself after spending time in wooded or brushy areas.

4. Be vigilant; look for unusual-appearing items in your stool.

Family:

5. Periodically disinfect your child's toys, particularly if he or she is in a group or institutional setting.

6. Thoroughly cook pork, beef, and fish.

7. Wash vegetables thoroughly.

8. Periodically check your child's head for lice, and while you're at it check yours, too.

9. Look for odd-looking items in your family's stool; be alert to any reports of lice infestation, pinworm outbreaks, or other public health considerations in your area. Pay particular attention to school outbreaks

10. When in doubt, ask your doctor or other health professional.

Pets:

11. Every now and then, visually examine pet droppings for tapeworm proglottids or other "unusual signs."

12. Keep sandboxes clean and covered; cats love to use sandboxes as their toilet.

13. Control your pet's tendency to roam, especially if it likes to hunt rabbits, mice, or other wildlife.

14. Practice proper pet sanitation; clean up after your pet.

15. Have your pet checked regularly by a veterinarian.

Chapter 8 Summary

Key Points

1. There are a number of interspecific relationships between animals. If the interaction is necessary for the survival of one of the participants, the relationship is obligatory, otherwise it is considered a facultative relationship.

2. Symbiosis is an intimate, interspecific relationship between two animals.

3. There are four types of symbiotic relationships: (1) mutualism, in which both parties benefit; (2) commensalism, whereby the symbiont benefits and the host is unaffected; (3) phoresis, where the symbiont uses the host for transport only; and (4) parasitism, a symbiotic relationship in which the symbiont (parasite) benefits at the expense of the host.

4. Parasites can be external (ectoparasites) or internal (endoparasites).

5. Three attractions of parasitism are: nutrition, protection, and reproductive effectiveness.

6. There are a number of barriers that a successful parasite must overcome. Among these are finding the host, attaching in the proper location, overcoming the host's defenses, and surviving outside of the host long enough to locate a new host.

7. Humans serve as final hosts for a wide array of parasites. The chapter presents the life cycles of a number of these.

8. Infestation of humans by the head louse, *Pediculus humanus*, is discussed in detail.

9. There are a number of things you can do to avoid parasitic infestation: personal hygiene, pet hygiene, common-sense interactions with pets, proper cooking techniques, be alert for signs of parasites.

Key Terms

autoinfection
commensalism
cryptobiosis
cysticercosis
definitive host
direct life cycle
ectoparasites
endoparasites
erratic parasite
facultative relationship
final host
host specificity
hydatid cyst
hyperparasitism
incidental parasite
indirect life cycle
intermediate host
interspecific relationship
intraspecific relationship
molecular mimicry
mutualism
obligatory relationship
parasitism/parasitoid
periodicity
periodic parasite
permanent parasite
phoresis
reservoir host
superparasitism
symbiosis/symbiont
tissue specificity
vector

Points to Ponder

1. Differentiate between an ectoparasite and an endoparasite. Give examples of each.

2. What is an obligatory relationship? Give an example.

3. Give, define, and provide examples of the four types of symbiosis.

4. Differentiate between direct and indirect life cycles of parasites. Give an example of each.

5. Differentiate between a definitive (final) host and an intermediate host. Give an example of each.

6. What is a vector? Give an example.

7. Why do you think the mutualistic relationship is so common?

8. What type of problems must parasites conquer in order to be successful? Give an example of how they overcome each.

9. How are lice transmitted to humans? What are the symptoms and how can lice be controlled?

10. For each of the parasites listed below, give the following information:

✓ location of adult parasite in the human
✓ what symptoms does the human show
✓ name an intermediate host (if there is one)
✓ how do humans contract the parasite how the parasite is detected (diagnosed) in humans

    Ascaris lumbricoides
    Enterobius vermicularis
    Wuchereria bancrofti
    Trichinella spiralis
    Necator americanus
    Taenia solium

11. Identify some parasites that can be conveyed from pets to humans. Give examples of how each is transmitted and how it can be avoided.

IV. Inside Readings

Bird, A. F. and J. Bird. 1991. The Structure of Nematodes. New York: Academic Press.

Bogitsh B.J. and T. C. Cheng. 1990. Human Parasitology. Philadelphia: Saunders Publishers.

Brooksmith, P. 1999. Bugs, Bloodsuckers, Bacteria: On Your Body and in Your Home. New York: Barnes & Noble.

Brusca, R. C., and G. J. Brusca. 1990. Invertebrates. Sunderland, MA: Sinauer Associates.

Cheng, T. C. 1986. Parasitology. New York: Academic Press.

Faust, E. C., et al. 1968. Animal Agents and Vectors of Human Disease. Philadelphia: Lea and Febiger.

Levine, N. D. 1980. Nematode Parasites of Domestic Animals and of Man. Minneapolis:Burgess.

Moore, J. 1984. Parasites that change the behavior of their hosts. Sci. Am. 250(5):108-115.

Schmidt, G. D., and L. S. Roberts. 1989. Foundations of Parasitology. St. Louis: C. V. Mosby and Company.

Pechenik, J. A. 1996. Biology of the Invertebrates. Dubuque, IA: Wm. C. Brown.

Chapter 8

Laboratory Exercise

Symbiosis

Animal Interactions

Laboratory Objectives

- ✔ compare the different life cycles of the parasites presented,
- ✔ determine the impact of various parasites on human health,
- ✔ review the role of mutualism and commensalism in the lives of animals.

Supplies Needed:

- ✔ description of several parasitic life cycles
- ✔ selected microscope slides of parasitic roundworms,
- ✔ live specimens of the vinegar eel

Chapter 8
Worksheet

(also on page 593)

Symbiosis

Mutualism

1. Define mutualism and give two examples among those animals that we have studied.
   Both members of relationship benefit and depend on each other for survival ex: monkey and deer fish and shrimp

Commensalism

2. Define commensalism and give two examples among those animals that we have studied.
   Only one member of relationship benefited the other is unaffected ex: crab and worm eagle and deer

Parasitism

3. Define parasitism and give two examples among animals other than humans.
   One member of relationship benefited the other is harmed but not killed. ex: ants & catepillar, Bee and insect.

4. Examine a microscope slide of the encysted form of larval *Trichinella spiralis* (the trichina worm).

   a. Note the spiraled position assumed by the many larvae on this slide.

   b. How is this stage reflected in the parasites' name? *Because it is spiral muscle*

   c. In what type of tissue is the cyst embedded? *muscle tissue*

   d. What disease does this organism cause in humans? *Trichinosis*

5. Next, select a slide showing male and female hookworms, *Necator americanus*.

   Note: The parasite is essentially identical to another species of hookworm, *Ancyclostoma duodenale*. In fact, the names are sometimes interchanged. For our purposes, we will use *Necator* as the hookworm example as it is the more common parasite of humans.

   a. How is *Necator* transmitted to humans?
   Larval penetrate human skin upon contact

   b. Why is it more common in Atlanta than in Chicago?
   Because of warm weather, as the soil is more moist

   c. Note the large anterior teeth (the source of its common name).
   What do you think is their function?
   Help them break the mucus lining

   d. Finally, examine the male hookworm and locate the posterior, flared copulatory bursa.
   How do you think he uses these structures?
   It hooks on the Female

6. The final microscope slides to examine are of male and female *Enterobius vermicularis*, the human pinworm.

   a. Note the very small size of this parasite. Which is larger, the male or female? *Female*

   b. Describe how this parasite may be transmitted to humans.
   Airborn or sticky contamination

Worksheet – chapter 8

Instructions: The table below contains information about selected human parasites; a separate parasite for each row. The table is partially complete in that enough information is contained in each row to identify that parasite. Using lecture notes and text information, fill in the missing data.

Parasite sci. & common name	Human Location	Intermediate Host	Transmission	Pathology & Diagnosis
Taenia solium (pork tapeworm)	Intestine	Pig	Eggs are swallowed undercooked Food	mild intestinal upset, loss of appetite, diarrhea eggs & gravid proglottids detected in stool
Ascaris (Human round worm)	Small intestine	None	eggs are swallowed, larva migrate to lungs then up trachea and are again swallowed to mature in small intestine	female lays 200,000 eggs per day
Entero bius vermi cularis (Human Pinworm)	Large Intestine	None	Scratch the area & Pick up eggs & swallowed as fingers in mouth	most widespread human roundworm in U.S., female lays up to 16,000 eggs/day; flashlight and scotch tape tests to detect worms that crawl out of anus at night; familial infection
			filariaform larvae penetrate exposed skin from soil and migrate through blood vessels to lungs and then to intestine	
Trichinella spiralis (Trichina worm)	Small intestine	Pig/ Human	under cooked pork	trichinosis; may cause severe muscular fatigue
oncho cerca volvulus (Blinding worm)	Beneath the skin layer	Boooooo	bite of the blackfly (*Simulium*, spp.)	Chemotherapy can help against this Fly. Netting window & doors
wucher ella bancrofti (Human elephan tiasis)	Lymphatic System	mosquitos		gross enlargement of the extremities

9 Chapter 9

THE MOLLUSKS

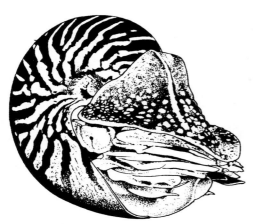

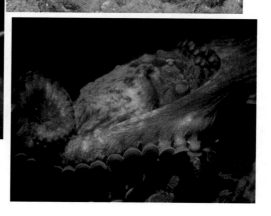

Adopt the character of the twisting octopus....now turn a different hue.
- Theognis

Phylum Mollusca

He was a bold man who first ate an oyster.
- Lesage

Introduction

*I*n one way or another most animals have an impact on humans. But rarely do we encounter a group that has influenced us as have the mollusks. We all , for instance, know of their culinary contributions to our lives: steamed clams, oyster Rockefeller, calamari, scallops, and escargot. But do we realize that throughout history, mollusks were an important, and at times, the primary, food source for many of our ancestors? Anthropologists can often trace the appearance and disappearance of entire civilizations through the piles of discarded molluscan shells. These kitchen heaps, called middens, can be huge. I recall on one of my research expeditions to San Clemente Island, off the coast of southern California, the huge mounds, some of them as high as 30 feet and stretching for hundreds of meters, of nothing but empty, discarded clam and abalone shells. Except for one burial site that we uncovered, these mounds were the only remnants of early native Americans who had occupied the island centuries before. Shellfish were then, and still are, a staple for many of earth's peoples.

But food is only one of many ways this animal has influenced man. From pearls and cameos for jewelry, to mother-of-pearl for buttons, shiny shells for barter (money), for hobbies (shell collections), communication (conch shells), to dyes, and inks, people have exploited and benefited from this invertebrate group. Mollusks have even contributed to the world of medicine. For instance, much of our knowledge of how the nervous system works is based upon experimentation with giant neurons of the squid and octopus. In addition, our understanding of animal behavior has been aided by the study of these same cephalopods. We have, indeed, profited by our interaction with mollusks. Even our feathered pets take advantage of their distant ancestors by nibbling on cuttlebones (the internal skeleton of cuttlefish, a relative of squid) for needed calcium.

The relationship, however, is not always so positive. Take, for example, the snails that serve as intermediate hosts for *Clonorchis*, the human liver fluke and *Schistosoma*, the human blood fluke. These snails aid in transferring two of the most potent human diseases. Or what about the damage that snails and slugs do to our carefully tilled gardens? How about the shipworm (a close relative of the clam) that has exacted incalculable damage to wooden-hulled ships, piers, and buildings by their incessant burrowing. Zebra mussels are, as I write this, vandalizing the aquatic landscape of the Great Lakes. And some mollusks are even venomous. The blue-ringed octopus is a small but deadly creature inhabiting Australian coastal waters, and some shelled cones can inject a painful, even fatal, neurotoxin. So, the man-mollusk connection is an old and multifaceted one...but not always welcomed. And it is this very give-and-take between the beneficial and the detrimental that I believe makes their study so rewarding.

Figure 9.1. Humans have benefitted from mollusks in a variety of ways from food to jewelry.

General Characteristics

One of the difficulties in studying such a diverse group of animals is determining what features each possesses to link it to all of the others. In short, what attributes are common denominators shared by all? It would be nice, for instance, if all had an easily recognizable body plan or if all had a series of uniquely obvious structures. But that's not the case with mollusks; there are very few features found only in this group, and even those are elusive. To add further to the problem of relationships is the fact that mollusks differ so much in size, feeding patterns, locomotion, and habitats. Put simply, mollusks inhabit a wide variety of niches. So whatever general plan may have existed is lost in the scramble to adapt to different lifestyles. Nonetheless, we will find in the phylum Mollusca that there are a number of important diagnostic characteristics, although only three seem to be enjoyed by nearly all of them: (1) a **foot**, (2) a **visceral mass,** and (3) a **mantle**.

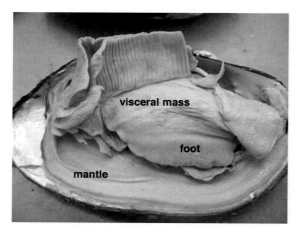

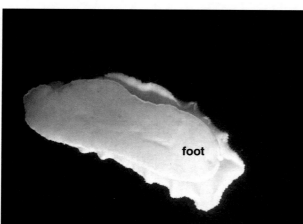

Figure 9.2. A. The general body plan of a mollusk showing the location of the mantle, foot, and visceral mass. The remaining plates show the structure of the foot in various mollusks.

The Foot

The foot, molluscan style, is a fleshy yet versatile mass of muscle. At times it is an oval, glandular pad used for slow ciliary, slimy locomotion as in snails or slugs. At other times we find it serving as a super suction disc to adhere a chiton to its rocky, wave-swept shoreline. The foot can be a muscular blade to burrow through mud-flats (clams) or to burrow into rocks and wooden surfaces (shipworms and their kin). In the cephalopods (e.g, squid, nautilus, and octopus), the foot has become a bundle of arms and tentacles for grabbing and handling prey. But whatever its shape or function, all mollusks seem to have a foot of some sort.

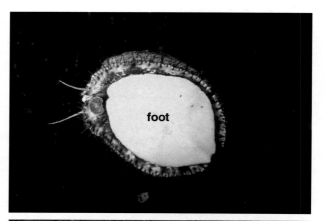

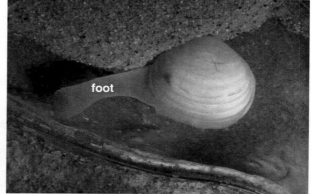

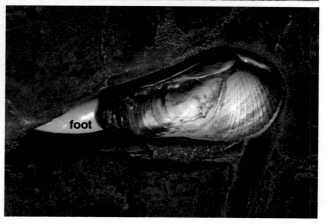

The Visceral Mass

You can think of the visceral mass as a sac to hold the major internal body organs. It is usually located dorsal to the foot, although in cephalopods it has shifted to a more posterior position. The visceral mass is easily seen in some taxa (e.g., chitons, clams, and cephalopods), but difficult to distinguish in others. In any event, the visceral mass, or *hump* as it is sometimes called, encloses the reproductive, digestive, circulatory, and excretory organs.

The Mantle

Perhaps the most unique feature of mollusks is the mantle. But even the mantle is difficult to track in its ever-changing guise. At times it is a membrane beneath the shell that drapes over the visceral mass in a capelike fashion (Figure 9.3). At other times it is an outer, fleshy skin that envelopes the body of a squid, cuttlefish, or octopus. Then again, it appears as a membrane wrapping the outside of a cowry's shell or winding inside of a snail's shell.

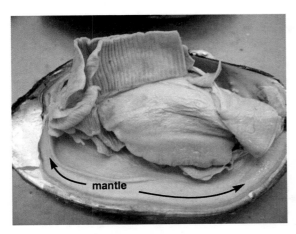

Figure 9.3. A dissected clam showing the location of the all-important mantle.

The importance of the mantle, however, is not only in its various designs, but also in the variety of functions it performs. And there are several:

THE MOLLUSCAN MANTLE

(1) In those mollusks that possess a shell (which is the majority of them), it's the lateral mantle surface that secretes it. The shell thickens along its interior surface in response to secretions of the mantle. Also, the shell increases in size, "grows," as the mantle adds new material to its leading edge. The result is a shell with rings very similar to those of a tree trunk, one ring for each growing season.

(2) If an irritant (e.g., sand, debris, etc.) becomes lodged between the shell and the mantle, the mantle will secrete concentric layers of **nacre** (the same material that lines the shells) around that substance in an effort to ease the irritation. After many years of gradual deposition, the result may turn out to be a perfectly round gem, a lustrous pearl.

(3) In many mollusks, the medial mantle surface forms a respiratory organ, either a gill or a lung, and provides a large surface area for the exchange of gases.

(4) In many species, the arrangement of the mantle forms a cavity, a "brood chamber". Within this space, larvae may be protected during their early development.

(5) The mantle may be supplied with sensory structures, such as eyes, touch receptors, and chemoreceptors.

(6) The muscular or ciliary action of the mantle causes a tidal flow of water to be pumped through its cavity. This flow of water is essential to many mollusks as a source of oxygen, food, and as a way to flush out wastes.

Functions of The Molluscan Mantle
~ A Summary ~

1. FORMS THE SHELL
2. MAY FORM A PEARL IN RESPONSE TO THE PRESENCE OF AN IRRITANT
3. MAY FUNCTION AS A RESPIRATORY SURFACE
4. FORMS A "BROOD" CHAMBER FOR LARVAL DEVELOPMENT
5. MAY BEAR A NUMBER OF SENSORY STRUCTURES
6. ITS CILIATED SURFACE MAY PUMP A CIRCULATING, INTERNAL STREAM OF WATER

Other Characteristics

In addition to a foot, a mantle, and a visceral mass, mollusks show several other important features.

A Shell

A calcium carbonate casing is present in most mollusks. As a matter of fact, many people know this group only from the shell; simply ask any beachcomber. The variations in shell size, coloration, pattern, shape, and design are truly dazzling. I think it's safe to say that a shell is the hallmark of the phylum.

However, not all mollusks have an external shell. Some, instead, have only a series of overlapping plates, others have but a pair of closing valves, while in others the shell is reduced to a small, saddlelike wedge. Then there are some mollusks that have lost the external shell altogether; it has sunk below the surface to become an internal supporting rod. Finally, some, such as the octopus, lack a shell altogether.

Nonetheless, most mollusks have one. In fact, all of the five major classes of mollusks have at least some members with a shell. Three classes (the polyplacophorans, scaphoporans, and the bivalves) are exclusively shelled (i.e., all representatives of this group have a shell). In another class, the gastropods, a large, conspicuous shell is present in most of its members although it's reduced in some (e.g, terrestrial and sea slugs, nudibranchs). And it's in only one case, the cephalopods, that a shell is not a prominent feature; however, even among cephalopods most representatives have at least an internal remnant of a shell.

The construction of the shell is fairly uniform. It's a triple-layered laminate composed primarily of calcium carbonate. The outside is covered by a black to greenish-black organic layer, the **periostracum**, thought to protect freshwater mussels from an acidic environment. The middle strengthening zone is the **prismatic layer**, and the innermost stratum is the **nacre**. It's the inner nacreous layer that gives the lustrous nature to mother-of-pearl, cameos, and, of course, to that most coveted of natural products, the true pearl.

Functionally, the shell is an excellent example of how evolutionary trade-offs occur in nature. On the one hand, this outer casing provides a protective sanctuary, a self-made coat of armor, so to speak. On the other hand, it's cumbersome to haul about. Carrying one about surely is an energy-demanding feat that limits its owner's maneuverability. Thus, shells are typically possessed by aquatic, slow-moving animals. Just consider the giant clam, for instance. Its shell is so heavy that once the clam has settled in a favored location it never moves again.

Shelled terrestrial forms, such as snails, are also famously slow of foot. It is no wonder then that when the external shell is finally eliminated, as in squid and octopuses, mobility increases tremendously. The trade-off is clearly one of speed and agility in favor of protection.

Figure 9. 4. A collection of molluscan shells showing the array of shapes and designs so coveted by shell collectors everywhere. The prismatic and nacre layers are indicated.

The Radula

A large number of molluscans possess a specialized feeding structure, the radula. A tongue-like device loaded with rasping teeth, plates, ribbons, or ridges, the radula is a unique feature of mollusks. No other animal group has such a device. It is especially prominent in gastropods, cephalopods, and chitons, although it is absent from bivalves (clams, mussels, scallops, etc.). Like a tongue, the radula can be protruded from the animal to scour algae or other material from hard surfaces or used to tear flesh. At times it is modified to bore holes into wood, into other mollusks, and even to scrape depressions on rock surfaces. Because of its constant action against hard surfaces, the radula is subject to continuous wear. As the surface wears away, it is continually replaced by newly formed tissue from its core.

An extreme example of the radula is seen in conus species (below). Here it is modified into a harpoon-like device that can inject a paralyzing toxin into its victim.

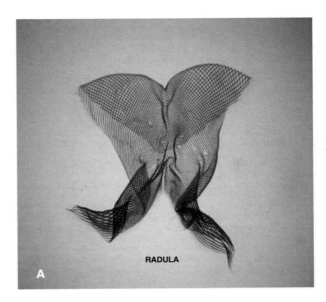

RADULA

A

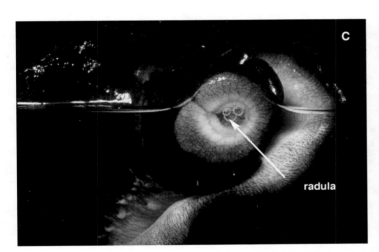

C

radula

B

D

modified radula into a venomous harpoon

Figure 9.5. Plate **A** is the surface of a snail radula displaying the ridges used to scrape food from hard surfaces. Plate **B** is a SEM of a similar radula showing the scraping teeth under much higher magnification. Plate **C** reveals the location of the radula on the undersurface of an abalone while a *Conus* (**D**) extends its radula concealing a deadly, toxin-laden harpoon.

Respiration

So far the animals we have studied in this course have exchanged respiratory gases (O_2 and CO_2) through the body surface. While cutaneous respiration of this type is satisfactory for relatively small, thin, or aquatic animals, it is decidedly unacceptable for larger, bulky, or purely terrestrial ones. Mollusks solved that problem by giving us the first dedicated respiratory surface, a gill. They've even gone an additional step in some terrestrial forms (the pulmonate mollusks) by developing a membranous respiratory sac, a lung. Most mollusks, however, breathe by way of gills.

Whatever the respiratory device might be, it must allow diffusion to occur at an efficient rate. To do that, the surface must satisfy a number of physiological and physical requirements. The better it satisfies these conditions the more efficient it will be in the delivery of oxygen and removal of carbon dioxide. See the box below for a detailed presentation of these requirements.

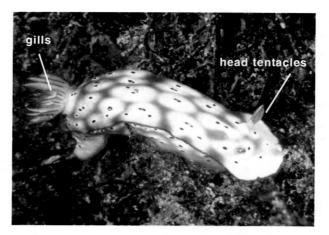

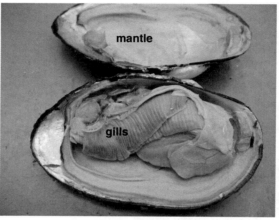

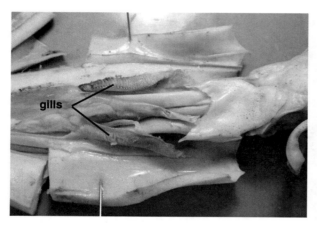

Figure 9.6. Three examples of the molluscan gill. To the left is a dissected squid revealing a pair of large internal gills. The photograph at the top is a nudibranch with its external finger-like gills. At the bottom is a dissected clam showing the location of the plate-like gills.

Requirements for Successful Respiration

FOR A STRUCTURE TO QUALIFY AS AN EFFECTIVE RESPIRATORY SURFACE, IT MUST SATISFY SIX REQUIREMENTS.

1. IT MUST BE LARGE; THE LARGER THE BETTER.

2. IT MUST BE PATENT; IT MUST REMAIN OPEN— ITS EFFECTIVENESS IS LOST IF THE SURFACE COLLAPSES.

3. IT MUST BE THIN; MOST RESPIRATORY SURFACES ARE MEMBRANES ONLY ONE CELL LAYER THICK.

4. IT MUST BE MOIST; DIFFUSION RATES CEASE WHEN THE SURFACE DRIES OUT AND DEATH QUICKLY FOLLOWS.

5. IT MUST BE WELL VENTILATED; O_2 MUST CONTACT THE RESPIRATORY SURFACE AND CO_2 MUST BE CARRIED AWAY.
   (Aquatic animals flush the surface with water to accomplish ventilation, while terrestrial animals aerate the respiratory surface either through a series of internal tubes or through various breathing motions.)

6. IT MUST BE IN CLOSE ASSOCIATION WITH THE CIRCULATORY SYSTEM

   (Once oxygen diffuses through the respiratory surface, it must then be transported to the various body cells. This is best accomplished by a branching network of blood vessels.)

A Circulatory System

Let's suppose that a potential respiratory surface satisfies requirements 1-5: it is large, patent, thin, moist, and well ventilated. This means that the surface is supplied with sufficient oxygen and that diffusion of that oxygen can readily occur. Yet the goal of respiration is not just the delivery of oxygen to the respiratory surface but, in actuality, it is the distribution of oxygen to, and removal of carbon dioxide from, every living cell in the body. So, you can see that without requirement number 6, respiration fails. Without a method to assure that oxygen from the gills, skin, or lungs is transported directly to a needy cell, respiration is futile, and the cell dies. Once cells die, entire organisms soon follow.

There is, fortunately, in higher animals, an organ system devoted to assuring that doesn't happen. I am referring, of course, to the circulatory system, a system of pumps and vessels dedicated to the traffic of materials from one place in the body to another. The circulatory system is designed to pick up oxygen at the skin, gill, lung, or other respiratory surface, and deliver it to individual cells where it is used in the release of energy— a process we call cell metabolism. There, the circulatory system seizes metabolic effluents, such as carbon dioxide and nitrogenous wastes, and transmits them to places in the body where they can be flushed to the outside environment. So it is no mere accident that along with the first dedicated respiratory surface, a gill, the first circulatory system also appeared. The two are inseparable partners in cell survival.

It is, however, important to keep in mind that, while the transport of oxygen, carbon dioxide, and metabolic wastes is a major function of the circulatory system, it's not its only function. This keystone system also ships nutrients to the cells, dispatches hormones and enzymes to targets throughout the body, and routes immune cells to places in need of protection. It is central to the regulation of fluid balance and, in higher animals, also helps to regulate body temperature.

I suppose this is an appropriate time to discuss the two types of circulatory systems: **open** and **closed**. In higher animals, especially the chordates, the circulatory pattern is closed and, as such, consists of five elements:

(1) a transporting fluid, **blood**, designed to carry oxygen, carbon dioxide, nutrients, and so on throughout the body;

(2) a pump, the **heart**, to provide the needed pressure to propel blood throughout the system;

(3) a series of vessels, the **arteries**, carrying blood away from the heart;

(4) a series of returning vessels, the **veins**, transporting blood back to the heart; and

(5) small, intervening vessels, the **capillaries**, situated between the arteries and veins.

Closed systems, by definition, require that blood remains confined to a labyrinth of tubules. As already indicated, closed systems are used primarily by "higher" animals, probably related to the challenges of an active and energy-demanding lifestyle. Nonetheless, some invertebrates, notably, the cephalopods, have also adopted this advanced pattern.

The other scheme, the *open system*, is identified by the lack of capillaries. As a result, blood must leave the arteries to enter sinuses or other open spaces usually called **hemocoels**. Once blood enters these spaces, it bathes or washes over the organs in a sloshing manner. The open pattern is less efficient and more primitive than the closed system and throughout the animal kingdom it is primarily used by invertebrate groups, including insects and the majority of the mollusks.

On the whole, mollusks have, by giving us simultaneously a matched respiratory and circulatory system, made an exceedingly important contribution to the further evolution of animals. In fact, it is hard to overstate the significance of these two contributions.

IT'S A FACT

CAPILLARIES ARE THE FUNCTIONAL UNIT OF THE CLOSED CIRCULATORY SYSTEM...THEY ARE WHERE MATERIAL EXCHANGE BETWEEN THE SYSTEM AND THE INDIVIDUAL BODY CELLS OCCUR. OXYGEN AND NUTRIENTS REACH THE CELLS *ONLY* THROUGH CAPILLARIES WHILE CARBON DIOXIDE AND WASTES ARE REMOVED, LIKEWISE, ONLY BY WAY OF CAPILLARIES.

A Parade of Mollusks

Mollusks consist of a wide-ranging assemblage of animals, some well known, such as clams and squid, others relatively unknown (limpets, chitons, and nudibranchs). There are roughly 85,000 known molluscan species. Mollusca is not the largest animal phylum on earth, but it is the largest phylum in the seas. This impressive group is divided taxonomically into four major classes.

Class Polyplacophora

Members of this class are known commonly as **chitons** (Figure 9.7). They are a small group of relatively primitive animals, characterized by a dorsal series of eight overlapping plates or valves. The animals are headless, somewhat flattened mollusks. Thus, to the casual eye, they have the appearance of a small, shielded or armor-plated "tank" slowly moving along rocky shorelines.

Chitons have a prominent ventral foot used for locomotion and for attachment. It can firmly adhere itself to rocks and other wave-swept outcroppings by creating a vacuum-like seal. This prevents the animal from being dislodged by waves or by prying predators. They also have, within the ventral mouth cavity, a rather large radula used to scrape algae from a variety of surfaces.

Class Bivalvia

Clams, mussels, oysters, and scallops are members of this familiar aquatic group (see Figures 9.7 - 9.9). As the name implies, these mollusks possess two protective, somewhat flattened, shells (valves) attached dorsally by a springlike **hinge ligament** and held closed by a pair of powerful **adductor muscles.** When a clam dies, the muscles decompose, allowing the hinge ligament to open the valves automatically. At the top of the shells, where they join together, is a small hump, the **umbo**. It represents the earliest stage of development. The seasonal growth lines seen on the exterior of the clam, for instance, denote the gradual growth and age of the animal. The newest shell is deposited by the mantle along the outer shell margin. Internally, the shell is lined with **nacre**, giving a smooth luster to the shell interior.

Lying against the shell internally is the thin, fleshy **mantle** enclosing a rather large mantle cavity. Flaplike outgrowths of the mantle form pairs of **gills** extending laterally into the cavity. The gills are supplied with cilia which beat to draw in a constant flow of water. Oxygen is drawn from the flow and carbon dioxide is flushed. At the same time food is filtered from the stream of water. The posterior edge of the mantle is modified into two tubelike siphons. One, the **incurrent siphon**, directs a flow of water into the mantle cavity, whereas the **excurrent siphon** channels water to the outside.

8 dorsal plates

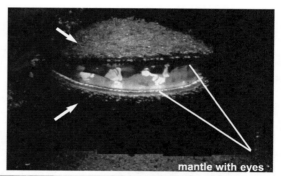

mantle with eyes

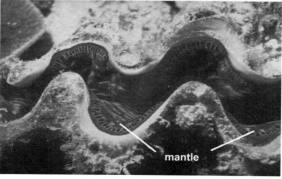

mantle

Figure 9.7. A chiton (above) shows the characteristic series of eight dorsal plates overlying a large muscular foot. The two bivalves to the right illustrate the typical upper and lower valves (arrows) of the class Bivalvia. Note also the mantle lining the shells.

The mantle may also possess sensory organs. **Photoreceptors** occur along the mantle margin, which, in scallops, form fairly well-developed eyes along the mantle edge. Also, the mantle may possess **osphradia**, special sensors to monitor the quality of incoming water. They may detect the presence of silt, toxins, pollutants, or even prowling predators.

A muscular blade-shaped **foot** extends ventrally and anteriorly from the body (Figure 9.5 below). It can be projected some distance into the mud or sand and used to "plow" the animal along or to burrow into the soft substrate. Locomotion is, thus, the role of the foot, although scallops can "clap" their shells together to effect an awkward swimming motion. Above the foot is a large **visceral mass** containing most of the internal organs. Leading into the mass are two pairs of **labial palps,** flaps that conduct food into the mouth.

Bivalves lack any semblance of the rasping radula. Food gathering, instead, occurs as a specialized form of filtration, called *deposit feeding* in which organic material is extracted from ingested sediments. Food passes from the mouth through an esophagus and into the stomach. From there it proceeds through a winding intestine to pass, eventually through a rectum and to exit by way of the anus. Note that the anus empties into the excurrent siphon and as a result digestive wastes are released above the incurrent siphon. This leads to a potential sanitation problem of mixing fecal material with incoming food, a situation we have come to know as *fouling*. A greenish-colored digestive gland is in close association with the stomach and intestine. All of the digestive features, except the labial palps, rectum, and anus, lie within the visceral mass.

Also dispersed within the visceral mass, and surrounding the intestines, are the **gonads**. Most bivalves are *dioecious*, but the sexes are not easily identified; the gonads require microscopic examination to distinguish male from female. Eggs are either shed into the surrounding water or, in some forms, the embryos are brooded within the mantle cavity. Marine bivalves produce small larvae called *trochophores,* while the freshwater forms produce a *glochidium* larva that becomes parasitic on fish gills.

Other organs located within the visceral mass are a pair of cerebral ganglia representing a rather simplified brain, especially important in controlling foot movement. Again, note that bivalves are considered noncephalic animals.

Dorsal to the visceral mass is a membranous sac enclosing the pericardial cavity and **heart**. The heart is a small muscular mass with two vessels (aortas) extending from it. It is rather unremarkable as hearts go, except for one little twist. In fact, perhaps the strangest feature of bivalve anatomy relates to the heart. Upon close examination you will see that, of all things, *the rectum passes through the center of the heart ventricle!* The adaptive significance, if any, of this strange arrangement is unknown. Lying ventral to the heart is a blackish **kidney**, the *metanephridium* used to excrete nitrogenous waste.

Figure 9.8. The internal anatomy of a freshwater mussel, a representative bivalve, is depicted below. The photo to the right shows several clam larvae, called glochidia, which are parasitic on fish gills.

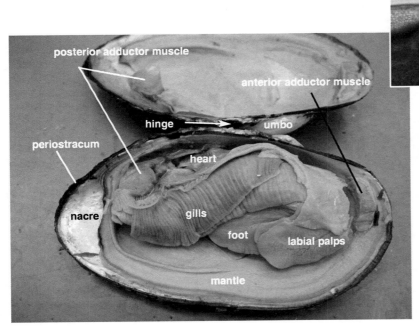

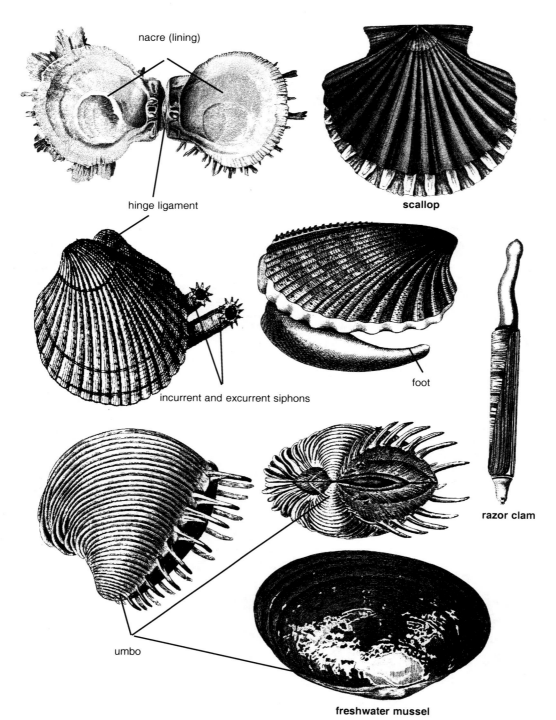

Figure 9.9. An assemblage of bivalves. Notice the location of the hinge ligament that helps to spring open the shells when the powerful adductor muscles are relaxed. Also note the position of the umbo, where the bivalves first started to develop, thus the oldest part of the animal.

Class Gastropoda

With roughly 70,000 species, most of the world's mollusks, by far, are slugs and snails. Some of the most splendid, elaborate, and elegant animals on earth are the gastropods (Figures 9.10 and 9.11 and color plate on page 246). The shapes, versatility, and sculptured designs of their shells, at times, defy description. And even some of the shell-less forms (i.e., nudibranchs) are endowed with their own brilliant colors and conspicuous displays. This is truly a large and colorful group of animals, which includes approximately 80% of all known mollusks.

The typical features of gastropods include (1) a well-developed head with eyes, tentacles, and other sensory structures; (2) a large, ventral foot often supplied with cilia for locomotion; (3) a single, twisted or coiled shell possessing but a single opening; (4) a visceral mass dorsal to the foot and occupying the coiled shell; (5) an aquatic habitat; (6) gill respiration; and (7) a twisting of the larvae resulting in what is known as torsion. For a detailed review of this process see the Inside View 9.1 on page 193.

There are, however, a number of gastropods that do not adhere to those "typical" features. For instance, there are shell-less gastropods, most notably the nudibranchs and slugs (see below). Also, not all gastropods are aquatic. A number of terrestrial slugs and snails exist. In the Pacific Northwest, there is a banana slug, a large, brightly colored animal found in the interior of redwood forests (see D below). As long as it remains moist, the slug can thrive some distance from open water.

In fact, staying moist is crucial for terrestrial mollusks, shelled or unshelled. Some terrestrial snails possess an **operculum**, a calcareous disc that plugs the entrance to the shell, thus limiting water loss. As the animal retreats into the shell passage, the operculum is the last to enter, thus sealing the opening. This is a valuable device to protect the animal from predators as well as to prevent desiccation.

Another variance is the method of respiration. By no means are all snails gill-breathers. There are a group of terrestrial (and a few aquatic) forms that use a saclike extension of the mantle as a lung.

Orange and speckled and fluted nudibranchs slide gracefully over the rocks, their skirts waving like the dresses of Spanish dancers.
- Steinbeck, *Cannery Row*

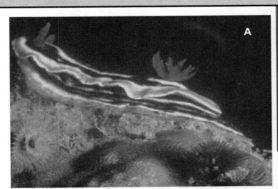

Figure 9.10. A number of shell-less gastropods. **A - C** are marine nudibranchs, among the most colorful of marine animals. **D** is a banana slug, a terrestrial gastropod of the northwestern United States. It is a large bright yellow animal reaching well over 6" in total length.

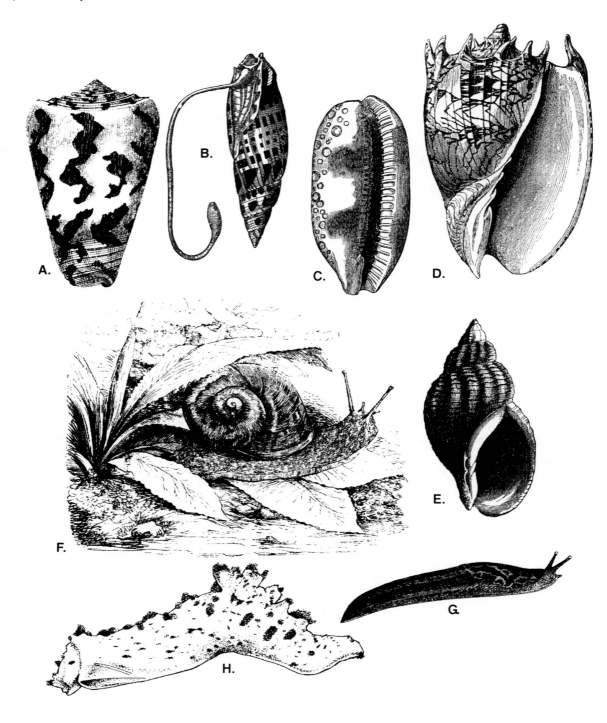

Figure 9.11. A representative collection of gastropods. **A**. tent, **B**. cone, **C**. cowry, **D**. conch, **E**. gosh, **F**. terrestrial snail, **G**. terrestrial slug, and **H**. a nudibranch (sea slug).

Class Cephalopoda

The cephalopods (Figures 9.2 and 9.13) are, to many zoology students, the most fascinating of the mollusks As a matter of fact, there are those of us who would rank the cephalopods among the most provocative of *all* animals. The oscillating octopus evokes a world of mystery— the unknown world of the abyss. Just mention the giant squid, all 60 feet of tentacles, arms, and crunching jaws, or speak of the small blue-ringed octopus, one of the world's most poisonous animals, and most people are captivated, beguiled to the bone. Cephalopods are, without doubt, a group worthy of our attention.

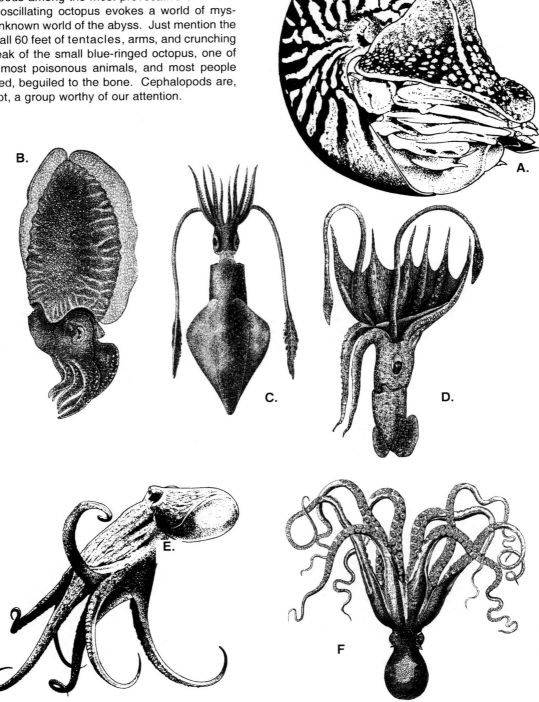

Figure 9.12. Several types of cephalopods. **A**. chambered nautilus, **B**. cuttlefish, **C-D**. squid, and **E-F**. octopuses.

The Ultimate Spineless Predator

The most obvious feature of cephalopods is the vast contrast in structure from that of other mollusks, the chitons, bivalves, and gastropods. They just look so much different than clams, snails, slugs, and nudibranchs, for in cephalopods, we see an enormous revision of the basic molluscan body plan. These are not soft-bodied, slow-moving, mucous-covered globs confined to a heavy shell. These are not filter feeders or algae scrapers. These are not headless, eyeless, defenseless lumps. Not the cephalopods. To the contrary, they are swift, agile, magnificent predators. In fact, their entire body design is but a series of adaptations to a predatory existence.

First of all, they have lost the shell. Well...that's not altogether true. The most primitive cephalopods, the nautiloids, do possess an impressive, multicolored shell (Figure 9.13). The chambered nautilus, for example, is a scavenger of the deep oceans capable of limited predation. Its shell is thin, lightweight, and gas filled to provide maximum buoyancy and a degree of mobility. In comparison to the other cephalopods, however, the nautiloids are slower-moving inhabitants of the deep ocean. They have not nearly reached the level of predation of their more advanced, shell-less cousins, the cuttlefish, squid, or octopus.

However, to say that cuttlefish and squid have no shell is also not altogether true. Granted, they have no visible shell, but buried in their flesh is a remnant. In the cuttlefish, for example, there is a fairly large dorsal, supporting rod, the **cuttlebone**. It consists primarily of calcium carbonate and is gas filled (again, to optimize buoyancy). By the way, this is the same cuttlebone used by caged canaries and parakeets as a convenient calcium source. The squid, like the cuttlefish, has an internal shell vestige, albeit a smaller version, the **pen**. It, too, is a dorsal structure, but much reduced over the cuttlebone. The pen is but a thin, almost transparent, flexible support for the agile squid.

It's up to the pinnacle of molluscan design to lose the shell altogether. The octopus has neither an external nor an internal shell. It has no skeleton, whatsoever. It is the consummate, boneless, shell-less, supple predator able to slide into the smallest of cracks or crevices.

The second important cephalopod adaptation to predation is a shift in the relationship between the foot and the visceral mass. You will recall that in other mollusks, the foot lies ventral to the visceral mass. The ventrally located foot then provides a means to crawl, adhere, or burrow, while the dorsal visceral mass is a hump filled with the body organs. This arrangement works well for a sedentary or slow-moving animal but is entirely insufficient for an animal wishing to become an active predator. Something drastic must change, and it did. In the cephalopods the foot shifts forward while the visceral mass moves posteriorly. The foot is transformed into a number of structures including a tube-like siphon, the arms, tentacles, and head while the visceral mass becomes a posterior sac encasing the internal organs.

So, the cephalopods have discarded the shell and have repositioned the foot to become a series of arms and tentacles located in the front of the animal. To assure that they can latch onto prey, the arms and tentacles are equipped with a series of suckers or hooks. Prey are grasped and then brought to the mouth where a beak, much like a parrot's beak, is present to rip the prey apart. A rasping radula is present just posterior to the beak to assist in further dismantling the prey. Some cephalopods have a toxic saliva that also helps to subdue a struggling victim. This method of food gathering, in which the prey is snatched and brought to the beak, is aptly termed **raptorial feeding.**

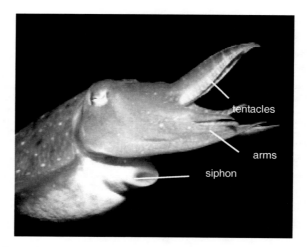

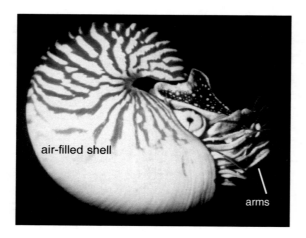

Figure 9.13. A chambered nautilus (left) with prominent external shell and cluster of up to 90 arms. The cuttlefish (above) has an internal shell, the cuttlebone. Note the large siphon used in jet propulsion.

We've now described a fairly formidable predator. But there's still a couple of important ingredients missing: **speed** and **agility**. We know that, among animals, the very best predators are fast and alert. The best predators can seek out, follow, and overcome a retreating quarry. In predation, active pursuit, or a swift lunge from ambush, is the name of the game. And cephalopods have solved this problem in a rather unique way. Rather than twisting the body from side to side or up and down as most predators do to gain swimming speed, they are able to accomplish swiftness in an entirely different manner. They move in a smooth, gliding, explosive fashion through **jet propulsion**. Water is pumped into a large body cavity and then forcibly expelled outward through a funnel, the **siphon**, to provide the force for powerful motion. To learn more about this type of locomotion, see the *Inside View* 9.1 (pg. 193).

You may recall from our earlier discussion that animals have two types of circulatory systems: open and closed. The majority of mollusks, such as the bivalves and gastropods, being slow movers, have opted for the open system. However, an open circulatory system is quite unsuitable for large, fast moving animals. Therefore, it's not at all surprising to discover that cephalopods have a closed system which quickly delivers oxygen and nutrients to individual cells.

But cephalopods, such as squid, have an unusual problem with their closed system. Their heart, called a **systemic heart**, provides blood pressure to send blood throughout the body delivering its life giving cargo. However, by the time blood finally reaches the lateral gills the pressure has dropped considerably. In fact, it is too low to push blood through the mass of blood vessels within the gills. To solve this problem, each gill is supplied with its own heart, a **branchial heart**. These "gill hearts" revitalize the blood pressure thus forcing blood through the gill blood vessels where valuable oxygen is picked up and carbon dioxide is released.

So, speed and agility are necessary components of the very best predators. And we now know that speed in cephalopods is achieved through jet propulsion and a closed circulatory system, but how does it become agile? In other words, how does a cephalopod best exploit its new found speed?

For one thing, an active predator must be able to sense and respond quickly to environmental stimuli. To do that, two things are invaluable: large eyes and a large brain. Cephalopods have both. In fact, and this is important, cephalopods have the largest and most advanced eyes among all of the invertebrates. Likewise, they have, by far, the largest brain of any invertebrate animal. With large, image-forming eyes, they can observe motion and detect shapes. With a large brain and jet propulsion, they can respond accordingly. Finally, add to that, the ability to profit from experience, that is, to learn (studies have indicated that many cephalopods are quick learners), and we have finally arrived at the ultimate invertebrate predator...the cephalopod!

But There's a Catch

There's one problem with all of this restructuring. And it's a relatively big problem. We have just learned that cephalopods sacrificed an external shell for speed. It seems to be a pretty good trade-off. But is it really? True, without a shell these animals have become efficient predators, many would say the ultimate invertebrate predator. But in losing the shell, they've also become vulnerable. They are now potential prey, themselves. There's no protective shell in which to hide. They can't just hunker down and wait out a predator as their relatives the clams or shelled gastropods do.

So how do cephalopods overcome this problem? Agility helps. Speed helps. So does *chicanery*! It's no coincidence that cephalopods have the means to escape predators through deception. After all, they have a large brain, big eyes, and a capacity to learn from experience. And they have something else...**sepia**...ink. By enshrouding a predator within a copious cloud of black ink, sometimes containing a slight sedative, the squid, cuttlefish, or octopus can stealthily make a clean getaway while the predator is confused and distracted.

But trickery involves more than an inky cloud. Cephalopods are also marvels at disappearing. Buried in their skin are special pigmented cells, **chromatophores,** that enable them to change color and patterns. They are masters of camouflage. It's just one more way to avoid being on the wrong end of the predator/prey dining table.

Oh, by the way, they have yet one more way to solve the vulnerability question. Through numbers. The truth is, cephalopods, especially squids, are tasty morsels. Huge numbers end up in the gullets of sharks, fish, whales, and even other cephalopods. Millions are harvested to be transformed into calamari, that rubbery but incredible appetizer. To withstand massive harvesting, squid have become prolific; they produce vast quantities of themselves. It seems that whenever there's a problem, there's a solution, and then another problem, and yet another solution...that's natural selection at its best.

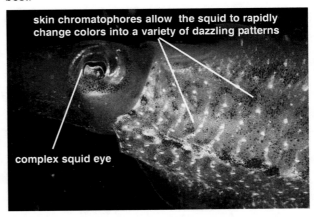

skin chromatophores allow the squid to rapidly change colors into a variety of dazzling patterns

complex squid eye

Figure 9.14. A squid combines a large brain, large eyes, and deceptive practices to become an effective predator and, at the same time, to avoid being a vulnerable prey.

Figure 9.15. A diverse assembly of mollusks. **A.** a terrestrial slug, **B.** another terrestrial slug, the banana slug, **C.** the internal structure of a nautilus shell showing the air-filled chambers that help to keep the animal afloat, **D.** a cone, or driller shell, **E.** a nudibranch showing external gills, **F.** oyster shells, **G - H** two gastropods bearing protective spines, and **I.** a close-up of gastropod coiling.

The labels within the figure read:

tentacles

B

pneumostome
(O_2 entrance for respiration)

gas-filled chambers provide buoyancy

C

D

proboscis

Inside of the probing proboscis is a venomous radular dart that is shot into the flesh of other gastropods (a cone's usual food source) or small fish. Certain species of cones can be dangerous to humans.

gills

tentacles

E

F

G

H

I

An Inside View 9.1

Twisting, Turning, Sliding, and Spurting

A Twisted Tale. If you have ever examined a gastropod shell, you have undoubtedly noticed that, in addition to its elaborate shape and overall beauty, there's a twist. That's right, a twist. In fact, it's a right-handed twist. If you hold the shell so that the pointed end faces you and the lip of the opening is facing downward, you will see that coiling is clockwise and the aperture (opening) is on the right side...it's a "right-handed" shell. So, what's so curious about that? The curiosity lies, not in being right-handed, but, instead, in the fact that about 12% of shells twist in the opposite direction... they're "left-handed." By the way, that's about the same percentage of humans who are left-handed! I'm sure there's no connection; it's just peculiar.

But why the twist in the first place? What purpose, if any, does a twisted shell provide? While we may not know for sure, we can make a pretty good guess. It probably has to do with the weight and bulkiness of the shell, and with balance. The animal's body is wound within the whorls; that is, the internal organs reside up inside of the coils. By winding the shell into a series of spirals, the center of gravity is shifted over the foot so that the animal is better balanced, and thus, better able to get around.

I think another factor in coiling is that it probably allows the animal to grow somewhat larger than otherwise possible. If the shell were uncoiled, the animal would soon reach a size beyond which it would tend to topple over.

Thus, by coiling, gastropods can better balance themselves and, at the same time, continue to grow.

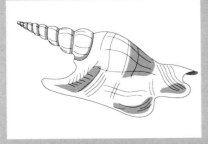

Another Twist. There's another curious phenomenon in gastropods that has to do with twisting. It seems that while still an embryo, the entire body turns on itself, 180 $^{\circ}$, so that what was once in the back is now up front. The mantle cavity, instead of being posterior, now lies over the head; the anus moves to a position over the mouth, and the gills, which were in the back, are now up front. The foot, which was up front, is now shifted posteriorly. Everything seems to become realigned—very confusing!

Furthermore, this embryonic twisting, or **torsion** as it's called, occurs in an anti-clockwise direction and thus has no relationship to the coiling of the shell. So, the question is...why the contortion? What possible advantage does torsion provide? Again, the answer is, we don't know. Perhaps as with coiling, it, too, helps in stability or balance. The shift in foot location seems to support that possibility. Perhaps it aids in respiration or in sensing the environment... we can only guess.

But one thing we do know as a result of torsion, the anus is now directly over the mouth. This relationship between mouth and anus introduces a very real sanitation dilemma, called **fouling**. As fecal material, and perhaps nitrogenous wastes, too, exit the anus, there's the prospect that they will become mixed with incoming food and fouling will occur. That's not good. This means that the animal now has to either tolerate feces mixed with food or...develop strategies to avoid it.

Some gastropods have done just that. They possess special pores through which wastes are removed; others have redirected the flow of water so that mixing doesn't occur, while still others go through a de-torsion during adult life. As I said, torsion is a very curious phenomenon.

A Slippery Progress. The terrestrial snails and slugs face a serious obstacle. If they are to remain truly terrestrial, they must find some way to prevent desiccation, to keep from drying out. Their flesh is bare and if not, in some way, protected will dehydrate causing an early, and I presume, not a very pleasant death.

Gastropods, and many other animals I might add, have solved this puzzle quite well. The answer is **mucus**. Glands on the body surface secrete this slimy, viscous material as a way to prevent losing water...it is a slugs' version of waterproofing. But it also serves as a lubricant. In gastropods, mucus does double duty; it helps to retain moisture and, at the same time, provides a slippery medium. Snails and slugs take advantage of mucus's slimy nature to locomote.

Located on the bottom of their foot are rows of cilia which beat in unison within the mucus sheath thereby effecting forward progress, albeit, a slow progress. So the next time you spot a slug's slimy trail, be duly impressed...it is evidence that slugs, too, can find one solution to two problems.

mucus trail

Spurting Forward. Earlier I described how important agility is to a predacious lifestyle. If a predator cannot capture its prey, it won't be much in the world of predators. I also discussed how cephalopods, by redesigning the body achieved swiftness through jet propulsion.

Now, let's look in greater detail at just how jet propulsion works in cephalopods. We will use the common squid, *Loligo*, as our example. First, some basic anatomy. The "skin" of a squid is really the mantle. It surrounds the body and in doing so creates a space, the mantle cavity. At the anterior end of the mantle is a collar, which surrounds the head. Between the collar and the head is a gap leading into the mantle cavity. On the ventral side of the collar is a funnel-shaped, hollow tube, the **siphon** .

When swimming, a squid always moves in the opposite direction of the water flow. If the funnel is facing forward (its usual position), the animal is propelled backward. For the animal to move forward, the funnel must point backward; to move to the left, the funnel must point to the right, and so on. Attached to the siphon are retractor muscles that move the siphon into its various positions.

The body fins are not normally used in propulsion but rather as stabilizers and balancers. As a result, the squid swims with a rather smooth and gliding action. It doesn't have to flap, wiggle, or contort its body as fish and mammals do. Nor does it have to push its body against the water; it just glides along propelled by a jet stream of water.

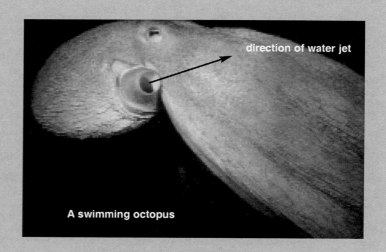

direction of water jet

A swimming octopus

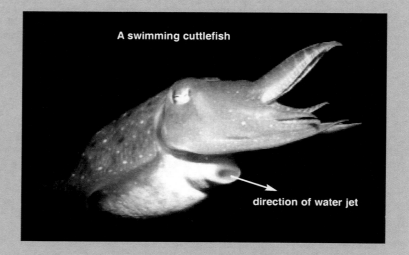

A swimming cuttlefish

direction of water jet

An Inside View 9.2

"A Light That Burns from Within"

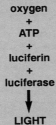

All work requires energy. In fact, the very definition of energy is the ability to do work. Whether it's the car in your garage, the toaster on your kitchen counter, the spider under your porch, or the cell buried deep within your *pectoralis major* muscle, energy is absolutely necessary if you want output. In fact, it's the most essential, indispensable ingredient in the machinery of life. And what do we know about energy?

(1) It cannot be created nor destroyed (at least that's what the physicists tell us); thus, it's not recyclable.

(2) The ultimate source of energy for living systems is the sun.

(3) Photosynthetic plants are needed to transform sunlight into chemical energy for their use and for ours.

(4) Cells, to stay alive, require a constant input of energy.

Oh, there's one more thing that we know about energy. Its use seems to be **inefficient,** in some cases, highly inefficient. Take, for example, the light bulb. Electric energy is needed to produce light from an incandescent bulb, but in so doing, around 98% of the energy input is lost as heat (just touch an angry light bulb and this fact is quickly evident).

Or, consider an entire ecosystem. As energy travels through one trophic level to another, heat is generated—about 90%, to be exact. That means that the producers (plants) and consumers (animals) use only about 10% of the ener-gy available to them. The rest is carried away on the warm breath of Mother Nature.

Even cells are not all that efficient. The average liver cell, nerve cell, or muscle cell rarely exceeds 25% proficiency; the rest (~75% or so) is, again, lost as heat. You only need to cuddle with somebody or touch a person after a long, hard workout to appreciate the radiant factor of metabolism.

But not all living energy systems are so inefficient. There's one process that eclipses this seemingly universal trend of poor energy conversion. And it's a fascinating one indeed.

Bioluminescence it's called...the ability of living organisms to produce light. Its efficiency rating is *over 92%*—one of the most effective energy converters known! In fact, bioluminescence is often termed "cool light" reflecting its efficiency.

As we now understand it, bioluminescence is a chemical process in which dietary energy is transformed into radiant energy... light. To accomplish that, four things are needed: oxygen, ATP (the energy source), luciferin (a phosphorus-containing compound), and the enzyme luciferase. Thus, the formula for bioluminescence goes something like this:

oxygen
+
ATP
+
luciferin
+
luciferase
↓
LIGHT

The types of organisms using bioluminescence are many and they come from a wide variety of groups:

bacteria	**amoeba**
sponges	**snails**
cnidarians	**bivalves**
cephalopods	**insects**
annelids	**fish**
crustaceans	**sharks**

A quick scan of the above list tells us that almost all bioluminescent organisms are aquatic and those that aren't are nocturnal in their terrestrial habits. To better understand the reason for that we need only to look at specific habitats.

First of all, approximately 90% of the earth is water. Of that, around 85% is below one mile, yet light only penetrates to about 600 feet. As a result, the vast majority of habitats on our planet are in total darkness, where no plants can exist.

The mesopelagic zone (below 600 feet, our largest environment) is a dark, uninviting, cold, pressurized, inhospitable place where the only food available to animals is each other or that which drifts down from above. Yet a wide variety of animals live there. And one of the factors that makes them able to do so is their ability to make light. Think about it. There aren't many ways to find food or each other in total darkness. But bioluminescence gives animals that ability.

There are two types of bioluminescence:

Extrinsic Bioluminescence

Many animals do not, or cannot, manufacture their own light and, thus, must "borrow" it from another organism. For instance, the flashlight fish has a pouch below each eye filled with bioluminescent bacteria. These symbiotic bacteria provide an illuminating glow that the fish uses to light up its surroundings.

Likewise, the angler fish has a dorsal spine that hangs in front of its large, viperlike mouth. The spine has a terminal bulb filled with glowing symbiotic bacteria that serve as a lure to attract unsuspecting prey. There's even a squid that has illuminating bacteria in its ink sac. Imagine the surprise of a predator that becomes engulfed in a cloud of shimmering ink!

Intrinsic Bioluminescence

Most bioluminescent animals, however, can produce their own light through special light-producing organs called *photophores*. These special photogenic organs actually produce light through the same mechanism already described (luciferin + luciferase + ATP + O_2 = light).

This light producing organ (the photophore) has a small lens for focusing, a diaphragm-like iris that can change the size of the beam from a wide angle to a narrow pinpoint, and a mirror-like reflector that lines the back of the organ to magnify the light. In other words, the photophore is very much like an eye... *except that it sends light rather than receives it.*

We have discussed the nature of bioluminescence, how it's produced, and what kinds of animals use it. Now let's find out why they do so. Although there may be a number of specific reasons for an animal to use light, there are really only three basic uses: *to find food, to escape becoming someone else's food, and to communicate.*

In the first category, that of food finding or preying on another animal, an individual may simply illuminate its environment to locate a food source. The flashlight fishes do just that with two beacons of light emanating from their cheek pouches to spot food on the ocean floor. On the other hand, bioluminescence may be used *not* to illuminate your food but to attract it to you. Both the angler fish and the viper fish lure prey in this manner. They wiggle a tiny light in front of their gaping jaws and any curious fish that wants to "check out" the strange light will more than likely discover more than he's bargained for.

A large number of animals use bioluminescence in defense. They may startle a potential predator by a sudden flash or they may confuse, or even temporarily blind, a predator by issuing a cloud of biolumines- cence ink as some squid are prone to do.

Another curious practice is seen in some animals that will autotomize a glow- ing body part when harassed. Some annelid (polynoid) worms, for instance, will purposely self-amputate a lumines- cent portion of their body when in the grasp of a pred- ator. This draws attention to the flashing part while the rest of the worm escapes.

One more defensive strate- gy is a type of countershad- ing called *ventral bolumi- nescence*. Located in the brain of some animals are special sensors termed **extraocular photorecep- tors**. These receptors are designed to detect ambient light intensity, that is, the level of light in the animals' immediate surroundings. This information is then passed to ventral pho- tophores, which mimic ambient light, thus, render- ing the animal virtually invis- ible to potential prey lurking beneath.

Here's how this might work. Imagine a prowling shark looking overhead to find its prey by relying upon a dark outline silhouetted against a bright sky. The shark can then ambush the unsuspecting prey from below. However, if the prey is able to match its ventral surface to the surrounding light, its outline will become obscured and difficult, if not impossible, for the shark to see. Even greater camou- flage is accomplished if the light flashes or glitters slightly to match the glim- mer of sunlight on the water's surface.

Although it's hard to over- estimate the importance of using light to escape ene- mies or to locate food, we, nonetheless, see even larg- er numbers of species using bioluminescence, not just for escape or food location, but to communicate. In fact, this may be its most valuable use.

Just recall summer evenings and the flashing of lightning bugs (fireflies) and you will realize that, even on land, bioluminescence has a role to play, and that role is communication.

To an animal that lives its life in darkness, either as a nocturnal terrestrial form or in the eternal blackness of the ocean floor, the ability to pass information quickly and unambiguously is criti- cal. And light does just that.

The communicative func- tions of bioluminescence are many. Here are three of the more important ones.

(1) **Species Recognition.** Many animals have their own specific patterns of bio- luminescence—lights unique to just them. As such they are easily recog- nized by other members of the same species. This can be important in establishing territories or in any type of group behavior.

(2) **Niche Distribution.** If emitting light enables an animal to establish a territo- ry, this in turn, may help to distribute a species within available habitat and thus to prevent overcrowding and the overuse of local resources.

(3) **Reproduction.** But per- haps the most common use of bioluminescence is in the realm of reproduction. A specific pattern, color, or pulsation can identify the sex of the "flasher" while at the same time serve to attract a mate. Furthermore, light of this sort may be a crucial part of the courtship ritual.

Another reproductive use of light is in the glowing eggs of certain fish and polychaete worms which makes them easier for the males to find and fertilize.

Of course, animals do not always subscribe to an either-or agenda in their use of bioluminescence. They, instead, intertwine the vari- ous uses as the need aris- es. An individual animal may, for example, use biolu- minescence to defend itself from a marauding predator one moment only to use it to hunt prey the next. Or, it may flash coded signals to induce group cohesion or to attract a mate.

In the darkness of night or the blackness of the marine abyss, light is employed in a variety of ways. To better understand its role in the life of animals, consider the fol- lowing example:

The little flashlight fish, *Anomalops katoptron*, is a nocturnal species found in the coral reefs of the Pacific Ocean. It has a pouch beneath each eye in which bioluminescent bacteria are embedded. A flap that cov- ers each pouch can be low- ered to reveal glowing bac- teria. In this way the fish uses the light as a beacon to locate food. However, it can also flash messages to other flashlight fish for reproductive purposes or just to help maintain proper distances from each other. In this way, *Anomalops* is using bioluminescence, not just for one function, but to achieve a combination of ends.

> THE RESULT OF THIS SIMPLE CHEMICAL PROCESS CAN BE DAZZLING. LIKE A NEON BILLBOARD ITS CURIOUS FLASHES, THROBBING PULSES, AND BRILLIANT BURSTS OF ORANGE, YELLOW, AND BLUE LIGHT ARE EMITTED FROM A VARIETY OF ORGANISMS AND FOR A VARIETY OF REASONS.

Chapter 9 Summary

Key Points

1. Mollusks are a multivaried assemblage of organisms that have had considerable impact on human development as a source of food, barter, jewelry, and for their aesthetic value.

2. All mollusks share three diagnostic characteristics: a foot, a visceral mass, and a mantle.

3. The foot is invariably involved in locomotion but is modified in the cephalopods into arms and tentacles.

4. The visceral mass contains the internal organs of digestion, reproduction, excretion, and circulation.

5. The mantle is a uniquely molluscan feature. It secretes the shell, acts as a respiratory surface, wafts water into the mantle cavity of bivalves, creates a brood chamber to house developing young, and possesses a number of sensory structures.

6. Most, but not all, mollusks bear an outer shell. At times, such as in the gastropods and bivalves, the shell can be large and heavy. In other forms, the shell is reduced to a series of external plates (chitons) or to internal supporting rods as in the cuttlebone of cuttlefish or the pen of cephalopods. All vestiges of a shell are missing from octopods.

7. Many mollusks have a tongue-like rasping structure, the radula. With such a device, the chitons, gastropods, and cephalopods can scrape food from hard surfaces (snails) or use it as a tool to shred flesh as in the cephalopods. The radula is absent from bivalves.

8. Mollusks are the first group to bring to the animal world a circulatory system. Theirs is an open system in which blood enters open spaces (sinuses) to bathe various tissues.

9. Mollusks are also the first group to introduce true respiratory structures. Most make use of gills, while a few terrestrial forms use a lunglike sac for gaseous exchange.

10. There are six requirements for any respiratory surface. It must be thin, moist, well ventilated, large, patent, and associated with the circulatory system.

11. Mollusks are spread among four major classes. The polyplacophorans are the primitive chitons. Those with two shells (valves) belong to the class Bivalvia while the largest class is the Gastropoda incorporating the snails, limpets, slugs, nudibranchs, and conchs. Finally, the cephalopods (squid, nautilus, cuttlefish, and octopods) are designed for an active predatory existence.

12. Among other characteristics, the majority of gastropods have a coiled shell to provide a compact, balanced body design.

13. Some gastropods display torsion in which the embryos become twisted about their body axis.

14. Cephalopods show a number of adaptations to predation: streamlined body, suctorial tentacles, raptorial feeding, and hydraulic jet propulsion.

15. Bioluminescence is an efficient form of energy use and is common in mollusks. It comes in two varieties: intrinsic and extrinsic, and serves a number of functions from food gathering to communication.

Key Terms

branchial hearts
closed circulatory system
coiling
cuttlebone
deposit feeding
excurrent siphon
foot
glochidia larva
hemocoel
hinge ligament
incurrent siphon
jet propulsion
mantle
metanephridium
nacre
open circulatory system
operculum
osphradia
pen
periostracum
prismatic layer
radula
raptorial feeding
sepia
siphon
systemic heart
torsion
trochophore larva
umbo
visceral mass

Points to Ponder

1. Describe locomotion in a squid.

2. What role do cilia play in clam feeding?

3. Describe how a squid feeds. What is that type of feeding called?

4. Describe the action of a radula in a snail and in a squid.

5. What role do osphradial play in the life of a clam?

6. Describe how the valves (shells) of a clam open and close.

7. Give an example of convergent (parallel) evolution between a squid and a human.

8. What is the function of sepia in a squid or octopus?

9. What role do chromotophores play in the life of a squid or octopus?

10. Define coiling in a gastropod. What is its significance?

11. How is a pearl formed?

12. Name several ways that mollusks are important to humans.

13. What is the function of a squid's branchial hearts?

14. Which molluscans are cephalic? Which ones aren't?

15. Give 3 ways that cephalopods differ from the other molluscan classes.

16. What is needed in order for a surface to be effective as a respiratory device?

17. What is a glochidium?

18. Define bioluminescence and describe how it is produced.

19. What are the two types of bioluminescence?

20. What are the functions of producing light and what kinds of animals use it?

Inside Readings

Brusca, R. C., and G. J. Brusca. 1990. Invertebrates. Sunderland, MA: Sinauer Associates.

Fiorito, G., and P. Scotto. 1992. Observational learning in Octopus vulgaris. Science 256: 545-547.

Gosline, J. M., and M. E. DeMont. 1985. Jet-propelled swimming in squids. Sci. Am. 252(1):96-103.

Gould, S. J. 1985. The Flamingo's Smile. New York: W. W. Norton.

Hyman, L. H. 1967. The invertebrates, Vol. 6 Mollusca. New York: McGraw Hill.

Linsley, R. M. 1978. Shell form and evolution of gastropods. Am. Sci. 66:432-441.

Ludyanskiy, M., et al. 1993. Impact of the zebra mussel, a bivalve invader. BioScience 43:533-545.

Morton, J. E. 1979. Mollusks. London: Hutchinson University Press.

Pechenik, J. A. 1996. Biology of the Invertebrates. Dubuque, IA: Wm. C. Brown.

Vermeij, G. 1993. A Natural History of Shells. Princeton: Princeton University Press.

Ward, P. D. 1987. The Natural History of Nautilus. Boston: Allen and Unwin.

Chapter 9

Laboratory Exercise

Phylum Mollusca

Laboratory Objectives

✓ to identify the distinguishing characteristics of the phylum Mollusca,
✓ to identify various specimens of mollusks on display,
✓ to give the taxonomic classification (i.e., phylum and class) of those mollusks studied,
✓ to identify external and internal anatomical features of the mussel,
✓ to identify the external and internal features of the common squid, *Loligo*, and
✓ to characterize the functional anatomy of the class Cephalopoda

Supplies needed:

✓ dissection microscope
✓ compound microscope
✓ prepared specimens of various mollusks
✓ freshwater mussels for dissection
✓ preserved squid for dissection

Chapter 9
Worksheet

Mollusca 1

I. Molluscan Survey: Display Specimens

There are several examples of mollusks on display. Study those characteristics that typify each class.

Class Polyplacophora (chitons): eight overlapping dorsal plates; large, ventral disc shaped foot, slightly cephalic

Class Gastropoda (snails, slugs, limpets, nudibranchs): either a one-piece, coiled shell; or shell-less; cephalic

Class Bivalvia (clams, oysters, scallops): two valves, no radula, a ventral foot and dorsal visceral mass; weakly cephalic

Class Cephalopoda (squid, cuttlefish, octopus, nautilus): foot modified into tentacles; shell reduced or absent; highly cephalic

II. Freshwater Mussel

Mussel: External Anatomy
Follow the dissection instructions given below to identify the highlighted features.

1. Examine your specimen's exterior (Figure 9.16 and below). Note first that there are **two valves**, or shells, held together at the top. Near the area where they attach is an elevated region, the **umbo**, which represents the very first shell of this animal. As the clam grew, it deposited new shell along the outer margins; thus, the most recent shell is along the leading or outer edge while the oldest is represented by the umbo. Examine the contour of each shell. You will note a series of **growth lines,** each indicating one season's growth. Therefore, by counting the lines you can establish the relative age of your specimen.

2. Note that the shell is covered with a greenish-blackish material that easily flakes off. This is an organic layer, the **periostracum.** It is thought to provide some protection for the animal in an acidic environment.

3. At the dorsum of the mussel, near the umbo, is the region of valve attachment. The two shells are held together by a **hinge ligament,** a springlike device that causes the shell to open automatically unless held shut by the two **adductor muscles.** You will need to cut these muscles in order to open the clam.

4. Finally, orient your specimen with the umbo facing you and the smaller blade of the shell to your right, as in the diagram below. The anterior area of the animal is now to your right, and the posterior region is to the left.

POSTERIOR

growth lines

ANTERIOR

umbo (oldest area)

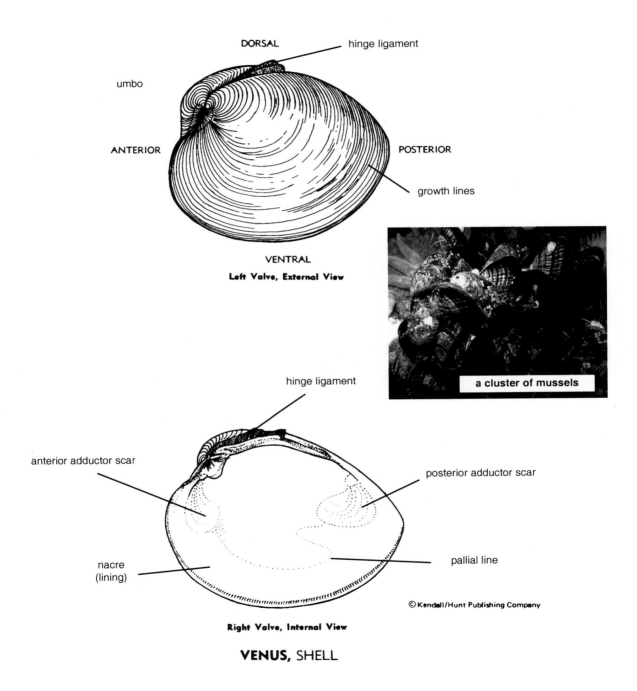

DORSAL — hinge ligament

umbo

ANTERIOR

POSTERIOR

growth lines

VENTRAL

Left Valve, External View

a cluster of mussels

hinge ligament

anterior adductor scar

posterior adductor scar

nacre
(lining)

pallial line

© Kendall/Hunt Publishing Company

Right Valve, Internal View

VENUS, SHELL

Figure 9.16. External and internal view of a clam shell.

Mussel: Internal Anatomy

Follow the directions given below and use diagrams and photos in the lab manual to locate the highlighted items.

1. Open your specimen. The easiest way to do that is to first study a diagram or photograph. Note that there are two large muscles, the **anterior** and **posterior adductor muscles,** that hold the shells closed but do not, by the way, open them. You will need to cut both of these muscles. To do so, slip a scalpel blade between the valves and slice through both the anterior and posterior muscles. When you do that the shells will open easily. Be careful not to twist the blade as it may break.

2. Try, for the time being, to keep the tissues in their natural location. Examine the membranous **mantle** lining each shell. Note that it is attached to the shell along its margin, about 1/2 inch from the edge. This attachment zone is the **pallial line** and is where new shell is deposited as the animal grows.

3. Next, examine the interior of the shell. The lustrous lining is **nacre,** also called *mother-of-pearl.* It is used to make jewelry and in the past was a major source of buttons as well as common currency (money). Also, note that attached to each shell is the remnant of the two adductor muscles that you cut.

4. Now examine the remaining anatomy of the mussel. Locate the **gills,** as large serrated flaps growing from the dorsal part of the mantle. Then find the large ventral, blade shaped **foot.** Note the direction it is pointing (is it anterior or posterior?).

5. The area of the mussel above the foot is the **visceral mass.** It contains some of the major body organs. You will dissect this later. But first locate, along the anterior area of the visceral mass, two fleshy flaps. These are the **labial palps.** Between the palps is the mouth. By using a probe you can determine where the mouth is located.

6. Examine the area where the two valves are attached (i.e., the **hinge ligament**). Just below the hinge is a thin sac that contains the heart. Do not open this yet; you will do so later.

7. Now open the visceral mass. To do so, grasp the foot and slice into the mass as though you were slicing a bagel. Ask your instructor if you have a question about how to do this. Once the visceral mass is properly opened, the internal organs are exposed. You should see the **intestine** as a tube winding through the mass. Trace it toward the palps and note a slight enlargement; this is the **stomach.** Surrounding the stomach is a greenish mass, the **digestive gland,** which assists the animal in food digestion. Enclosing the intestine throughout the visceral mass are the **gonads.** Although the mussel is dioecious, you cannot differentiate between the sexes without experience and the aid of a microscope.

8. The final step in the dissection is to expose the **heart** and **kidney.** To do that, return to the dorsal area where the two valves are attached. Locate the thin **pericardial sac** that contains the heart and open it carefully. A sharp pin probably works better here than a scalpel. Examine the small muscular heart externally. Note the openings on the heart surface leading to the pericardial sinus. Then carefully cut open the heart to reveal, of all things, the **rectum** inside! Can you think of a reason that the rectum passes through the heart?

9. Finally, examine the blackish area ventral to the heart. This is the **kidney.**

You need to be able to locate and give functions of the following mussel features:

✓ umbo	✓ visceral mass
✓ growth lines	✓ stomach
✓ anterior and posterior regions	✓ intestines
✓ periostracum	✓ gonads
✓ hinge ligament	✓ digestive gland
✓ nacre	✓ heart
✓ mantle	✓ rectum
✓ foot	✓ kidney
✓ gills	✓ labial palps
✓ pallial line	✓ posterior and anterior adductor muscles

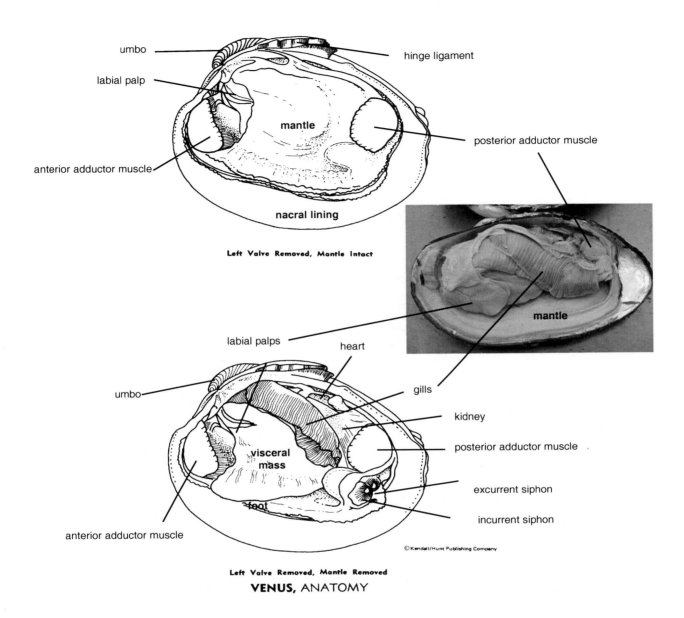

Left Valve Removed, Mantle Intact

Left Valve Removed, Mantle Removed

VENUS, ANATOMY

© Kendall/Hunt Publishing Company

Figure 9.18. Internal anatomy of a clam.

1. anterior adductor muscle
2. posterior adductor muscle
3. gill
4. mantle
5. foot
6. visceral mass
7. labial palps
8. heart within pericardial sac
9. hinge ligament
10. umbo

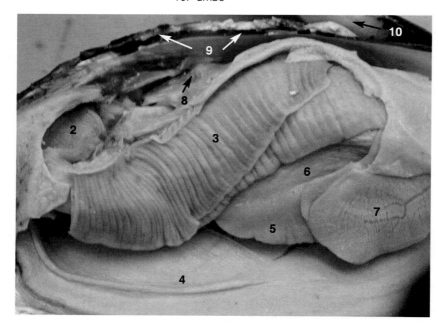

→ ANTERIOR

Figure 9.17. Upper photograph is of a freshwater mussel with its right valve removed. The lower photo is with the visceral mass dissected to show internal features. The animal's anterior is to the right.

1. anterior adductor muscle
2. posterior adductor muscle
3. gills
4. mantle
5. foot
6. digestive gland
7. stomach
8. intestines
9. gonads
10. heart with interior rectum
11. kidney
12. umbo
13. hinge ligament

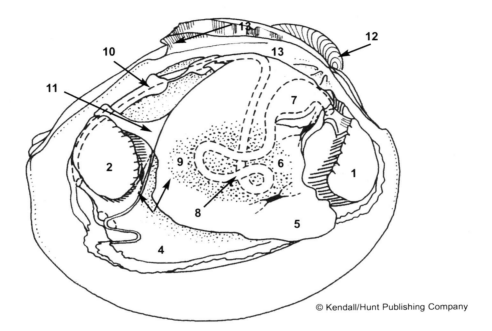

© Kendall/Hunt Publishing Company

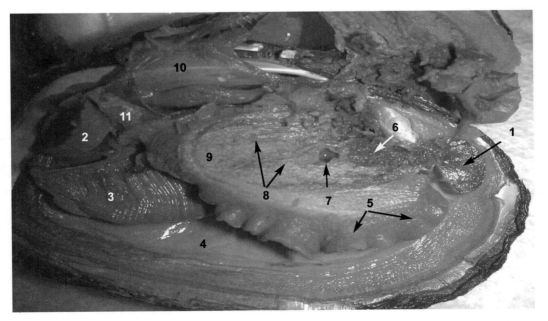

Figure 9.19. Internal anatomy of a clam.

Chapter 9
Worksheet

(also on page 595)

Mollusca ll

***Loligo* (squid) Dissection**:

Squid: External Anatomy
(Figures 9.20–9.24)

1. Examine the head of a preserved squid and note the large **eyes** (you will dissect one later), the anterior **arms** and **tentacles**. How many are there of each? Closely examine an arm and a tentacle noting the **suckers** used to grasp prey. View a sucker under a dissecting microscope. Pull the arms apart to locate the **mouth** surrounded by the **periostomial membrane**. You will dissect the contents of the mouth area later.

> Describe how a squid feeds:
>
> 8 arms, 2 tentacles; uses its arms and tentacles to attack the prey and bring the prey to his mouth

2. Now examine the rest of the exterior. Note that the "skin" of the squid is really the **mantle**. The squid has two large lateral **fins** along its posterior border.

What is the function of the fins?   to balance and stabilation

3. Now find the anterior **collar,** which encloses the internal organs. Place a finger (or a probe) between the mantle and the internal organs into the **mantle cavity**. Now locate the funnel-shaped **siphon**.

> Describe the role of the siphon, the mantle cavity, and the collar in squid locomotion. Explain how a squid makes a **right** turn.
>
> The role of the siphon is to bring in water movement like the jet machine.
> mantle cavity - water goes through in the gills
> collar - closes the mantle cavity
> siphon - Release the water.
> It will close off one side of the collar and then get the intake of water in and get out through siphon
> move the siphon to left

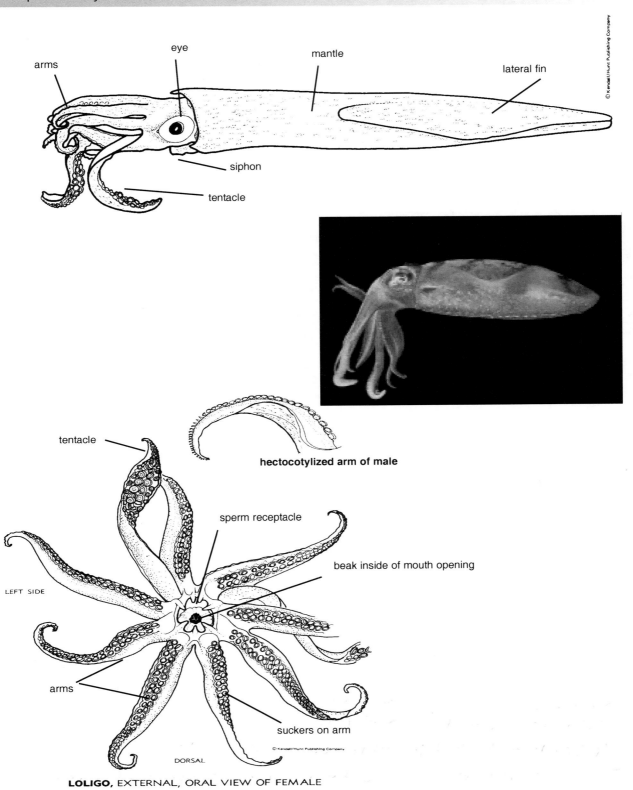

LEFT SIDE

DORSAL

LOLIGO, EXTERNAL, ORAL VIEW OF FEMALE

Figure 9.20. External features of *Loligo*, the Pacific Squid. The photograph above is a lateral view of a squid showing arms and tentacles.

Squid: Internal Anatomy
(Figures 9.21–9.24)

1. Place the squid so that its ventral side is facing you. The siphon will be exposed in this position. Use scissors or a scalpel to open the specimen along the ventral midline. Do not cut too deeply or you will destroy underlying tissue. Cut all the way to the posterior end of the animal. Now make one or two lateral cuts on each side of the animal in the region of the fins and away from the midline. This will allow you to pin back the mantle easily to expose the internal organs.

2. Do not cut into the organs until instructed to do so. You should at this point be able to locate the **pen** as a reddish colored supporting rod lying just beneath the dorsal mantle.

3. First determine if your specimen is a *male* or *female*. If it is a female you will see two large glands located in the midregion of the specimen. These are **nidamental glands** used to produce a gelatinous coating for the eggs. Also, if the specimen is a female you should see masses of small **eggs** in the **ovaries** located in the posterior region of the animal. Finally, to complete the female reproductive system, locate the **oviduct** on the animal's left side. Note its opening through which eggs pass on their way to the mantle cavity.

4. If your specimen is a mature male, you should first locate the small tubular **penis** on the animal's left side. Then find the centrally positioned **testes** and bulb-like **spermatophoric gland** situated on the animals' left side. Sperm are packaged into bundles and stored in the spermatophoric gland until passed to the female when mating. The male uses a modified (hectocotylized) arm for this purpose. Although the male does have a penis, it is far too small to be used as a true copulatory organ.

5. Locate the **ink sac** lying beneath the **rectum** (actually it's dorsal to the rectum, since your animal is on its back). The ink sac may be black or somewhat metallic in color. Be careful not to rupture the sac as the ink (sepia) can be rather messy. Note that both the **anus** and ink sac empty into the siphon.

> Explain why the ink sac and rectum both empty into the siphon:
>
> *To shoot it out of the system*

6. Next locate the large, straplike **retractor muscles**, one attached to each side of the siphon.

> What do you think is the function of the siphon retractor muscles?
>
> *To open and close off the siphon; use to monitor the directions*

7. Examine the large, lobed **gills** located on each side of the animal. Note their large surface area. Why are they so big? *So they can absorb more water, and do the gas exchange*

Process oxygen fast to proc so more surface area   *use the O₂*

8. Locate the **systemic heart.** To do that, you will need to carefully remove overlying membranous tissue between the base of the gills. The heart is easy to recognize as a fairly large, triangular, muscular mass with several vessels emanating from it. Now locate the two **branchial hearts** situated at the base of each gill.

> Explain the function of the systemic and branchial hearts.
>
> *Systemic heart provide blood pressure to send blood through out the system*
> *Branchial heart + Heart of the gills that force blood through blood vessels forcing O_2 intake, CO_2 release*

9. The squid's **kidney** can be located as a rather diffuse whitish mass located just anterior to the systemic heart.

10. There are three more internal items you need to examine, all associated with the digestive system. First, find the very large **digestive gland** located beneath the ink sac and retractor muscles. Then locate the **stomach**. This is a little tricky since you will have to dig around somewhat to expose it. It is thumb shaped and located on the animal's right side just posterior to the gills. It's usually about the size of your little finger, but may be larger if the squid has recently fed. If food is present, you may want to open the stomach to determine its contents. The final internal structure to locate is an extension of the stomach, the **cecum**. The squid doesn't have a true intestine, so the cecum is used as a large auxiliary sac for digestion. It is often rather gel-like in texture.

  * Quite often there are tapeworms in the squid caecum. Search your specimen for the white, ~1/4-inch-long parasites.

11. Now you will dissect the head contents. To do so, place the squid so the dorsal surface faces you. Then make a sagittal incision through the base of the arms, between the eyes, to expose a round muscular body about the size of your thumb. This is the **buccal bulb.** Notice a tube, the **esophagus**, leading posteriorly from the bulb to the stomach.

12. Open the bulb with a scalpel. Inside locate the large parrotlike **beak**. Pry open the beak to expose the tongue shaped **radula** inside equipped with bristlelike teeth.

> Why is the buccal bulb so large and tough?
>
> *So it can intake large and tough prey*

13. Next dissect one of the eyes. Be careful when you open it as fluid may squirt out. The squid eye is constructed very much like yours; in fact, the squid eye and the human eye are classic examples of *parallel evolution.* There is an opening, the **pupil,** that allows light to enter the eye proper. Light passes through the round, marblelike **lens** for focusing and then strikes the **retina**, a dark brown to blackish layer that transforms light into the visual signal sent to the brain. If cow eyes are available you may wish to dissect one and compare its anatomy to the squid eye. See page 214 for instructions on how to dissect a mammalian eye.

14. Your final dissection is to expose the **brain**. Continue the cut posteriorly that you made to dissect the buccal bulb. Cut deeply between the eyes until you see a large white mass, which is the brain. Squids have the largest eyes and brain of any invertebrate animal. Next locate the pair of stellate ganglia (fig. 9.21). Just pull the body mass to one side which will expose these star-shaped nervous structures. They function as relay centers to control the powerful mantle muscles responsible for jet propulsion.

> Can you think of any relationship between predation and the squids' tentacles, siphon, eyes, and brain?
>
> *Tentacles allow them to reach out the prey, Siphon allows them to go bast, eyes help to see the predator and brain helps to think*

You need to be able to locate and give functions of the following squid features:

✓ arms	✓ tentacles	✓ fins
✓ mouth	✓ ~~periostomial membrane~~	✓ ink sac
✓ mantle	✓ mantle cavity	~~✓ kidney~~
✓ collar	✓ siphon	✓ siphon retractor muscles
~~✓ nidamental glands~~	✓ eggs	✓ ovaries
~~✓ oviduct~~	✓ penis	✓ testes
~~✓ spermatophoric gland~~	✓ buccal bulb	✓ beak
✓ radula	✓ esophagus	✓ stomach
✓ cecum	✓ rectum	✓ anus
~~✓ digestive gland~~	✓ gills	✓ systemic heart
✓ branchial hearts	~~✓ brain~~	~~✓ stellate ganglia~~
✓ eyes (lens, pupil, retina)	✓ pen	

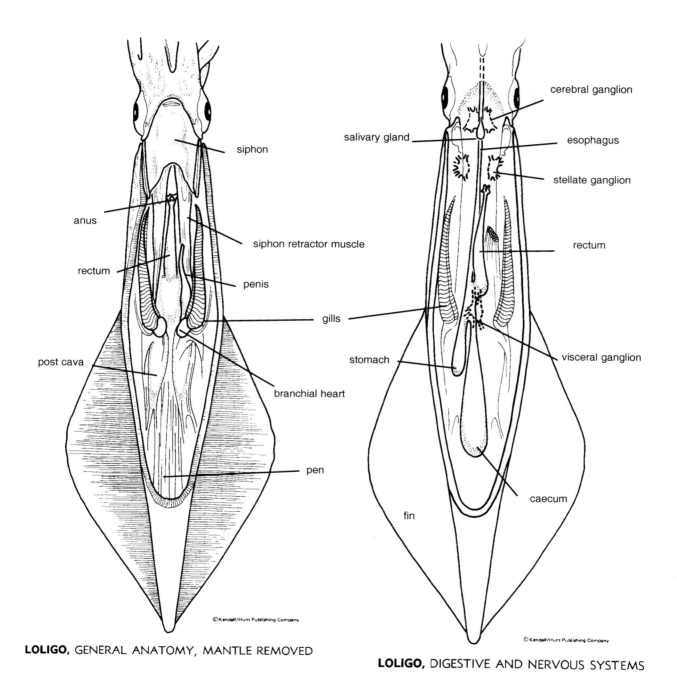

LOLIGO, GENERAL ANATOMY, MANTLE REMOVED

LOLIGO, DIGESTIVE AND NERVOUS SYSTEMS

Figure 9.21. Internal anatomy of a squid.

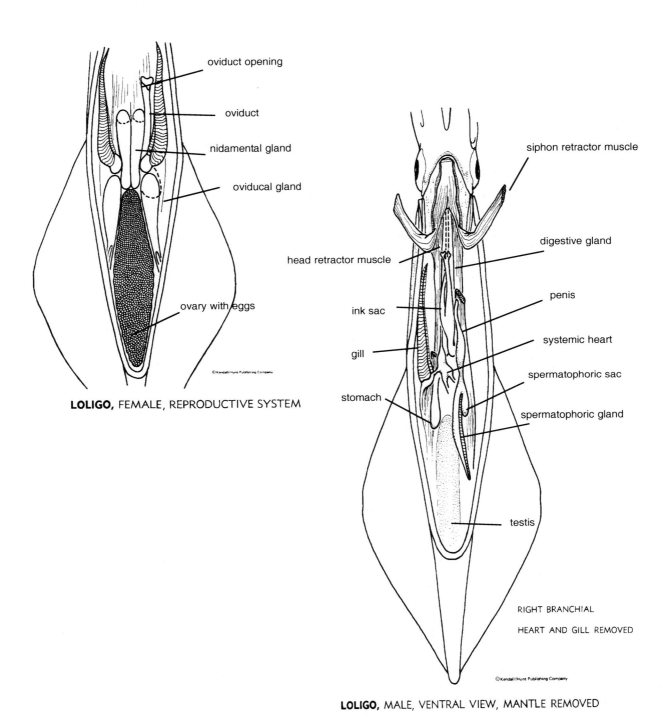

LOLIGO, FEMALE, REPRODUCTIVE SYSTEM

LOLIGO, MALE, VENTRAL VIEW, MANTLE REMOVED

Figure 9.22. Internal anatomy of a squid showing the female reproductive system.

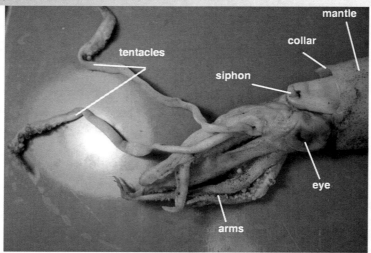

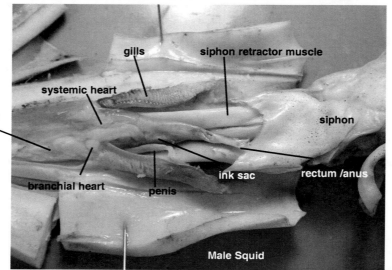

Male Squid

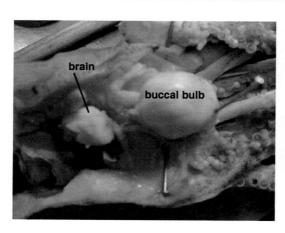

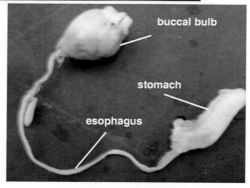

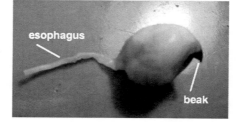

Figure 9.23. External and internal anatomy of the squid.

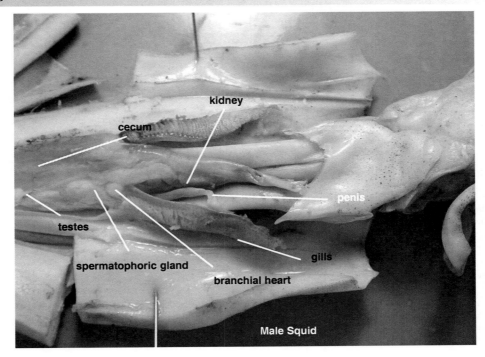

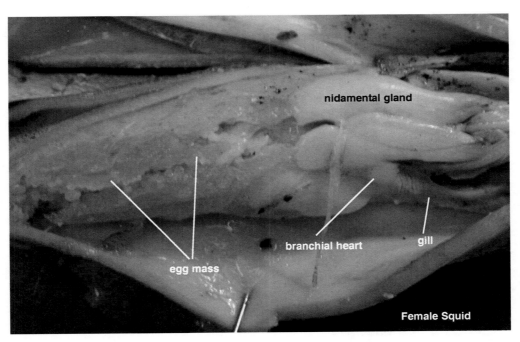

Figure 9.24. Internal anatomy of a male (top) and female (bottom) squid.

Mammalian Eye

Cow's Eye Dissection

A. External Anatomy

1. Examine the exterior of a cow's eye. Note the large amount of **adipose tissue** (fat) on its surface. Fat is important as it provides important padding to protect the eye from impact injuries.

2. Exiting from the rear of the eye is the **optic nerve.** You will find the nerve as a whitish, rather round, and hard cablelike structure. The optic nerve consists of over a million individual axons that carry visual information from the eye to the brain.

3. There are six muscles, known as **extraocular muscles,** that move the eye. These are seen as a number of reddish slips that attach at various points on the eye surface.

4. Covering the front of the eye is the **cornea.** In life it is a clear layer; however, in your specimen it will appear somewhat opaque due to the preservative. At the center of the cornea is an opening, the **pupil.** Light passes through the cornea and pupil to enter the interior of the eye. The colored ring around the pupil is the **iris,** which can dilate or contract to control the amount of light entering the interior.

B. Internal Anatomy

1. To enter the eye, you will need to cut through the tough outer coat, the **sclera,** the "white" of the eye. This is best done with scissors. However, be careful not to squeeze the eye as you cut into it so that you don't squirt the internal fluid onto yourself.

iris   pupil

A.

2. Once the eye is opened (see diagrams **B** and **C** below), you can begin to locate the following important internal features:

  a. The **lens** is a hard, roundish object just behind the pupil. It's held in place by microscopic, threadlike **suspensory ligaments**.

  b. Surrounding the lens is a ring of dark tissue, the **ciliary body.** This exerts force on the lens (by way of the suspensory ligaments) to change the shape of the lens during focusing. As a person ages, the ability of the lens to alter its shape diminishes, and as a result, eyesight declines.

  c. Within the "eyeball" is a mass of gelatinous material, the **vitreous humor** which helps to maintain overall eye shape.

  d. Lining the inside of the eye is a thin, dark layer, the **retina.** It consists of several layers of nerve cells that collectively receive, transform, and send light signals to the **optic nerve** that exits the back of the eye.

  e. You will note that beneath the retina is a shiny, opalescent layer. This is the **tapetum lucidim,** which allows a greater level of night vision in nocturnal animals. The tapetum causes the familiar reflective "eyeshine" from the eye of nocturnal animals; *it is absent from a human eye.*

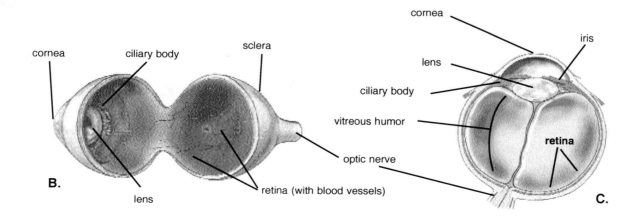

cornea   ciliary body   sclera   cornea   iris   lens   ciliary body   vitreous humor   optic nerve   retina

B.   lens   retina (with blood vessels)   **C.**

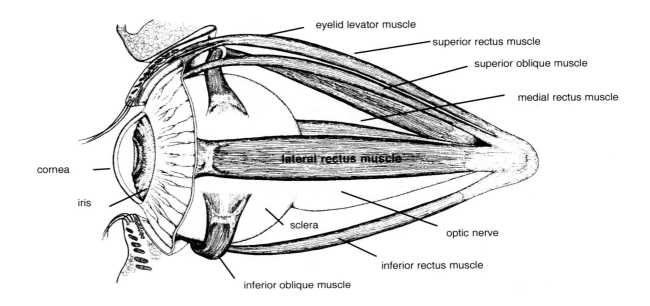

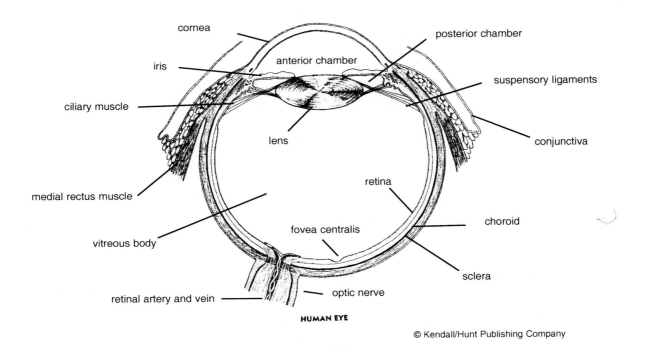

HUMAN EYE

Figure 9.25. The external and internal anatomy of the human eye.

10 Chapter 10
PHYLUM ANNELIDA

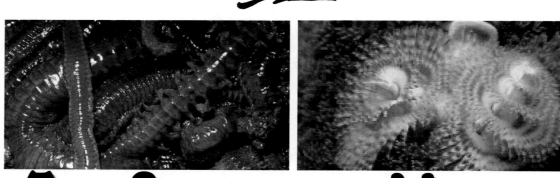

THE SEGMENTED WORMS

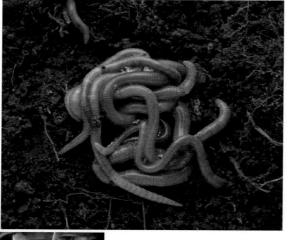

From earthworms to the strange Christmas tree worm (upper left) to the even stranger fan worms (lower left), all annelids share the same basic feature... segmentation.

PHYLUM ANNELIDA

The Segmented Worms

The plow is one of the most ancient and most valuable of man's inventions; but long before he existed the land was, in fact, regularly plowed, and still continues to be thus plowed...by earthworms. It may be doubted that there are any other animals which have played so important a part in the history of the world, as have these lowly organized creatures.

- Charles Darwin

Introduction:
Darwin, the earthworm's champion

*I*n addition to his notoriety over the theory of evolution, Charles Darwin was also well known as an expert on annelids, especially earthworms. But his assertion, in the above quotation, that earthworms are man's most important ally is a little tough to take. We all know that a dog is man's best friend...but then, ever since our study of parasites, that, too, may come into serious question. However, I'm not sure that I'm quite ready to replace Hunter with a squirmy, mucus-covered, unthinking, unknowing, miniature earth plow!

On second thought, I guess we shouldn't be too hasty here. Perhaps Darwin had a point. What is it about annelids that had evolution's architect so captivated?

Metamerism . A simple glance at any member of this phylum, say an earthworm or a leech, will reveal its singular most important feature: annelids are **segmented**, or in the jargon of biologists, they are metameric. The animal is divided, longitudinally, into a series of identical rings. In fact, the word annelid refers to "little rings" and each ring is termed a **metamere**.

Segmentation is such an important contribution that virtually all higher animals take advantage of it. For instance, the two most successful animal phyla, the arthropods and the chordates, are segmented, although each probably derived it separately. We are quite certain that arthropods evolved from annelids, and thus received segmentation directly from them. Chordates, on the other hand, being far removed from the annelids, most likely developed metamerism on their own. Nonetheless, you can see evidence of it in both phyla: in the segmented body of a caterpillar, the abdomen of a grasshopper or crayfish, in the arrangement of their appendages, in

the repetition of fish ribs and snake vertebrae, and in the serial organization of human nervous and muscular systems (e.g., the "six pack" pattern of the abdominal muscles of weight lifters).

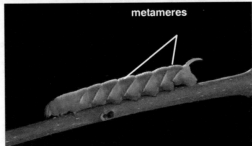

Figure 10.1. Metamerism, otherwise known as segmentation, occurs first in annelids as in the earthworm (top photo) but is repeated throughout higher animals as in the caterpillar as well as in the chordates (fish, reptiles, birds, mammals).

The Significance of Metamerism

As you might guess, since segmentation is so prominent, it must confer some advantage to an animal. It does. As a matter of fact, it bestows several benefits.

First of all, each segment is isolated, internally, from its neighbors by an encircling, membranous septum, resulting in a series of individual compartments. Each compartment, filled with its own coelomic fluid, then becomes a distinct hydrostatic skeleton. As a result, muscular contractions of the body wall allow one or more segments to become rigid. This rigidity is a great aid in locomotion. It permits an earthworm, for example, to push forcibly as it burrows through relatively hard soil, a process called peristalsis. As one part of the body is anchored in place by special spines, called **setae**, the rigid segment plows through the soil. Crawling and swimming are accomplished in a similar manner.

A second advantage of metamerism is also related to compartmentalization. Because segments are separated from each other, injury to one segment does not necessarily affect others. Thus, metamerism serves to **lessen the impact of a possibly debilitating injury**.

Segmentation also permits what is generally called **regional specialization**. Adjoining metameres can fuse together to form a functional unit. For example, local segments may unite to form a head to bear sensory structures, or they may form a functional thorax to bear appendages, or an abdomen specialized to house internal organs. Specialization is exhibited by insects and other arthropods as seen below in Figure 10.2

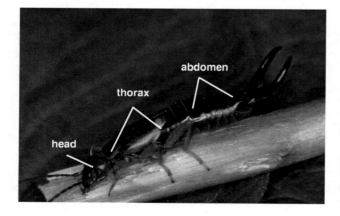

Figure 10.2. An earwig showing the basic body plan of an arthropod. Here there are three specialized regions, the head, thorax, and abdomen.

Beyond Metamerism

Segmentation is certainly the most pronounced feature of the annelid body plan, but not the only one.

External Anatomy

The most anterior segment of the annelid body is the **prostomium**. It's a fleshy knob that overhangs the mouth, and internally contains the brain and a variety of sensory structures (eyes, palps, and tentacles), depending upon the animal in question. At the other end of the animal is the **pygidium**, another fleshy knob, which bears the anus (Figure 10.8). Between these two extremes are the numerous trunk metameres. Each trunk segment may bear a pair of oarlike lateral appendages, as in the polychaete **parapodia** (Figure 10.5), or they may lack appendages altogether, as in earthworms and leeches. Furthermore, several trunk segments may fuse into a glandular band-like structure encircling the animal as in the **clitellum** of earthworms.

Finally, the majority of annelids possess **setae**, bristles or spinelike outgrowths of the body wall. Although the number of setae may vary from one annelid to another (e.g., there are none in leeches, a few in earthworms, and are numerous in the sandworms), their function is always *to provide traction* while the animal is burrowing or crawling. They may also anchor the animal within its burrow to impede extraction by a predator such as when a robin plays "tug-of-war" with an unwilling earthworm.

Nervous System

Without doubt, annelids are cephalic, although head formation is more pronounced in some species than in others. Located toward the front of each animal and just above or the pharynx, are a pair of bulbs, the **suprapharyngeal ganglia**. This primitive "brain" receives incoming messages from sensory structures in the head as well as signals from the posterior body. The ganglia are connected to an elongated ventral nerve cord which runs below the digestive tube. The nerve cord, in turn, has swellings at each body segment from which pairs of small lateral nerves branch to supply the body wall and other structures.

Sensory structures are much more advanced in the bristleworms than in other annelid groups. The sandworm *Nereis*, for example, has a number of well-developed **eyes** that are bilaterally arranged in pairs on the head to detect light intensity and direction, motion, and may very well be able to form crude images. **Touch** (tactile) and **chemo-receptor** are located on tentacles and palps that extend from the sandworm's head, as are **statocysts** for spatial orientation. Other sensory structures, such as photoreceptors, thermoreceptors, and even pH sensors, are found throughout the phylum, often scattered along the animal's general body surface.

There are two annelid lifestyles in which special

sensors are of particular importance. First, we find that in burrowing annelids (e.g., sandworms, tubeworms, lugworms, and earthworms) *touch receptors* are widely distributed over the entire body. These sensors receive and convey important tactile information about the worm's immediate surroundings and are, thus, of considerable survival value. Second, in the parasitic leeches *chemo-receptors* play an equally vital role. Leeches rely, at least in part, on chemical secretions to locate potential hosts. The ability to detect even minute amounts of host emissions is one of the hallmarks of parasitic annelids, such as leeches. Thus, we find that chemosensors are prominent features among the leech sensory apparatus.

Digestive System

Annelids take advantage of a number of feeding strategies. One pattern is direct **deposit feeding** as in the burrowing earthworms. Here the animals ingest soil as they tunnel, digest the organic contents, and eliminate everything else. A variation of this pattern, called **selective deposit feeding**, is seen in some polychaetes in which they extract organic material from sediments *before* ingesting it. Yet another pattern is followed wherein polychaetes filter food from the surrounding water using mucous-covered **radioles** (spaghetti-like tentacles that fan out from the head, see below). Yet other polychaetes are predacious and, in fact, some even inject venom into their prey. Finally, some leeches, it seems, can alternate between an ectoparasitic and a free-living existence.

Regardless of the feeding pattern they might use, all annelids have a complete digestive system divided into an anterior **foregut**, a **midgut**, and a posterior **hindgut**. However, the annelid digestive system is more than a simple tube with a hole at each end. It actually is a fairly sophisticated apparatus exhibiting considerable regional specialization. The **mouth**, for example, is specialized for the intake of food. Just behind it is a muscular **pharynx** to "suck" the food into the mouth. Next is an elongated tube, the **esophagus**, that transports food to a storage chamber, the **stomach**. Upon leaving the stomach, food enters the **intestine** where digestion and absorption occur. Finally, undigested contents are eliminated through the **anus**.

The above pattern is basic for the phylum. However, there are many deviations from this plan. For example, the pharynx of many polychaetes may be everted as a **proboscis**, which, in turn, may bear powerful jaws or feathery food-gathering radioles. The stomach, in earthworms, is divided into an enlarged storage chamber, the crop and a muscular gizzard. Food from the crop is passed to the gizzard where it is pulverized by powerful muscular action. In leeches, the crop is invested with numerous pouches or cecae that can expand immensely during a blood meal.

To further enhance digestion, many annelids have developed mechanisms to increase surface area. For instance, the earthworm intestine has, along its dorsal face, a straplike structure, the **typhlosole**, which hangs into the intestinal lumen. The typhlosole greatly increases the absorptive surface area.

Respiration

The interchange of oxygen and carbon dioxide is accomplished in two ways. Foremost is the passage of gases through the general body surface (i.e., **cutaneous respiration**). Secondly, there may be extensions of the body wall into **gills** specialized for gas exchange. In either case, the animal must remain moist for diffusion to work. Thus, most annelids are aquatic. Those forms that happen to be terrestrial secrete mucus to prevent dehydration; but they, too, are restricted to a moist environment.

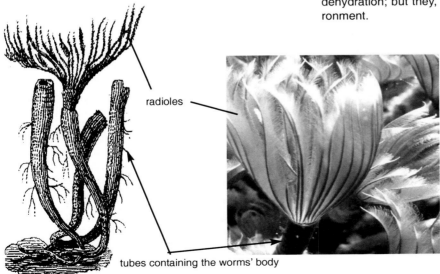

radioles

tubes containing the worms' body

Figure 10.3. Annelid tubeworms showing their feathery feeding radioles emerging from the top of the tube.

Excretory System

One of the unique features of annelids is that they possess a pair of "kidneys" or **nephridia** for each body segment.   The arrangement and function of a nephridium is quite simple.   It consists of three parts:

(1) a ciliated funnel, the **nephrostome**, that collects fluid from one coelomic cavity and passes it to,

(2) a coiled tube that lies within the coelom of the next segment. The coiled portion is supplied with numerous blood vessels into which ions from body fluid are resorbed. The tube may expand into a bladder where the concentrated fluid collects and then empties to the outside through a pore,

(3) the **nephridiopore** (Figure 10.11).

Circulatory System

The annelids, as a group, have a closed circulatory system as diagrammed in the figure below and shown in Figure 10.12.   Blood flows toward the head through a large longitudinal **dorsal vessel** which can be seen lying atop the digestive system. Blood then passes laterally and ventrally, through a series of **aortic arches**, to collect below in a **ventral vessel**, another large tube lying along the underside of the digestive system. Blood is transported posteriorly through the ventral vessel and is exchanged between the ventral and dorsal vessels by a series of capillaries, usually a pair for each segment. *Thus, blood makes a loop, going toward the head in the dorsal vessel and toward the tail in the ventral vessel.* Valves located on the interior wall of the ventral vessel prevent backwash to ensure the correct direction of blood flow. Finally, there are special arteries that supply the various body parts: nervous system , digestive system, body wall, gills (if present), and nephridia.

A Question of Hearts

Some authors regard the dorsal vessel in the earthworm as the "true" heart, while others give the aortic arches that distinction. In actuality, however, all vessels contract to propel blood, although the dorsal vessel, being the largest, undoubtedly provides the most power. Thus, while there is no centralized heart in annelids, the dorsal vessel comes the closest.

Reproduction

Some polychaete annelids are dioecious (male and female reproductive organs occur in separate individuals). However, the phylum, as a rule, is monoecious; all earthworms and all leeches are *simultaneous hermaphrodites* (both partners exchange sperm while copulating). Subsequent fertilization, may take place in the surrounding water (external fertilization), as in many polychaetes, or will most likely occur internally as in all earthworms and leeches. Some aquatic annelids such as the polychaetes, have a free-swimming, ciliated larva called a **trochophore**. This small larva grows by adding segments until it gradually attains adult appearance. Most annelids, however, lack a free larval stage and embryos grow, instead, within a sealed **cocoon**. They hatch from the cocoon as small versions of the adult.

The reproductive and developmental pattern varies considerably among the annelid taxa. In fact, details can be somewhat confusing to sort out. Therefore, rather than giving an in-depth account of each group here, reproductive specifics will instead be furnished as we discuss the three major classes.

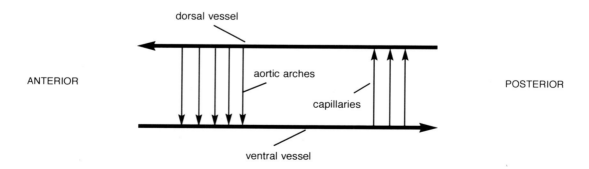

Figure 10.4. The flow of blood through the annelid circulatory system. Blood makes a loop by flowing toward the head in the dorsal vessel and toward the tail in the ventral vessel.

Class Polychaeta
(bristleworms)

Of the over 17,000 known species of annelids, approximately 70%, are actually marine polychaetes, the bristleworms. This is a primitive yet intriguing group that varies in a number of ways from the other annelid taxa: they look different, they have atypical feeding patterns, and their reproductive biology is unique. They range in size from a few millimeters to over three feet in length, and fall into one of two categories: free-living, swimming (errant) forms, and those that occupy burrows, the sedentary bristleworms.

There are three structural features that distinguish polychaetes from the earthworms and leeches. First, the *prostomium is prominent,* usually forming a definite "head." The head often has a protrusible pharynx bearing chitinous **jaws** or fangs that the animals can use to impale their prey. Located on the pharynx are also a variety of sensory structures, including eyes (at least two, often four or more), tactile and feeding tentacles, and sensory palps.

The second anatomical feature unique to polychaetes are fleshy outgrowths of the body wall called **parapodia** (Figure 10.5). Typically, each body segment bears one pair of parapodia. These are highly vascularized and, thus, form a large surface area for exchange of respiratory gases. In addition to being used as a respiratory device, the parapodia are also used as paddles for swimming, as braces for crawling, and as levers for burrowing.

Finally, this group is named for the bunches of conspicuous **bristles** that protrude from each parapodium. The bristles, actually **setae**, are stiff, chitinous projections that are used to anchor the animal while crawling and/or burrowing. As the animal progresses forward, the setae grip the ground, thus preventing the animal from sliding backwards.

Feeding Behavior

Polychaete annelids exploit a variety of feeding strategies. Some, such as the common sandworm, *Nereis*, are carnivorous. Powerful muscles manipulate the sharp jaws to grasp small worms, crustacea, and a host of other invertebrate species. In some species, the jaws are even accompanied by poison glands. There are other polychaetes that use feathery tentacles, **radioles**, extending from the prostomium to filter organic material from the surrounding water (Figure 10.6 C, D, E). These curious suspension feeders, often occupying burrows or tubes, look very much as though they have swallowed a mop, handle first! Sometimes the radioles occur in whorls as in the christmas-tree worms.

Another nutritive pattern involves deposit feeding. Here the animal (e.g., lugworm) simply ingests large amounts of sediments and extracts, from it, any organic matter. Still other polychaetes have adopted a scavenging lifestyle in which they feed on dead animals and other organic debris. Finally, there are some polychaete annelids that have undertaken a symbiotic relationship with another animal. For instance, there have been symbiotic polychaetes found in the gill chambers of chitons, in the mantle cavity of large mollusks, and even within the shell of hermit crabs. The symbiotic forms typically feed on debris or food scraps from their host.

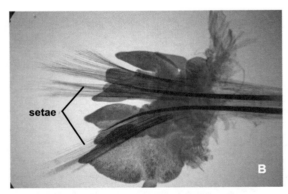

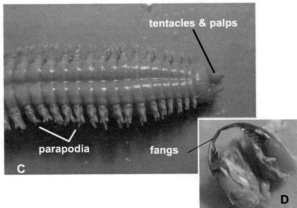

Figure 10.5. A, the sandworm, *Nereis*, a carnivorous polychaete; **B**, a microphotograph of a single, paddlelike parapodium with the clusters of setae, illustrating why this group is often known as the bristleworms; **C**, a ventral view of the anterior portion of a sandworm; **D**, a pair of fangs dissected from a sandworm.

Figure 10.6. Representative polychaete worms. **A.** a bristleworm ("fireworm") showing pronounced segmentation and lateral projections called parapodia; **B**, a christmas-tree worm affixed to a bit of coral (note the radioles arranged in a whorled spiral characteristic of christmas tree worms); **C-E**, fanworms showing feathery radioles extended from the body tube.

Reproductive Biology

Although some polychaetes can reproduce asexually through budding or fragmentation, the general reproductive pattern is through mating by separate sexes. However, in spite of the fact that there are male and female individuals, distinct gonads (testes and ovaries) do not form. Instead, sperm and eggs are produced by specialized areas of the peritoneum (i.e, the membrane that lines the coelom) and then shed into the general body cavity. Gametes are eventually released into the surrounding water where external fertilization occurs. In most polychaetes, a free-swimming, ciliated larva, the trochophore, develops, as described on the previous page. It lives for some time as a larva, continuously elongating by budding new segments just in front of the pygidium (tail). As the trochophore continues to enlarge, it eventually reaches a critical size, after which it undergoes a type of metamorphosis into an adult.

There are many variations on this basic reproductive theme of polychaetes. But the strangest occurs in certain tube-dwelling species. These annelids devote a portion of their body as a separate, sexually mature, reproductive entity called the **epitoke**. One or more of these mature sections bud off from the adult and actively swim toward clusters of other epitokes. In a swarming mass, they all release gametes into the surrounding water where fertilization occurs. The epitoke then dies, but the adult from which it sprouted, lives on!

Figure 10.7. A polychaete annelid with the posterior reproductive section, the epitoke.

Class Clitellata

Members of phylum Annelida that have a clitellum belong to class Clitellata. This class contains around 8,000 species in two familiar subclasses, Oligochaeta (earthworms) and Hirudinea (leeches).

Subclass Oligochaeta
(earthworms)

If there is anything such as a typical, proverbial annelid, it's the earthworm. To most people, just the mention of the word "worm" conjures up the image of a squirming, slimy earthworm. Indeed, these annelids are part of the human lexicon: "The early bird gets the worm"; "Let him squirm, like a worm on a hook"; "The grave will make worm's meat of me"; "I wormed my way into her heart." All you have to do is mention biology and the lowly earthworm immediately becomes the center of conversation; most of us were introduced to dissections by this omnivorous annelid. Some of us are repulsed, while others are entranced. Many a young lad or lady was disgusted by the inner workings of a slippery earthworm, while the very same view irretrievably hooked others into a life of zoology.

Oligochaetes are truly among the most identifiable of animals. They are found in both freshwater and marine habitats, and are abundant on land. Indeed, it's the terrestrial earthworms that we know best. Although living on land (actually, in burrows), they, like their aquatic cohorts, breathe through the body surface and thus must remain moist; drying out being the mortal enemy of an earthworm. To prevent dehydration, the earthworm's skin is literally blanketed by mucous-secreting glands, as anyone who has held a mass of squirming earthworms well knows. The mucilaginous secretion helps to retain moisture and thereby prevent desiccation; a parched earthworm is, frankly, a dead earthworm.

As a group, oligochaetes are highly segmented, both inside and out. They have no parapodia and are only very slightly cephalic, lacking a distinct head. Nonetheless, they do have an enlarged pair of anterior ganglia serving as a brain as well as a number of receptors located anteriorly. In addition, they are well equipped with sensors scattered over the body surface. They possess diffuse photoreceptors and are negatively phototactic, especially repelled by bright light. Likewise, they have acute touch and chemo-receptors throughout the epidermis.

Unlike the bristleworms (polychaetes) that have large numbers of setae on each body segment, oligochaete annelids have but a few. In fact, the very name Oligochaeta refers to having only a "few bristles." Like their polychaete cousins, the setae serve essentially the same purpose; they anchor the animal within its burrow and assist in forward locomotion by preventing slipping or backsliding.

Reproduction

Some oligochaetes can reproduce through budding or fragmentation, however the primary pattern is sexual. Oligochaetes are monoecious; they are known as simultaneous hermaphrodites since sex partners concurrently donate sperm to each other. Sperm, manufactured by the testes, are stored within large, paired **seminal vesicles** until mating when they are passed to the partner by way of **sperm ducts**. Once transferred, they are stored within paired elements of the female system, the **seminal receptacles**.

As eggs are deposited, the stored sperm become mobilized for fertilization, which occurs either internally or more commonly, within a **cocoon** as it passes over the sperm ducts. The cocoon is a chitinous envelop, formed by the large band-like gland, the **clitellum**. The cocoon slips over the head of the worm, and as it does, sperm and eggs are deposited within. Oligochaetes develop directly, having no larval stage.

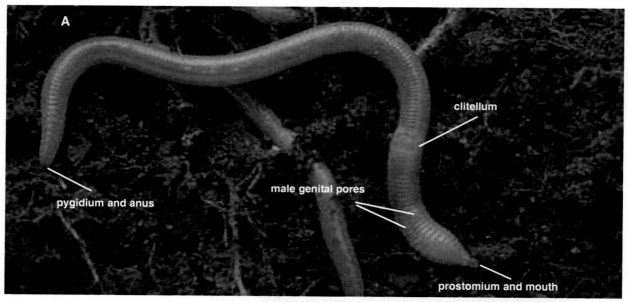

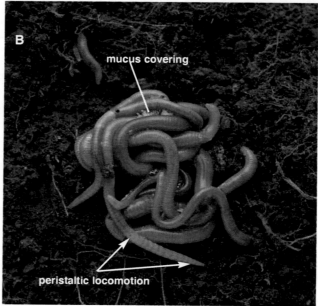

Figure 10.8. The common oligochaete, *Lumbricus terrestris*. Note the overall vermiform shape and the high degree of segmentation. **A**, a ventral view of an adult. Male genital pores and the glandular clitellum are indicated; **B**, a mass of adults. Note the mucus coating (arrow) and the indication of peristalsis in the bottom individual (double arrows).

Subclass Hirudinea
(leeches)

While it is true that, to some people, earthworms are unwelcome creatures, they are no match when compared to leeches. There are few, if any, animals more reviled than the leech. The notion of being covered by black lumps of bloodthirsty annelids makes our skin crawl. And if our language is littered with foul epithets for earthworms, it is utterly overflowing with vile references to leeches. Who wants to be considered a "bloodsucking leech"? ...a worthless bloodsucker of others?...a lowly freeloader on one's friends, family, society? Yes, leeches are near the bottom of our list of animal companions.

All members of the subclass Hirudinea are known as leeches. They are primarily organisms of freshwater environments, although some are marine and a few, such as the giant land leech of South America that reaches 25 cm (~ 10 inches), occur in moist terrestrial habitats. But not all leeches are bloodsuckers. Some are predatory and others are scavengers.

As a group, leeches are thought to have evolved from an oligochaete (earthworm) ancestor. Thus, it is not surprising that the two share a number of similar features. They both lack parapodia and are only slightly cephalic. The head lacks eyes, tentacles, or palps. Likewise, they are both hermaphroditic, possess a clitellum, and use cocoons to house developing young. Also, both oligochaetes and hirudinians lack a free-living larval stage. But they differ from other annelids in a number of respects.

Leech Features

(1) They are dorso-ventrally flattened.

(2) They are the only annelid class with members who are ectoparasitic.

(3) They possess enlarged digestive ceca, corresponding to their blood-sucking habits, which can expand tremendously as they become engorged with blood.

(4) Leeches lack the setae that are so characteristic of other annelids.

(5) Their epidermis is well supplied with chromatophores, allowing many to change colors to match their surroundings.

(6) They lack internal compartmentalization.

(7) They have a unique nervous system. Instead of one enlarged anterior ganglion (brain) as in most animals, leeches actually possess 34 individual brains distributed along their ventral nerve cord.

(8) Their method of locomotion varies from the o t h e r annelids. Without separate internal compartments, leeches cannot rely on the alternating rigid blocks of hydrostatic segments as the polychaetes and oligochaetes do for burrowing, or for creating peristaltic waves.   Leeches, instead, use either a wriggling swimming motion or an inchworm type of locomotion   in which a posterior sucker latches onto the substrate, allowing the anterior end to move for ward   and   take hold. The posterior portion is then brought forward to again attach. The animal thus moves forward in gradual, pincherlike pulsations.

Figure 10.9. A medicinal leech, *Hirudo medicinalis*, attached to the author's arm.

An Inside View 10.1

Leeches, Fingerprints, and Rattlesnakes

How do they find me?

For each of the past 15 summers, I have taken an odyssey to the Boundary Waters Canoe Area on the border of Minnesota and Canada. There, in the great outdoors, I seek nature's antidote to the urban murrain. I escape into a world of eagles, campfires, purple sunsets, and loons. However, on a number of trips I have had the misfortune of meeting another presence, a miniature vampire of sorts. For on many occasions, I have served as leech-lunch. I have found them securely fastened to my arms, legs, and other abstruse locations. Without exception I am always surprised at how quickly they find me. It seems that no sooner do I step into a marsh, a pitch-black lake, or a quaking beaver pond than I see a shadowy clump hunkered down on my calf or arm. They seem to arrive out of nowhere with me as their hapless target. Just how can they be so expeditious?

Well, the truth is, hungry leeches have at their disposal various means to locate a warm-blooded host. First of all, they have, scattered over their body, special tactile sensors able to discern slight vibrations in the water. So, by the time I step into the cold darkness, they are already on the alert. Then, by using thermoreceptors, they are able to sense my body heat. I guess even on the coldest, rainiest day, I am still cuddly enough to merit their celebrated kiss. Finally, they smell me; and after a week in the back country, I'm not too surprised. They have epidermal chemosensors that detect substances oozing from a potential host. As a matter of fact, you can perform a simple test to demonstrate their sensitivity to faint human secretions. Just touch the inside of a beaker containing hungry leeches. They will immediately detect the oils on your fingerprint, become agitated, and quickly move toward that spot seeking a bloody meal. So, how do they find me? That's easy; to a leech, I am a huge, warm, vibrating, smelly target. (KH)

Why do I tolerate them?

Frankly, I endure them because, for the most part, I don't even know they're there. From the leeches' viewpoint, once they've located a meal, they must be able to attach and feed relatively unmolested. But how can they extract blood, of all things, without being detected? To do so, they have a couple of tricks up their slimy sleeves. First of all, leeches have three razor-sharp teeth that can quickly slice through flesh. This they can do unannounced because they have, in their saliva, an anesthetic that numbs the immediate area.

Thus, I can be attacked without my knowledge, often by a relatively large number of leeches. In addition to the anesthetic, they also inject an anticoagulant, which delays blood clotting, sometimes for up to 48 hours. And if that's not enough, they have one more secretion, a vasodilator that expands my small blood vessels. These three salivary secretions allow the leech to gorge itself with comparative ease, and without much interference. After a complete blood meal, lasting from 10 minutes to an hour, the sated leech drops off, not to feed again for about 12 months. That's a relief... but come to think of it, that's just about the time I'll return for my annual getaway.(KH)

Medicine's Bloody Little Secret

Leeches and medicine have coexisted for a long, long time. The earliest records out of ancient Egypt refer to the medical use of leeches. I'm certain that if records existed from our cave-dwelling ancestors they would tell stories of the power of leeches.

The reason is linked to the early interpretation of disease. For most of our history as a species, it was thought that diseases were due to an imbalance in the body's "humors." In other words, the medical community believed that human afflictions were due primarily to the presence of "bad blood." The remedy for such a simple diagnosis was equally uncomplicated. Just remove the offending fluid, bloodletting it was called, or more precisely, leeching. Barber-surgeons (that's right, barbers; the traditional red-and-white barber pole reflects their one-time status as surgeons!) would just attach several leeches to the body and drain away the foul blood. Virtually any human ailment was thought curable through bloodletting. Fevers, aches, obesity, tumors, boils, laryngitis, acne, pneumonia, nausea, headaches, and even mental illnesses were among the afflictions for which leeches were the favored treatment.

Bloodletting was so prevalent that a thriving industry of leech collectors developed. Collectors simply walked slowly through infested waters and then picked the feeding leeches from their legs or torso. During the 18th and 19th centuries, in Europe alone, millions upon millions of leeches were used annually. Collection became so intense that the number of leeches finally dwindled and, in fact, in France and England, they virtually disappeared. Even today the leech is an endangered species in England. The practice of leeching peaked in the 1830s and then fell into disrepute as the medical world began to uncover the real causes of disease. By the end of the 19th century, leeches were all but eliminated from the physicians little black bag. (KH)

However, one question remains...did it work? Did bloodletting really help the patient? What do you think? Would draining a soda-can full of your blood cure your headache, resorb that pesky pimple, or take off a few pounds? The fact is, bloodletting offered little relief to the ails of our forefathers. On the contrary, leeching, in many instances, gave the patient a bitter dose of anemia, or even worse.

A note of interest here is that George Washington, after contracting a severe sore throat, was bled three times on December 14, 1799, although with instruments, not leeches. He died that very day. While any connection between bloodletting and his death is speculative, it's quite obvious that bleeding sure didn't help. Another interesting side note is that it wasn't always the patient who was victimized. It is well known, for example, that leech collectors, too, often became pasty-faced anemics.

The ebb and flow of medicine is, at times, strange to behold. What is common practice one day, is phased out, only to return later as a promising new elixir. Leeches are a good example. They were, for most of human existence, a popular treatment for our maladies. Then they lost favor, their use became ridiculed and subverted.

But they were down, not out. As a matter of fact, they have recently resurfaced; they have again raised their thirsty little lips to re-enter the world of medicine.

The modern physician recognizes that *Hirudo medicinalis*, the medicinal leech, can, indeed, be an ally if used properly. They are no longer applied indiscriminately to relieve a patient of a cup or so of blood. Today, they are used quite specifically. For example, in those cases where blood becomes pooled just beneath the skin, such as around the site of a suture, or even a blackeye, leeches can be quite effective. By placing a few hungry ones near a subcutaneous wound, the blood can be safely and completely removed without the need for further surgical intervention. Also, as the leech feeds it injects *hirudin*, a powerful anticoagulant that is helpful in restoring blood flow following reconstructive surgery.

Because of hirudin, leeches may help in re-establishing blood flow following attachment of severed fingers, hands, ears, or even disseevered penises.

They are also used to treat varicose veins, cauliflower ears, and glaucoma as well as aiding in the bonding of tissue grafts. Lately, leeches are being used to prevent, or even dissolve, blood clots. In fact, researchers have recently isolated compounds in leech saliva that show promise in the treatment of cardiovascular disease.

Snakebites, Anemia and AIDS

Leeches have also been used to treat snakebites. The venom injected by poisonous pit vipers, such as a rattlesnake, is a complicated cocktail of toxins and proteins (see page 471). A snakebite may evoke a number of symptoms, usually involving tissue deterioration and subcutaneous hemorrhage at the injection site. In some cases, leeches may be effectively used to remove blood collected beneath the skin following snakebites.

There are, however, some risks involved in using medicinal leeches. Anemia is experienced by approximately 50% of patients following prolonged leeching. And the physician must take care that leeches do not roam. For instance, if they are applied to the facial area and are not sufficiently confined, leeches may migrate into the nasal passages or bronchial airways causing serious respiratory blockage or dysfunction.

Then there's the question of HIV and AIDS. While leeches have never been implicated in the transfer of HIV, the possibility does exist. Thus, leeches that have been used in a medical procedure are discarded as biohazard waste; they are never reused.

In spite of these few side effects, the leech, like Phoenix, has risen from obscurity, renewed and reinvigorated, to resume its place among medicine's ever-expanding arsenal of MRIs, CAT scans, lasers, electron beam scanners, laparoscopes, and heart transplants.

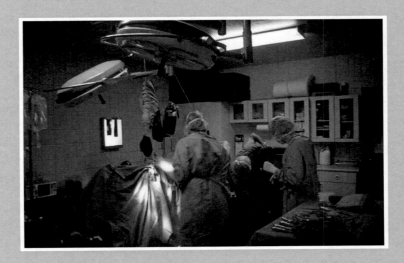

Chapter 10 Summary

Key Points

1. The most characteristic feature of the phylum Annelida is metamerism, also known as segmentation.

2. Metamerism has a number of advantages, among them are compartmentalization; segmented hydrostatic skeletons, which enhance burrowing, swimming, and other means of locomotion; reduction in the effect of injury; and the possibility of regional specialization.

3. External features of most annelids include an anterior prostomium, posterior pygidium, and a pronounced clitellum.

4. Most annelids are weakly cephalic; nonetheless, they are equipped with a variety of sensors including tactile (touch), chemo-receptor, photoreceptors, thermoreceptors, statocysts, and pH sensors.

5. Annelids show diverse feeding patterns ranging from simple filter feeding, direct and selective deposit feeding, predation, and parasitic.

6. As a group, annelids have a complete digestive system consisting of a mouth, foregut, midgut, hindgut, and anus. These regions may be variously modified depending on feeding style and food type. For example, annelids may possess a prominent proboscis with predatory jaws; a sucking pharynx; a storage chamber (crop); an internal muscular gizzard for food maceration; elongated tubes (ceca) or an intestinal typhlosole, both to increase surface area for food absorption.

7. Respiration is accomplished either through the general body surface (cutaneous) or by way of gills.

8. Annelids have a closed circulatory system consisting of a pumping device (heart), arteries, veins, and interconnecting capillaries. Blood basically passes toward the head through dorsal vessels, passes ventrally through a series of aortic arches, is then propelled rearward through ventral vessels, and finally passes dorsally through peripheral and other vessels to complete a circulatory loop.

9. Most annelids are monoecious although a few have separate sexes (dioecious). As a rule, members of the phylum practice simultaneous hermaphroditism in which both individuals are simultaneously impregnated. Some aquatic annelids have a free-swimming larva called a trochophore, however, the embryos of most annelids grow within a sealed cocoon.

10. The annelid phylum contains three major classes.

Polychaeta. The bristleworms constitute the bulk of the phylum. All possess an effusion of setae (bristles) and are highly segmented, marine annelids that bear a pair of appendages (parapodia) on each segment. They also display an array of feeding styles ranging from filtration to predation.

Oligochaeta. The earthworms and their kin are also highly segmented but bear few setae and they lack the paired appendages seen in the bristleworms. They are weakly cephalic and most are effective burrowers using their hydrostatic skeleton to plow their way through the soil.

Hirudinea. Known as leeches, member of this group are dorso-ventrally flattened, lack setae, and bear chromatophores in their skin to enable them to change colors. The most characteristic feature of leeches, however, is their blood-sucking habit. All are ectoparasites on a variety of animals, including humans. Leeches locate their host through detection of vibrations, chemoreceptor, and heat detectors Upon feeding, they inject an anesthetic, a vasodilator, and an anticoagulant to prevent detection and to permit a continuous blood flow.

Key Terms

anticoagulant
aortic arches
clitellum
crop
direct deposit feeding
dorsal aorta
gizzard
hirudin
metamerism
nephridiopore
nephridium
nephrostome
parapodium
prostomium
pygidium
radioles
segmentation
selective deposit feeding
seminal receptacles
seminal vesicles
setae
simultaneous hermaphroditism
suprapharyngeal ganglia
trochophore larvae
typhlosole
vasodilator

Points to Ponder

1. What is metamerism?

2. Metamerism has a number of functions; give four.

3. Compare feeding patterns of an earthworm with those of a sandworm (i.e., bristleworm).

4. Compare direct deposit feeding with selective deposit feeding.

5. Sketch and label the earthworm digestive system and give the function of each part.

6. Describe how the typhlosole functions.

7. Sketch the circulatory pattern of blood in an earth worm. Label each part.

8. Describe how leeches feed. How do they locate their hosts (e.g., humans)? How do they overcome the problem of blood clotting in the host? How do they control the level of blood flow in the host? How do they begin feeding without the host's knowledge?

9. Describe how the hydrostatic skeleton aids an earthworm in burrowing.

10. Describe earthworm reproduction. What is the role of the clitellum?

11. What is a trochophore larva?

12. Define simultaneous hermaphroditism.

13. What is an epitoke?

14. Describe the structure and function of the earthworm excretory system.

Inside Readings

Brusca, R. C., and G. J. Brusca. 1990. Invertebrates. Sunderland, MA: Sinauer Associates.

Dales, R. P. 1967. Annelids. London: Hutchinson University Library.

Lee, K. E. 1985. Earthworms: Their Ecology and Relationships with Soils and Land Use. Orlando: Academic Press.

Lent, C. M. , and M. H. Dickinson. 1988. The neurobiology of feeding in leeches. Sci. Am. 258(6):98-103.

Mann, K. H. 1962. Leeches: Their Structure, Ecology, and Embryology. New York: Pergamon Press.

Mill, P. J. ed. 1978. Physiology of Annelids. New York: Academic Press.

Newell, R. C. 1970. Biology of Intertidal Animals. New York: American Elsevier.

Pechenik, J. A. 1996. Biology of the Invertebrates. Dubuque, IA: Wm. C. Brown.

Satchell, J. E. ed. 1983. Earthworm Ecology. New York: Chapman and Hall.

Sawyer, R. T. 1986. Leech Biology and Behaviour. New York: Oxford University Press.

Thorpe, J. H. , and A. P. Covich, eds. 1991. Ecology and Classification of North American Freshwater Invertebrates. New York: Academic Press.

Wallwork, J. A. 1983. Earthworm Biology. Baltimore: University Park Press.

Chapter 10

Laboratory Exercise

Phylum Annelida

Laboratory Objectives

✓ to reproduce the classification of the specimens studied,
✓ to identify external and internal features of annelids and explain their respective functions,
✓ to explain the adaptations of an earthworm for life in the soil, and,
✓ to differentiate among the various forms of the phylum via preserved specimens.

Supplies needed:

✓ Dissection microscope
✓ Compound microscope
✓ Prepared microscope slides of the earthworm
✓ Live *Lumbricus* (the earthworm) and *Hirudo* (the medicinal leech)
✓ Preserved *Lumbricus* and *Nereis* (the sandworm) specimens
✓ Representative preserved annelid specimens, models, and biomounts.

Chapter 10
Worksheet

(also on page 599)

Phylum Annelida
Class Polychaeta

One of the common polychaetes of seashores is the sandworm *Nereis*. It is a predator of the intertidal zone where it feeds on a variety of invertebrates. *Nereis* is an active swimmer and can crawl quickly in search of prey or to escape danger.

I. *Nereis*, external anatomy

1. Place a sandworm specimen so that the dorsal side is facing you. Observe its basic shape and external features. Note the pronounced segmentation and the paired, lateral appendages, the **parapodia**, on each body segment. View these under a dissecting microscope and observe the bundle of bristlelike **setae** emerging from each parapodium.

What are three functions of parapodia?

Swimming, burrowing, crawling

---

2. Sandworms have a well-developed head and pharynx, although the pharynx is often withdrawn into the animal. If your specimen has an extended pharynx, view its numerous appendages: eyes, tentacles, palps, and anterior jaws. If not, then observe the specimen on display.

*Do not dissect the sandworm.
When you are finished with the external examination, return the specimen to its container.

II. Display Specimens

Study the various other polychaetes on display. Examine the specimen labeled *Amphitrite*. The feathery structures extending from the anterior segment are called radioles.

What are the functions of the radioles?
They filter organic material surrounding water.

Subclass Oligochaeta

Perhaps the most common biological specimen is *Lumbricus*, the common earthworm. They are used in virtually all beginning lab courses. However, even if you have already dissected one (or more), keep in mind that earthworms possess a number of significant annelid structures.

III. Living Earthworm, *Lumbricus*

1. Place a live earthworm on a moist paper towel. Determine the dorsal, ventral, anterior, and posterior surfaces of your specimen. (**Note**: The dorsal side is darker and setae are located only on the ventral surface.) You can feel the setae (bristles) by grasping the worm in one hand and pulling your fingers first toward the tail and then toward the head. In what direction do they point?

How do you explain this?

Help anchor the worm within its burrow and assist in forward locomation by preventing slipping.

---

2. What are the two functions of the setae?
— Help anchor the worm
— preventing slipping backward

3. Now observe your worm's locomotion. This type of movement is called **peristalsis**.

How do muscles and coelomic compartments combine to produce this type of locomotion?

they contracts

---

4. Does the earthworm have a preferred orientation? Place it on its dorsal surface; does it right itself? Virtually all animals will behave in the same manner.

How do you explain this apparent "need" to remain upright?

Yes, they will right itself, they used contraction to do. They have touch receptor, photo receptor which allow them to right itself.

---

5. Describe your animal's response to:

   a. **Touch** - gently touch the animal with a soft object, then poke it with a needle probe.
      Record your observations by comparing the worm's reactions to these two stimuli.

      It will coil itself

   b. **NaCl** - place a small amount of salt directly on the worm. What is the response?

      It Freakes out.

   > Explain at the cellular level.
   >
   > The salt dries out the worm because the salt absorb the moisture which cause the worm to freak out.

6. Farmers consider earthworms to be highly beneficial. Why is that?

   > Because earthworms can dig the soil and it gives air to the soil and after that the feces of earthworm isalso benefical.

IV. Preserved Earthworm

External Anatomy

Examine a preserved earthworm. Note its general shape and distinctive segmentation. Examine the anterior end to find the mouth located between the most anterior segment, the prostomium, and the first true segment. Likewise, find the anus on the pygidium (the last segment). Next locate the ventral opening of the vas deferens on the 15th segment and the small oviduct openings on segment 14. Finally, examine the glandular, ringlike **clitellum**. In the space below, describe the function of the clitellum.

   > Clitellum - secretes the muscus during copulation and Forms the cocoon to protect the developing embryos.

V. Earthworm Dissection

Internal Anatomy

(Figures 10.10 - 10.13)

1. Place the earthworm so the dorsal surface is facing you. Using a sharp scalpel or a needle probe, make a shallow incision through the epidermis from the prostomium (head) back to about 2 inches posterior to the clitellum. Be sure not to cut too deep as you will damage the underlying structures.

2. Use dissection pins to spread the body wall so that the internal structures are adequately exposed. However, avoid pinning the most anterior segments as this may obscure the brain and related structures.

3. Note that each internal compartment is separated by a thin tissue, the **septa**, so that each segment has its own internal coelomic cavity.

4. Now examine the circulatory system. Find the **dorsal blood vessel** lying atop the digestive system. You will see this as a dark line running the length of the animal. Trace it forward until it terminates at the series of five **aortic arches**. These are identified as small, blackish, fingerlike objects that project ventrally. The arches join the **ventral vessel,** another black line that can be located along the ventral surface of the intestine.

5. Next locate and examine the nervous system. It consists of a pair of **suprapharyngeal ganglia** (brain) that appear as tiny white "pinheads" at the anterior end of the pharynx. These form a loop around the pharynx to unite below as the **ventral nerve cord**. The cord runs the length of the animal beneath the ventral blood vessel. Find the whitish-colored cord by gently moving the intestine aside.

6. The earthworm has both male and female reproductive organs. The large whitish objects lateral to the aortic arches are the **seminal vesicles**. Buried within the vesicles are the small testes, which to observe must be carefully dissected. Exposure of the testes is optional. The ovaries, likewise, are too small to view without the aid of a microscope (finding them, too, is optional). The remaining female structures are the small paired **seminal receptacles** situated antero-lateral to the vesicles. Sperm from the mating partner are stored in the seminal receptacles until needed.

7. The most complicated apparatus in the earthworm is the digestive system. Carefully examine it and locate each of the following features. First find the muscular **pharynx**. Note how it is attached to the body wall by a series of stringy muscles. Posterior to the pharynx is the elongated **esophagus**, which leads to the saclike **crop**. The crop is a soft enlargement of the esophagus. Behind the crop is the muscular **gizzard** leading to the **intestine**, which passes to the **anus**.

8. Carefully open the intestine to reveal the straplike **typhlosole** running the length of the intestine.

What is the function of the earthworm's typhlosole?

Increase absorptive surface area

VI. Earthworm, Cross-Section Slide

Select a microscope slide of the earthworm in cross section. Starting at the outside (the outer edge of the specimen) and working your way internally, find the following structures:

- ✓ **epidermis** - most exterior tissue
- ✓ **circular muscles** - fibers lying just beneath the epidermis
- ✓ **longitudinal muscles -** fibers facing the coelomic cavity
- ✓ **coelomic cavity** - the open space in each segment that houses the internal organs
- ✓ **intestine** - the elongated, centrally located tube
- ✓ **typhlosole** - the round tube within the intestine
- ✓ **dorsal blood vessel** - lying above the intestine
- ✓ **ventral blood vessel** - lying beneath the intestine
- ✓ **ventral nerve cord** - lying beneath the ventral blood vessel

Subclass Hirudinea

Leeches

1. Living Specimens. Observe the swimming and crawling patterns of a leech.

2. Preserved Specimens. Observe the various leeches on display.

Note the following features in the preserved specimens and in the photograph below:

> ✓ the dorso-ventrally flattened body
> ✓ the anterior and posterior suckers
> ✓ lack of distinct head
> ✓ absence of parapodia
> ✓ absence of setae

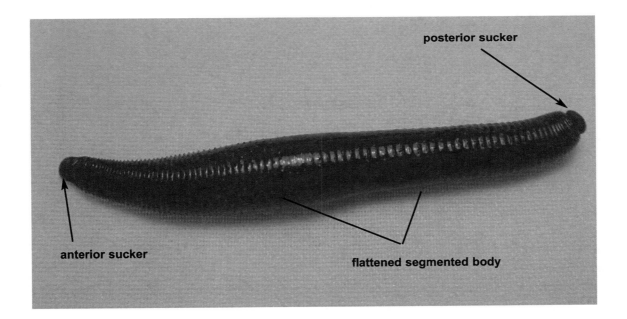

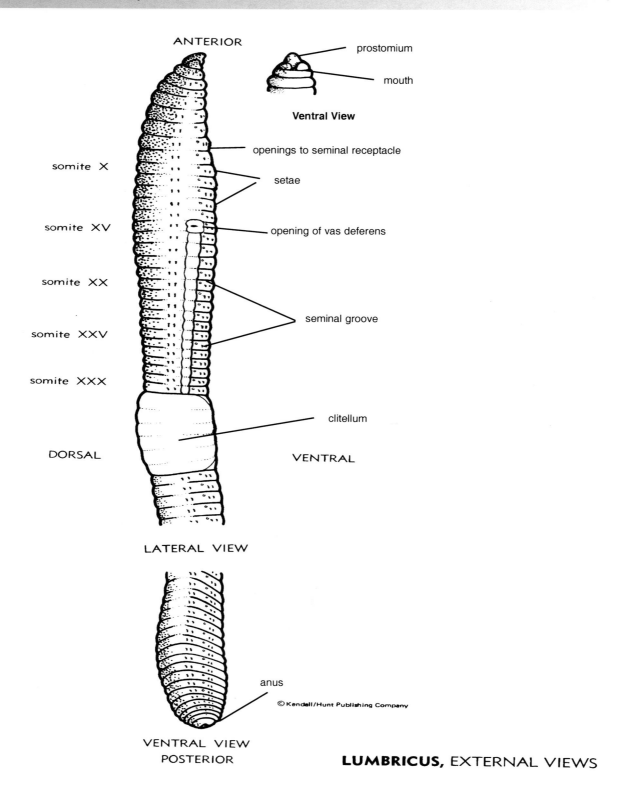

ANTERIOR

prostomium

mouth

Ventral View

openings to seminal receptacle

setae

somite X

somite XV

opening of vas deferens

somite XX

seminal groove

somite XXV

somite XXX

clitellum

DORSAL

VENTRAL

LATERAL VIEW

anus

© Kendall/Hunt Publishing Company

VENTRAL VIEW
POSTERIOR

LUMBRICUS, EXTERNAL VIEWS

Figure 10.10. The common earthworm, *Lumbricus*, external features.

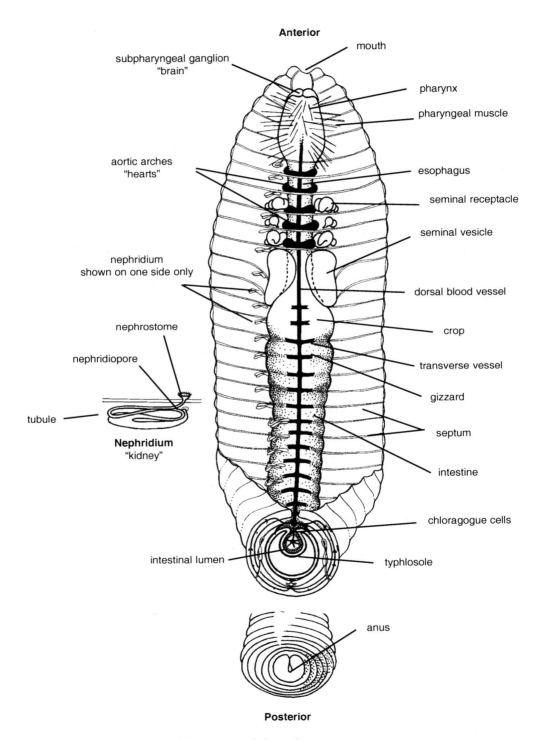

Figure 10.11. The common earthworm, *Lumbricus*, general dissection.

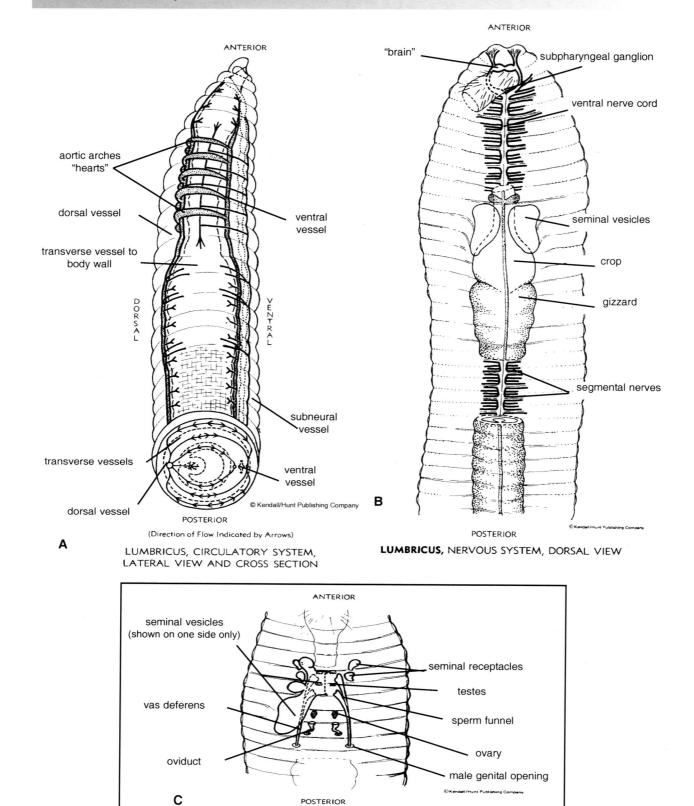

ANTERIOR

aortic arches
"hearts"

dorsal vessel

transverse vessel to
body wall

DORSAL

ventral
vessel

VENTRAL

subneural
vessel

transverse vessels

ventral
vessel

dorsal vessel

© Kendall/Hunt Publishing Company

POSTERIOR

(Direction of Flow Indicated by Arrows)

A

LUMBRICUS, CIRCULATORY SYSTEM,
LATERAL VIEW AND CROSS SECTION

ANTERIOR

"brain"

subpharyngeal ganglion

ventral nerve cord

seminal vesicles

crop

gizzard

segmental nerves

B

POSTERIOR

© Kendall/Hunt Publishing Company

LUMBRICUS, NERVOUS SYSTEM, DORSAL VIEW

ANTERIOR

seminal vesicles
(shown on one side only)

seminal receptacles

testes

vas deferens

sperm funnel

ovary

oviduct

male genital opening

© Kendall/Hunt Publishing Company

C

POSTERIOR

LUMBRICUS, REPRODUCTIVE SYSTEM, DORSAL VIEW

Figure 10.12. The common earthworm, *Lumbricus*, circulatory (A), nervous (B). and reproductive (C) systems.

ANTERIOR

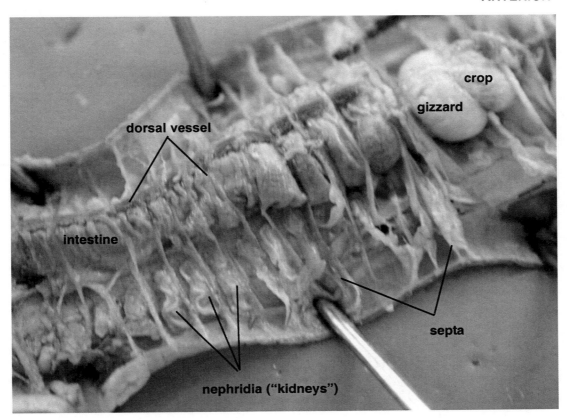

Figure 10.13. Dissection of the common earthworm, *Lumbricus*.

Clitellum - Secretes coocon
Setae - prevents slipping when moving
Gizzard - Grind boods
crop - store food
Dorsal blood vessels - carries blood vessels
Seminal vesicles → Big white
            receptacles → Small white

Annelids

CHRISTMAS TREE WORMS

LEECHES

FANWORMS

SANDWORMS

EARTHWORMS

A COLOR PORTRAIT – PART I
...FROM SPONGES TO BUGS

SPONGES

sponge gemmule

green anemones

stinging tentacles

corals

CNIDARIANS

surface polyps

coral

anemones

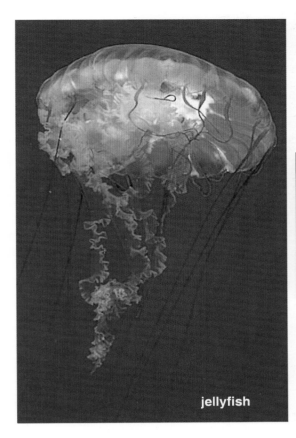

jellyfish

sea anemone

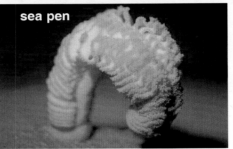

sea pen

anemones

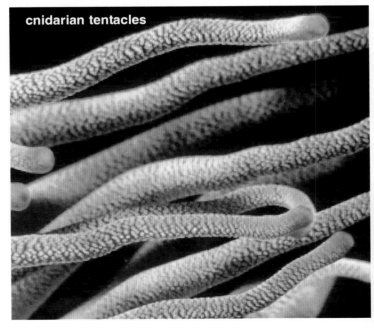

cnidarian tentacles

Portuguese Man-o-War

an attached anemone

FLATWORMS

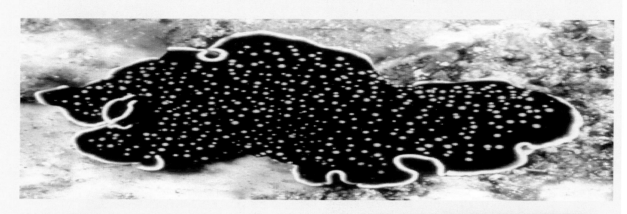

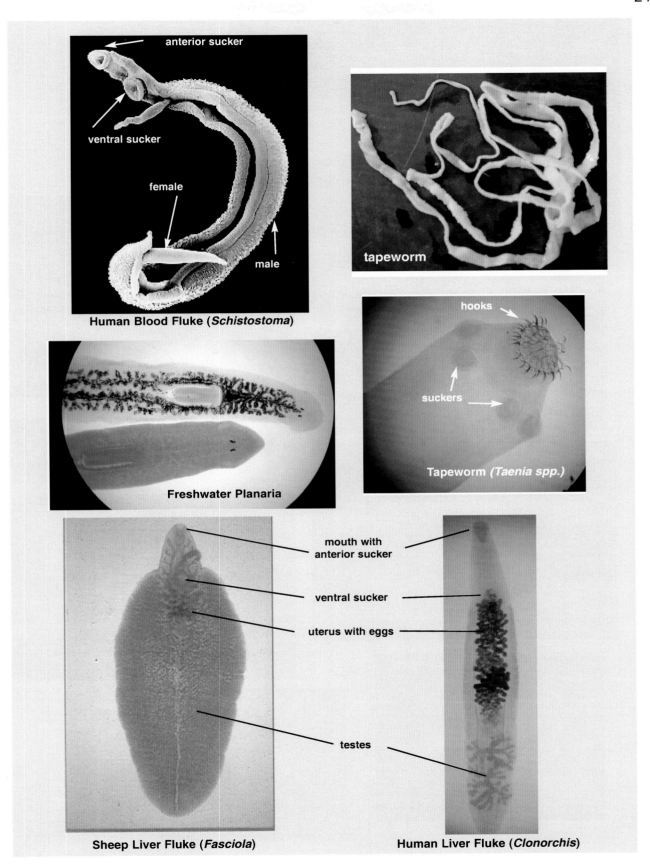

anterior sucker

ventral sucker

female

male

Human Blood Fluke (*Schistostoma*)

tapeworm

hooks

suckers

Tapeworm (*Taenia spp.*)

Freshwater Planaria

mouth with anterior sucker

ventral sucker

uterus with eggs

testes

Sheep Liver Fluke (*Fasciola*)

Human Liver Fluke (*Clonorchis*)

banana slug

chiton

nudibranch

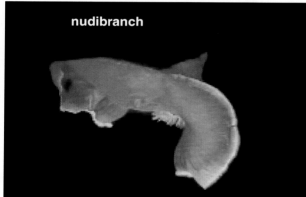

nudibranch

MOLLUSKS

nudibranch

nudibranch

cone

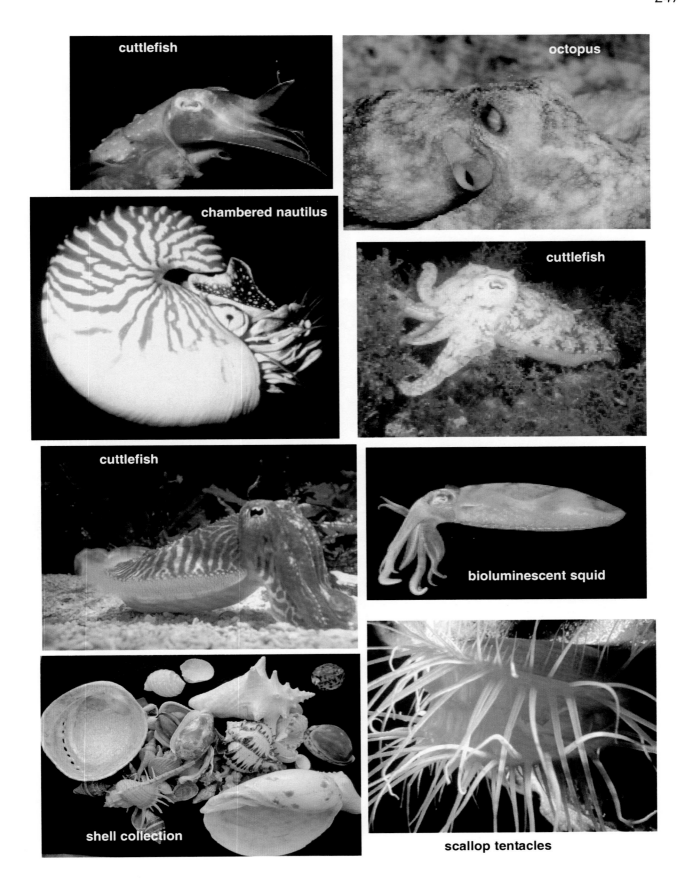

cuttlefish

octopus

chambered nautilus

cuttlefish

cuttlefish

bioluminescent squid

shell collection

scallop tentacles

earthworms

sandworms

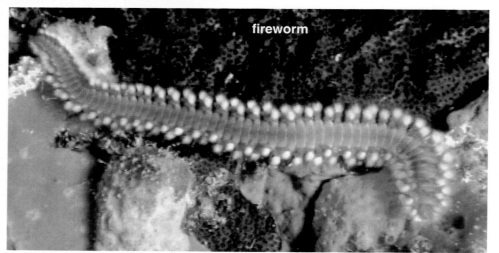

fireworm

ANNELIDS

medicinal leech

Christmas-tree worms

Christmas-tree worms

fanworm

cluster of fanworms

cluster of fanworms

ARACHNIDS

female black widow spider

Black widow spider, Copyright by John Jackman. Reprinted by permission.

brown recluse spider

Brown recluse spider, Copyright by Rick Vetter. Reprinted by permission.

garden spider

lynx spider

tarantula

female scorpion carrying young

wolf spider carrying an egg case

CRUSTACEANS

cleaning shrimp

orange spider crab

a large species of spider crab

a land crab

a hermit crab

arrow crabs

barnacles

crayfish

land crab

sponge crab

red banded coral shrimp

INSECTS

dragonfly

black swallowtail butterfly

praying mantis

white admiral butterfly

stick insect among leaves

cicada

flower beetle

grasshopper

beetle

giant water bug

velvet ant

honeybee

flower beetle

dragonfly

wasp, ventral view

MORE INSECTS

INSECTS & SOME RELATIVES

giant centipede

hissing cockroach

millipede

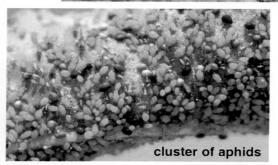
cluster of aphids

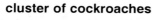

cricket

emerging insect

cluster of cockroaches

centipede

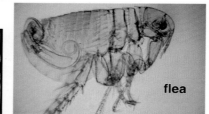
flea

mosquito in action

termite soldier

11 Chapter 11

THE ARTHROPODS

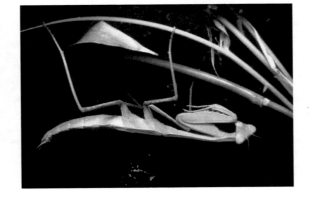

Chapter 11

Phylum Arthropoda:

INCREDIBLE SUCCESS

Introduction

There's few who would argue that of all the animals on earth arthropods are the most successful. Regardless of how success might be measured, arthropods prevail. If diversity is the yardstick, arthropods win; 8 out of every 10 known animals is an arthropod. If habitat usage is the barometer, arthropods dominate; they have invaded virtually every known environment and monopolize an incomprehensible number of ecological niches. If sheer numbers is the standard, arthropods win again. They are among the most prolific of animals. They "spread like flies." In fact, using the housefly as an example, it seems that more of them hatch each day than there are humans on earth. Or consider the termite or the ant. It is estimated that there are 1,500 pounds of termites and one million ants for every person alive! A single termite queen lays 30,000 eggs per day (that's one every three seconds). She lies in her underground boudoir, fed, cleansed, pampered, and all the while incessantly pumping out eggs. And what about locusts? They emerge in sun-eclipsing clouds of over 50 thousand million individuals, eating and reproducing their way to ignominy. One swarm can consume over 3,000 tons of vegetation per day!

Arthropods have, indeed, discovered what is needed to survive, and they do it better than anything else. I've studied insects for a long time and I'm amazed at their versatility, resourcefulness, and resilience. No matter what we do to control or eliminate them, they rebound as tough as ever, ready, it seems, to tackle any assault we might deliver. In the U.S. alone, we spend $250 million each year to unsuccessfully annihilate cockroaches. They continue to thrive while we, on the other hand, seem intent on destroying the very natural bounty on which we depend. We despoil natural habitats at an appalling rate. Eliminating, all the while, the very diversity that gives earth its vitality. And who will survive the onslaught? Why, arthropods, of course.

This recurring idea of human extinction followed by the takeover by swarming hordes of arthropods was graphically expressed years ago:

it won't be long now, it won't be long
man is making deserts of the earth
it won't be long now
before man will have it used up
so that nothing but ants
and centipedes and scorpions
can find a living on it.
                                        -Marquis

We've all heard about an impending revolt by spiders, scorpions, and insects. But is it real, or just science fiction hype? Is it possible that arthropods will take over, and if so, when will it happen? Finding the answer to that question is much like untying the Gordian knot; it's hard to know where to begin and you're a little afraid of the outcome once the question is unraveled. However, according to famed biologist, E.O. Wilson, who has studied this question perhaps more than anyone, it's not a matter of when arthropods will inherit the wind for, in his words, "they already have!"

Figure 11.1. Ants are among the most successful of all animals. They exist in huge, social colonies that number well into the millions of individuals.

Changing Earth's Tapestry

Before we discuss what makes arthropods so successful, there's one important event to consider. Many millions of years ago animals accomplished a feat so momentous that it changed the very nature of our planet: they invaded land. And it didn't happen just once, but twice. Much has been made of amphibians as the source of land animals; of how they evolved from lungfish and eventually gave rise to the reptiles and then onward to birds and mammals. We marvel at their skills, their initiative, their boldness. We applaud their success. But what we seem to forget is that they weren't the first to creep on shore...arthropods were.

The best we can figure is that amphibians began the process about 350 million years ago (mya), during the late Devonian period. But long before there were amphibians, long before the first dinosaur left its scaly footprint in shallow seas, arthropods had made that same treacherous leap...from water to land. Fossil records show that approximately 70 million years before the amphibian explosion (i.e, ~ 420 mya), some type of arachnid (a spider or, more likely, a scorpion) had already become terrestrial. The commanding question then, the one we cannot escape, centers on the challenge confronting these pioneers. What did it take to make that giant step that transformed life on our planet?

Keep in mind that most of the earth was under water, and still is for that matter. Keep in mind, too, that the colonizing animal (whatever it might have been) was already adapted for life in the water. Also, we must realize that when the invasion occurred no other animal existed on land; there were no butterflies flirting from flower to flower (indeed, there were no flowers); there were no snakes slithering through the grass, no birds issuing their noisy protest. Land was, indeed, unoccupied by animals. Well, to be honest, there may have been a snail or two sliding through the underbrush. There may have been a few annelids burrowing here or there along the seashores. But truly terrestrial animals? There were none. So I wonder. What compelled those first creatures to leave the comfort of their watery sanctuary to lurch, wriggle, or crawl onto virgin soil?

Figure 11.2. What was the early landscape like, and what did it take for early arthropods (such as centipedes or maybe scorpions) to invade this virgin terrain?

THE ATTRACTION...
...THE CHALLENGE

*S*o, what was the attraction? Why did some antediluvian ancestor crawl onto land? In retrospect, we can propose at least three possible reasons: (1) food, (2) an abundance of open niches, and (3) the absence of predators. A combination of those three factors most likely provided an overwhelming pressure to exit an aquatic domain in favor of a terrestrial one. We know, of course, that one more component was involved. Any colonist must have already possessed the appropriate and necessary features to successfully invade land. In other words, the immigrant must have been *preadapted* in some way that enabled it to thrive on land.

However, it is important to remember that in spite of the attractions, terrestrial life is much different than that in water. In many respects, it's much more difficult. To truly become a terrestrial animal, there were several very real impediments to overcome. Among these were:

(1) **Respiration** on land is much different than that in water. Gills, the traditional aquatic respiratory device, doesn't work well on land. First of all, out of water, the gill collapses. As a result, the surface area is reduced below the capacity to support life. Another possible route is to exchange gases directly through the body surface. However, cutaneous respiration, which can work in moist terrestrial environments but not in dry ones, does, by that very factor, restrict the full invasion of arid terrestrial niches. To overcome these two limitations, the respiratory surface had to be altered: moved internally to remain moist, and kept patent to overcome the tendency to collapse.

(2) Respiratory surfaces are not the only thing subject to dessication. The entire body surface, if not protected, will also dry out if exposed to atmospheric conditions. If an animal dehydrates, it perishes. So, the ability to **prevent dehydration** is another vital step in the total colonization of land.

(3) Because gravity is less of a factor in an aquatic environment, an animal living in water tends to float instead of scraping along the bottom. It doesn't face the degree of friction as does its terrestrial counterparts. An aquatic animal is buoyant. Thus, to colonize land, an animal must overcome friction caused by the pull of **gravity**. Special appendages must be developed that will lift the animal off the ground and, at the same time, propel it forward.

(4) Another advantage an aquatic animal has over terrestrial ones is in the discharge of metabolic wastes and the management of ion and water balance, i.e., **osmoregulation**. In water, an animal can excrete nitrogenous wastes as ammonia which is diffused into the surrounding water. However, on land this is impossible. A land animal must find a way to release metabolic waste, not as ammonia, but in some other form, such as urea or uric acid...what we generally call urine. At the same time, a land animal must balance ion concentrations (especially NaCl) and control water loss. This necessitates a reconstruction of the organ involved (i.e., the kidney).

(5) One of the primary differences between life on land and life in the water is ambient **temperature**. In water, temperature is relatively stable; there's just not much variation. However, on land it's another story; temperature seems to be constantly fluctuating. Shifts occur throughout a given day, season, year, and even within different microhabitats. Life on land is, truly, one of thermal extremes. Thus, a terrestrial animal must be able to compensate for a constantly changing temperature.

(6) All animals require a constant supply of energy. **Acquiring food** posed, at least initially, a problem for land invaders. First of all, since there were no other animals on land during the first wave of colonization a predator would have significant problems finding food. This, of course, changed quickly as animals poured into terrestrial ecosystems. On the other hand, for the first animals, the pioneers, there was an abundance of plant life in the form of ferns, cycads, and mosses. Taking advantage of these new foods posed yet another challenge to land colonization.

(7) To survive, an animal must have appropriate mechanisms to **sense its environment**. The better equipped an animal is to receive and process incoming stimuli, the better its chances of survival. However, waterborne stimuli travel differently than airborne ones. For example, waterborne chemicals are perceived in different ways than those carried by air currents— light travels in distinct pathways on land and in water; and the same is true for touch, sound, and temperature. Thus, to be successful, a terrestrial animal must alter its sensory apparatus accordingly.

(8) Perhaps the singular most important thing an animal does in its lifetime is to **reproduce**. Leaving behind that precious genetic message is an animal's one true legacy. Through reproduction, a populations' inheritable code is made available to the next generation. An important business, indeed. To meet this demand, animals have developed a number of strategies. One that works exceedingly well in water is *broadcast spawning,* the release of large number of gametes into the surrounding environment where fertilization and development occurs. This is the method used by the majority of aquatic animals, although a number of other strategies are also practiced.

There is, however, a real problem with this reproductive strategy on land. To reproduce properly, sperm and eggs must remain moist...they cannot dry out. The same is true for the developing embryo. Thus, external fertilization does not work for land-dwelling animals. To solve the problem of dehydration, land animals have opted for internal fertilization. As a matter of fact, internal fertilization has become the hallmark of terrestrial reproduction.

Internal fertilization requires a pivotal change in reproductive structures and behavior. No longer will spreading one's gametes into the water suffice. Thus, special devices for introduction of sperm into the female's reproductive system must be developed. As a result, a number of intromittent organs (e.g., penises) have evolved. Secondly, like us, each individual animal has a "personal space" that is guarded against encroachment. Since internal fertilization requires that sperm enter the female this cloudlike personal space must be abridged. To do that, animals have developed proper courtship behaviors to break down a partner's natural aggressiveness. Finally, internal fertilization and terrestrial reproduction have demanded a different scheme for development of the embryo. Some animals opt to retain eggs internally (e.g., mammals), while others lay some type of external, shelled egg (insects, reptiles, and birds). Such land-adapted eggs allow embryological development within a controlled, nutrient-rich environment (Figure 11.3 and also chapter 14). It is no accident, therefore, that a terrestrial egg has evolved.

The transition from water to land must have been an incredibly challenging and risk-filled venture. But for those animals that achieved it, an unbelievable opportunity awaited. Countless empty habitats were there for the taking. A whole panorama of unoccupied ecological niches spread before them. It wasn't long before those early pioneers gained a toehold and soon arthropods began a massive radiation. They crept, climbed, sprinted, flitted, scurried, darted, tunneled, jumped, and flew in every possible direction until they eventually carpeted the earth.

Animal colonization of land, some 420 million years ago, provided the spark for the diversity explosion that followed, and which continues to the present day. Emancipation from the aquatic environment did, truly, change our planet's landscape. And it was the arthropods that first lit that spark.

Figure 11.3. To thrive and reproduce on land, animals had to find a way to protect the precious developing embryo without having to return to water. They did this by developing a shelled egg.

Barriers to Terrestrialism

1. **Must be able to breathe on land**
2. **Must keep from dehydrating**
3. **Must overcome pull of gravity**
4. **Must be able to osmoregulate on land**
5. **Must be able to deal with temperature fluctuations**
6. **Must be able to live on terrestrial food sources**
7. **Must cope with terrestrial sensory stimuli**
8. **Must be able to reproduce on land**

Protostome vs. Deuterostome

Figure 11.4 below shows a possible phylogenetic tree of the major animal phyla. You will note that the great majority of animals belong to the Bilateria line. These animals are all bilaterally symmetrical, although echinoderms are secondarily radial. You will also note that bilateral animals are subdivided into two lineages: the **protostomes** and the **deuterostomes**. To understand what these two terms mean, we must look to animal development. Following fertilization, the zygote soon begins to undergo successive mitotic divisions called cleavage, forming organisms with cell numbers of 2, 4, 8, 16, 32, 64, and so on. At some point a hollow sphere, the blastula, develops. Soon a portion of the blastula begins to fold inward, i.e., to invaginate, in a process called gastrulation. The point of initial invagination is called the blastopore. Gastrulation rearranges cells to form the three germ layers; ectoderm, endoderm, and mesoderm. It also results in a tube-within-a-tube arrangement with one end of the inner tube being the primordial mouth and the other end, the anus. The way in which an embryo undergoes cleavage and gastrulation determines whether it's a protostome or a deuterostome. If cleavage is spiral (i.e., occurs at an acute angle to the embryo) and the blastopore forms the mouth, then the animal belongs to the protostome branch (proto = first). If, however, cleavage is radial (i.e., occurs in a parallel fashion) and the mouth forms opposite the blastopore, then that animal is a deuterostome.

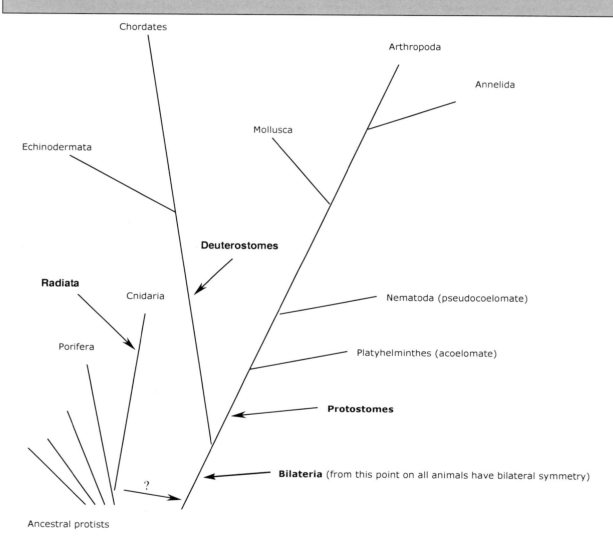

Figure 11.4. A cladogram (evolutionary tree) showing possible relationships between selected animal phyla. Note that all animals from cnidarians onward have bilateral symmetry. This is true even for the echinoderms, although as adults, they are secondarily radial. Also, note that the two main evolutionary branches, the protostomes leading to the arthropods, and the deuterostomes, giving rise, eventually, to the chordates. See the discussion given above of these two major lines.

So, What Is an Arthropod?

The irony surrounding arthropods is that, on the one hand, they are the most successful animals that have ever existed, while on the other hand they contribute very little, if anything, to higher animals. Unlike some of the groups we have studied so far, such as the flatworms, molluscs and annelids, from which other animal groups evolved, arthropods give rise to no other taxa. They, instead, constitute a dead-end group in what is called the protostome lineage (Figure 11.4 on the preceding page).

So does that mean that arthropods haven't developed any meaningful adaptations? On the contrary, they have concocted quite a few, each furnishing something to the incredible success that arthropods enjoy. Seven of the most important are

(1) an exoskeleton
(2) paired, jointed appendages
(3) tagmatization
(4) metamorphosis
(5) prolific reproduction
(6) advanced sensory structures
(7) complex behavior

The Exoskeleton

The chitinous exoskeleton has probably been more responsible for arthropod success than any other single feature. It provides an excellent armorlike protective covering as anyone who has tried to "squash" a flea can attest. It is a tough yet flexible covering that gives needed support to the animal, provides anchorage, and a series of levers for muscle attachment; thus, it is important in arthropod movement and overall locomotion. Finally, the exoskeleton, if impregnated with a waxy substance, as is the case in most arthropods, provides a waterproof covering. Because of this decreased permeability to water, arthropods are protected from both waterlogging and excessive evaporation at the same time. In other words, the arthropod exoskeleton prevents dehydration, one of the important barriers to life on land.

The exoskeleton consists of two layers: an outer, thin, **epicuticle**, and an inner, thicker, **procuticle**. The epicuticle contains a waxy lipoprotein giving it the waterproofing properties. The waxy epicuticle is also impervious to chemicals, such as pesticides, which helps to explain why we have such a difficult time controlling insects. The inner procuticle is a hard, tough, polysaccharide layer consisting primarily of chitin. Because of the abundance of arthropods and the use of chitin by many other invertebrates, chitin is, next to cellulose, the most bountiful organic material on earth.

Although the exoskeleton is the major contributor to arthropod success, it is not without its disadvantages. For example, it can be quite heavy, which is one factor in limiting the overall size of terrestrial arthropods. Also, since the exoskeleton is made of non-living material it cannot grow. For an arthropod to gain size, it must periodically shed the exoskeleton, a molting process called **ecdysis**. The fossil record indicates that arthropods possessed an exoskeleton long before they invaded land. In other words, the exoskeleton was a preadaptive feature that greatly aided the invasion of terrestrial environments.

Jointed Appendages

For an animal to move efficiently on land, its body needs to be raised off of the surface. The less contact with the ground, the less friction an animal encounters and the more efficient is its locomotion. Paired appendages, one located opposite the other, provides sufficient support and elevation. However, to improve mobility, the limb needs to be flexible. Strategically placed joints along the appendages accomplishes just that. Thus, the **paired, jointed appendage** of arthropods is an invaluable addition on the path toward terrestrialism. As with the exoskeleton, aquatic arthropods had evolved the jointed appendage long before land invasion occurred; the appendage, too, was preadapted.

Tagmatization

Arthropods, like their annelid predecessors, are metameric. Segmentation, in itself, confers several advantages to an animal (see chapter 10), but it also may lead to the fusion of neighboring segments. This fusion, in turn, leads to regional specialization. An area consisting of co-joined segments can be specialized to perform specific functions for the animal. This fusion process is termed tagmatization. Arthropods usually have two or three tagmata. For example, in the insects there are three: (1) a head is designed expressly for feeding and sensory functions, (2) the thorax is specialized for locomotion (it always bears the legs and/or wings, and (3) the abdomen carries out visceral (bodily) functions. The degree of tagmatization differs among the various arthropods and will be discussed along with the specific groups. In any event, the role tagmatization plays is to increase an animal's overall efficiency.

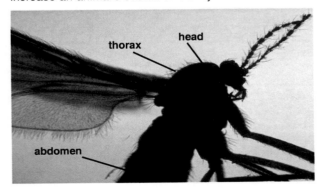

Figure 11.5. Insects exhibit three areas of tagmatization: a head, thorax, and an abdomen.

Metamorphosis

The basic advantage of metamorphosis, in which a larva differs, sometimes radically, from the adult, is that each assumes separate lifestyles. The adult invariably requires a different food source than the larva, they occupy distinctly different habitats, and engage in different activities. For example, the larval housefly (maggot) lives and feeds in garbage, excrement, or other filth. The adult, on the other hand, is a flying insect that feeds primarily on sugary substances. As a result, they don't compete with each other. It's as though they were two separate species occupying individual ecological niches. In fact, there are many instances among arthropods (and other animals, as well) wherein the larva and adult were first identified as distinct species before it was discovered that they were merely different stages of the same animal. The various types of metamorphosis will be discussed in detail in chapter 14.

Prolific Reproduction

The reproductive ability of arthropods is epic. All you have to do is to think of the massive emergences of locusts, gnats, mosquitoes, and even spiders and you realize their great capacity to reproduce. Arthropods do, indeed, multiply rapidly and prolifically. And that's one of the primary foundations for their ecological success. Any species that combines a relatively short life span (as is virtually the case with all arthropods) with a high reproductive potential has the power to evolve rapidly. It's much easier for such a species to take advantage of mutations, competition, and natural selection than a long-lived animal that produces few offspring.

Here's an example of how these two factors, short life span and high fertility, can lead to rapid evolutionary success. Suppose that a given arthropod, let's say a carrion beetle (i.e., an insect that feeds on dead animals), is exposed to a change in its environment. For argument's sake, suppose that a pathogenic toxin suddenly appears in all carcasses that this beetle normally uses as food. Feeding on these dead bodies can be fatal, and is for the vast majority of the population. However, since the beetle's population is extremely large and varied, there is a good chance that some members might harbor a genetic resistance to the toxin. They alone can eat without mishap. They survive, reproduce, and pass on their genetic resistance. Soon, the population of carrion beetles are all "immune" to the action of the toxin. The population has shifted from a vulnerable one to one now unaffected. The unprotected beetles succumbed to the toxin while their immune brethren survived to reproduce. So, because of the quick turnover (i.e., short lifespan) and high fertility, natural selection can act swiftly to produce a newly adapted, toxin-resistant population.

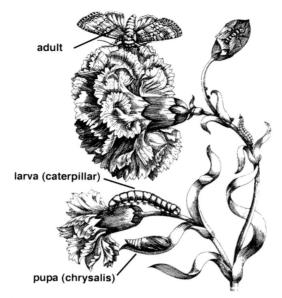

Figure 11.6 A number of life stages can occur on different parts of the same plant, thus reducing competition between larva and adults.

Sensory Structures

Terrestrial stimuli differ from those in aquatic environments. Visual, auditory, tactile, and chemical signals, for instance travel in different fashion on land. Therefore, to maximize their senses, land arthropods must rely upon a suite of sensory modalities beyond that of most other invertebrates. Vision, touch, olfaction, and hearing are among their most important.

Most arthropods possess two types of photoreceptors. The so-called simple eye, the **ocellus**, is designed to detect light intensity and motion. It may also play a part in regulating daily and seasonal rhythms. The **compound eye**, on the other hand, is designed specifically to detect movement. It probably forms primitive images, as well. As the name implies, the compound eye consists of individual receptors, numbering into the thousands in some forms. Apparently, in certain arthropods, the compound eye is also able to discern colors.

Tactile receptors (hairs, setae, palps, pectines) may be scattered over the body surface, often concentrated ventrally on legs, feet, mouth parts, and the abdomen. In addition, most arthropods have antennae that are equipped with tactile receptors.

Chemoreceptors can respond to a variety of stimuli, all related to what we know as the sense of smell. These sensors are found over much of the animal's surface as hairs, pegs, plates, pectines, and pores. They are often concentrated in the mouthparts, antennae, or on reproductive structures. The ability of arthropods to detect the smallest amounts of airborne or waterborne chemicals is extraordinary.

While arthropods do not receive sound stimuli in the same manner that we do, they, nonetheless, are able to detect a wide range of airborne and waterborne vibrations. Special **mechanoreceptors**, such as hairs and setae, are widely distributed over the animal's surface. Pressure waves created by sound frequencies displace or agitate the sensors forming a signal that is sent to the brain. Spiders, for example, are well equipped with sensory hairs to receive the slightest of vibrations. In addition, many arthropods, such as crickets, grasshoppers, and cicadas, have special organs for sound reception placed on legs or the abdomen.

To accommodate such a wide field of sensations, arthropods have evolved a relatively advanced nervous system. They have a pair of large **cerebral ganglia** ("brain") located above the pharynx, which is a control center for the head and related sensors. Issuing from the ganglia is a **ventral nerve cord** running the length of the animal. Nerves leave the cord at intervals to receive sensory information and to otherwise innervate the various body organs and related structures.

Behavior

Thus far in our study of zoology, we have encountered a vast spectrum of animal behaviors. From the simple trial-and-error learning seen in the Platyhelminthes (Planaria) to the advanced courtship and communicative patterns of cephalopods, animals have evolved a striking diversity of adaptive behaviors.

However, nowhere among invertebrates do we see the degree of behavioral sophistication as that displayed by arthropods. They have evolved a system of interactions with their environment, and with each other, that surpasses all other invertebrates. Among their repertoire of behaviors are elaborate courtship rituals, defensive displays, intricate web-building performances, habitat alteration, care of young, and complex intraspecific communications.

A rich and impressive list, indeed, but nothing compared to their crowning achievement. It's in the realm of **social interactions** that arthropods reach a level of refined behavior far beyond any other invertebrate animal. The social order of ants, bees, termites, and others is as intricate as any in the animal kingdom. They have developed a caste system wherein each animal has a specific role to play. For example, in termite colonies (Figure 11.7), which may number in the millions, there are **workers** to build the intricate tunnel systems and care for the queen, **soldiers** to defend the colony, and a **queen** to ensure a continual supply of eggs. Each is specialized to perform a crucial task; each contributes to the survival of the whole. As a result, the colony functions as one individual— a mammoth, mega-individual. This is, truly, division of labor at its most extreme.

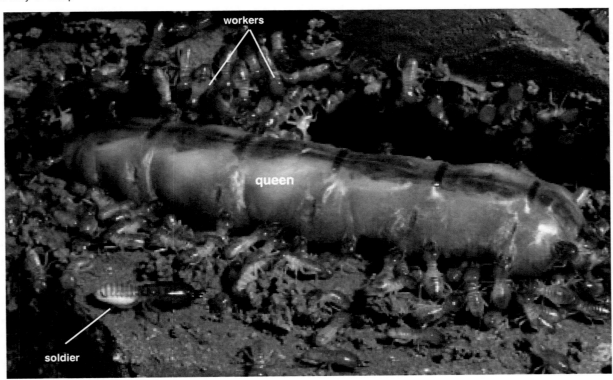

Figure 11.7. A termite colony with the huge queen tended by an array of workers and their large-headed soldier protectors.

Chapter 11 Summary

Key Points

1. Arthropods, including the spiders, crustaceans, centipedes, millipedes, and, especially, insects, have become the most successful animals on earth.

2. Arthropods were the first animals to truly invade land, being attracted to the abundance of available niches, lack of predation, and the profusion of food.

3. But to become terrestrial, arthropods had to overcome a number of obstacles brought on by dehydration, gravity, temperature extremes, and new food sources.

4. Several features have contributed to arthropod's unprecedented success: an exoskeleton, tagmatization, paired jointed appendages, metamorphosis, prolific reproduction, advanced sensory structures, and complex behavior.

5. The majority of animals are allocated to either the protostome or the deuterostome taxonomic branch.

Key Terms

Bilateria
blastopore
blastula
broadcast spawning
chitin
cladogram
cleavage
compound eye
deuterostome
ecdysis
ectoderm
endoderm
epicuticle
exoskeleton
gastrula
gastrulation
internal fertilization
mesoderm
metamorphosis
ocelli
patency (respiratory)
phylogenetic tree
procuticle
protostome
radial cleavage
Radiata
spiral cleavage
tagmata
tagmatization
zygote

Points to Ponder

1. What kind of animal first invaded land?

2. What was the attraction to invade land?

3. What were the eight barriers to terrestrialism that arthropods faced?

4. How did arthropods solve each barrier?

5. What is an exoskeleton and why is it so important?

6. What role does wax play in the success of the exoskeleton?

7. What are two disadvantages of the exoskeleton?

8. What is ecdysis and how does it contribute to the success of the exoskeleton?

9. Most animals belong to the Bilateria group. What features place them into this group?

10. What are the differences between a protostome and a deuterostome animal?

11. What is a cladogram?

12. What role do jointed appendages play in the overall success of arthropods?

13. What is tagmatization?

14. What is metamorphosis and what is its significance?

15. What role does a short life span coupled with high reproductive rates play in the success of arthropods?

16. Arthropods have two types of photosensors (i.e., eyes). What are they and how do they differ?

17. What type of behavior is seen in arthropods that surpasses all other invertebrate animals?

Inside Readings

Brusca, R. C., and G. J. Brusca. 1990. Invertebrates. Sunderland, MA: Sinauer Associates.

Gray, J., and W. Shear. 1992. Early life on land. Am. Sci. 80:444-456.

Little, C. 1990. The Terrestrial Invasion: An Ecophysiological Approach to the Origin of Land Animals. New York: Cambridge University Press.

Mayr, E. 1997. This Is Biology: The Science of the Living World. Cambridge: Harvard University Press.

Pechenik, J. A. 1996. Biology of the Invertebrates. Dubuque, IA: Wm. C. Brown.

Shear, W. A. 1993. One small step for an arthropod. Nat. Hist. 102: 46-51.

Wilson, E. O. 1992. The Diversity of Life. New York: W. W. Norton & Company.

12 Chapter 12

PHYLUM ARTHROPODA

THE CHELICERATES

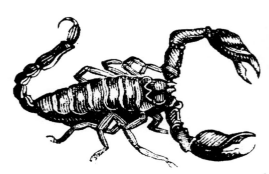

horseshoe crabs
spiders
ticks
scorpions
pseudoscorpions
sea spiders

Chapter 12

THE CHELICERATES:
Spiders...and more

Arthropod Variety

*I*t's hard to imagine the vastness of the phylum Arthropoda. The number of species (almost 1,200,000 known) is so immense that a classification scheme alone could literally fill a book. While we need to understand something of their taxonomy, we certainly cannot tackle anything like a complete scheme. So, to simplify matters we will restrict our discussion to extant (living) forms and limit the number of groups to those most representative of the phylum as a whole.

The phylum Arthropoda is normally subdivided into three subphyla: the chelicerates, the crustacea, and the uniramians. Each of the following three chapters presents a synopsis of one of these three subphyla.

Subphylum Chelicerata

The most primitive of the arthropods are, in many respects, also the most intriguing. To this taxon belongs the horseshoe crabs, the scorpions, the pseudoscorpions, the spiders, the sea spiders, the ticks, and the mites, among others. Truly, a picturesque troop if there ever was one. However, as different as they seem on the surface, they all share four special features: (1) no antennae, (2) no mandibles (i.e., chewing mouthparts), (3) modification of the first two appendages into structures called chelicera and pedipalps, and (4) a body organized into two regions: a cephalothorax (the combined head and thorax) and an abdomen. There are two important classes in this subphylum.

Class Merostomata

Early each summer, along the eastern coast of the United States, a vast pilgrimage occurs. Thousands upon thousands of tanklike horseshoe crabs (*Limulus polyphemus)* congregate to perform their annual breeding ritual. Males and females, like little armored combatants, grapple in their amorous embrace along miles of sandy beaches. As the female deposits and

buries her 30,000 or so eggs, the male floods them with his sperm. Now the real scramble begins as hordes of gulls, shorebirds, crows, jays, and other scavengers gorge themselves on the nutritive-rich eggs. The survival, in fact, of some migratory shorebirds may very well depend upon the abundance of mating horseshoe crabs.

However, horseshoe crabs (also known as king crabs) are not crabs at all. They are, instead, one of the most ancient of animals, more closely related to spiders and scorpions than to lobsters, fiddler crabs, or the many other true crabs. As such, they are primitive-looking creatures, covered dorsally with two armored casings, one comprising the head and thorax, called the cephalothorax or prosoma, itself covered by a shieldlike carapace. The other casing, the opisthosoma, envelopes the abdomen. Adding to their primitive appearance are a number of movable, protective spines located on the abdomen. Posteriorly, there is a spine-like tail, the telson, that the animal uses as a lever during locomotion and to right itself if placed on its back. Horseshoe crabs are the only member of the chelicerate subphylum to possess a pair of compound eyes, which are located on the prosoma.

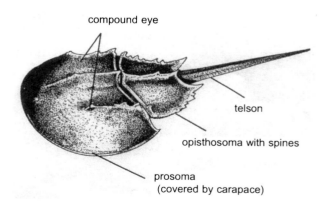

compound eye

telson

opisthosoma with spines

prosoma
(covered by carapace)

Figure 12.1. External anatomy of a horseshoe crab.

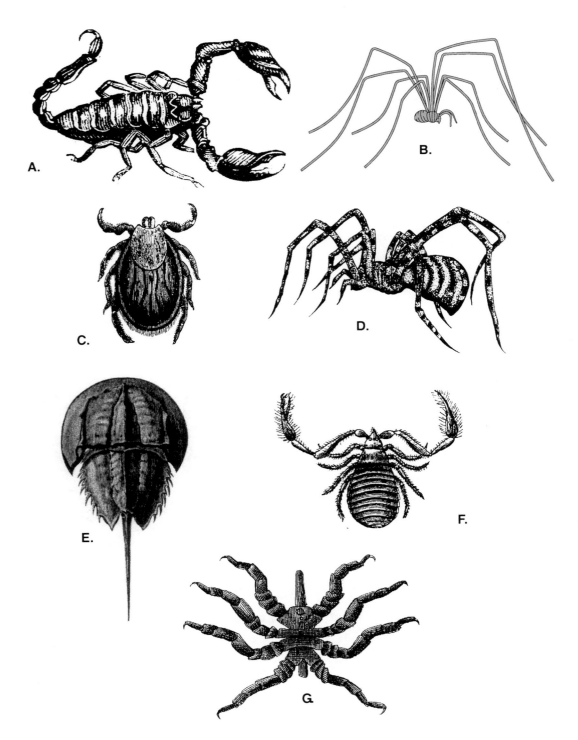

Figure 12.2. Representative members of the subphylum Chelicerata, not drawn to scale. **A**, scorpion. **B**, daddy longlegs. **C**, tick. **D**, spider. **E**, horseshoe crab. **F**, pseudoscorpion. **G**, sea spider.

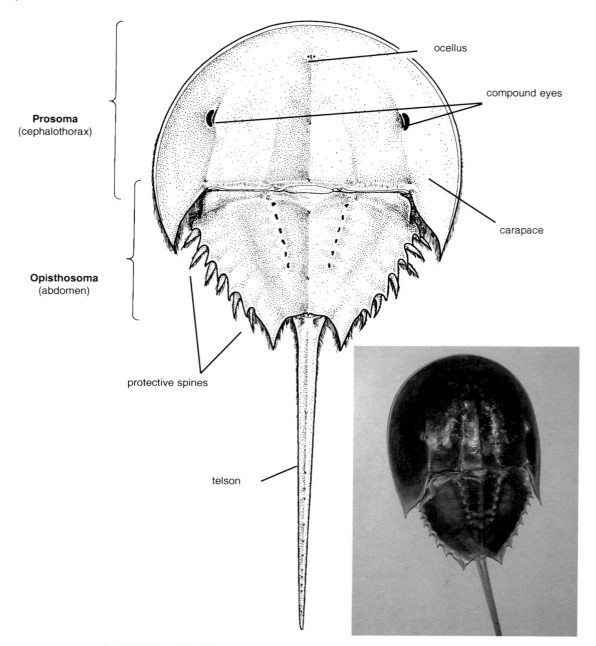

LIMULUS (HORSESHOE CRAB), EXTERNAL, DORSAL VIEW

Figure 12.3. A dorsal view of the horseshoe crab, *Limulus*.

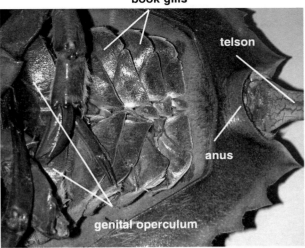

The primitive nature of *Limulus* is even more evident if we turn the animal over (Figure 12.4 & 12.5). Ventrally, the **cephalothorax** bears six pairs of appendages surrounding the mouth. The first pair, the **chelicera**, for which the subphylum takes its name, are small and possess terminal pincers (**chela**) to help manipulate food. The second pair, called **pedipalps**, are identical to the remaining appendages, except in males when they may be modified as sperm transfer devices. The next three pair, the **walking legs**, are used in locomotion and all bear terminal pincers for food handling. The last pair are modified into a spatula-like blade to remove debris from the gills. At the base of each appendage is a **gnathosome**, a device apparently used to crush and grind prey before swallowing. Posterior to the appendages, and located on the abdomen, is a series of flat, platelike appendages. The first plate, the **genital operculum**, functions in reproduction while the remaining five, the **book gills**, are respiratory structures. The anus is located at the junction of the abdomen and the telson.

Horseshoe crabs have remained virtually unchanged for 500 million years. Early fossils from the Paleozoic era are nearly identical to today's living forms in size and basic structure. They are marine animals living on the ocean floor searching or plowing about for mollusks, worms, or other small animals upon which they feed. Their body design and lifestyle constitutes one of evolution's true success stories.

Figure 12.4. Ventral view of *Limulus*, the horseshoe crab.

It's A Fact

Horseshoe Crab Blood Saves Lives!

There's a curious thing about *Limulus*...it has blue blood. For many years the pharmaceutical community has taken advantage of this oddity. If fact, today there is a $50 million industry centered around this blue blood, which has a special property...it can verify the purity of commercial drugs, a property no other blood has. When extracted and freeze-dried as Limulus Amebocyte Lysate (LAL), it is used to determine the presence of contaminants in virtually all drugs used in this country and elsewhere. Horseshoe crab blood is the only known source of this compound. Each year over 300,000 crabs are collected, bled, and then returned to their ocean habitat.

* ANOTHER INTERESTING FACT ABOUT HORSESHOE CRABS IS THAT CHITIN FROM THEIR SHELLS IS USED IN THE MANUFACTURE OF SURGICAL SUTURES.

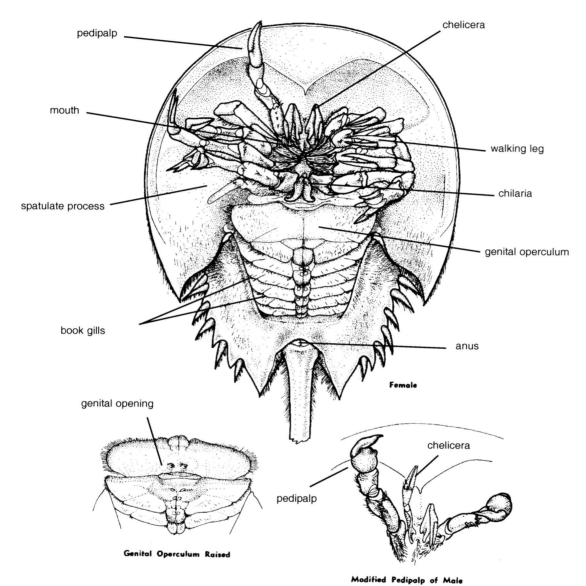

pedipalp

mouth

spatulate process

book gills

chelicera

walking leg

chilaria

genital operculum

anus

Female

genital opening

Genital Operculum Raised

chelicera

pedipalp

Modified Pedipalp of Male

LIMULUS (HORSESHOE CRAB), EXTERNAL, VENTRAL VIEW

© Kendall/Hunt Publishing Company

Figure 12.5. A ventral view of the horseshoe crab, *Limulus*.

Class Arachnida

There are few animals that generate more basic fear than spiders and scorpions. The over 100,000 species of spiders, scorpions, ticks, mites, daddy longlegs (harvestmen), whip scorpions, sun spiders, and pseudoscorpions comprise the class of arthropods known as arachnids. To many of us, they are the most loathsome and fearsome of creatures. Hairy, mysterious, poisonous, and noxious, they represent the underworld of the animal kingdom. Nightmares are filled with these creepy demons. Our popular jargon even has a special name for fear..."arachnophobia" it's called. What's so special about arachnids to stir up such interest?

First of all, they are primarily terrestrial animals. There are only a few aquatic species, and even those are thought to be derived from terrestrial ancestors. They possess eight legs and the body, like that of the horseshoe crab, is divided into two regions. Anteriorly is a large **cephalothorax** (prosoma), covered by a carapace, and posteriorly is an **abdomen** (opisthosoma). In spiders, the prosoma and abdomen are joined by a narrow stalk, the **pedicel**. This allows the abdomen a greater range of movement, especially in laying down silk.

In ticks and scorpions, however, the two body parts are broadly connected. The abdomen may be segmented, as in scorpions, or fused, as in most spiders and ticks. In scorpions, the abdomen ends in a stinging apparatus (Figure 12.6) that can be quite dangerous, even fatal, to humans.

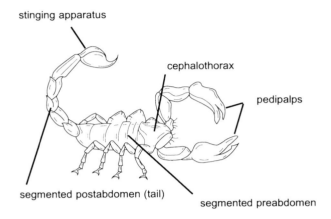

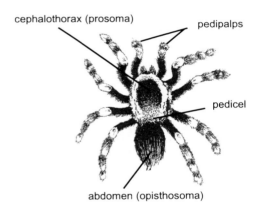

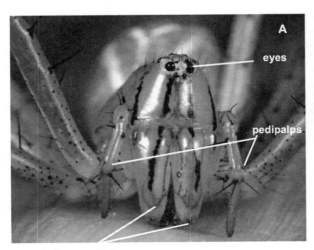

chelicerae with fangs

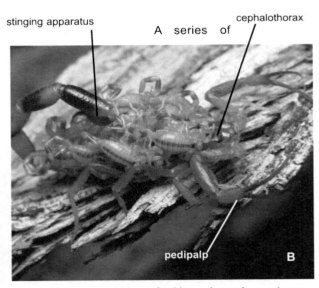

Figure 12.6. A comparison of spider and scorpion anatomy. The frontal view of an adult spider (A) shows compound eyes and mouthparts. A female scorpion (B) carrying young on her back. Note pincerlike pedipalps and segmented abdomen with terminal stinger.

A series of appendages projects from the cephalothorax. The first, a pair of **chelicera**, are used by these arachnids to process food. In scorpions and ticks, the chelicera are rather small and used to manipulate food into the mouth, while in spiders they bear an internal poison gland and terminal fangs for delivery of venom (Figure 12.6).

Virtually all adult spiders are venomous although only a dozen or so species worldwide are dangerous to humans. Two that occur in the midwestern United States are the notorious black widow (*Latrodectus mactans)* and the brown recluse (*Loxosceles reclusa*). The black widow, recognized by the black, shiny exoskeleton and red hourglass shaped design on the abdomen usually occurs in shaded areas such as debris heaps or woodpiles. The brown recluse is more inclined to inhabit basements or attics, thus, people have a greater chance of encountering it than the black widow. The brown recluse can be identified by a dorsal violin-shaped marking on its thorax. Contrary to popular myth, the female black widow (and other female spiders, for that matter) seldom devour the male following mating.

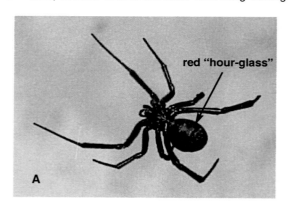

red "hour-glass"

A

B

Figure 12.7. The above photographs are of the two most poisonous spiders in the U.S., the black widow (top) and the brown recluse (lower). Note the red "hour-glass" shape on the ventral surface of the black widow and the "fiddle" design on the back of the brown recluse (arrow)...also refer to the following page.

The second pair of arachnid appendages are the **pedipalps**. In spiders, ticks, and mites, these paired structures appear as miniature legs used either as tactile sensors, to process food, or by the male to assist in sperm transfer. In scorpions, however, the pedipalps are large and elongated with grasping claws, the chela, similar in appearance to lobster chelipeds. Posterior to the pedipalps are four pairs of walking legs.

Respiration in arachnids is accomplished by two means. Spiders and scorpions use internal **book lungs**, similar to the book gills of horseshoe crabs, although located internally. These leafy respiratory surfaces communicate to the outside through small slotted ventral openings, the **spiracles**. Air enters the spiracles, floods the gill surfaces, and diffuses into the circulatory system. In other arachnids, a system of branching tubules, the **trachea**, originate at the spiracular openings to carry oxygen directly to internal tissues (Fig. 12.8). Trachea of this type are found in ticks, mites, insects, and on the abdomen of some spiders.

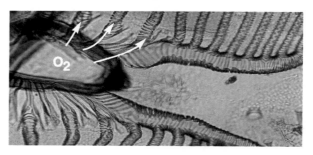

O_2

Figure 12.8 An example of the tracheal respiratory system used by various arthropods.

The sensory apparatus of arachnids is relatively well developed. While some have no eyes at all, most have four pairs of **simple eyes** able to detect variations in light intensity and gross movements. These eyes occur in two forms: *primary eyes* for day vision and *secondary eyes* for night or evening vision. It is doubtful that arachnid eyes form discrete images, although some specialists are of the opinion that hunting spiders can locate suitable prey based upon shape.

Scattered over the surface of arachnids are a number of other sensory structures. Setae, hair, bristles, or spines are used as **mechanoreceptors** to detect sound vibrations. Some arachnids also possess special **slit organs** distributed along their ventral surface. A membrane covering the slit vibrates in response to sound waves and transmits information to the nervous system. All arachnids possess **taste** and **olfactory sensors** scattered over the epidermis, especially concentrated around the mouth and head. Finally, scorpions bear unique comblike appendages, the **pectinate organs**, suspended ventrally from their thorax. These are thought to be either chemoreceptors, capable of tasting/smelling the environment, or mechanoreceptors, which can sense subtle vibrations.

red "hour-glass"

The Black Widow Spider
(ventral view)

What Arachnids Eat

Spiders. Arachnids, as a group, use a variety of feeding strategies. Spiders, for instance, are carnivorous, dispatching prey with a **neurotoxic venom**. Poison glands buried at the base of the fangs pump venom into the prey, partially digesting it from the inside. Since spiders lack chewing mouthparts, they must convert or pulverize their prey (usually insects) into a semiliquid pulp, which is then ingested. There is also strong evidence that many spiderlings, before they become predacious, survive by eating *pollen granules* that have become trapped on the sticky threads of webs. This makes perfect sense since many spiderlings are so tiny that subduing even the smallest insect would be a considerable feat.

Scorpions. Scorpions, too, are predacious, killing their quarry with poison from the abdominal stinger. Although insects are the primary food of scorpions, they have been known to feed on other, often larger, animals, such as lizards and snakes. Prey are restrained by the large chela (pincers), the tail is lowered, and the venom is then forcibly injected. The illustration below depicts a scorpion in its feeding posture. Once the prey has been immobilized, external digestion begins. As the food is partially dissolving, the small clawed chelicerae crush or macerate it into a mushy brew that is then ingested.

While most scorpions are not considered dangerous to humans, stings being similar to that of a wasp or bee, a few species, including one that occurs in Mexico and the southwestern United States, can inflict debilitating, even fatal stings. As with any toxic animal, if encountered it is best to leave them undisturbed.

Ticks. Unlike spiders and scorpions, ticks do not prey n other animals. They, instead have developed slicing and piercing mouthparts adapted to feeding on blood. Virtually any vertebrate animal can be host to ectoparasitic ticks. Once attached, a feeding tick may remain for several days until it becomes engorged with blood, becoming two to three times larger than when it started. When sated, they randomly drop off to begin their reproductive cycle. Although ticks, themselves, are relatively harmless, their bite being barely noticeable, they can serve as vectors of several viral human diseases. Lyme disease, Rocky Mountain spotted fever, tick fever, and tick-borne typhus can all be transferred through a feeding tick.

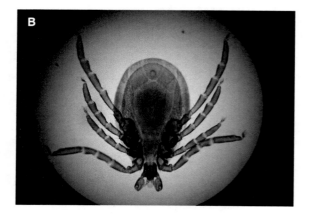

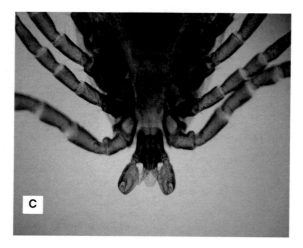

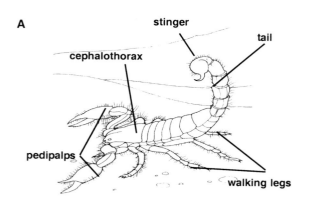

Figure 12.9. A, a scorpion in defensive posture; **B** & **C**, the common wood tick, *Dermacentor*. Note the mouthparts modified for cutting and slashing through flesh and sucking on blood.

Mites. Mites seem to have adopted a number of different feeding patterns. Some are herbivores, others are scavengers, a number are predacious, and still others are parasitic on invertebrate and vertebrate animals, as well as on plants. Regardless of the feeding method used, all mites have one thing in common: they are diminutive, rarely exceeding 1 mm in length. It seems mites have discovered that being tiny has certain advantages. They have found that there are plenty of recesses, nooks, and caverns, for extremely small arthropods. As a result, they have invaded just about every miniature ecological niche. Consider the following kinds of mites: dust mites, spider mites, pillow mites, mange mites, scabies mites, skin mites, and chigger mites...just a few of the many minuscule pests known as mites.

Even though tiny, they can be serious economic or medical pests. They destroy crops, reduce the vitality of domestic animals, and may serve as disease vectors (e.g., they carry various plant viruses, and at least one mite transmits the disease scrub typhus to humans). *Sarcoptes scabei,* the scabies mite, may be best known for causing mange in animals, but it also is common on humans, worldwide. See *Inside View 12.1,* page 281, for a more detailed look at the relationship between mites and humans.

Other Arachnids. *Pseudo-* or *false scorpions* resemble true scorpions but lack the elongated telson and stinging apparatus. They are small (< 7 mm) and reside in a variety of habitats such as under rocks, in leaf litter, under bark, and in loose soil. They feed on other invertebrates. The *harvestmen* (daddy longlegs) are primarily predators on other arthropods, although they will, at times, revert to scavenging of decaying vegetable matter and flesh.

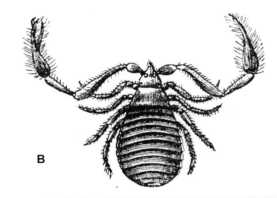

Figure 12.10. Various arachnids. **A.**, the mange mite, *Sarcoptes scabei,* causative agent of mange (scabies) in animals. Note the many bristles and spines that serve to anchor the parasite among the fur of its host. Plate **B** shows a pseudoscorpion. A scorpion look-alike that lacks the stinging apparatus. A daddy longlegs is shown in plate **C**.

Silk

Many spiders use silken threads to assist in locating and capturing their prey. **Silk** is formed by special silk glands situated in the abdomen and deposited by small, fingerlike **spinnerets** located at the abdomen's tip. There are six known types of silk, differing in thickness and function. For example, silk may serve as a cocoon to enclose the eggs, to form a mat on the floor of a nest, or even for locomotion. By ejecting a long, thin, single thread, a young spiderling can be carried by the wind from one location to another in a process called "ballooning." Another function of silk is as a safety line. As a wandering spider roams about, it constantly attaches threads to the substrate. If it slips or is dislodged, it is caught by the drag line instead of falling to the ground.

However, the most common use of silk is as an aide in feeding. The threads may form a tunnel or a trip line that signals to a hidden spider that prey is nearby, or the silk can be woven into **webs** varying from simple designs to extremely elaborate constructions.

One of the more complex is the **orb web** in which various zones are used for different purposes. For instance, the central hub serves as a waiting platform for the resident spider, while the bulk of the web, the **sticky spiral**, contains a glue to entrap unwary prey. Once the prey is entangled in the adhesive threads, the spider paralyzes it with a quick bite and then enwraps the stupefied victim in a shroud of silk. The victim may be eaten immediately or saved for later.

Many orb weaving spiders build a new web each day, and have been clocked to complete construction in less than 30 minutes. In many cases, the spider actually eats the old web before constructing the new one. "Waste not" is apparently the motto of these frugal arachnids.

Figure 12.11. Uses of spider silk. On the lower left is an orb web showing the central hub and surrounding sticky spiral that entraps the prey. The spider on the upper right photo is wrapping a paralyzed bee within a shroud of silk for a later meal. The other two photographs shows the attack position of garden spiders on their webs.

Sea Spiders
(Class Pycnogonida)

Some of the strangest creatures on earth are the sea spiders. They are all slow-moving arthropods found throughout marine ecosystems. Some patrol the tidal zones for small prey, while others search the benthic depths for worms, or other invertebrates. Although they are superficially similar to spiders, they belong to an altogether separate group, the class Pycnogonida. Nonetheless the two animals do share enough features (e.g., presence of chelifores [chelicera] and palps [pedipalps], liquid feeding habits, and parallel structures of their brains) to be considered close relatives.

In addition to the above structures, sea spiders possess several distinct structures of their own: (1) both sexes possess special grooming appendages, the **ovigers**, emanating from the head (in the male, the ovigers are used as special brooding chambers for eggs during embryonic development), (2) the head is projected anteriorly as an elongated "proboscis" bearing the **chelifores** and **palps**, (3) the body proper ("trunk") is greatly reduced, so much so that the gonads and digestive system *penetrate into the legs,* (4) the trunk has lateral projections, called **pedestals**, to which the legs are attached, (5) each leg contains numerous ventral **gonopores** as exit channels for egg or sperm release, (6) the abdomen is reduced to a stubby, vestigial stump that contains the anus, and (7) the legs are often extremely long, at times, more than 16 times the length of the body. Also, the legs differ from all other arthropods in having nine segments or joints instead of the standard four.   This body design, with a reduced, pedestaled trunk, a vestigial abdomen, nine-jointed legs, and appendages modified in the male for egg attachment, is different than that of any other known animal.

Sea spiders are certainly arthropods, since they have an exoskeleton and paired jointed appendages, but they are so distinct as to rate a separate class, and in the eyes of some taxonomists, even a separate subphylum.

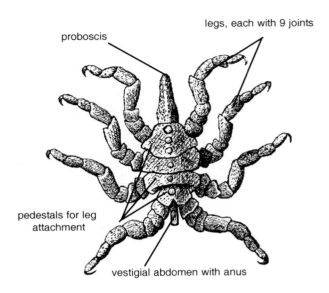

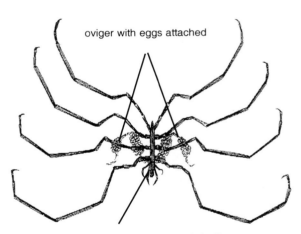

Figure 12.12. Two examples of pycnogonids. Note the large proboscis-bearing palps (pedipalps) and chelifores (chelicera), the reduced trunk with lateral pedestals for leg attachment, the reduced abdomen, the nine-jointed legs, and the grooming appendages (ovigers) with eggs attached in the male.

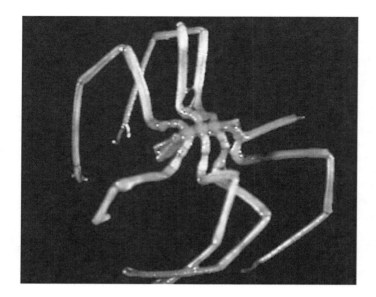

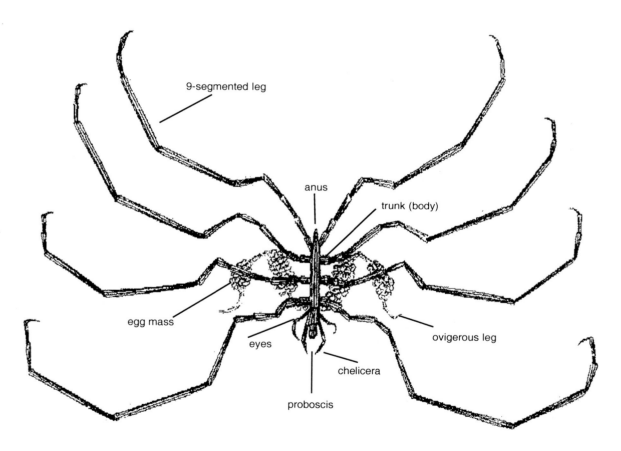

Figure 12.13. The anatomy of a male sea spider belonging to the genus *Nymphon*. Note the extremely small body, the elongated 9-segmented legs, and the egg mass attached to the oviger.

An Inside View 12.1

Mighty Mites

*...the average living room is a hive of activity and the bedroom is probably busier still,
where even the most virtuous of us share our beds with thousands of other creatures.*

\- Brookesmith

Facial Warriors

My guess is that this morning you bathed. Or, perhaps it was yesterday? Maybe you used one of those liquid, antibacterial soaps; after all, the primary aim of bathing is to rid ourselves of external microbes and the smell they spawn, body odor we call it. The problem is that you, like everyone else, probably weren't very successful in combating these ubiquitous foes. But rest easy, you have a comrade, an ally, in this fight. Unknown to you, day in and day out, your skin is a battleground where tiny warriors contend for supremacy. Bacteria are feeding on your surface debris and mites, in turn, are feeding on the bacteria. That's right, mites, tiny arachnids, minuscule parasites are in residence on your body's surface. You have them, I do, we all do. They are skin mites with the impressive names of *Demodex folliculorum* and *Demodex brevis*.

The first type (*folliculorum*) resides in pits at the base of hair follicles, and, for some reason, they prefer the hair of your eyebrows. The other species, *brevis*, lives in the adjacent sebaceous glands. Both like to gobble up skin flakes, body oil (sebum), makeup, or anything else lingering on your wrinkled brow. But what they really, truly relish are bacteria. These mighty combatants escape the friendly confines of hair follicles or sebaceous glands, to live for a few days on your face, feeding on the abundant bacteria, and then crawl back into their crypts to lay eggs. The newly hatched mites eventually make their way to the surface, as their parents did before them, and the whole cycle repeats.

However, unlike some parasites, skin mites are harmless and, in fact, may actually contribute to the health of your skin! But where did they come from? How did you become contaminated with these miniature hedge hogs? Well, it seems that you can pick them up in a variety of ways, usually through contact with an infected person. They are transferred by casual touching, by wearing someone else's hat, using a friend's face cloth, or perhaps through kissing. There's even evidence that good ole mom may have delivered them to you during breast feeding! And just in case you would like to view these miniature gluttons crawling about your face, here's how. Pluck a few hairs from your eyebrows, place them on a microscope slide, and take a peek under low power.

You will discover that they are quite cute. Like little stubby cigars, these tiny mites, with tiny legs, patrol the cosmic moonscape of your forehead searching for a feast of bacteria and dead skin.

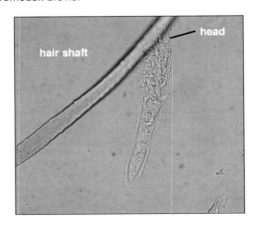

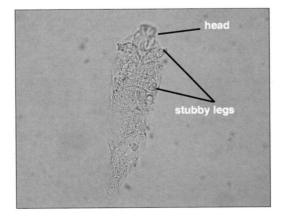

Figure 12.13. The cigar-shaped skin mite, *Demodex*, lives in association with hair follicles where it feeds on skin flakes and bacteria. The photograph to the left shows a skin mite attached to a hair shaft.

Pillow Fights and Dust Storms

One day not too long ago, while reading near a window, I became captivated by the flow of dust particles streaming in the sunlight. I knew that people of science were supposed to be inquisitive so, in keeping with that pioneer spirit, I began to contemplate dust. Just what is it? Where does it come from? What's it doing in my house? Why is it streaming in my sunshine? Well, after some searching and rummaging in one dust-covered book after another, here's what I found out. First of all, about 80% of household dust is dead human skin. That's right, skin, our own skin. After some rather unsettling reflection on that little fact, I finally concluded that it's really not too surprising since skin is our largest organ and it's replaced every month or so. New cells, by the millions, constantly, incessantly, push dead ones up to the surface where they fall away in dusty little clouds. Imagine, millions of dead, scaly, skin flakes shed by each of us every day. They take up an airborne existence to travel around our homes, offices, schools, and restaurants like tiny snowflakes drifting aimlessly here and there. I discovered that they float on currents of air outward from the center of a room toward its edges where they fall in a veritable blizzard. That's why dust concentrates on picture frames, fireplace mantles, TV sets, and unopened zoology books.

If roving skin makes up around 80%, what's in the other 20%? What's lurking there? Experts in dust content tell us that there is a patchwork of items in that remaining one-fifth. It seems, for instance, that in addition to skin flakes, dust also contains a share of pollutants, pollen, human hair, carpet fibers, insect and spider excrement and their exoskeleton fragments, and, oh yes, the mites. Carpet mites, otherwise known as *Dermatophagoides pteronyssimus,* and these, even more than their *Demodex* cousins, relish dead skin. They drift from floor, to ceiling, to window sill, to grandfather clock, in a constant tide of dust, feeding all the while on those gliding flakes of skin. Come to think of it, I'm rather glad they're here among us. Who wants to wake up each morning covered by a film of dusty scales?

Now that brings us to the bedroom. You would think that there's at least one pest-free place in our lives, one sanctuary where we can escape Mother Nature and her mischievous moods. Although there may very well be such a place somewhere, it's not your bedroom, not your bed, and certainly not your trusty, cherished pillow. In fact, a bed is a perfect incubator for mites; it's warm, it's humid, and it has a nocturnal, endless source of food. It is estimated that a single bed is a breeding ground for millions of dust mites. There we emit our body heat, breathe and sweat moisture into the mattress, toss and turn constantly scraping off dead flakes like falling snow, all the while providing nourishment for our tiny bed companions.

Oh, there's one more thing. About that pillow of yours, and who doesn't like to have a nice fluffy pillow on which to rest one's weary, fact-filled head? Pillows, one of life's true comforts. But, you guessed it! Mites adore pillows; it's their Garden of Eden. Hundred of thousands tumble out when you "plump" one, and washing doesn't seem to offer much control. There, on our warm pillows, mites feed, defecate, lay their eggs, raise their young, and die. And while they don't seem to harm us any (unless you suffer from asthma), somehow knowing they are there makes me a bit uncomfortable. Somehow, my psychological well-being doesn't relish the idea of sleeping on a mat of mites, rolling about in their dead carcasses, wallowing in their copious excrement. As a matter of fact, it bothers me a lot. And I suppose the next time I drift off to sleep in that hotel room, I'll be thinking of the many squatters who have preceded me offering their flaky feast to the legions of mites nestled somewhere below, keeping them fat, healthy, and happy.

Chapter 12 Summary

Key Points

1. The phylum Arthropoda contains an immense number of species assigned to three major subphyla.

2. The subphylum Chelicerata contains the mites, ticks, true scorpions, pseudoscorpions, true spiders, sea spiders, and horseshoe crabs.

3. All members of this subphylum possess the following characteristics: no antennae, no mandibles, a pair of chelicera, a pair of pedipalps, and a two-part body (the cephalothorax and abdomen).

4. Horseshoe crabs (class Merostomata) are primitive arthropods having existed virtually unchanged for over 500 million years.

5. Horseshoe crabs have a two-part body (prosoma and opisthosoma) and a posterior tail-like telson. They possess paired walking legs with basal crushing devices known as gnathosomes, and members of the group use platelike structures, the book gills, for respiration.

6. The largest group of chelicerates are the arachnids: spiders, ticks, mites, daddy longlegs, scorpions, and pseudoscorpions.

7. Arachnids are essentially all terrestrial animals whose body is divided into an anterior prosoma and a posterior opisthosoma. The anterior appendages, the chelicera, are food-processing structures that, in spiders, are modified into poison fangs. Pediplaps are usually sensory structures (although are modified into large claws in scorpions). All arachnids also possess four pairs of walking legs.

8. Respiration is accomplished either by book gills or through an interlacing series of internal tubes, the trachea. The tubes allow air to penetrate throughout the animal's internal body.

9. Most arachnids as adults are predacious, although some (mites and ticks) are parasitic. Some of the parasitic forms can transmit a number of diseases to humans (e.g., Lyme disease, Rocky Mountain spotted fever, typhus, and tick fever).

10. Silken threads are extruded from the posterior abdomen of spiders through a series of fingerlike spinnerets. Silk is used for a variety of purposes.

11. Mites are cosmopolitan parasites that may be found on or associated with humans.

12. Sea spiders (class Pycnogonida) are strange marine arthropods. They are long-legged arachnids with an elongated proboscis, nine-segmented legs bearing gamete openings, the gonopores, reduced abdomen, and special grooming devices called ovigers.

Key Terms

book gills (lungs)
carapace
chelicera
chelifores
genital operculum
gnathostome
gonopores
horseshoe crabs
mites
opisthosoma
ovigers
pectinate organs
pedicel
pedipalps
prosoma
slit organs
spiracles
telson
ticks
trachea

Points to Ponder

1. What four features do all members of the subphylum Chelicerata share?

2. Describe a horseshoe crab.

3. Name the different kinds of arachnids.

4. Name three ways that arachnids differ from all other arthropods.

5. Describe the two ways that arachnids breathe.

6. What is the function of slit and pectinate organs?

7. What is the function of spinnerets? silk glands? What are some of the ways that spiders use silk?

8. What two Illinois spiders are potentially dangerous to humans?

9. How do ticks feed?

10. What are *Demodex* mites? Where are they found? What do they feed on? How does a person become contaminated with them?

11. Where can you find dust mites? What do they feed on?

12. What are sea spiders? What features make them distinct from other arthropods? On what do they feed?

Inside Readings

Arnaud, F., and R. N. Bamber. 1987. The biology of Pycnogonida. Adv. Marine Biol. 24:1-96.

Brownell, P. H. 1984. Prey detection by the sand scorpion. Sci. Am. 251(6):86-97.

Brusca, R. C., and G. J. Brusca. 1990. I invertebrates.Sunderland, MA: Sinauer Associates.

Eberhard, W. G. 1990. Functions and phylogeny of spider webs. Annu. Rev. Ecol. Syst. 21:341-372.

Emerson, M. J., and F. R. Shram. 1990. The origin of crustacean biramous appendages and the evolution of Arthropoda. Science 250:667-669.

Foelix. 1982. Biology of Spiders. Cambridge: Harvard University Press.

Hopkin, S. P., and H. J. Read. 1992. The Biology of Millipedes. New York: Oxford University Press.

Kantor, F. S. 1994. Disarming Lyme disease. Sci. Am. 271(3):34-39.

Kaston, B. J. 1978. How to Know the Spiders. Dubuque, IA: Wm. C. Brown.

McDaniel, B. 1979. How to Know the Ticks and Mites. Dubuque: Wm. C. Brown.

Pechenik, J. A. 1996. Biology of the invertebrates. Dubuque, IA: Wm. C. Brown.

Polis, G. A. ed. 1990. The Biology of Scorpions. Stanford, CA: Stanford University Press.

Rudloe, A., and J. Rudloe. 1981. The changeless horseshoe crab. Natl. Geogr. pp. 562-572 (Apr).

Savory, T. 1977. Arachnida. New York: Academic Press.

Shear, W. A. 1994. Untangling the evolution of the web. Am. Sci. 82:256-266.

Sonenshine, D. E. 1991. Biology of Ticks. New York: Oxford University Press.

Vollrath, F. 1992. Spider webs and silks. Sci. Am. 266(3):70-76.

Weygoldt, P. 1969. The Biology of Pseudoscorpions. Cambridge: Harvard University Press.

Witt, P. N., and J. S. Rovner, eds. 1982. Spider Communication: Mechanisms and Ecological Significance. Princeton: Princeton University Press.

A Female Wolf Spider Carrying an Egg Case

Chapter 12

Laboratory Exercise

The Chelicerates

Laboratory Objectives

✓ to identify the three subphyla of arthropods that we have studied,
✓ to briefly characterize each subphylum through observation of living and preserved specimens,
✓ to identify anatomical features of the horseshoe crab,
✓ to identify the major external features of a spider from preserved specimens, and
✓ to identify the major external features of a sea spider from preserved specimens.

Needed Supplies

✓ preserved specimens of *Limulus*, the horseshoe crab
✓ preserved specimens of spiders, mites, ticks, scorpions, and sea spiders

Chapter 12
Worksheet

Phylum Arthropoda
Subphylum Chelicerata

Class Merostomata

1. *Limulus*, the horseshoe crab, is a living example of a primitive arthropod. Refer to Figures 12.3 and 12.5 for dorsal and ventral views. Examine the dorsal surface of a specimen and locate both simple and compound eyes, the posterior telson, and the shieldlike carapace covering the cephalothorax. Note that segmentation is indicated by the double line of pores on top of the abdomen as well as by the movable spines along its rim.

The abdomen shows obvious fusion of neighboring segments. What is fusion of this sort called?

How do you think the movable abdominal spines are used?

2. Now turn the specimen over and observe the ventral structures. Locate:
   - ✓ mouth
   - ✓ genital operculum
   - ✓ chelicera
   - ✓ anus
   - ✓ book gills
   - ✓ pedipalps (can you tell if your adult specimen is male or female?)
   - ✓ walking legs
   - ✓ grooming leg (the last of the walking legs)

Class Arachnida

3. Examine the scorpions, spiders, ticks, and mites on display. Note the narrow waist in spiders and the fanged chelicera. Compare the leglike pedipalps of spiders to the pedipalp pincers of scorpions. Try to locate the spinnerets of spiders (what is their function?) and the segmented tail of scorpions.

4. Observe a *Demodex* mite under low power of a microscope. Compare the body shape of a follicular (*Demodex*) mite with a scabies mite. To what do you attribute the difference in shape?

5. Finally, observe the tick specimen on display. Note its ovoid shape (it looks a little like a watermelon seed). What is its food?

Class Pycnogonoida

6. Examine the sea spiders on display. Note the elongated legs and small barrel-shaped body. How many legs does the specimen have? How many joints are there in each leg? What are ovigers?

13 Chapter 13
PHYLUM ARTHROPODA

THE CRUSTACEANS

barnacles
shrimp
crabs
lobsters
copepods
isopods

Chapter 13

The Crustaceans

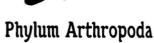

You cannot teach a crab to walk straight.

- Aristophanes

Phylum Arthropoda
Subphylum Crustacea

A number of representative crustaceans are shown in Figure 13.2 on page 290. These are, for the most part, common and recognizable animals contributing significantly to the world's food supply. They are an extremely widespread, diverse, and prolific group of organisms including such forms as planktonic krill, crabs of various sorts, lobsters, crayfish, shrimp (prawns), copepods, isopods (roly-polys), and barnacles. They range in size from the very tiny (<1 mm) planktonic forms to giant Alaskan crabs with a leg span of over 13 feet.

Except for the terrestrial isopods (roly-polys and wood lice), crustaceans are all aquatic, found throughout both marine and freshwater ecosystems. They may occur at the shoreline, in the tidal zones, in brackish water, in ponds, rivers, ditches, or deep in the oceans. Some even live in water-filled tunnels.

Characteristics of the Crustaceans include:

(1) **two pairs of antennae;** the first pair is usually called antennules;

(2) a body normally arranged into **two tagmata**: the *cephalothorax* (often covered by a shieldlike carapace) and the *abdomen*;

(3) **biramous appendages**, demonstrating the concept of *serial homology* (see the *Inside View 16* on page 302);

(4) special food-preparing appendages, the **mandibles**;

(5) predominately **gill respiration**, although a few cases of cutaneous respiration exist;

(6) osmoregulation is usually accomplished by way of **antennal** (green) **glands**;

(7) the **exoskeleton** is generally impregnated *with calcareous deposits,* resulting in a tough, hard, outer shell,

(8) for the animal to grow, the exoskeleton is periodically shed, a process called **ecdysis**.

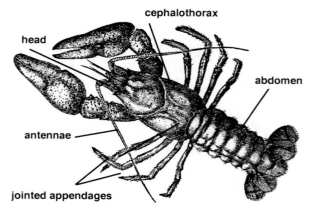

Figure 13.1. External features of a representative crustacean, the crayfish.

head — cephalothorax — abdomen — antennae — jointed appendages

As you might suspect with such a large, diverse group, the classification system is quite extensive. In fact, it's rather difficult to fully appreciate crustaceans without a glimpse into their taxonomy as given below.

Phylum Arthropoda

Subphylum Crustacea

Omit **Class Branchiopoda** - small freshwater animals with leaflike appendages: brine shrimp, clam shrimp, water fleas. The common crustacean, *Daphnia* is a member of this class.

(there are ~ 800 described species)

Omit **Class Maxillopoda–Subclass Ostracoda** - minute crustaceans with a hinged bivalve carapace enclosing head and body; a single median eye; limbs reduced; benthic, planktonic and a few terrestrial forms: bean shrimp

(there are ~ 8,000 described species)

Omit **Class Maxillopoda–Subclass Copepoda** - a single, median eye; no carapace; most are tiny <10 mm long; often carry eggs trailing in special appendages, the ovisacs: copepods

(there are ~ 9,000 described species)

Omit **Class Maxillopoda–Subclass Branchiura** - oval; with mouthparts modified for parasitism on various fish: fish lice

(there are ~ 130 described species)

Class Maxillopoda–Subclass Cirripedia - body greatly modified; limbs absent; bivalve carapace; compound eyes lost; reduced head: barnacles

(there are ~ 1,000 species)

Class Malacostraca - 19-20 segments each bearing a pair of primitive biramous appendages.
(there are more than 20,000 described species). There are four important orders:

Omit {
Order Isopoda - legs similar, terrestrial habitats: pillbugs, wood lice
Order Amphiopoda - body laterally flattened: sand fleas
Order Euphausiacea - makes up much of oceans' plankton: krill
Order Decapoda - 10 walking legs: shrimp, lobsters, crayfish, crabs] learn this

To keep our discussion of the crustaceans under some control, we will limit ourselves to just two classes: the diverse and well-known **malacostracans** (lobsters, shrimp, crayfish, and crabs) and the more atypical, lesser-known, **cirripedians** (barnacles).

Class Malacostraca

To the upper crust of society, a gourmet meal is incomplete without a malacostracan. Dinner tables of the rich and famous are festooned with lobster, shrimp, and crab—a feast appreciated by kings, presidents, and CEOs. However, it's not only the elite that appreciate delicacies; the rest of us indulge as well. For example, every cajun in Louisiana knows of a delight equal to that of any lobster. Drive the country roads of just about any Louisiana parish on a warm spring afternoon and you will see entire families picnicking along bayous, ponds, and ditches. Look closely and you will see baited traps scattered throughout the roadside waterways. They're practicing the age-old custom of "crawfishing"—catching buckets of crayfish for sumptuous spreads of etouffee or bisque. In both instances, through the fancy lobster plate and the simple crayfish boil, crustaceans are doing their part to feed a hungry world.

But, humans are not the only ones to appreciate the nutritional pre-eminence of malacostracans. Small fish, large fish, squid, sharks, seals, seabirds, and whales gorge themselves on everything from planktonic krill to crabs and lobsters. One authority claims that a single whale may consume four tons of krill (a small marine crustacean) per day. Raccoons, mink, otters, and a variety of game fish, from large-mouth bass to northern pike, dine eagerly on crayfish. Truly, an enormous variety of marine and freshwater animals depend upon this taxon for their daily fare. It doesn't take much to appreciate the economic value of this group of animals.

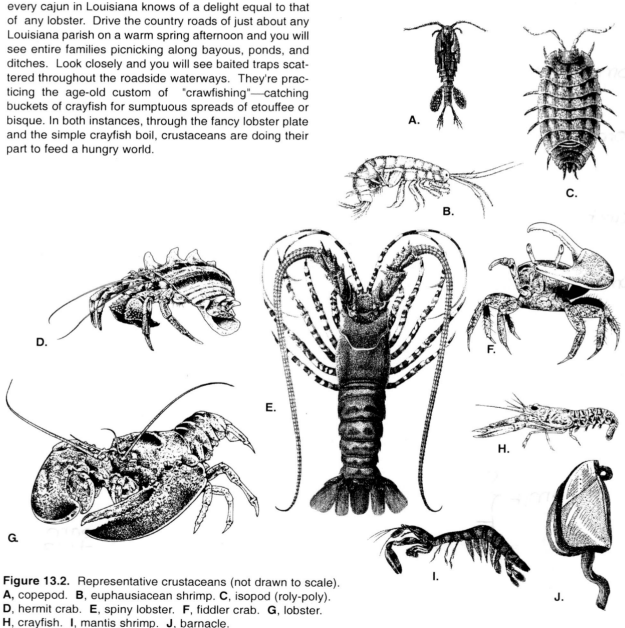

Figure 13.2. Representative crustaceans (not drawn to scale).
A, copepod. **B,** euphausiacean shrimp. **C,** isopod (roly-poly).
D, hermit crab. **E,** spiny lobster. **F,** fiddler crab. **G,** lobster.
H, crayfish. **I,** mantis shrimp. **J,** barnacle.

The demand for shrimp, lobster, crayfish, and crabs make them among the most relentlessly hunted of any animal. The numbers harvested from the wild runs well into the millions of tons per year. To compensate for this level of depletion, there are ongoing attempts to commercially farm both shrimp and crayfish. So far these attempts have yielded promising results that will hopefully ease the pressure on natural populations.

Malacostracans represent somewhere around 75 - 80% of all crustaceans, and while there is considerable diversity within the group, the crayfish and lobster are similar enough that either can be used as an example of a typical crustacean. During the following discussion of the crayfish, keep in mind the general characteristics of the subphylum given at the beginning of the chapter. Many of these diagnostic features are found, not only in the crayfish, but throughout all crustaceans.

Malacostracan (Crayfish) Anatomy

Appendages

Much of the scientific interest in crayfish is centered on their appendages. There are two prevailing reasons for that. First, they show a fundamental, primitive design, that of the **biramous appendage** (see the *Inside View 13.1* on page 294) in which there are two branches (rami) radiating from a central base. Second, the crayfish appendage is a classic example of what zoologists call **serial homology.** Homologous structures of this type occur when a basic pattern (e.g., the biramous appendage) is used to mold a variety of adaptive features. Each crayfish appendage starts its embryonic life as a two-branched (biramous) structure. As the animal matures, the biramous appendage is molded into a number of different designs, each with a different function. This remarkable evolutionary plasticity is apparent as a series of appendages ranging from the anterior-most **antenna** to the posterior-most **telson.** As we discuss each of the three body regions of the crayfish, we will review respective appendages.

There is one other dimension of the crustacean appendages that deserves comment. These animals, and many others I might add, have the distinct ability to sacrifice all or part of an appendage when threatened. If a predator happens to grasp a crayfish leg, the crayfish can voluntarily sever the appendage at a predetermined fracture plane located near its base. As a result, the predator dines on a small portion of crayfish while the bulk of the animal escapes. A valve closes the opening at the fracture plane so that body fluids do not escape.

Animals that practice **autotomy,** as the process is called, include crustaceans, some insects, echinoderms, and even many lizards. For such self-amputation to work, the lost structure must be replaced. Regeneration of the lost part is one thing all autotomous animal have in common. Each animal that has lost an appendage or other body part, is able, within a short time, to regrow the missing part.

Figure 13.3. A crayfish with a regenerating right cheliped. Note also the compound eyes, antennae, antennules, and rostrum.

Head

The head of the crayfish, as with all crustaceans, is unique in that there are two pairs of **antennae.** These are sensory structures equipped with tactile and olfactory receptors. The most anterior pair, the **antennules,** are *biramous* (two branched), reflecting the primitive condition. Posterior to the antennae, and surrounding the mouth, are three pairs of specialized head appendages designed to handle and process food. First, there are paired **mandibles** with serrated edges to partially masticate or crush food before it is swallowed. Immediately posterior to the mandibles are two pairs of leaflike **maxilla** that assist in passing food toward the mandibles or mouth. The second maxillae help to process food and also serve as "bailers" to create a respiratory current of water over the gills.

On the tip of prominent stalks are two large **compound eyes,** which in some animals can be rotated independently on the stalk. It is doubtful that members of this group can detect anything more than light intensity and movement, although there is some evidence that the lobster, and maybe the crayfish, can form rough visual images. Between the eyes is a forward projection **rostrum,** which serves as a protective shield for the eyes.

The head bears, internally, two important structures. First, it shouldn't be surprising that the cephalic nature of crayfish (i.e, antennae and eyes located on the head) would require a prominent **brain**. Located just behind the eyes, it sends, posteriorly, a pair of fibers that unite beneath the esophagus as a **ventral nerve cord**. The cord then extends the length of the animal as a whitish cable enervating organs, muscles, and other tissues along the way.

The head also houses the osmoregulatory organ, the **green gland**, sometimes called the antennal or maxillary gland. It functions much like a kidney by regulating ion balance and fluid levels. It does this by filtering blood through an elongated tubule and then allowing the filtered fluid to exit through an opening, the **nephridiopore**, located at the base of each antennae. It's important to note, however, that most of the nitrogenous waste from cell metabolism is excreted as ammonia, *not* by way of the green gland, but, instead, across the gill surface.

Thorax

In the crayfish, the head and its adjacent region, the thorax, are fused into a single **cephalothorax**. The thoracic portion bears the external gills and eight pairs of appendages. The first three are maxillipeds, surrounding the mouth as food-processing structures. Each maxilliped is biramous in that it has a base and two branchlike extensions. The next five pairs of appendages are **walking legs**; however, the first of these is modified into the large clawed **chelipeds** used in defense, feeding, and to control the female during copulation.

Internally, the thorax houses several visceral organs. First is the large **digestive gland** that serves as a liver to aid in the digestive process. Also, the **gonads** and associated reproductive structures are located in the thorax as is an enlarged **stomach**. The stomach is situated behind the head and, in fact, is often called the "head stomach" because of its unique location. Lining the inside of the stomach is a masticating apparatus, the **gastric mill**, consisting of three rows of calcareous teeth. Food that is not adequately pulverized by the mandibles is further macerated by the gastric mill. Exiting the stomach, and extending throughout the abdomen, is a tubular **intestine** that terminates at the **anus** on the last segment, the **telson**. Finally, the **heart** and related vessels are located in the thorax directly posterior to the stomach. Blood leaves the heart through a series of vessels that terminate as hemocoels to bathe the various tissues and organs. This is a classic case of an *open circulatory system*.

Abdomen

Each of the six abdominal segments bears a pair of appendages. The first five are the **swimmerets**, which function to create respiratory currents and, in females, to carry the eggs. They are not, as their name implies, involved in the act of locomotion. In the male, the first pair of swimmerets, and sometimes the second pair as well, are modified into sperm transfer devices, called **gonopods**. The last abdominal segment bears the flattened tail-like **uropods** used in swimming. Large flexor and extensor muscles provide the force for the powerful swimming action of the crayfish tail.

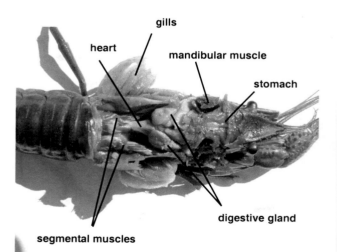

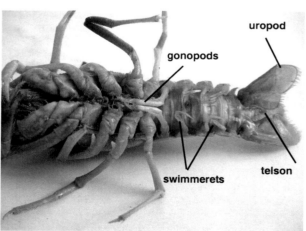

Figure 13.4. Left, the internal anatomy of a crayfish; right, ventral external structures of a male crayfish showing gonopods, the sperm transfer devices.

Ecdysis

If you have ever tried to extract meat from a lobster claw or have felt the painful grasp of a crayfish cheliped, you know how formidable the crustacean exoskeleton can be. As a tough, protective outer covering, it has few equals in the animal world. But it has other functions as well. For example, it's an impervious anti-dehydration cloak preventing excessive water loss. It also serves as the attachment site for muscles, much like an internal skeleton. However, for all of its positive qualities, the exoskeleton does have some drawbacks, two, to be exact.

The first problem with an exoskeleton is its *weight*. It's heavy to tote about. This is especially true for land-dwelling species. In water, bulk is not so much of a problem, but on land it can become a limiting factor. As a matter of fact, the reason that terrestrial arthropods are so much smaller than their aquatic counterparts is due to the weight of their exoskeleton. A king crab, with a span of well over 12 feet, would be completely immobilized on land.

The second problem has to do with growth. An exoskeleton is made of nonliving material, and thus *cannot grow incrementally* as does living tissue. This, of course, presents a dilemma for any animal wearing an exoskeleton. How is it to get any bigger if its chitinous coat can't? The solution is to completely shed the old covering periodically, then add a new, larger one into which the animal can grow, a process called ecdysis. And that's exactly what arthropods do. They regularly molt the old exoskeleton and add a new, soft one, which gradually hardens.

In point of fact, a new exoskeleton is deposited before the old one is shed. The animal then "puffs" itself up to fit into its new shell. While increase in arthropod size occurs in spurts, its tissues actually grow constantly, becoming wrinkled until molting occurs. Then before the new exoskeleton hardens the animal's tissues expand to accommodate the new cover.

However, ecdysis, too, presents a problem. There's a period during which an animal is without its protective shell, and thus, extremely vulnerable. This is the "soft-shelled" stage, and apparently a stage much sought by predators.

The most straightforward explanation of molting (**ecdysis**) is that it's a physiological process controlled by the action of two counterbalancing hormones. A *Y-gland*, located near the base of the antennae, secretes **ecdysone**, a molting hormone (MH). Its presence in the blood causes molting to occur. A similar gland, *X-gland*, found near the base of the eye-stalk, secretes molt-inhibiting hormone (MIH). The presence of this hormone, obviously, impedes molting.

The regulation of ecdysis then depends upon the respective levels of the two hormones in the blood. If MIH > MH, molting will not occur; if, on the other hand, MIH < MH, then ecdysis will proceed. Evidence indicates that the blood level of MH actually stays relatively unchanged, while MIH tends to fluctuate. If something interferes with the production of MIH, then molting will naturally occur. We think that the triggering mechanisms are a combination of internal and external factors. *Temperature*, for instance, may be a triggering mechanism. An increase in environmental temperature may elicit a decreased production in MIH and molting results. The same may be true of *photoperiod*. As light intensity or even periods of daylight increase, MIH production decreases, causing ecdysis. Thus, changes in temperature or photoperiod may serve as the triggering event to stimulate ecdysis. On the other hand, an internal stimulus might be the trigger. Something as simple as the growth of the individual inside of its exoskeleton may be enough to initiate molting. As the animal reaches a certain size, the central nervous system may inhibit the secretion of MIH, causing molting to proceed.

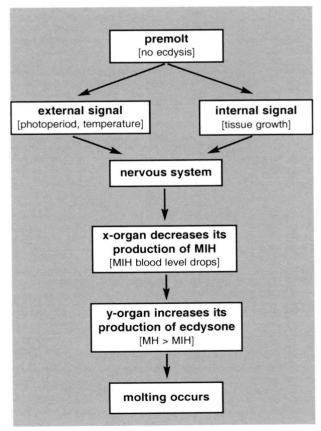

Figure 13.5. The sequence of external and internal events leading to ecdysis (molting) in crustaceans.

An Inside View 13.1

The Crustacean Appendage
-A Study in Plasticity-

One of the most attractive features about crustaceans is that they're a leggy bunch. Some, such as the spider crab, the Alaskan king crab, and the gleaner shrimp, seem to be nothing but legs. And of course, who doesn't appreciate the bountiful donation of lobsters to the world's table? Where would Epicureans of the world be without those meaty limbs? However, there's more than mere dining luxury that earns my affection for the crustacean appendage. To me, it's the biological intrigue more than the gustatorial delight. For the crustacean appendage has an important story to tell...*a story of structural plasticity.*

A quick glance at a live crayfish and one might conclude that evolution's grand designer pulled out all stops when charting the crustacean limb. Natural selection has fashioned no fewer than 15 different architectural patterns, 10 of which are found on the crayfish alone. What's even more provocative is that each pattern has its own purpose or function; each limb design has its own role to play. If we use the crayfish (or lobster) as an example we can more readily appreciate this relationship between structure and function.

Appendage	Structure	Function
head appendages		
antennules	double branched, elongated filaments equipped with hairs and other sensors	detects chemical signals (taste/smell), vibrations (touch) and helps to orient the animal (equilibrium)
antennae	an elongated, single filament equipped with sensory receptors	assists antennules with taste and touch reception
mandibles	chitinous slabs with serrated edges	used to crush and partially dismantle food
first maxilla	small leaf-like blades fringed with sensory hairs	food processing; helps to move food toward the mouth
second maxilla	similar to 1st maxilla in construction but also has a blade-like "bailer"	assists 1st maxilla in food handling and also pulls (bails) water from the gill chamber, thus helping in respiration
thoracic appendages		
1st maxilliped	similar to maxilla but has a pointed spine and a blade-like base	food handling and also serves as chemoreceptor (taste) and in touch reception
2nd maxilliped	has flattened base with two spines and leaf-like gills on interior surface	food handling, touch, taste, and respiration
3rd maxilliped	same as 2nd maxilliped only larger	same as above
cheliped	this is the first of 5 pairs of walking legs; large with claw-like chela	food handling, defense and offense; the chelipeds are also used by the male to manipulate the female during mating
walking legs 2 - 5	consists of single, strong branch; some may have small pincers and gills	used in locomotion (walking) and if chela are present, in food handling and respiration
abdominal appendages		
gonopod (male only)	first of 5 pairs of swimmerets; it is modified into a prong-like rod	used, along with the 2nd swimmerets, to transfer sperm from the male to the female during copulation
swimmerets	small, biramous appendages fringed with spines	enlarged in females to carry the huge volume of eggs; in both sexes, they circulate water to gills; do not aid in swimming
uropod	a large appendage with two expansive flattened blades	used as a propulsive blade during swimming (especially backward) for escape; also used to protect egg mass

The Crustacean Appendage
-A Study in Plasticity-

(the story continued)

Now that we have seen, up close, the structure and function of the various crustacean limbs, let's turn our attention to two fundamental questions. First, what's the reason for the rich variety in the crustacean appendage...why are there so many types? Second, and more importantly, *how* does such richness occur? The answer to the first question is rather straight-forward: each specific type of appendage has a different job to perform. It's the old tale of division of labor. Each appendage contributes to the overall efficiency of the entire animal; the chores, duties, responsibilities, and needs of the organism are met by separate designs. In other words, if each part, in this case, appendages, does its job, then the whole animal is more efficient, better able to compete, and more likely to survive.

However, the answer to the second question is not so simple; indeed, it is quite a bit more complex. How does such structural diversity happen...where do all of these different appendages come from? To answer that we need to look more closely at a basic crustacean limb, say, the swimmerets or the antennules (Figure 13.9 and 13.11). Note that there are three parts to the appendage, a base and two outward-extending branches. The base is termed the **protopodite**, the inner (medial) branch, facing the animal's midline, is the **endopodite**, while the outer (lateral) branch is the **exopodite**. This Y-shape design, in which two branches project from a base, is termed a **biramous appendage**. The fact is, all 10 types of crayfish appendages are embryonically based upon this biramous plan. In the adult, however, several limbs have become secondarily uniramous as an adaptation to its particular function (e.g., the walking legs, antennae, cheliped). In other appendages, one or the other branch has been reduced, again as a result of specific roles (e.g., 3rd maxilliped). In certain cases, the base or inner branch may, itself, carry a lateral projection, for example, the gills of walking legs and the palps of maxillipeds.

What are we to make of all of this? Well, it appears that each crustacean appendage has evolved from a primitive biramous form; indeed, *each* appendage passes through this biramous condition during embryonic development. Then as development proceeds, the basic model (biramous plan) is modified, serially, to perform different functions. Put another way, a series of different structures is molded from a basic source. The cheliped, the antennae, the swimmerets, and so on, all had their origin in the biramous appendage but have been modified to perform their specific functions. We call this condition **serial homology**. This is a basic concept in biology and the crustacean is a classic example of the concept.

Figure 13.6. Representative crustaceans showing a variety of appendages. An arrow crab (**A**) nestled among the stinging tentacles of an anemone; note its extremely long legs which are also seen in the banded coral shrimp (**B**). A species of spider crab (**C**) illustrates the protective nature of body and leg spines.

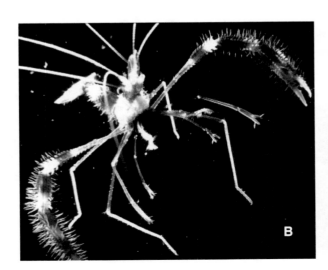

Crustacean Coloration

Some crustaceans have bright, ornamental colors that apparently serve reproductive and recognition functions. Others maintain more cryptic patterns, and still others can shuttle back and forth between the two (Figure 13.7). The ability to match a particular substrate when needed is an important survival strategy for many animals, no less so in crustaceans. The actual color change is due to fluctuations of special epithelial pigments: reds, yellows, blues, white, browns, and black, that are clustered within chromatophore cells of the exoskeleton.

We have seen this type of camouflage in a number of animals, especially the squid. However, color change in the squid was of neuronal (i.e., nervous) control and, therefore, almost instantaneous. Not so in crustaceans. Here the color conversions are due to hormones and, as such, is a much slower process. It may take a minute or so for a crustacean to change from one pattern to another. The hormones, themselves, are produced by secretory cells located in the eyestalk. For each pigment, there are two controlling hormones, one countering the other. One of the two serves to maximize a color and the other to minimize it. The comparative levels of the two hormones thus determine the color pattern produced.

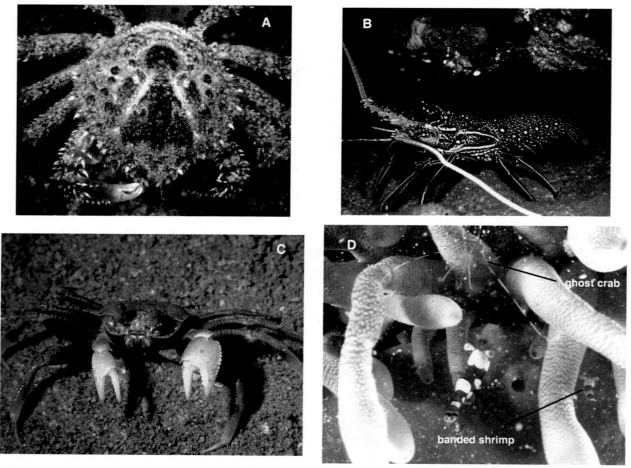

Figure 13.7. Camouflage in crustacea. **A**, a large spider crab uses a multicolored body to match its substrate; **B**, a spiny lobster glitters with a spotted background to blend and break up its outline; **C**, a shore crab in defensive posture, but blending well into its surroundings; **D**, a ghost crab and two barber pole shrimp are well disguised within the protective embrace of an anemone's tentacles.

Class Maxillopoda–Subclass Cirripedia
~ the barnacles ~

The greatest deviation of all from the typical crustacean body plan is seen in the barnacles. These strange-looking animals appear to be more closely aligned with mollusks than with arthropods, let alone with the crustaceans. In fact, it wasn't until the latter half of the 19th century that they were correctly assigned to the arthropod and crustacean line.

The reason for their oddity is that the barnacles have taken up either a sedentary or parasitic lifestyle, and have modified their appearance accordingly. They have *no head*, have r*educed the first pair of antennae*, and have *lost the second pair;* indeed, they use the first antennae to glue themselves to a substrate and then flip over so that, as adults, they spend their entire lives upside down! Barnacles have *no gills, no heart, the abdomen is greatly reduced, their thoracic appendages are either modified, reduced, or absent, and most of them possess a molluscan-like bivalve shell.* Perched atop the two shells are two or more smaller valves that are movable and can open or close as needed for feeding. Nonetheless, the presence of jointed feeding appendages, called **cirri**, as well as a **nauplius** larval stage show the relationship of barnacles to other crustaceans.

Free-living barnacles secrete a powerful gluelike substance that attaches them to some object such as floating debris, a ship hull, or another living animal (Figure 13.8). This secretion is, in fact, the strongest adhesive known. Any locomotion is generally accomplished by larval forms only; adults are entirely sessile. An interesting side note is that it seems larvae of attached species are attracted to adults of the same species by pheromones (special chemicals) issued by the adults. As a result, large clusters of attached barnacles of varying ages tend to commingle.

As a group, barnacles are monoecious but cross fertilize by using a male penis that extends from the shell (often reaching longer than its owner) to find a nearby partner.

Feeding by free-living barnacles is accomplished with six pairs of modified feathery appendages, the **cirri**, which filter plankton from the passing water. The parasitic barnacles, on the other hand, have a mouth modified into piercing and sucking parts to withdraw fluid from their hosts. They may occur as ectoparasites of sea stars, urchins, or brittle stars. There are also some cirripedians that are endoparasitic on other crustacea, usually crabs. They initially parasitize the gut but slowly send out branching roots that essentially invade every part of the host's body.

Figure 13.8. Whale barnacles (probably of the genus *Choncoderma*) clustered on the head of a gray whale, a favorite site for barnacle attachment. The insert shows a closer view of a cluster of barnacles. Note the shell-like appearance and the inner valves which open during feeding.

Chapter 13 Summary

Key Points

1. Crustaceans possess two pairs of antennae, biramous appendages arranged in a series from anterior to posterior, food-processing mandibles, and gills for respiration.

2. A pair of special structures called green glands are located in the ventral head area and used for osmoregulation and excretion.

3. Most crustaceans belong to the class Malacostraca which includes the crayfish, lobster, shrimp, and crabs.

4. Crustaceans have a variety of appendages that begin their development with two branches, or rami.

5. These biramous appendages are then modified in the adult to perform a variety of functions from grasping (the pincher-like chelipeds), to feeding (maxilla and maxillipeds), to locomotion (the walking legs and the powerful tail-like uropod), to sperm transfer (the male gonopods), to carrying eggs (the female swimmerets).

6. Modification of a basic plan (i.e., the biramous appendage) into a number of different structures with different functions is called serial homology.

7. Crustaceans, and a number of other groups as well, show a phenomenon known as autotomy in which a limb or other structure is voluntarily sacrificed to escape predators or other dangerous situations.

8. Located within the crayfish stomach is a series of calcified teeth, the gastric mill, used to masticate or pulverize any large food items that may have been ingested.

9. The crustacean body is covered by a calcified, hard exoskeleton.

10. While the exoskeleton serves as an important protective covering, it does have two drawbacks: it's heavy and it cannot grow.

11. Thus, the exoskeleton must be periodically cast off in a process called molting or, more correctly, ecdysis.

12. The shedding (molting) process is controlled through a complex feedback mechanism between the animal's external and internal environment. Two glands, the X-gland and Y-gland, control the actual shedding process by way of their hormonal secretions.

13. Many crustaceans can alter their coloration through hormonal secretions as a defense against predation or for sexual recognition.

14. Barnacles, class Cirripedia, are highly modified crustaceans adapted to a sedentary, filter-feeding, or parasitic existence.

Key Terms

antennae
antennules
barnacles
biramous appendage
chelipeds
cirri
ecdysis
ecdysone
endopodite
exopodite
gastric mill
gonopods
green glands
isopods
krill
maxilla
maxillipeds
molting hormone
molt-inhibiting hormone
nephridiopore
premolt
protopodite
rostrum
serial homology
swimmerets
telson
uniramous appendage
uropods
X-gland
Y-gland

Points to Ponder

1. List the major features of the crustacea.

2. Give two anatomical differences between the crustaceans and other arthropods.

3. What is a biramous appendage?

4. What is serial homology and how does it relate to the biramous appendage of crustaceans?

5. Name a terrestrial crustacean.

6. Name the kinds of arthropods that belong to the class Malacostraca.

7. What is the green gland, where is it located, and what is its function?

8. What are chelipeds?

9. What is the function of the gastric mill and where is it located?

10. Give the various functions of crayfish swimmerets, including the gonopods.

11. What is ecdysis? How is it controlled in an animal such as the crayfish? What hormones are involved in the process? (Be sure to include the role of the X-gland, the Y-gland, as well as types of both internal and external signals.)

12. Name the head appendages found on a crustacean and give their respective functions.

13. What appendages are located on the thorax of a crustacean? What is the function of each?

14. What is autotomy and how is it important in the life of a crustacean?

15. How does the ability to change color aid survival in a crustacean? How is color change accomplished?

16. Why is a barnacle considered a crustacean?

Inside Readings

Abele, L. G. ed. 1982. The Biology of Crustacea.
New York: Academic Press.

Bliss, D. E. 1982. Shrimps, Lobsters and Crabs. Piscataway, New Jersey: New Century.

Bliss, D. E. , ed. 1982-1987. The Biology of Crustacea. vols. 1-9. New York: Academic Press.

Brusca, R. C., and G. J. Brusca. Invertebrates.   Sunderland, MA: Sinauer Associates.

Cameron, J. N. 1985. Molting in the blue crab. Sci. AM. 252(3):102-109.

Fitzpatrick, J. F. Jr., 1983. How to Know the Freshwater Crustacea. Dubuque, IA: Wm. C. Brown.

Pechenik, J. A. 1996. Biology of the Invertebrates. Dubuque, IA: Wm. C. Brown.

Schram, F. R. 1986. Crustacea. New York: Oxford University Press.

Chapter 13

Laboratory Exercise

The Crustaceans

Laboratory Objectives

✓ to identify major external and internal features of crustaceans through dissection of the crayfish,
✓ to identify various types of crayfish appendages,
✓ to understand the relationship between the biramous appendage and serial homology, and
✓ to describe respiration in the crayfish.

Needed Supplies

✓ preserved specimens of crayfish for dissection
✓ preserved samples of crabs, shrimps, lobsters, and barnacles

Chapter 13
Worksheet

(also see page 603)

Phylum Arthropoda
Subphylum Crustacea

1. Examine the various crustaceans on display. Try to locate the following crustacean features

   ✓ the presence of two pairs of antennae
   ✓ body divided into two tagmata; the cephalothorax and abdomen (sometimes covered by carapace)
   ✓ biramous appendage
   ✓ mandibles
   ✓ presence of gills
   ✓ thick, hardened exoskeleton

Class Malacostraca
The Crayfish: External Anatomy
(Figures 13.9-13.11)

2. Examine the external features of both a male and female crayfish and locate:

   ✓ eyes and eyestalks
   ✓ antennules
   ✓ antennae
   ✓ rostrum
   ✓ mandibles
   ✓ maxillae & maxillipeds
   ✓ carapace

   ✓ abdomen
   ✓ telson
   ✓ anus
   ✓ chelipeds (first pair of walking legs, modified into a large pincer)
   ✓ walking legs
   ✓ gonopods (on male only)
   ✓ swimmerets

3. Use the features below to identify specimens as either male or female.

   Males: large chelipeds used to control female during mating, narrow abdomen, large first swimmeret called gonopods, openings from the two male sex ducts located at the base of legs # 5.

   Females: chelipeds proportionally smaller than males, broad abdomen, small or missing first swimmeret, remaining swimmerets relatively large to carry eggs, opening into the seminal receptacle is centrally located between legs # 4 & #5, the two openings from the oviducts are located at base of walking legs # 3.

Notice how hard the exoskeleton is in your specimens. What four functions does it perform?

_____

_____

_____

_____

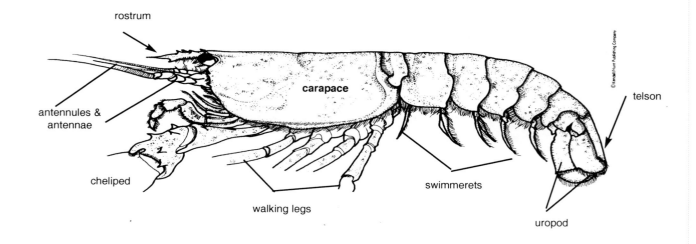

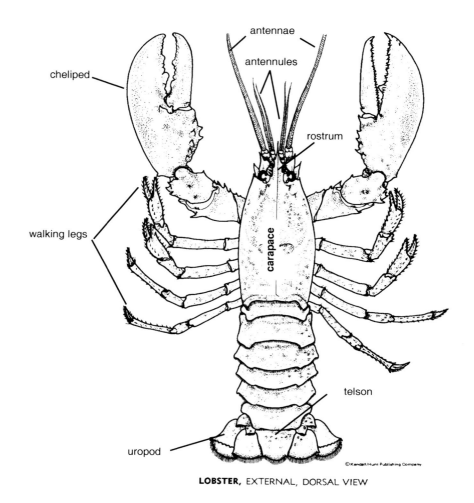

LOBSTER, EXTERNAL, DORSAL VIEW

Figure 13.9. External, lateral and dorsal view of a crayfish or lobster.

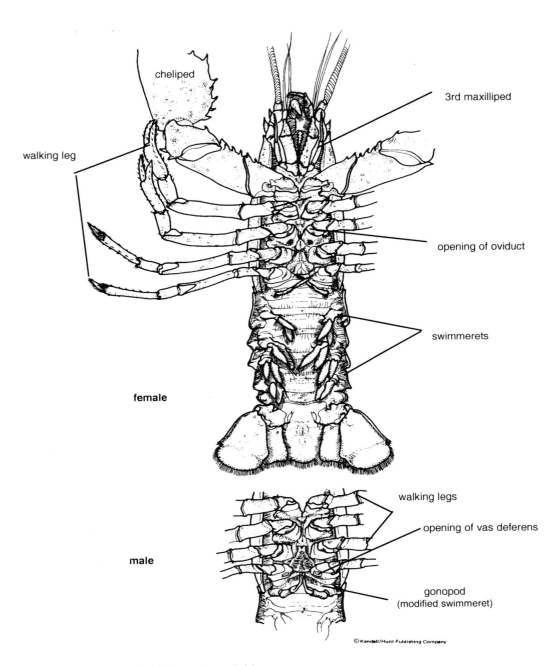

cheliped

3rd maxilliped

walking leg

opening of oviduct

swimmerets

female

walking legs

opening of vas deferens

male

gonopod
(modified swimmeret)

© Kendall/Hunt Publishing Company

Figure 13.10. Ventral view of a lobster (crayfish).

4. The primitive crayfish appendage is **biramous**, consisting of a base and two branches. The embryonic appendage is modified into 10 adaptive, functional variations. This arrangement is an example of serial homology in which a basic form (here, the biramous appendage) is modified into a series of functional structures. From one of your specimens, extract one of each general type of appendage, sketch, and give its major function below.

Function   **Sketch**

antennules: Reflect primitive   (also on page 604)
conditions

antennae: Equipped with sensory
structure with tactile
& olfactory receptors

mandibles: Help to magilate or
crush food before swallowing

maxillae: Assist in passing food
(either one of two)
towards the mouth

maxillipeds: Food handling
(any one of three)

chelipeds: Used for defense, feeding
(pincers) and control female during
copulation

walking legs 2 - 5: used in locomotion
(any one)

male swimmeret #1 used for sperm
(gonopod)   transfer

remaining swimmerets: Enlarged in female
to carry huge volume of eggs/
circulate water to gills (no swimming)

uropod: Use for swimming

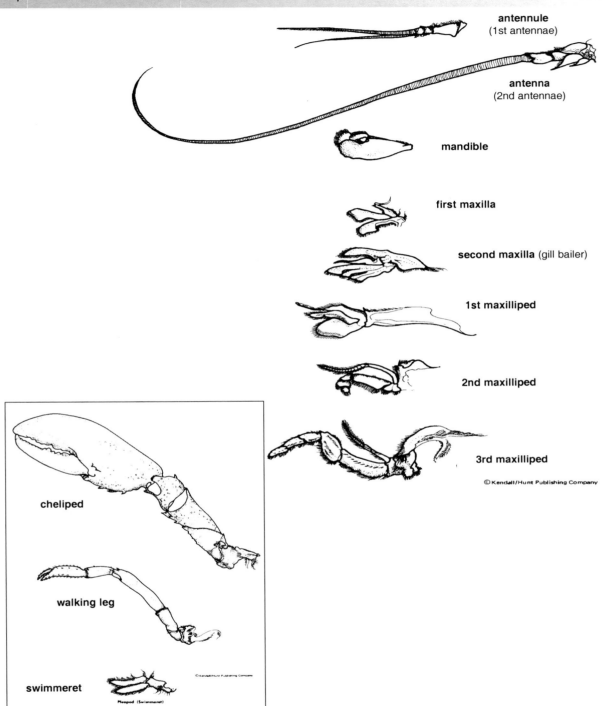

antennule
(1st antennae)

antenna
(2nd antennae)

mandible

first maxilla

second maxilla (gill bailer)

1st maxilliped

2nd maxilliped

3rd maxilliped

© Kendall/Hunt Publishing Company

cheliped

walking leg

swimmeret

Pleopod (Swimmeret)

© Kendall/Hunt Publishing Company

Figure 13.11.   Appendages of a lobster or crayfish.

The Crayfish: Internal Anatomy

- dissection -

Note: Dissect the specimen from which you removed the appendages.
However, be sure to examine the internal anatomy of both males and females, by viewing the dissection of classmates.

1. This is a relatively simple dissection. All it requires is the careful removal of the **carapace**. This is done by placing your thumb under the edge of the carapace, near the gill chamber, and peeling away the **exoskeleton**. You can do this either in one complete piece or by chipping small pieces until the carapace is removed. The key here is to be careful so that you do *not* disturb underlying structures, especially the heart.

2. Once you have removed the carapace, you can see, from a dorsal view, the various internal organs. Examine the large feathery **gills** that occupy gill chambers on both sides of the animal. Next, note a large whitish mass occupying much of the body cavity. This is the **digestive gland**. Anterior and slightly dorsal to the digestive gland is the **stomach**, also known as the proventriculus. In crayfish, it is huge and sometimes called the "head stomach" due to its anterior location. You will be asked to remove and examine it later.

omit 3. The **heart** is a medium-sized organ easily identified because of the openings, or **ostia**, through which blood passes. It is located immediately under the carapace a little posterior to the stomach. Although the circulatory system is open, there are, nonetheless, a number of vessels present. The largest is the **dorsal aorta**, which may be seen running posteriorly from the heart. While you are examining this area, carefully lift the aorta and locate the **intestine** that lies beneath it. The intestine, of course, passes from the stomach all the way to the **anus**.

omit 4. You are asked to identify two muscles at this time. First, locate the two round **mandibular muscles** found on each side of the stomach. These are attached ventrally to the mandibles and provide the powerful force for crushing food. Next, locate the large muscle masses in the tail. These **segmental muscles** serve as flexors and extensors to move the powerful tail in rapid, flipping actions.

omit 5. Unless your specimen is sexually mature, the reproductive organs will be small and difficult to locate. In any event, carefully lift the heart and look for **ovaries** or **testes** that lie just beneath it. Ovaries are invariably "spotted" due to the presence of ova. If your specimen is gravid, the entire posterior body cavity will be filled with eggs. A careful examination of a male crayfish will reveal two small, elongated, saclike testes attached to a coiled tube, the **ductus deferens**, which carries sperm to the outside.

6. Now return to the stomach. Grasp it firmly and remove the entire structure. Notice that as you remove it there is an attachment point ventrally. This is the **esophagus** leading from the mouth. Note also that another tube, the intestine, leads dorso-posteriorly from the stomach. Next open the stomach and flush it with water. Examine the three internal chitinous teeth. This is the **gastric mill** used to grind and pulverize any large food items. The chitinous lining of the stomach, including the gastric mill, is shed during molting.

7. Now carefully examine the cavity left by the removed stomach. There are several items here to identify. First, two large **green glands** (sometimes called antennal glands) can be seen lying in pockets at the base of the cavity. These are excretory structures responsible for osmoregulation and ion regulation. Carefully lift one and you will see a small tube leading from it to the excretory pore at the base of the antennae.

8. Finally, examine the nervous system. You should locate two whitish, threadlike fibers passing around the esophagus. These are extensions of the **nerve cord**, which unite ventrally as the ventral nerve cord and pass posteriorly toward the anus. Trace the two fibers anteriorly and you will see that they unite to form the small supra-esophageal ganglion, better known as the **brain**. If you look carefully, you will see fibers radiating out from the brain to the eyes, antennae, etc.

In summary, you should be able to identify the following structures on a preserved crayfish:

✓ gills	✓ digestive gland
✓ stomach	✓ gastric mill — underneath teeth
✓ esophagus	✓ intestine
✓ heart, ostia	✓ dorsal aorta
✓ ovaries	✓ testis
✓ ductus deferens	✓ green gland → Big
✓ brain	✓ ventral nerve cord
✓ mandibular muscles	✓ segmental muscles → Flexor muscle of abdomen

ob intestine

curls tail under

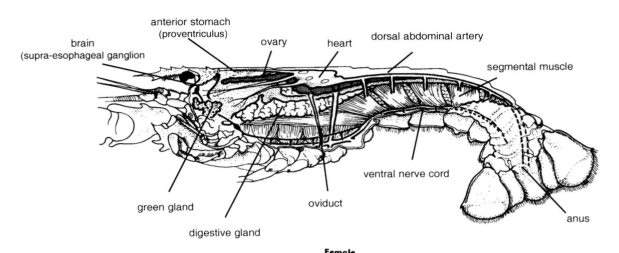

anterior stomach
(proventriculus)

brain
(supra-esophageal ganglion)

ovary   heart   dorsal abdominal artery

segmental muscle

ventral nerve cord

green gland

oviduct

anus

digestive gland

Female
LEFT DIGESTIVE GLAND REMOVED TAIL TURNED TO SHOW VENTRAL SURFACE

Antenna - pair of
mobile
appendages
Antennules -
Serve as
Feelers
sensitive

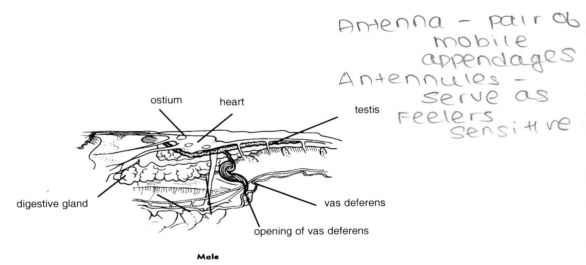

ostium   heart   testis

digestive gland   vas deferens

opening of vas deferens

Male
AREA OF GONADS LEFT DIGESTIVE GLAND REMOVED

© Kendall/Hunt Publishing Company

LOBSTER, DISSECTION, LATERAL VIEW

Figure 13.12. Lobster (crayfish) dissection, lateral view..

Gastric mill - underneath
Intestine - to grind food

Crelipeds - Defense, mating
Gills - Breathing
Green Glands → Kidneys
of the crayfish
Telson - movement
uropods - provide back
wards
movement
Swimmeets - use for
locomotion &
reproduction

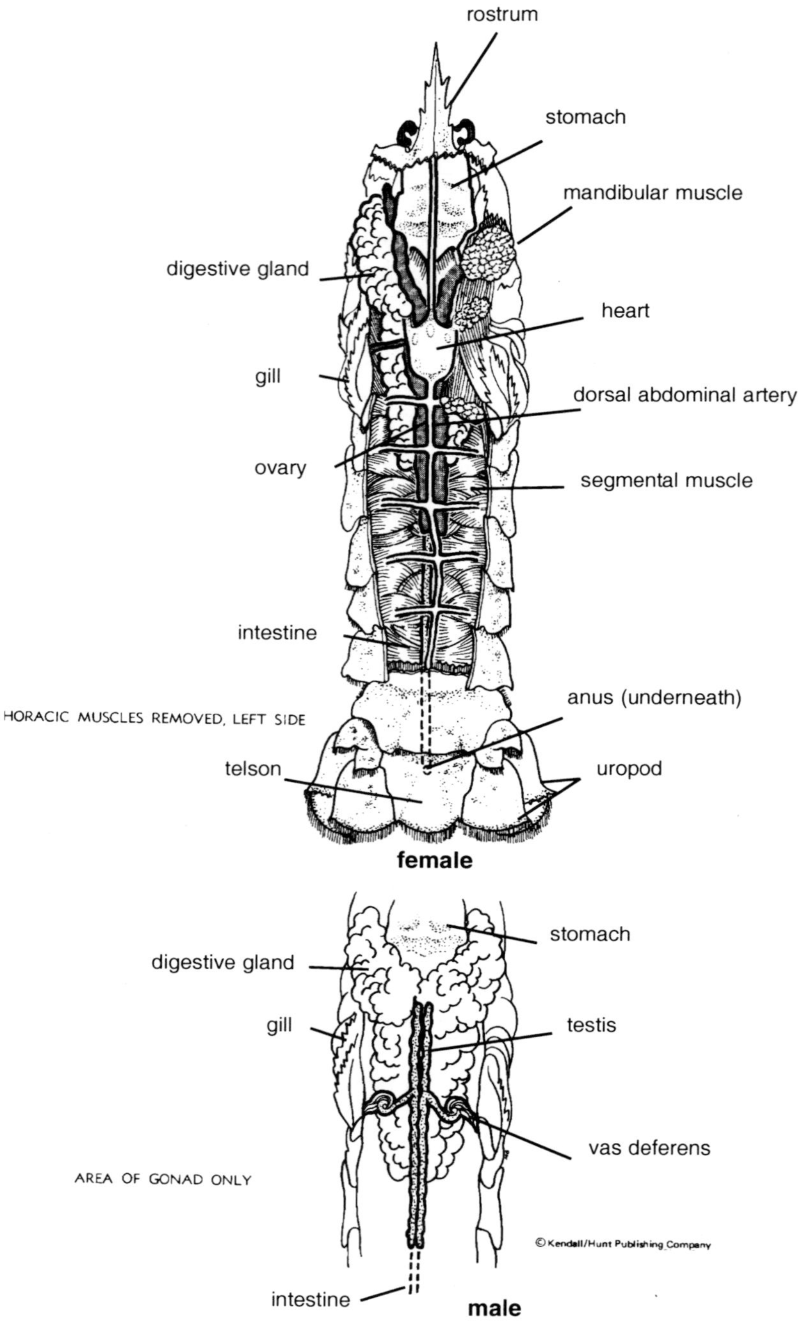

Figure 13.13.   Lobster (crayfish) dissection, dorsal view.

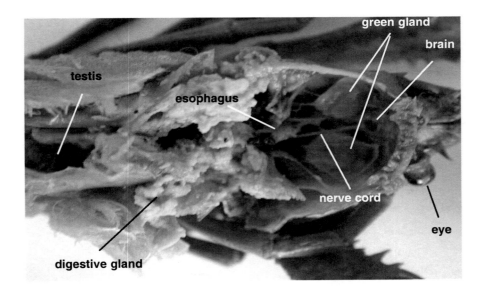

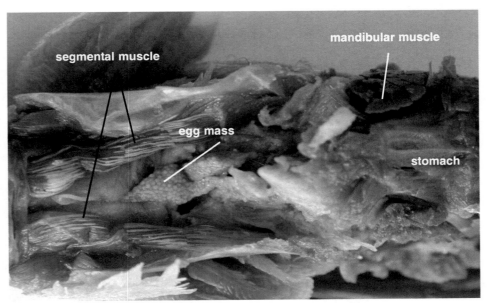

Figure 13.14. Internal structures of a crayfish. The upper photograph shows the contents of the cavity when the stomach is removed. The lower plate (head to the right) is of a female with eggs beginning to form.

14 Chapter 14

PHYLUM ARTHROPODA

THE UNIRAMIANS

centipedes
millipedes
insects

The Uniramians

*So important are insects and other land-dwelling arthropods,
that if all were to disappear,
humanity would probably be doomed within a few months.*
-E.O. Wilson

Introduction

A uniramian is an organism whose appendages have only one branch, or to be more exact, are unbranched. Unlike the crustaceans, with their biramous appendages, the limbs of uniramians are never "two branched." But there's much more than just a single-branched appendage that identifies this impressive group of animals. For example, they all have one pair of antennae, chewing or modified mandibles, tracheal respiration, Malpighian tubules for excretion, an elongated multisegmented trunk with a pair of appendages per segment, and a ventral nerve cord.

Sometimes obvious features, such as these, tell only part of a story, and this is one of those occasions. Uniramians are so much more than mandibular animals with unbranched appendages breathing through a tubular network. On the contrary, in this taxon, what lies beneath the surface is nothing short of incredible. Their distribution, diversity, social organization, reproduction, relationship to plants, and impact on humans is utterly staggering.

Although I will detail this idea later when we discuss the insects, for now let's look at just one colorful example. There is a longstanding relationship between plants and members of this subphylum. Without this group of animals, there would be virtually no fruit and no flowers. Insects and angiosperms (e.g., flowering plants) have co-evolved in a mutualistic relationship over the last 150 million years or so. Each depends upon the other. Plants rely upon insects for pollination and for recycling of nutrients. Flower color, flower shape, fruits, and scents attract insects. Insects, in turn, eat plants, reproduce there, and use them for shelter and as hunting ground. Neither could survive without the other. And just what would our earth be like without flowers, trees, bugs, and butterflies? This planet's very biodiversity depends, in large measure, on the action of this branch of the arthropod tree.

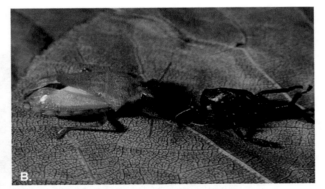

Figure 14.1. Selected relationships between insects and flowering plants. **A**, a grasshopper lies cryptically concealed upon a leaf; **B**, a pentastomid bug converts a leaf into a combat zone as it attacks an unwary beetle; **C**, a jeweled cluster of beetles feed on a flower head; **D**, a honeybee extracts nectar as it inadvertently pollinates a flower.

Diversity

The subphylum Uniramia contains five taxonomic classes, two of which are so insignificant that we will do little more than mention them. The remaining three classes are much more abundant and are discussed separately, and in more detail.

One small class, the Symphyla, consists of tiny (< 1 cm) eyeless, plant-feeding, centipede-like animals. They forage among plant litter, often feeding on decaying vegetation. Another uncommon taxon, the Pauropoda, are smaller yet (< 2 mm). They, too, are sightless, and possess strange Y-shaped antennae. Pauropods inhabit moist soils, woodlands, and rotting vegetation upon which they feed.

Class Chilopoda:

~ the centipedes ~

Centipedes are common nocturnal predators of small animals throughout the world. They range in size from a few centimeters to the large Vietnamese centipede that grows to over 20 cm (~ 8 inches; Figure 14.3). Their prey usually consists of invertebrates, although some of the larger centipedes will prey on frogs, lizards, and small rodents. Centipedes have even been observed rearing back and snatching flying insects from the air!

They differ anatomically from all other arthropods in several respects. First of all, their first pair of trunk appendages (maxillipeds) are modified into a pair of **poison fangs** that inject venom from a basal gland.

The venom in some of the larger species can be painful but is rarely fatal to humans. It is potent enough, however, to disable prey animals. Once immobilized, the prey is crushed and torn apart by mandibles and maxillae that surround the mouth. Food is generally swallowed in small, somewhat liquefied, pieces.

Even though centipedes possess an exoskeleton, as do all arthropods, it lacks a waxy coating. This, plus the fact that the openings (spiracles) leading to the tracheal respiratory system cannot close, limit centipedes to moist habitats or to settings where desiccation is reduced, such as under rocks, logs, in leaf litter, and so on. Since it is difficult for centipedes to control water loss, they restrict their activity to shaded areas or tend to be nocturnal so as to escape the dehydrating effects of the sun.

A further examination of centipede anatomy reveals a strongly segmented, dorso-ventrally flattened trunk (Figures 14.2 and 3). The number of trunk segments varies from around 15 to 173 (the maximum so far encountered), and each segment generally bears a single pair of clawed appendages. The last trunk segment often supports elongated sensory appendages, the **cerci**, which probably detect vibrations or may serve as chemosensors. Centipedes also have one pair of antennae and simple ocelli (light-sensing organs). Compound eyes are present in only a few centipede species, and others are completely blind. Finally, some centipedes possess special defensive devices called **repugnatorial glands** to repel potential predators. These glands secrete a noxious or poisonous chemical that is often "kicked" in the direction of an adversary.

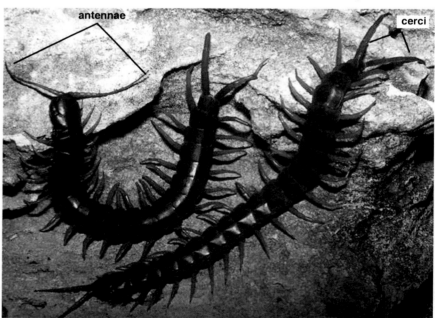

Figure 14.2. Common centipedes of the class Chilopoda. Note the flattened body, a single pair of legs per body segment, one pair of antennae, and a pair of posterior sensory appendages, the cerci.

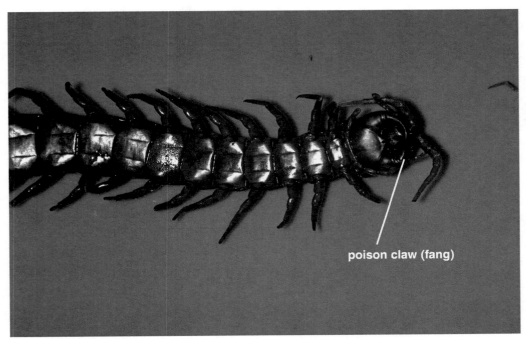

Figure 14.3. A dorsal view of a giant centipede (top). Note its flattened shape and one pair of legs per body segment. Below, a ventral view showing the enlarged maxillipeds modified into poison fangs.

Class Diplopoda
~ the millipedes ~

Like the centipedes, these "thousand leggers," better known as millipedes, have a wax-free exoskeleton. Thus, they, too, are limited to moist habitats where they slowly plow along feeding on decomposing vegetation. During embryonic development, adjacent segments of the body fuse into double units, called **diplosegments**. As a result, each visible segment bears, not one, but *two pairs of legs.* With some millipedes consisting of 200 segments, an animal may possess as many as 400 legs, far from the thousands that its name implies.

Millipedes are round in cross section and possess one pair of antennae. The first maxilla is enlarged into a plate that the animal uses to ram through loose soil or litter. They may possess a number of ocelli, but never compound eyes. Millipedes are absolutely harmless, but not defenseless. When molested they do have two tactics at their disposal. Their first line of defense is to contort their body into a springlike coil. The hard exoskeleton then serves as an impregnable barrier.

Secondly, many millipedes, like their centipede counterparts, also have repugnatorial glands. A disturbed millipede may eject caustic phenols, aromatic quinones, and some species can even spray a jet of hydrogen cyanide up to a foot or more to discourage harassment!

IT'S A FACT

Nature's Insecticide

NEW WORLD MONKEYS HAVE RECENTLY BEEN OBSERVED PARTICIPATING IN A BIZARRE RITUAL. UPON CATCHING A HANDFUL OF CENTIPEDES AND MILLIPEDES, THE MONKEYS CRUSH THEM TOGETHER, AND THEN RUB THE MASHED BODIES ONTO THEIR FUR. IT IS THOUGHT THAT SECRETIONS FROM THE CRUSHED REPUGNATORIAL GLANDS PROVIDE PROTECTION FROM NOXIOUS PESTS SUCH AS BITING FLIES, MOSQUITOES, FLEAS, TICKS, AND OTHER ECTOPARASITES. THIS IS A RARE EXAMPLE OF WILD ANIMALS USING ANOTHER ANIMAL TO WARD OFF YET A THIRD (A PEST).

defensive posture

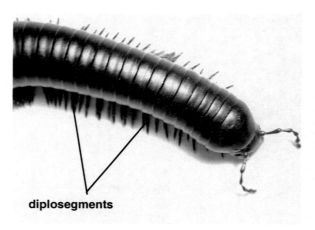

diplosegments

two pairs of legs
per segment

Figure 14.4. Three views of millipedes. The top left clearly shows the rounded shape consisting of diplosegments. The upper right photograph depicts a millipede rolled into a defensive posture. The ventral view of the animal on the bottom right gives a good illustration of the two pairs of legs per segment.

The Insects

Most animals are aquatic. Of those major phyla we have studied in this class so far, the Protista, Porifera, Cnidaria, Platyhelminthes, Nematoda, Annelida, and Mollusca are all predominately aquatic animals. We have yet to investigate the Echinodermata (next chapter) but they, too, are aquatic. Yet, even though most kinds of animals are aquatic, the great majority of individual animals live on land, not in water.

To claim that most of this planet's animals are terrestrial is absolutely true. How is that possible? The answer to this apparent contradiction may be found in the diagram below. Of the 1.3 million species so far identified, over 75% are insects. If you lump all arthropods together, the number jumps to around 85%. Arthropods are manifestly terrestrial animals. So even though most of the major kinds of animals prefer to live in water, arthropods don't, and since they are disproportionally abundant, they dominate the zoological landscape.

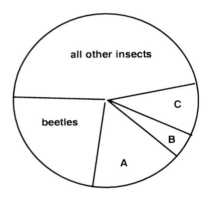

Figure 14.5. The relative distribution of known animals. **A**, arachnids and other noninsect arthropods. **B**, chordates (fish, amphibians, reptiles, birds, and mammals). **C**, all remaining phyla combined. Insects constitute approximately 75% of all animals on earth and most of these are beetles. Compared to the 0.05% of identified chordates, insects dominate by sheer variety alone. Flight (photo to the right) is one of the primary factors why insects have been so successful.

To put this into perspective, consider the other predominately terrestrial phylum, the chordates (i.e., fish, frogs, reptiles, birds, and mammals). There are only around 43,000 of them. That's less than 0.05% of the total number of the known animals on earth compared to the 75% represented by insects. To compound matters, keep in mind that most experts maintain that we have only identified a mere fraction (~ 1/50) of the insects that actually exist! We can only conclude that they are truly the masters of the terrestrial world.

More than any other invertebrate, the insect embodies the very idea of success. A small, compact body covered with a protective exoskeleton, well-developed sensory apparatus (especially their compound eyes), unparalleled ability to reproduce quickly and incessantly, a terrestrial egg, metamorphosis, modifiable mouthparts, complex instinctive behaviors, elaborate social orders, and an ability to fly have combined to produce the animal kingdom's most successful experiment. Stated more succinctly, the overall body design and lifestyle of insects have forged the earth's most highly adaptive animal.

Insects are, truly, evolution's success story. Take flight, for instance. Insects are the only invertebrate that has become airborne. By doing so they can more easily escape danger, locate food far out of reach of their landlocked colleagues, find otherwise inaccessible shelter, and relocate with relative ease. Flight has enabled insects to exploit every possible ecological niche; there's few places on earth that wings cannot reach. The development of winged flight has given insects a great, and by most accounts, an unfair advantage over all other invertebrates. Little wonder that most adult insects fly.

How do I know an insect if I see one?

In addition to the characteristics given on the preceding pages (e.g., small size, prolific reproduction, metamorphosis, and flight), insects are also characterized by a number of other features. They are mandibulate arthropods (i.e., they possess mandibles) and, as such, are most closely related to the crustaceans. As adults, they have a body covered by an exoskeleton and divided into three tagmata: the head, which bears feeding and sensory structures; the thorax, bearing locomotor appendages; and the segmented abdomen, enclosing the primary visceral organs of digestion, reproduction, excretion, and circulation. The abdomen also bears the external genitalia. Adults possess one pair of antennae and three pairs of legs. Most adults have, in addition, two pairs of functional wings.

hardened forewings covering hindwings

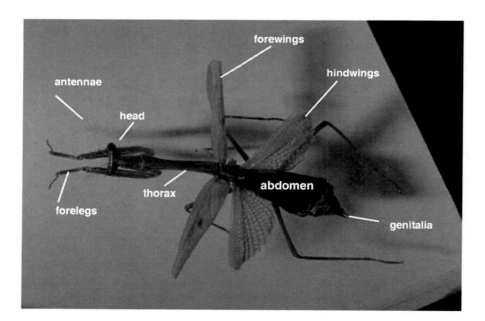

Figure 14.6. The typical insect body plan consisting of a head, thorax, and an abdomen is evident in this praying mantis photograph. The head bears the sensory structures and mouthparts, the thorax supports the legs and wings, and the abdomen contains most of the internal organs as well as the external genitalia. The forewings are often modified into shell-like protective coverings as in the beetle in the upper photograph.

How do insects sense their environment?

Although we can never know exactly what any animal sees, feels, or smells, we can make some pretty good guesses. Basically, we think that insects glean information about their surroundings through three pathways: auditory (sensing airborne or substrate vibrations), olfactory (smelling or tasting chemicals in their environment) and visual (detecting light variations, motion, or forming images).

Mechanoreception. Airborne and substrate (i.e., ground) vibrations are detected by a variety of receptors. Hair-like projections on the legs, setae scattered over the body, modifications in the exoskeleton, hairs on the antennae, and large air sacs covered by membranes (tympanic membranes) can all be deformed, deflected, or otherwise distorted by pressure waves. What this means is that insects are endowed with a well-defined, well-tuned sense of hearing, which they use to detect predators, prey, and each other. We have all heard the pulsating chirps of a cricket or the perpetual buzz of a cicada as they seek to attract a mate. Cockroaches can detect the bare whisper of air currents caused by a person walking into a darkened room. They are already up to full speed and sneaking under the sink, or scurrying beneath the refrigerator by the time you switch on the light.

Chemoreception. Insects are apparently able to detect molecules of airborne chemicals in much the same way that we detect odors. However, an insect's power of discrimination is much, much more refined than ours. For instance, some male moths can sense even a few molecules of a chemical signal (pheromone) issued by a female a half mile away. In the same way, a mosquito can discern the presence of a host animal by the carbon dioxide released during respiration. You attract mosquitoes by simply breathing. Social insects keep in contact through chemical signals and some (e.g., ants) can deposit a chemical trail leading to a food source. Most of the chemoreceptors are located on an insect's mouthparts, legs, or antennae (which may be long and thin, club-shaped, or bear featherlike tufts).

Photoreception. Adult insects have two kinds of eyes; most have a pair of large **compound eyes** (shown in the photo to the right) and some possess, in addition, simple eyes called **ocelli**. The simple eyes are best suited to detect shades of light intensity and, according to some entomologists, may help regulate diurnal (daily) rhythms. The compound eyes, on the other hand, have much different capabilities. They are designed to form images, see color, and, especially, to detect movement.

Compound eyes can most likely form a blurred or coarse-grained image, perhaps a mosaic of sorts, but certainly not the type of sharp image our eyes form. It is also unlikely that they can see over a long distance. However, the compound eye sees well into the ultraviolet spectrum, something we cannot do. They can also detect polarized light, which they apparently use as a tool in navigation and orientation.

But the compound eye's real triumph is as a motion detector. To best understand why that is so, we need to briefly examine the structure of this type of eye. In most cases, the eyes are large, convex, and occupy much of the lateral side of an insect's head (see below). Each eye consists of up to 30,000 individual units called **ommatidia**. These receptors send separate visual images to the brain, and it's for that reason we think the image may be a mosaic. Although each ommatidium functions independently, all are linked together to form a curved visual unit looking much like a humpbacked **honeycomb**.

Each ommatidium can sense slight changes in light intensity. As light passes over the eye, such as when a nearby object moves, each of the thousands of ommatidia records the change. It is estimated that the insect eye can thus detect light movement of less than $0.10°$. In simple terms, that means an insect can detect the slightest twitch, the barest shiver, the merest tremble.

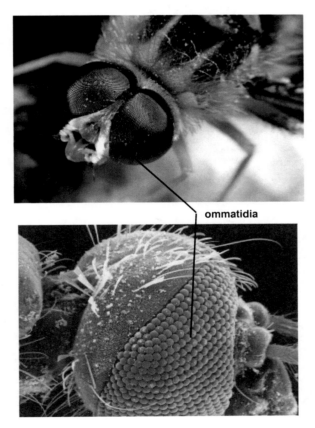

ommatidia

Figure 14.7. Some of the sensors of a fly. In addition to the photoreceptors (compound and simple eyes) and chemoreceptors (antennae), note the many body hairs that function as mechanoreceptors. The scanning electron micrograph (SEM) in the lower plate shows a fly eye under extreme magnification.

How do insects locomote?

Adult insects use a variety of ways to get from one place to another. They walk, run, jump, crawl, climb, burrow, swim, and fly. However, in all cases, the locomotor appendages, the six legs and two or, more likely, four wings, are borne on the middle body section, the thorax.

In general, the speed of travel is directly related to three factors: diet, method of feeding, and escape. Predacious insects are usually swift-moving animals; whereas, plant-eating or scavenging insects are often slow moving. Compare, for instance, the relative speeds of a scavenging millipede or maybe a flower-sipping butterfly with that of a prowling tiger beetle or a dragonfly (both predators). Tigers beetles and dragonflies are, in the realm of arthropods, world-class sprinters whereas butterflies and millipedes move through life at what seems to be a slow-motion pace.

However, plant-eating or scavenging insects are not always slow moving. In fact, some are swift, and need to be, to escape danger. For example, a deerfly that is prey for a dragonfly relies upon its awareness and quickness to flee. The same is true with cockroaches who can serve as dinner for a number of animals. They are impressive sprinters even though they, themselves, are scavengers. Houseflies, too, are famous for their ability to elude even the most diligent swatter, as are grasshoppers.

Figure 14.8. Examples of arthropod locomotion in relation to food source. **A**, millipedes and butterflies are relatively slow-moving herbivores, whereas tiger beetles (**C**) are swift predators. Note the impressive jaws and long legs that make the tiger beetle a formidable maurader easily able to overtake and subdue its prey. Although cockroaches are scavengers, they are nonetheless, extremely fast. Cockroaches (**D**) are relatively quick moving scavangers while grasshoppers (**E**), like butterflies, are slow-moving plant feeders, yet they can quickly escape using their powerful jumping hind legs. The well-camouflaged mantis (on the next page) is another potent predator that lies in wait to ambush any passing prey. Note the large raptorial forelegs equipped with spines as well as the powerful hind legs used to quickly explode at an unwary passerby. Otherwise, mantids are rather deliberate, slow-moving insects.

On the other hand, the method of obtaining a meal may belie an animal's speed. Praying mantids (below), for example, are the insect equivalent of a military sniper, lying quietly in wait until an unsuspecting victim comes within range, then they strike with laser-like speed.   Mantids, themselves, however, are laboriously slow-moving animals. They walk leisurely and they fly clumsily.   Yet beneath the surface is supreme quickness. They are sluggish, until a target approaches; then they pounce like a thunderbolt.

Figure 14.9. Praying mantids are the quintessential insect predator. They lie in wait until an unsuspecting victim approaches, and then with their raptorial forelegs, they impale their hapless victim with lightning-like speed.

The basic design of an insect leg is shown in Figures 14.9 and 14.10, below. The leg is attached to the thorax by a rotating junction, the **coxa**, which, in turn, articulates with an oval-shaped **trochanter** much like a ball-and-socket joint. Both of these segments allow for wide-angle movement of the entire limb. The **femur** and **tibia** possess internal muscles for the movement of distal sections, especially the **tarsus**, which contacts the substrate surface. It (the tarsus) usually has claws, hooks, or hairs to increase the area of contact. In certain parasitic insects (e.g., fleas and ticks), the tarsus has additional modifications for attachment to their host's body.

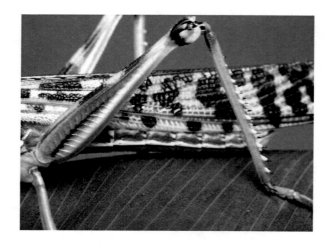

Also shown in Figure 14.10 are two highly specialized appendages. A limb adapted for pollen collection is shown in the **A** panel. It is somewhat flattened and bears baskets, hairs, rakes, and combs for pollen attachment. Figure 14.10 **B** is a mantid leg modified for quick capture of prey. Note the powerful **tibia** capable of explosive maneuvering of the tarsus to entrap prey. Both the tibia and tarsus are also equipped with spines to better grasp a struggling quarry.

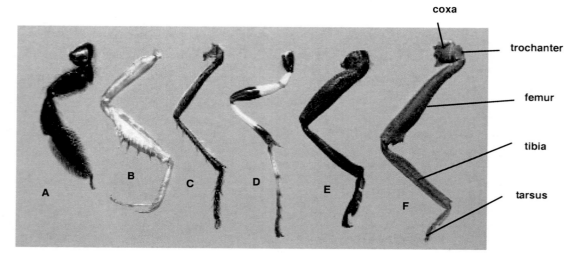

Figure 14.10. Representative insect legs are exhibited in the lower photograph. **A**, a leg, such as that of a bee, modified with flanges and hairs to carry pollen; **B,** raptorial leg, such as found on mantids, used to capture prey (note spines and claws for impaling the prey); **C - F** are elongated, rather generalized, legs adapted for walking and running. The segments of a typical leg are indicated on panel **F**. The upper photograph shows the powerful hindlegs of a grasshopper which enable it to quickly escape an attacker.

Insect Flight

Of all the ways insects locomote, none is more impressive or important than flight; indeed, the very essence of insect locomotion is winged flight. First appearing some 350 million years ago, perhaps as a modified gill-like outgrowth of the exoskeleton, or maybe as a flapping device to help control body temperature, the wing has given insects an advantage not afforded other invertebrates, and very few vertebrates, for that matter. The ability to escape danger, to locate food and shelter, to find mates, to migrate long distances, or to simply relocate locally are all made possible by winged flight. Without it, insect success would surely be far below what it is today.

Most adult insects have wings, and most winged insects have two pairs. Some groups, such as the dipterans (flies, mosquitoes, and relatives) have but one pair, but the great majority have a pair of **forewings** and a pair of **hind wings**. Those insects that lack wings (e.g., silverfish, fleas, lice) either evolved before wings appeared or else secondarily lost their wings as a lifestyle adaptation. At times, the front wings have become modified into a shell-like cover (as in beetles or true bugs) and the hind wings bear the burden of flight. But even in these cases, four wings are present.

The general anatomy of a wing is shown below. Note the four membranous wings with their network of cross veins serving as blood vessels and airways.

Figure 14.11. Wings are one of the true marvels of insect anatomy. The dragonfly in panel **C** shows the primitive condition with four equally sized wings. Note the numerous cross veins for blood and air passage. The butterfly above (**A**) has forewings and hind wings covered with scales and of unequal size. A number of insects have transformed the forewings into leathery or hard plates called elytra as seen in the beetle above (**B**). These cover and protect the more delicate hind wings, which are usually folded beneath the elytra.

How and what do insects eat?

The variety of food that insects consume is reflected by the variety of their mouthparts. In effect, insects, as a group, can devour anything edible, living or dead. They are carnivorous, herbivorous, ectoparasitic, and many are scavengers. Some even feed exclusively on animal dung while others "grow" fungi in subterranean gardens. Still others manufacture their own food, honey, from flower nectar. Each strategy requires special mouthparts, sometimes modified only slightly from a primitive condition while at other times requiring elaborate alterations. Although the types of mouthparts are, indeed, extensive, they fall into just three broad categories. The most primitive, and most widespread, is the **biting-chewing** type seen in grasshoppers, crickets, mantids, dragonflies, beetles, biting lice, and others. Here the serrated mandibles, under control of powerful muscles, work together to tear off chunks of food, which is crushed and then swallowed.

The second pattern, the **piercing-sucking** type is used by mosquitoes, biting flies, fleas, ticks, true bugs (i.e., hemipterans), and sucking lice, among others. This category involves adaptation to a liquid diet of some type, such as blood, plant sap, or nectar. In each case, the mouthparts form an elongated tube through which the liquid is sucked into the mouth. In those insects that withdraw blood or plant "juice," the sucking tube is surrounded by a stylet for piercing flesh or tissue. A modification of this type is seen in butterflies, moths, and some flies in which piercing is unnecessary and a simple strawlike tube is all that is needed. The tube may be extremely long and coiled when not in use.

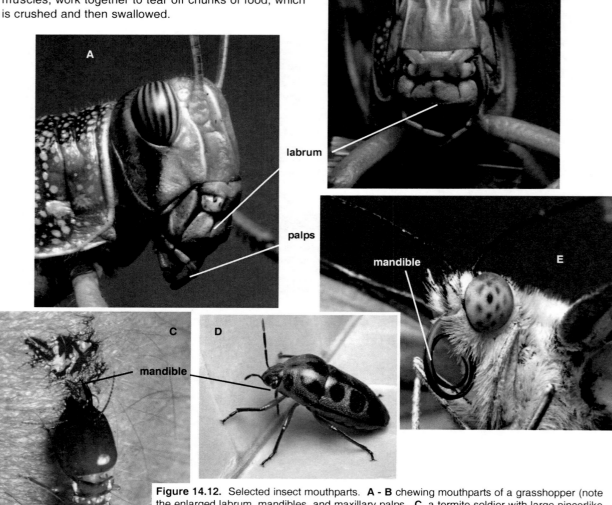

Figure 14.12. Selected insect mouthparts. **A - B** chewing mouthparts of a grasshopper (note the enlarged labrum, mandibles, and maxillary palps. **C**, a termite soldier with large pincerlike mandibles powerful enough to draw blood as seen here. **D**, a beetle with piercing-sucking mouthparts extracting sap from a leaf (note the probing mandible). **E**, the coiled sucking mouthparts of a moth. Note also the large compound eyes in the moth and grasshopper.

The third and final mouthpart classification is the **sponging** type. Houseflies and many other dipterans have developed a padded extension on the end of the labium, the "lower lip," to sop up a liquid diet (below). They actually convert solid or semi-solid food into a liquid form by first secreting saliva onto the food and then sponging it up as it liquifies (see *Inside View* 14.1, page 328 for a detailed discussion of the housefly). The food is then transported by capillary action to the mouth. Some "biting" flies deviate from this pattern of salivating and sopping. They, instead, slash through the epithelium and then soak up a blood meal. You may have been victim to horseflies, deerflies, or other biting flies of this "slash-and-sop" sort.

sponging pad (labium)

Why is their reproduction so special?

I guess there's only one way to describe the reproductive capability of insects: *prolific*. It is estimated that there are around 200 million insects on earth for every person alive! Certainly, their capacity to propagate is simply enormous—an output that's staggering in its abundance and overwhelming in its diversity. And why is that? While there are probably numerous reasons, two stand out above all others:

The Terrestrial Egg

Insect embryos develop within a fluid-filled, shelled egg deposited by the female. It is this **cleidoic egg**, perhaps more than anything else, that has made terrestrial life possible. On the whole, land reproduction is a double-edged, perilous process. First, external union of egg and sperm is virtually impossible on land. Sperm and eggs cannot simply be broadcast into the atmosphere and expected to undergo fertilization. Therefore, for fertilization to occur it must be *internal*. The male must find a way to introduce sperm *into* the female; he must somehow impregnate her. This means that he must

find her and then get close to her, real close. Over the eons of insect evolution, insects have developed many ways to locate each other, but none more effective than chemical attractants...pheromones. One sex or the other (usually the female) releases these romantic clouds and males are drawn in, often from relatively long distances. It is reported that a male moth can detect just a few molecules of a respective pheromone over one-half mile away!

But finding each other is just half the battle. Once together they must overcome natural shyness, or natural aggressiveness, whatever the case may be. To do that, insects have stumbled on a clever device, a stylized courtship process that leads each participant, step-by-step toward copulation.

Once internal fertilization is accomplished, the embryos must be provided with a place to develop. There are only two options available: internally (inside of the female) or externally. *Internal development*, while safe and dependable, exacts a stiff price. The cost is in numbers. The female can only carry a limited number of developing embryos internally, therefore, **viviparity** (live young) or **ovoviviparity** (holding shelled eggs internally) results in relatively few offspring. The other option, *external development*, has the potential to produce high numbers but it, too, has drawbacks. For one, it can be hazardous; the young, if exposed, may very well fall prey to prowling predators. However, desiccation is an even more dangerous problem. Embryos, if unprotected, will dehydrate. To counter dehydration, a shelled egg has been developed. In fact, a chitinous shelled egg goes a long way toward solving both problems. It can be hidden, buried, camouflaged, or even actively defended as is the case in many social insects (termites, ants, bees, wasps). But more importantly, the shelled egg protects against desiccation. It is this feature that makes the cleidoic egg so valuable. In fact, the land-adapted egg has been one of the most important adaptations in arthropod evolution. Without it, land-based reproduction would most likely have never occurred among invertebrate animals.

Figure 14.13. A giant water bug with shelled eggs fastened to her back.

Metamorphosis

All insects undergo some growth from hatching to adulthood; adults are invariably larger than immatures. The insect developmental pattern is as follows: the young hatch and begin to grow, but to do so they must periodically shed (molt) their nonliving exoskeleton, a process called **ecdysis**. There are a number of molts that may occur, usually six, before adulthood is reached. In the periods between molts, the developing individual is called an **instar**. So the normal sequence is from egg to instar to molt (ecdysis) to instar and so on until the adult stage is reached following the fifth instar.

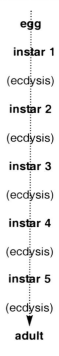

egg

instar 1

(ecdysis)

instar 2

(ecdysis)

instar 3

(ecdysis)

instar 4

(ecdysis)

instar 5

(ecdysis)

▼

adult

There are a few insects that hatch from an egg and grow to adulthood without any discernible physical change. For instance, the primitive springtails and silverfish emerge from the egg as a small version of the adult, pass through a number of molts, and become sexually mature adults. Maturation of that sort, in which there is no measurable change from juvenile to adulthood, is termed **ametabolous** or *simple development* and only occurs among primitive, wingless insects. In fact, this nonmetamorphic pattern is used by less than 1% of insects.

However, all other insects display some degree of distinct change during their post-egg development. We call this type of marked change **metamorphosis** and there are two major types. In a number of species, the immature insect resembles the adult in that it has compound eyes, similar mouthparts, antennae, and similar walking legs. In this early stage, called a **nymph** (or if aquatic, a naiad), wings and sexual organs are always absent. Development occurs in gradual stages in which the nymph becomes more and more like the adult through a series of molts until wings and sexual organs develop. This gradational pattern is known as **hemimetabolous** development (the terms *gradual* or *incomplete* metamorphosis can also be applied). Approximately 9% of insects are hemimetabolous, including true bugs, lice, earwigs, termites, dragonflies, mayflies, grasshoppers, cockroaches, cicadas, praying mantids, and crickets. One final note about hemimetabolous insects: both the *immatures and adults occupy similar habitats*.

The remaining insects (~ 90%) show *complete* or **holometabolous** development. Beetles, butterflies, moths, flies, bees, lacewings, fleas, mosquitoes, and ants are some of the species that undertake this, the most striking of all insect development. In this pattern, a wormlike **larva** (also known as a maggot, caterpillar, or grub) emerges from the egg. It *bears no resemblance whatsoever to adults;* indeed, both live in far different habitats with far different habits.

The larva is basically a feeding and growing stage while the adult is the reproductive unit. Larvae go through a number of instars and molts before entering a dormant state, as **pupae**. The pupa then undergoes an altogether dramatic transformation, **pupation**, into the adult form. Pupation occurs in special casings (a *cocoon* or *chrysalis,* below), or it may take place in plant stems, twigs, tree trunks, decaying logs, in prepared cells (honeybees), within the bodies of living and dead animals (parasitoids), or in other secluded locations.

Figure 14.14. A monarch butterfly just emerging from the puparium, here called a chrysalis.

Insect Development
~ A Summary ~

Insects use three developmental patterns:

(1) **Ametabolous** (simple) - used by primitive, wingless insects in which the juveniles are small versions of the adult. A number of molts allows the animal to grow into adulthood. Less than 1% of insects are ametabolous: silverfish and springtails (below) are examples.

Sequence: egg → juvenile → adult

springtail

(2) **Hemimetabolous** (incomplete) - the immature insect has some resemblance to the adult and is gradually transformed, through progressive molts, into a sexually mature individual. About 9% of all insects use this pattern: dragonflies, roaches, mantids, crickets, and grasshoppers.

Sequence: egg → nymph (naiad) → adult

dragonfly adult

(3) **Holometabolous** (complete) - this is the most common developmental pattern and is used by about 90% of all insects. The immature insect is grossly different in appearance and habits from adults and undergoes an abrupt, radical change into adulthood. Examples are beetles, flies, butterflies, moths, ants, and bees.

Sequence: egg → larva → pupa → adult

Significance: Since so many insects use the more advanced form of metamorphosis, especially the holometabolous pattern (~ 10/1 ratio), there must be some pragmatic value to the practice. Most entomologists would agree that the advantage lies in ecological diversity. That is, the immature, and adults, being dissimilar in structure and physiology, do not compete with each other for environmental resources; the more dissimilar they are, the less they compete. In other words, the immature insect and the adult virtually occupy different ecological niches. In addition, the immature stage is basically a growth form, while the adults are primarily reproductive units; some adults don't even feed, but are designed only to propagate.

adult

pupa

larva (caterpillar)

eggs

How are social orders constructed?

It seems that very early in their evolution some insects discovered one of the true secrets of success: togetherness. We see it in their elaborate courtship interactions, in care for eggs, and we see it, most explicitly, in their social organizations. A few species, and only a relatively few (around 13,000 out of the vast number of insects), are genetically programmed to work together in a complex and systematic way. In the terminology of behavioral biology, they are **eusocial**. All species of ants and termites are eusocial as are some bees and a few wasps. And what makes this phenomenon even more intriguing is that among all of the invertebrate animals, only insects form these elaborate social units.

A truly eusocial arrangement is invariably *colonial*. The colonies can be rather small, consisting of a few hundred individuals, such as some wasp aggregations, or they may be huge, with millions of swarming individuals as in ants and termite colonies. However, the common feature in all insect social orders is the division of labor among different castes. Within the colony are individuals of various types (castes), each type adapted to perform a specific function. For instance, there is a **primary queen** who is the reproductive source of the colony. Secondary and even tertiary queens may exist to replace the primary queen, if necessary, or new queens may be cultured from an undesignated female if the old queen dies. Another member of the caste system is the sterile **worker** who carries on the major building of the colony and cares for the queen, eggs, and larvae. **Soldiers** are invariably present as defenders of the colony, and stout defenders they are as illustrated by the termite in the photo below drawing blood with its powerful mandibles. The vast majority of insect societies are females; males, sometimes called **drones**, are infrequent in these colonies and are only used to fertilize new queens.

Ants are a perfect example of how effective social organization can be. By the number of species (~12,000), they make up less than 0.02% of the insect world yet, since their colonies are so enormous and so complex, *they actually constitute over 50% of total insect biomass.* A single ant colony may possess 10-12 million individuals. And we can only guess how many colonies exist...the number must surely be astronomical.

The structure of their colonies revolves around the queen. A virgin, winged female undergoes a nuptial flight in which she is impregnated by one or more winged males. She then initiates a new colony and all of the males die. The queen builds a nest and thereafter produces a constant stream of eggs using the sperm stored from the lone mating flight. All fertilized eggs develop into female workers who tend to the needs of the queen and care for developing eggs.

This means that all members of the colony, the queen, the workers, and the soldiers are, in reality, sisters. Males are only produced after several years when the colony attains sufficient size. The queen creates males by releasing unfertilized eggs. They are winged and, unlike the females, who perform all of the colony's duties, males do no work. They are soon ushered from the colony to locate virgin females and engage in mating flights. Males then die and the impregnated females go off as queens to start new colonies.

Figure 14.15. A soldier termite is charged with the responsibility of guarding the society, especially the queen. When provoked, soldiers can be formidable as seen in the photo to the left. The queen (above) is constantly tended by a cluster of workers. The worker in the foreground (arrow) can be seen cleaning the queen's antennae.

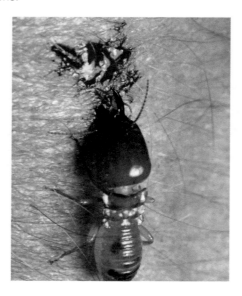

Insects and People

Throughout the last 350 million years, insects have exerted an awesome impact on planet earth. Like a ubiquitous whirlwind, they have wielded their massive armies of tiny warriors to influence the biosphere as has no other. All organisms have been affected, and we too, as one of the latest animals to appear, have shared...at times suffering but, in the main, benefiting.

Negative Impact

Pests: There's no denying it, some insects are tormentors. Gnats, cockroaches, houseflies, ants, clothes moths, noisy cicadas, and stinging bees are all bothersome. They take their toll on us, primarily a psychological one.

Parasites: Other insects are even more troublesome since they rely directly on us for food. Mosquitoes, biting flies, bed bugs, lice, fleas, and kissing bugs can be little tyrants as they search our hides for a helping of blood or flesh.

Disease vectors: Still other insects can exact a more devastating price as carriers of disease organisms. Malaria alone, carried by mosquitoes, has killed more people than all human hostilities combined. And it still rampages. Insects also transmit typhus, encephalitis, yellow fever, bubonic plague, anthrax, leprosy, and many other diseases. The common housefly, alone, is vector of over 60 disease organisms (see *Inside View* 14.1 on next page.

Crop destruction: Each year approximately 10-12% of the world's food supply is either eaten or despoiled by insects. The cost in dollars and human misery is unfathomable.

General damage: Termites tunneling through a house's foundation, carpenter ants boring into wood for a nest, long-horned beetles killing shade trees, clothes moths pilfering a closet of fine garments, carpet beetles fragmenting your new oriental rug, and dermestid beetles turning my bird and mammal museum skins into a shambles are all examples of general damage done by insects.

Benefits

The relationship with insects can be bothersome and, at times, life threatening. They are, indeed, a force to be reckoned with. *But, on balance, they do far, far more good than bad.* The interaction between insects and the rest of us, especially us, is pre-eminently beneficial.

Products: Insects provide us with wax, silk, dye, honey, and even pleasure for those who admire their intriguing richness.

Research: The study of insects, termed "entomology," has have given us a greater understanding of these organisms. From insect control to basic physiology, particularly, how nervous systems work, entomologists have added significantly to our knowledge of animals. The entire science of genetics is built on the tiny wings, eye color, legs, and antennae of the fruit fly, *Drosophila melanogaster.*

Recycling: Today, in a nearby woodland a deer will die of old age. Or he may succumb to some parasite or to some disease. As he stumbles and falls, the echo is heard throughout ecosystems large and small as millions of his kin, furry, feathery, scaly, or chitinous, do likewise. Have you ever wondered what happens to the legions of carcasses that litter our wildlands and roadways each day? Why aren't we up to our armpits in rotting corpses? Because of insects, that's why. Most dead animals are recycled by insects. Maggots, carrion beetles, w a s p s , and other insects attack and devour a carcass within days. The value done by this nutrient recycling is inestimable.

Food: Some people eat insects. A chocolate-covered ant gulped at a party or a termite mound torn apart by a nomadic tribe can serve a gustatory function. But think of the nutritional value that insects provide daily to billions of birds, bats, frogs, shrews, skunks, lizards, and snakes. They provide an irreplaceable base to food webs everywhere.

Biocontrol: Insects eat other insects. Sometimes they even eat insects that pester us or eat our crops or destroy our forests. They are nature's own insect control network.

Pollination: But by far the most important contribution of insects is the transferral of plant sex cells. It is estimated that around 65% of all plant species are pollinated by insects. From flowers to towering trees, insects keep the reproductive process going. Without insects we would live in a world of ferns, mosses, and evergreens. Most of our fruits and vegetables are insect pollinated as are our cultivated ornamentals. My flower garden wouldn't exist without visits from bees, beetles, and butterflies.

IT'S A FACT

Without insects, life on earth would be a shallow, barren landscape of rotting, putrefying carcasses. True, we might be free from malaria and yellow fever, we probably wouldn't have to combat plague or typhus, we could rest easily without fear of attacks by bed bugs. Picnics would be missing the uninvited mosquito or ants marching in cadence. We wouldn't need to worry so much about crop devastation; but then, there would, of course, be few crops. Without insects our food supply would dwindle to virtually nothing. Our world would truly be impoverished beyond recognition without these tiny swarming...so successful...arthropods.

An Inside View 14.1

Musca domestica
The Common Housefly:
Order Diptera

✪ Houseflies are hardy, ubiquitous pests found wherever humans live. Although they cannot survive temperatures below freezing, they can, because of their close association with people, overwinter in cracks and crevices in our homes. In fact, because of central heating, they can, at times, breed the year around right beside us.

✪ Each compound eye contains approximately 4,000 lenses; thus, flies have a wide-angle, omnidirectional visual field. That's why it's so difficult to swat one.

✪ Unlike most insects, flies only have one pair of wings; the rear ones are reduced to club-shaped structures, the halteres, which help to maintain balance while flying. The wings beat at a speed of almost 20,000 beats per minute (that's about 330 beats per second). Compared to the hummingbird's pace of 80 beats per second, the flutter of fly wings is a virtual blur!

✪ Their food consists primarily of sugary materials, although females feed on protein when forming eggs. Prior to feeding, houseflies regurgitate (i.e., vomit) on their food to liquefy it, and then sponge up the sugary broth. On their tiny feet are special receptors that they use to taste potential food.

✪ Houseflies have the habit of defecating every four to five minutes, usually on their food.

✪ Copulation, in houseflies, takes approximately one hour. After mating, the male injects a fluid into the female that makes her unwilling to mate with any other male; thus, he assures that his sperm will be the ones to fertilize the eggs. Following copulation, the female deposits eggs in other animals' excrement, in rotting carcasses, or on garbage.

✪ Development, from egg to pupa to adult, occurs in about 18 days. Adults normally have a life-span of approximately 30 days, but may live as long as 5 months.

✪ Houseflies have an enormously high rate of production: a single batch of 120 eggs from one gravid female would, in five months, yield about 191,000,000,000,000,000,000 flies if they all survived. Some experts claim that this equals enough flies to cover the earth's landmass to a depth of 18 feet! In point of fact, more flies hatch each day than there are humans on earth.

✪ The housefly is a vector of approximately 60 disease-causing organisms such as typhoid fever, leprosy, gangrene, bubonic plague, dysentery, tuberculosis, anthrax, and cholera.

✪ They may even spread ulcer-causing bacteria. Most of these disease organisms become attached to the feet of the fly following a visit to feces, carcasses, or other waste material. The disease organisms may then become transferred to human food as the fly is feeding. Houseflies can also harbor ~ 6.0 million bacteria on the outside of their body, and another 40 million internally.

✪ Just a few decades ago, houseflies were susceptible to pesticides such as DDT; however, they are now highly resistant to this and many other chemicals used in our battle to control household pests.

adult fly

larva (maggot) in a puparium

eggs on a dead caterpillar carcass

The Terrestrial Conundrum
~ A Riddle Solved By Arthropods ~

There's a puzzle, a riddle of sorts, that has perplexed zoologists for decades. It goes something like this: How can a small, unassuming, inconsequential creature like a centipede, a spider, or a scorpion afford to leave the comforts of the sea and overcome insurmountable odds to survive on land? Some 350 million years ago some arthropod did just that. Not only did arthropods survive, but they have since conquered terrestrial environments to become the most successful animals ever seen. How they accomplished this feat is revealed by the eight problems and the solutions outlined below.

Problem # 1 Animals must be able to breathe on land.

> Solution: Although today's aquatic arthropods still use a gill for gaseous exchange, their terrestrial kin developed two new internal structures for the purpose of breathing. The first is a series of internal flaps, the book lung, of the chelicerates. This can be seen in horseshoe crabs and in spiders. The second respiratory innovation is the tracheal system in which a series of tubes leading from slitlike spiracles transmits oxygen to the internal tissues and carries away carbon dioxide. By internalizing the respiratory surface, it remains moist and protected.

Problem # 2 Dessication is every animal's mortal enemy; they must solve the problem of dehydration.

> Solution: The remedy here was fairly simple. Preterrestrial arthropods had already developed the exoskeleton. All that is needed for true terrestrialism is some way to waterproof this hard cover. This was accomplished by permeating the exoskeleton with a waxy substance that prevented water from entering and exiting the animal.

Problem # 3 The pull of gravity on land is much greater than in water; so, how is friction prevented?

> Solution: To effectively move on land, an animal must resist friction caused by gravitational forces. The paired, jointed appendage provided the solution. The appendage not only lifted the organism off the surface, but, since it was jointed, it also improved overall motion for active, forward locomotion.

Problem # 4 Regulating water and salt levels is more difficult on land; how did early arthropods achieve terrestrial osmoregulation?

> Solution: Terrestrial arthropods convert metabolic waste from ammonia (in aquatic forms) to urea or urine. To do that, they developed a new kidney, the Malpighian tubule.

Problem # 5 How does an animal conform to the constantly changing temperatures of terrestrial habitats?

> Solution: This was a tough problem for arthropods to solve, and they were not altogether successful. The best they could do was to modify behavior to accommodate temperature fluxes. They slow down or hide when it's cold, emerge and speed up when it's warm. In fact, during long cold or dry spells, some arthropods can enter into a prolonged dormant state referred to as diapause.

Problem # 6 How can an aquatic animal take advantage of the new food sources available on land?

> <u>Solution:</u> The remedy here lies in the relative plasticity of the arthropod appendage. Through a process termed serial homology, a basic structure is modified into a number of functional units. Several of these units surround the mouth (e.g., mandibles, maxillae, maxillipeds, pedipalps, etc.). These adaptable mouthparts were transformed into a variety of different feeding structures: sponging, chewing, slashing, sucking, biting. Thus, arthropods could well exploit the various foods available on land.

Problem # 7 A successful land dweller must be able to receive airborne stimuli.

> <u>Solution:</u> Again, fortunately, aquatic arthropods had already developed the infrastructure for terrestrial responsiveness. The compound eye already existed, as did antennae. Both of these are essential to the reception of visual and chemical stimuli. Receiving sound waves posed a bigger challenge. This was eventually solved in two ways. First, the body could be covered by sensory hairs or setae to detect ground vibrations, and, second, a membrane, the tympanum, was developed to detect airborne vibrations.

Problem # 8 To be truly successful, they must be able to reproduce on land rather than in water.

> <u>Solution:</u> To be truly successful on land, arthropods must protect the embryo and keep it moist without having to return to aquatic environments to do so. This was, perhaps, the biggest challenge facing arthropods as they contemplated a terrestrial invasion. It was accomplished by the advent of two things. First, they must do away with external fertilization...it doesn't work on land. So, arthropods developed the necessary structures so that eggs can be fertilized <u>within</u> the female. By the way, this also required a change in behavior allowing the two partners to "get together."
>
> Secondly, there must be some way to overcome the drying effect that a terrestrial environment would have on the embryo. This was solved, rather ingeniously, by the appearance of a <u>shelled egg.</u> Thus, the embryo could develop unaided within its own protected, aquatic sphere.

> **Note**: Many of the necessary features for terrestrialism already existed among aquatic arthropods. This meant that they were already poised to invade land...they were, in many respects, **preadapted** to do so.

Chapter 14 Summary

Key Points

1. The subphylum Uniramia includes the centipedes, millipedes, and the insects as the major groups.

2. Uniramians share several features including unbranched appendages, a single pair of antennae, chewing, sponging, or sucking mouthparts, a ventral nerve cord, tracheal respiration, and Malpighian tubules for excretion and osmoregulation.

3. Centipedes are nocturnal, flattened, elongated arthropods that rely upon poison fangs to overcome their prey. They possess a single pair of legs per body segment.

4. Millipedes (class Diplopoda) are diplosegmented, worm-shaped arthropods that feed on decomposing vegetation.

5. Both centipedes and millipedes possess specialized repugnatorial glands that secrete a noxious discharge to discourage would-be predators.

6. Approximately 75% of all known animals are insects and about 40% of these are beetles. Thus, beetles (order Coleoptera) constitute the largest assemblage of animals on earth.

7. Insect success is due to a number of factors: small size, high reproductive rate, exoskeleton, terrestrial egg, metamorphosis, well-developed sensory apparatus, flight, and complex behaviors.

8. An insect's body is divided into three parts: the head, which bears the feeding and sensory structures; the thorax, which bears the locomotory appendages (legs, wings); and the abdomen, which houses the major internal organs and bears the external genitalia.

9. Insects possess powerful senses: the tympanic membrane detects vibrations (mechanoreception); chemosensors are scattered along the body surface; and simple and compound eyes serve as photosensors.

10. Insects use legs and wings for locomotion. Legs can be modified for walking, jumping, digging, swimming and running. Insects were the first animals to develop wings and most adult insects have two pairs.

11. The basic mouthparts of an insect include a pair of mandibles for chewing and associated appendages to assist in that process. However, the basic mouthparts have been modified for piercing, sucking, and sponging of food.

12. The reproductive mechanism of insects shows adaptations to a terrestrial life. Specifically, courtship behaviors, internal fertilization, and the development of the terrestrial egg are important terrestrial adaptations.

13. The terrestrial, shelled egg, in particular, is a significant adaptation to life on land. It allows the embryo to develop in a protective, dehydration-resistant chamber outside of the female.

14. To grow, an insect must periodically shed its exoskeleton in a process called ecdysis. Periods of ecdysis alternate with periods of development and growth during which the insect is termed an instar. Thus, insect growth involves an alternating pattern of ecdysis, instar, ecdysis, instar, and so forth until adulthood is attained.

15. Most insects also undergo some transformation in body form from immature to adulthood...a process termed metamorphosis.

16. Some insects undergo only minimal change (ametabolous development); others show more moderate modification (hemimetabolous development). But most insects undergo dramatic changes (holometabolous development) from egg through larval and pupal stages before adulthood is reached.

17. Certain types of insects (ants, termites, bees) show a tremendous degree of social interactions. Such eusocial insects have caste systems consisting of workers, soldiers, and queens, each carrying out important roles in the society.

18. Insects, as a group, have both negative and positive impacts on humans. They can be vectors of disease, damage crops, sting, serve as parasites, and cause general damage to our homes and facilities. On the other hand, they serve as vital elements in the recycling of materials, pollination of plants, and in the control of harmful insects.

19. We have used a number of approaches to control insect pests. From mechanical fly swatters and netting to the use of powerful insecticides, we have waged a losing battle against insects. Recently, a more environmentally friendly method of integrated pest management has shown favorable results.

20. The common housefly is a good example of an insect pest that has withstood our onslaught. Its behavior, reproduction, and association with humans has made it one of the most ubiquitous of pests.

21. Insects are the most successful of terrestrial animals. To achieve this success, they have overcome a number of obstacles such as terrestrial respiration, dehydration, gravity, land osmoregulation, temperature fluctuations, airborne stimuli, and terrestrial reproduction. The key to their amazing success lies in how they have overcome each of these obstacles.

Key Terms

ametabolous development
biting-chewing mouthparts
biocontrol
cerci
cleidoic egg
complete development
compound eye
coxa
diplosegments
drone
ecdysis
entomology
eusocial behavior
femur
gradual development
halteres
hemimetabolous development
holometabolous development
instar
integrated pest management
juvenile stage
Malpighian tubules
metamorphosis
naiad
nymph
ocelli
ommatidia
ovoviparous
pheromone
piercing-sucking mouthparts
primary queen
pupa
puparium
repugnatorial glands
royal jelly
simple development
soldiers
sponging mouthparts
sterile worker
tagmata
tarsus
terrestrial egg
tibia
trochanter
tympanic membrane
uniramous appendage
viviparity

Points to Ponder

1. What is a "uniramian"?

2. Give five features of a uniramian.

3. Describe a centipede.

4. Describe a millipede.

5. What is the function of repugnatorial glands found in centipedes and millipedes?

6. Insects are the most numerous animals on earth. What *type* of insect is the most numerous?

7. Give three features of the class Insecta that distinguish them from the other arthropods.

8. Sketch a "typical" insect leg.

9. Name three types of insect mouthparts and indicate how each is used.

10. What is special about the insect egg that has led to their success?

11. What is the relationship between ecdysis and the instar stage of insects?

12. Name and describe the three different types of metamorphosis found in insects. Give an example of an insect for each of the three types.

13. What advantages does insect metamorphosis give?

14. Describe the social order found in termite colonies. Give the types of individuals and their functions.

15. What is entomology?

16. Humans and insects enjoy a variety of interactions. Give five ways that we benefit from the relationship and five ways in which we are adversely affected.

17. What is integrated pest management?

18. Describe how a housefly feeds.

19. What are some of the diseases a housefly can spread to humans?

20. How did insects solve the problem of terrestrial respiration? dehydration? terrestrial locomotion? land temperature changes? terrestrial reproduction?

Inside Readings

Arnett, R. H., Jr. 1986. American Insects: A Handbook of the Insects of America North of Mexico. New York: Van Nostrand Reinhold.

Berenbaum, M. R. 1994. Bugs in the System: Insects and Their Impact on Human Affairs. Reading: Addison-Wesley.

Borror, D. J., and D. M. DeLong. 1964. An Introduction to the Study of Insects. New York: Holt, Rinehart and Winston.

Brusca, R. C., and G. J. Brusca. Invertebrates. Sunderland, MA: Sinauer Associates.

Coleman, N. 1991. Encyclopedia of Marine Animals. New York: HarperCollins Publishers.

Davies, R. G. 1988. Outlines of Entomology. 7th. ed. London: Chapman & Hall.

Ellis, R. 1996. Deep Atlantic: Life, Death, and Exploration in the Abyss. New York: Alfred A. Knopf.

Franks, N. R. 1989. Army ants: A collective intelligence. Am. Sci. 77:139-145.

Gordon, D. M. 1995. The development of organization in an ant colony. Am. Sci. 83:50-57.

Grisham, J. 1994. Attack of the fire ant. BioScience 587-590.

Holldobler, B. and E. O. Wilson. 1990. The Ants. Cambridge: Harvard University Press.

Klowden, M. J. 1995. Blood, sex, and the mosquito. BioScience 45:326-331.

Papaj, D. R., and A. C. Lewis. 1993. Insect Learning: Ecological and Evolutionary Perspectives. New York: Chapman & Hall.

Pechenik, J. A. 1996. Biology of the Invertebrates. Dubuque, IA: Wm. C. Brown Publishers.

Pimental, D., et al. 1991. Environmental and economic effects of reducing pesticide use. BioScience 41:402-408.

Rosmoser, W. S. 1981. The Science of Entomology. 2nd Ed. New York: MacMillan Publishing Co.

Wilson, E. O. 1971. The Insect Societies. Cambridge: Harvard University Press.

Wooton, R. J. 1990. The mechanical design of insect wings. Sci. Am. 263(5):114-120.

Chapter 14
Laboratory Exercise
The Uniramians

Laboratory Objectives

✓ to distinguish between members of the three uniramian classes,
✓ to use a dichotomous key of insect orders and correctly identify representative insects,
✓ to identify external anatomy of male and female grasshoppers,
✓ to dissect and identify external and internal structures of a female grasshopper.

Supplies needed:

✓ prepared and living specimens of the three uniramian classes
✓ specimens of 12 insect orders
✓ prepared specimens of both male and female grasshoppers
✓ dissecting equipment

Chapter 14
Worksheet

Phylum Arthropoda
The Uniramians

Subphylum Uniramia

Class Chilopoda (centipedes)

Examine those centipedes on display. Note the flattened trunk with *a pair of appendages for each segment*. Also observe the large **fangs** located on the maxillipeds just ventral to the elongated antennae. Finally, examine the sensory **cerci** projecting from the final segment.

Class Diplopoda (millipedes)

Compare the basic structure of a millipede with that of a centipede. Note the rounded shape, lack of poison fangs, and presence of *two pairs of appendages* for each body diplosegment.

Class Insecta (insects)

There are a number of insects on display. Examine each to learn its specific properties. Note the three body regions (**head, thorax,** and **abdomen**), *three pairs of legs* borne on the thorax, and wings (if present). Use the key provided to identify each of the orders portrayed below. Using a dissecting microscope, or a magnifying glass, compare each to the description given.

Representative Insect Orders

three thread-like cerci

Ephemeroptera (mayflies)
Mayflies are primitive insects with a prominent, aquatic larval stage. The adult part of life cycle is reduced. The adults live for only a few days, at most, during which time they don't feed, spending most of their time mating in flight. When at rest, the wings are held vertically over the back. The abdomen has three long thread-like cerci. Their legs are usually small and mouthparts are pronounced in larva but vestigial (nonfunctional) in adults. Mayflies are an important food source for many fish.

Odonata (dragonflies and damselflies)
The aquatic larvae of these insects are predacious, as are the adults, which prey on flying insects. These primitive insects have large compound eyes, reduced, bristlelike antennae, long and slender abdomens, chewing mouthparts, and two pairs of net-veined wings held out to the side when at rest (dragonflies) or vertically over the back (damselflies).

Orthoptera (grasshoppers, cockroaches, katydids, crickets, walking sticks, mantids)
The legs of orthopterans are adapted for running (cockroaches), jumping (grasshoppers, crickets, katydids), or for raptorial feeding (mantids). Their forewings are narrow and leathery, and the hind wings are usually large and fanlike,. They possess the following features: chewing mouthparts, cerci, antennae moderate to very long, female with pronounced terminal ovipositor, sound reception is a tympanum on legs or abdomen, and parts of the exoskeleton may stridulate to produce sound.

termite queen

Isoptera (termites)
All termites are eusocial with a well-defined caste system of blind workers, blind soldiers with enlarged mandibles, and reproductives (males are winged but live for only a brief period, queens are initially winged, but shed them and become transformed into large repro ductive individuals cared for by workers).   These are small, soft-bodied insects. The abdomen is broadly joined to the thorax. Termites live in huge swarming colonies and can be highly destructive when they occur in human residences.

forceps-like cerci

Dermaptera (earwigs)
In some locations, earwigs are common household residents, but they are scavengers and present no health threat. Although they possess enlarged forceplike abdominal cerci used in defense they are harmless, hind wings are large and folded under short, horny forewings.

Anoplura (sucking lice)
These are blood-sucking ectoparasites of birds and mammals, including humans.
Head lice (cooties) and pubic lice (crabs) have infested mankind throughout history.
They have piercing-sucking mouthparts, wings are absent, legs are adapted for attachment, and compound eyes are missing. They have a broad body capable of great distention during a blood meal.

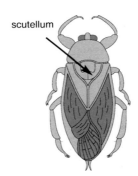

scutellum

Hemiptera (true bugs)
True bugs possess piercing-sucking mouthparts forming a jointed needlelike beak arising from the front portion of the head. The beak is held between the forelegs when not in use. Both sets of wings lie flat against the abdomen when at rest; forewings thick and leathery at the base and membranous toward the tip, hind wings large and membranous but folded beneath the forewings. A prominent triangular plate (the scutellum) is located between the bases of the forewings. Most bugs are herbivores, although many are predacious, and some are ectoparasites (kissing bugs and bed bugs both infest humans). As a group, hemipterans are of great economic importance as destroyers of crops.

Homoptera (cicadas, leafhoppers, plant hoppers, aphids)
Homopterans feed on plant sap with mouthparts that are modified for piercing and sucking. Some forms are wingless but most have two pair of membranous wings held tentlike or rooflike over the abdomen when at rest.

Hymenoptera (ants, bees, and wasps)
This insect order is characterized by the presence of a narrow waist joining the abdomen to the thorax. They possess chewing mouthparts and many have a venomous stinger derived from the ovipositor. Their hind wings are small and joined to larger forewings by a hook. There is a tendency in this group to form eusocial units. Some species are nuisances, others are highly beneficial in pollinating plants.

Siphonoptera (fleas)
These small, wingless insects have a laterally compressed body, mouthparts modified to feed exclusively on blood, and hindlegs adapted for jumping (other legs are adapted for clinging). All are ectoparasites and some are transmitters of serious disease (i.e., bubonic plague). Adults are not highly specific and will switch from one host species to another; humans are often infested with dog, cat, or rat fleas, as well as their own.

halteres

Diptera (flies, mosquitoes, and gnats)
Only one pair of wings is usually present, the other pair reduced to knobby structures called halteres which are used as gyroscopes for balance while flying. Dipterans have large eyes located on a highly mobile head. Mouthparts are variable and may be adapted for sponging, piercing, or chewing. These insects are of great economic and medical importance as vectors of disease and also are of considerable nuisance. However, they serve important functions in carrion removal and their maggots are also used in medicine to remove dead flesh from infected wounds.

Lepidoptera (butterflies and moths).
These insects have large compound eyes, long antennae and mouthparts modified into a long tube (coiled when not feeding) to extract nectar. The wings and body are covered with overlapping, dense, pigmented scales. Butterflies hold their wings vertically at rest, while moths rest with wings positioned horizontally. Larvae are of great economic importance as destroyers of crops and cloth. Most adults, however are important pollinators.

Coleoptera (beetles)
This is the largest order of insects; there are more beetles than all non-arthropods combined. The head bears well-developed antennae, eyes and chewing mouthparts. There are two pairs of wings with the first pair consisting of a hard shell-like covering, the elytra, that meets in a straight line down the back. The hind wings are membranous and folded under forewings when at rest. The body is compact and hard. Some beetles are serious agricultural pests, while others are predators of harmful pests.

elytra

Key to Selected Insect Orders*

1 a. Wings present and well developed 2
   b. Wings absent 20

2. a. Forewings thickened or leathery (at least at base); hind wings membranous, may be hidden beneath forewings...... 3
   b. All wings membranous throughout 7

3. a. Mouthparts beaklike (somewhat like a syringe and often held between legs when not in use) 4
   b. Mouthparts mandibulate, adapted for chewing........ 5

4. a. Beak arises from front of head, forewings leathery at base but membranous at tips (true bugs) .... **Hemiptera**
   b. Beak arising from rear of head, forewings uniform and held tentlike over abdomen (plant hoppers)........ **Homoptera**

5. a. Abdomen with forcepslike cerci, forewings short and covering folded hind wings (earwigs).... **Dermaptera**
   b. Not as above 6

6. a. Forewings hard, veinless, shell-like meeting in a straight line; hind wings folded under forewings (beetles).... **Coleoptera**
   b. Forewings not as above, hind wings broad and usually shorter than forewings; forelegs may be raptorial, hind legs may be
       modified for jumping, body may resemble sticks (grasshoppers, mantids, crickets, cockroaches, stick insects)... **Orthoptera**

7. a. One pair of wings only 8
   b. Two pair of wings 11

8. a. Body grasshopper-like, hind legs adapted for jumping ..... **Orthoptera**
   b. Not as above 9

9. a. Mouthparts vestigial (missing); three long, threadlike cerci (mayflies) ...... **Ephemeroptera**
   b. Mouthparts chewing, no cerci 10

10. a. Sucking mouthparts; hind wings reduced to clublike halteres, tarsi five segmented (flies) ..... **Diptera**
    b. Mouthparts chewing, tarsi two or three segmented ...... **Psocoptera**

11. a. Wings covered by overlapping, pigmented scales; mouthparts coiled (butterflies/moths) .. **Lepidoptera**
    b. Not as above ... 12

12. a. Wings long and narrow, fringed with extremely long hairs (edges appears fuzzy) .... **Thysanoptera**
    b. Not as above 13

13. a. Forewings large and triangular; hind wings small and rounded; wings held vertically over back; wings heavily veined;
        soft-bodied insects with two or three long, threadlike tails (mayflies) .... **Ephemeroptera**
    b. Not as above 14

14. a. Tarsi five segmented (just count the segments on the distal leg region [tarsus]) 15
    b. Tarsi consisting of four or fewer segments 16

15. a. Rather hard-bodied, wasplike insects; usually a narrow waist attaches abdomen to thorax; hind wing smaller than
        forewing ...... (bees, wasps, yellowjackets) ..... **Hymenoptera**
    b. Not as above 16

16. a. Hind wings and forewings of equal size; wings with many veins; antennae short and bristlelike; abdomen long and slender; wings outstretched at rest (dragonflies) or held vertically (damselflies) ..... **Odonata**

    b. Not as above 17

17. a. Sucking mouthparts (mouthparts basically a tube or stylet) 18
    b. Chewing mouthparts 19

18. a. Beak arising from front of head (true bugs) .... **Hemiptera**
    b. Beak arising from hind part of head (cicadas, aphids) .... **Homoptera**

19. a. Wings similar to each other; soft bodied; cerci small, winged termites .... **Isoptera**
    b. Not as above 20

20. a. Ectoparasites of birds and mammals; body flattened laterally or dorso-ventrally 21
    b. Not as above; soft bodied, usually blind insects, may have large pincerlike mandibles (wingless termites) .. **Isoptera**

21. a. Body flattened laterally; jumping hind legs; body sparsely covered with spines or hairs (fleas) **Siphonaptera**
    b. Body flattened dorso-ventrally; large tarsal claws for attachment to host (sucking lice) ... **Anoplura**

*Modified from Borrer and DeLong, An Introduction to the Study of Insects. Holt, Rinehart & Winston.

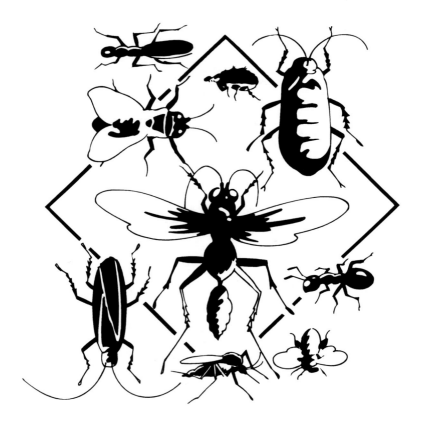

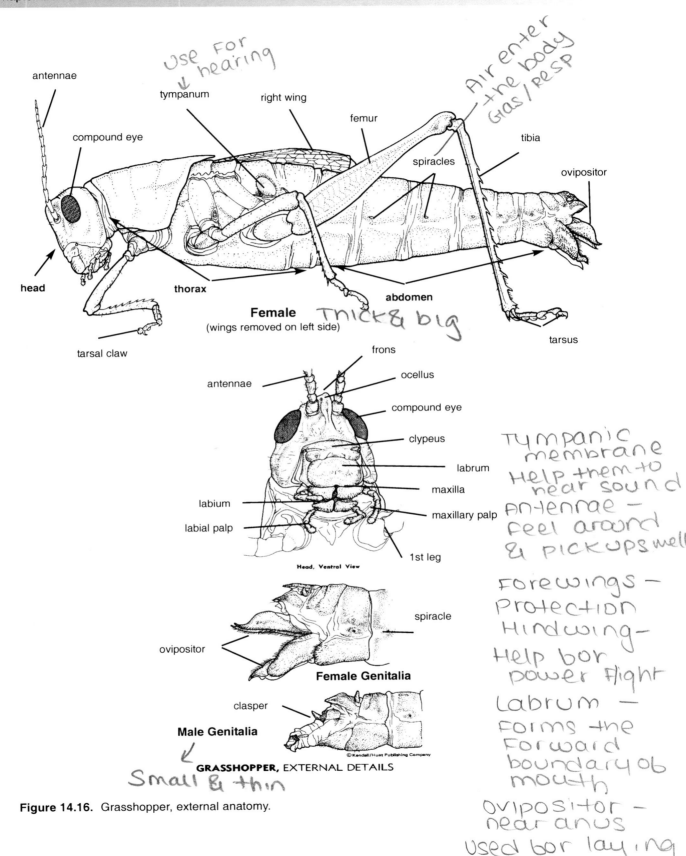

Figure 14.16. Grasshopper, external anatomy.

[Handwritten annotations:]

use For hearing

Air enter the body Gas/Resp

Tnick & big

Small & thin

Tympanic membrane Help them to hear sound
Antennae – feel around & pickup smell

Forewings – Protection

Hindwing – Help bor power Flight

Labrum – Forms the Forward boundary ob mouth

Ovipositor – near anus used bor laying bertilized egg

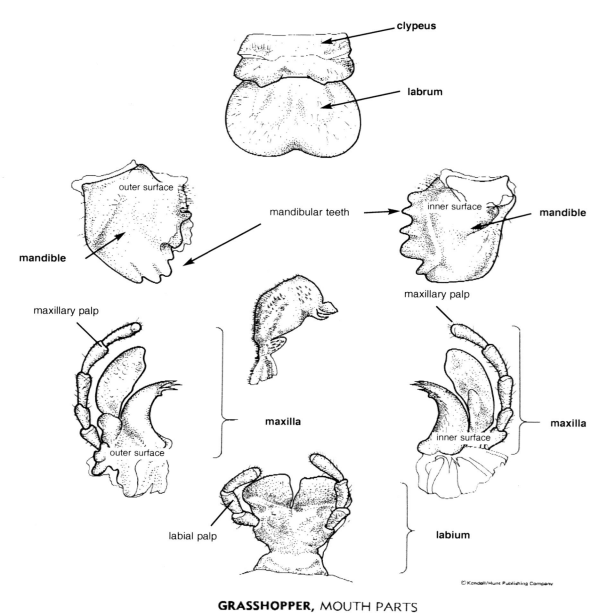

GRASSHOPPER, MOUTH PARTS

Figure 14.17. Grasshopper mouthparts, exploded view.

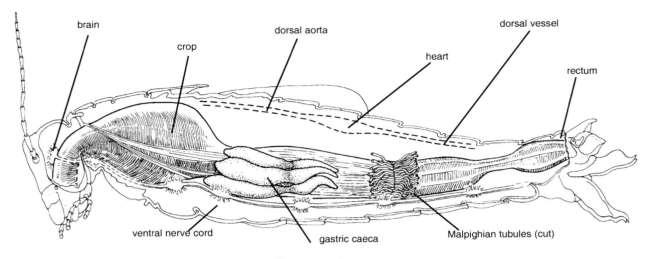

GENERAL ANATOMY

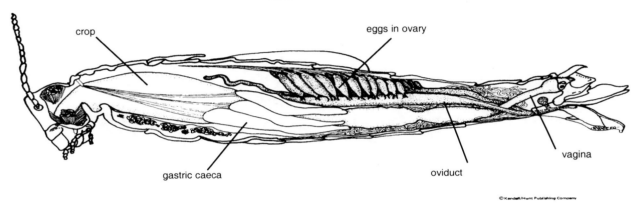

GRASSHOPPER, FEMALE REPRODUCTIVE SYSTEM

FEMALE ANATOMY

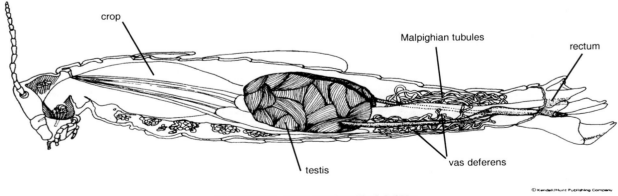

GRASSHOPPER, MALE REPRODUCTIVE SYSTEM

MALE ANATOMY

Figure 14.18. Grasshopper, internal anatomy.

15

Chapter 15

THE ECHINODERMS

Phylum Echinodermata

The weathered pearl-diver, enraged over starfish's appetite for oysters,
madly tore his five-armed antagonists into tiny fragments.
Imagine his dismay when he realized that instead of destroying his enemy,
he had only managed to create scores of newly regenerated rivals!

-Japanese folk tale

Introduction

To the right is a simplified cladogram showing the two main branches of animal evolution. The left branch of the "tree" bears two major phyla, the Echinodermata and the Chordata, on what is known as the deuterostome line (refer to Figure 11.4 page 262 for a discussion of protostomes and deuterostomes). Although echinoderms (sea stars, sea cucumbers, sea biscuits, sand dollars, sea lilies, and urchins) have no superficial resemblance to the chordates (fish, amphibians, reptiles, birds, and mammals), these two groups are, nonetheless, placed next to each other phylogenetically. What may appear ridiculous on the surface (a mountain lion would never be mistaken for a sand dollar), nevertheless, has a morphological and embryological basis. The justification for placing these two groups together is that (1) embryos of both groups use radial rather than spiral cleavage; (2) the anus of both echinoderms and chordates is derived, embryonically, from the blastopore; (3) the endoskeleton of both arose from mesoderm; and (4) in both groups the coelomic cavity develops similarly. For these reasons, you and I and other chordates have echinoderms as one of our nearest relatives, even though you look nothing like a starfish, and I bear little resemblance to an urchin!

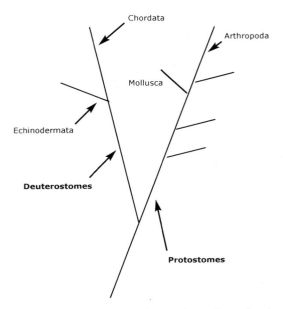

Chordata

Arthropoda

Mollusca

Echinodermata

Deuterostomes

Protostomes

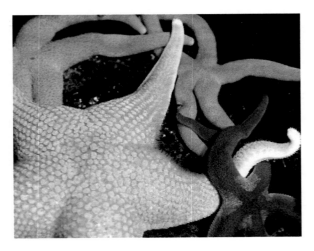

Figure 15.1. The phylogenetic tree above shows the close relationship between the echinoderms and the chordates (fish, amphibians, reptiles, birds, mammals). Both phyla are borne on the deuterostome branch of animal evolution even though they bear little resemblance to each other. The photo to the left contains a cluster of sea stars, and the photograph above shows two pencil urchins.

Echinoderm Characteristics

External Features

Echinoderms are headless animals and, for the most part, lack the anterior-posterior orientation seen in most other animals. In the echinoderm body plan, there is no front or rear, no right or left, no dorsal or ventral; instead, the animals are arranged along an **oral-aboral** axis. The surface bearing the mouth is automatically termed the oral face and is usually directed toward the substratum. The surface opposite the mouth is called aboral, and normally faces upward. The anus of echinoderms may exit either from the oral or aboral surface.

In sea cucumbers (Figure 15.2 **D**), the pattern seems to be more along the anterior-posterior line. But even in these wormlike echinoderms, there is an oral and an aboral organization. The oral surface actually wraps around most of the animal leaving a small "posterior" portion, containing the anus, as the aboral surface. The elongated form is secondary in sea cucumbers and, thus, tends to obscure its oral-aboral nature. In sea lilies (Figure 15.2 **G**, also page 350), the oral surface extends upward from a trailing, aboral stalk.

Most echinoderms are pentamerous in that they consist of body parts arranged in fives or multiples of five. The typical condition is seen in the sea star, *Asterias*, for example, as five arms radiating from a central disc (Figure 15.2 **A**). The centrally positioned disc bears the mouth and the anus as well as a unique structure, the **madreporite** (more on that later).

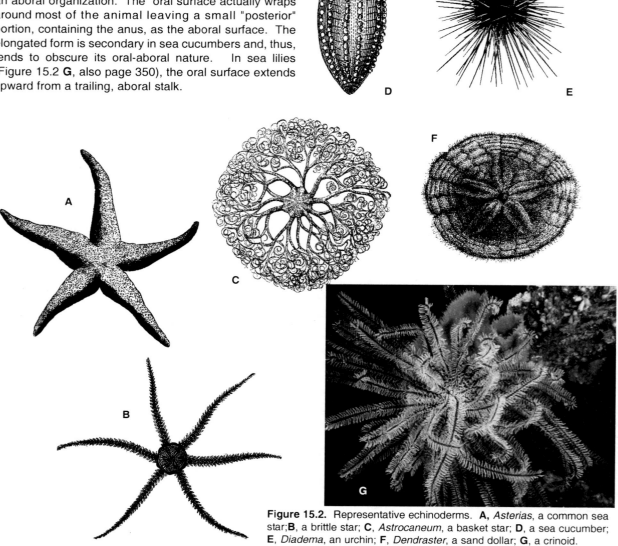

Figure 15.2. Representative echinoderms. **A,** *Asterias*, a common sea star; **B**, a brittle star; **C**, *Astrocaneum*, a basket star; **D**, a sea cucumber; **E**, *Diadema*, an urchin; **F**, *Dendraster*, a sand dollar; **G**, a crinoid.

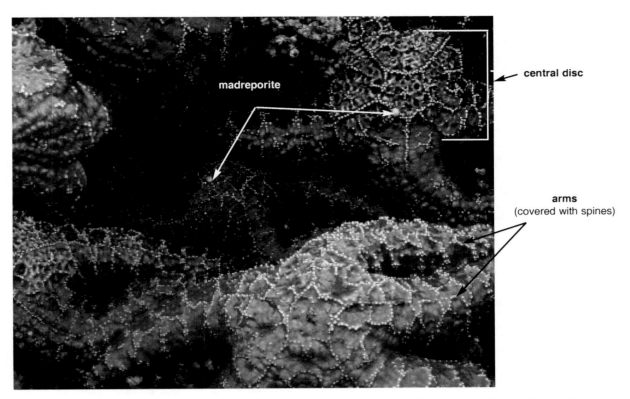

Figure 15.3. External features of sea stars. The madreporite is an aboral scablike device that is part of the water vascular system (WVS). It serves to filter the water as it enters the WVS. The surface of sea stars is studded with obvious spines, but what is obscured from view are the numerous respiratory papulae, the slitlike anus, and the miniature pedicellaria.

The epidermis (skin) of a sea star overlies a hard, calcareous endoskeleton constructed of separate plates or **ossicles.** Like chordates, who also possess an internal skeleton, the endoskeleton of echinoderms is *derived from embryonic mesoderm,* signifying an evolutionary relationship between the two groups. But unlike chordates, the endoskeleton of most echinoderms is essentially immovable, which means that the animals cannot use them as articulating surfaces for locomotion. This forces most echinoderms to rely on other means of movement. Sand dollars, urchins, and sea stars, for example, walk on the tips of their spines or use specialized tube feet to get around. In asteroids, such as the sea star *Asterias,* the plates allow only a slight gliding action; movement of its arms is so limited that they must depend solely on their many tube feet. On the other hand, an overlapping arrangement of ossicles in brittle stars is such that extensive mobility is possible, allowing these active animals to crawl, spider-like, on their five arms. As a result, brittle stars are the fastest moving of any echinoderms.

The term "echinoderm" means spiny skin. This is never more obvious than in sea stars, sand dollars, and especially, in the sea urchins. The skin is covered with spines of different lengths—some moveable, others fixed in place. These give the stars their spiny character and serve a defensive role for the animals. In fact, in some of the urchins the spines are venomous and can be quite painful or even fatal if stepped on. Also, scattered over the aboral surface of sea stars, sand dollars, and sea urchins are minute, clawed appendages, the **pedicellaria** (Fig 15.6, pg 351). These tiny pincers are thought to protect the skin surface from being smothered by drifting debris or from the attachment of sponge larvae. Thus, pedicellaria serve an important cleansing function by dislodging unwanted hitchhikers. Other pedicellaria have been known to purposefully attach debris to the skin surface to affect some sort of camouflage. Large pedicellaria may also serve to catch small fish, while others are venomous and, thus, seem to provide a defensive function.

Headless and Heartless

Echinoderms are the only major phylum that is exclusively marine...there are no freshwater echinoderms. In their seawater home they slowly creep along the bottom sediment. Because of their habitat and slow-moving ways, they lack a number of structures that we have become accustomed to in most other invertebrates.

First of all, most adult echinoderms have radial symmetry. However, it seems they start their lives as bilateral animals but become radial by the time they reach adulthood. Thus, echinoderms have what is called **secondary radial symmetry**. The question is, why?

Radial symmetry, as seen in the cnidarians (jellyfish), is a way to compensate for an otherwise slow, aquatic existence. Food and sensory stimuli, coming from all directions, can best be processed by a radial body form. As a result, echinoderms (and cnidarians) have no use for many conventional organ systems. For instance:

(1) **Echinoderms never use gills, lungs, or trachea for respiration.** Instead, clustered about the aboral surface are minuscule, naked, epidermal outfoldings called papulae. These membranes, along with tube feet, serve as respiratory surfaces (Fig. 15.5). Tube feet protrude orally through special ambulacral grooves, that run the length of each arm. These two surfaces (papulae and tube feet) combine to provide a large area for cutaneous respiration.

(2) **Echinoderms have no head.** Except for sea cucumbers, which appear to have something akin to an anterior head, all echinoderms, being radial, are noncephalic for those reasons described above.

(3) **Echinoderms lack a brain.** The nervous system, what there is of it, is diffuse (i.e., scattered). There is no compact brain, nor is there a ventral nerve cord as seen in other invertebrates. There are a few dispersed sensory structures, such as the simple eyes at the tips of certain sea star arms, but overall the nervous apparatus of these invertebrates is quite limited.

(4) **Echinoderms are heartless.** There is no central pumping device, there are no blood vessels, there is no blood, indeed, there is no circulatory system in echinoderms.

(5) **Echinoderms have no kidneys.** They have no proto-nephridium, no nephridium, or Malpighian tubules. Excretion is accomplished cutaneously through the body surface, primarily through the tube feet.

(6) **Echinoderms have no penis.** As a rule, echinoderms are dioecious animals and because of their aquatic existence, the vast majority use external fertilization. Eggs are dispersed into the surrounding water and sperm are generally released nearby. There are only a few reported cases of internal fertilization among echinoderms (e.g., the sea daisies).

A number of echinoderms reproduce asexually. Some of the sea stars, for instance, can split the central disc into pieces and each piece will subsequently regenerate the lost part. Also, some stars can voluntarily detach individual arms or even pieces of arms that will, likewise, regenerate. Autotomy is practiced by some echinoderms so that when harassed an animal will surrender an arm to escape. Naturally, the sacrificed limb will be replaced as shown below.

Figure 15.4. Two sea stars in the process of regeneration. Note the growth zone at the extreme tips of each newly regenerating arm. Also note that the animal in the lower photo has six rather than the standard five arms.

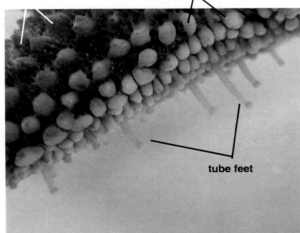

Figure 15.5. The upper photograph shows the ventral view of a sea star revealing the ambulacral grooves through which the tube feet project. The tube feet are clearly seen in the lower photo as are the fixed, immovable spines.

A Unique Plumbing System

From what you have just read, it seems that sea stars and their relatives are known more for what they aren't than for what they are. To some extent that may be true, but hidden within the confines of every echinoderm is one feature that is theirs and theirs alone. They have a system of interconnected, fluid-filled tubes and canals that is found in no other group of animals. They have a **water vascular system** (WVS). Much of echinoderm success, and they are a mightily successful group, is due to this intricate and unique system which is used in locomotion and for a number of other functions.

The system is filled throughout with slightly modified seawater which enters the system through a filtering plate, the **madreporite**. Water then passes directly into a short tube, the **stone canal**. This, in turn, leads to a circular or **ring canal** buried deep within the central disc. Extending outward from the ring canal are several elongated **radial canals**, one into each of the five arms. Fluid from the ring canals passes into a series of radial canals and then finally flows into an enlarged bulb (**ampulla**) at the top of each tube foot. It is estimated that some sea stars have in excess of 2,000 of these tube feet. The water vascular system is shown in Figure 15.15 on page 363.

The WVS works very much like a hydraulic pump. As fluid passes throughout the various tubes, pressure is ultimately applied within the bulb-like ampulla. This increased pressure, in turn, causes the tube foot to become erect. Relaxation of the ampulla reduces pressure as the foot is retracted. Each tube foot can be independently controlled allowing the animal to slowly move along as integrated waves of podia extend, contact the surface, and then withdraw. In many species, the tip of each foot is equipped with a suction-like device allowing it to grip the surface. Other echinoderms have adhesive discs that can adhere momentarily to the substratum. In any case, this is an extremely slow, yet persistent, mode of travel.

Tube feet have other functions as well. For instance they serve, as mentioned earlier, as respiratory and excretory surfaces. They may also be used in feeding. One of the preferred foods of sea stars are bivalves. A feeding star wraps around a clam, gripping each valve with numerous tube feet. Although the pressure applied by a single tube foot is minimal, and even the combined pressure of all "pulling" feet is not great, the sea star persists until the clam gapes ever so slightly. The star then everts its membranous stomach into the gap and digests the clam inside its own shell!

Echinoderm Variation

Living echinoderms are divided into six taxonomic classes, each with a different body type. One, the Concentricycloidea, known as the sea daisies, has only one species and is not included in the following discussion.

Sea Lilies, Feather Stars
class Crinoidea

cirri (arms)

crown

arms

stalk

The greatest assembly of crinoids (sea lilies and feathers stars) are found in Australian waters in and around the Great Barrier Reef. They are the most primitive of living echinoderms and differ conspicuously from others in the phylum. Although the sea lilies (shown above) are stalked and permanently attached to the substrate, the feather stars are mobile and free swimming. In both cases, the many arms (**cirri**) form a crown arrayed outward from the centrally located, upward-directed mouth, much like the arrangement of tentacles around the mouth in a sea anemone. A sea lily may possess in excess of 200 arms, each studded with tube feet that secrete mucus to entrap drifting food particles as the arms wave about. Thus, all crinoids are suspension (filter) feeders.

Although they do possess a water vascular system, they have no madreporite. Nor do they possess the spines so characteristic of the phylum. Pedicellariae, too, are absent.

Sea Stars
class Asteroidea

The best known echinoderms, the sea stars, belong to the class Asteroidea. They share the common star shape consisting of relatively immobile arms radiating out from a rather indistinct central disc. They are usually pentamerous, but variations are found. The mouth is directed downward on the oral surface and tube feet are distributed in ambulacral grooves along each arm (Figure 15.5). The madreporite and anus are aboral (directed upward) and that surface is replete with spines, respiratory papulae, and tiny clawlike pedicellariae.

Locomotion is accomplished by the slow deliberate progression of tube feet alternately gripping and releasing the substrate. Individual arms can move slightly but are not instrumental in overall locomotion.

The digestive system differs from that of other invertebrates. It consists of a mouth, esophagus, a double stomach, a series of pouches, and an anus. Food passes from the mouth to a short esophagus and into a two-part stomach. The lower portion is called the **cardiac stomach**, which first receives the food and passes it on to the upper, **pyloric stomach**, where most digestion and food absorption occurs. The cardiac stomach can be extruded from the mouth for external digestion, as in feeding on clams, while the pyloric stomach is confined internally. Branching from the pyloric stomach into each arm is a large pouch, the pyloric cecum (also called the digestive gland). The combined surface area of the ceca provide an extensive region for digestion and absorption of food. Undigested food then exits through the anus located on the aboral surface opposite the mouth.

Asteroids possess a series of gonads, typically a pair housed within each arm. The gonads are located above the digestive glands and each communicates to the outside through a tubular gonoduct. The sexes are separate and fertilization in asteroids is external.

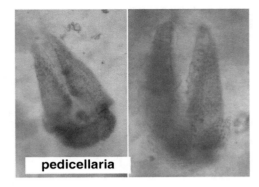

pedicellaria

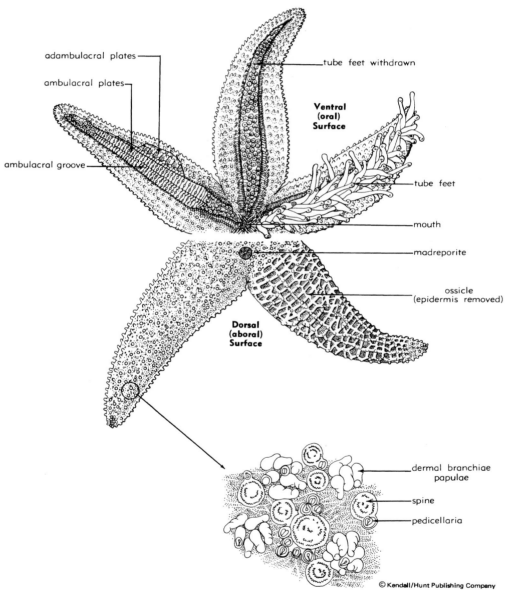

adambulacral plates

ambulacral plates

ambulacral groove

tube feet withdrawn

Ventral (oral) Surface

tube feet

mouth

madreporite

ossicle (epidermis removed)

Dorsal (aboral) Surface

dermal branchiae papulae

spine

pedicellaria

© Kendall/Hunt Publishing Company

Figure 15.6 External anatomy of a sea star.

Brittle & Basket Stars
class Ophiuroidea

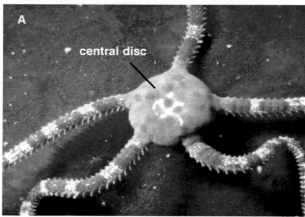

central disc

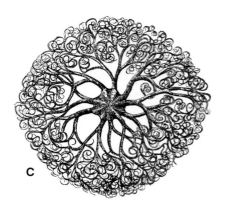

Figure 15.7. Members of the class Ophiuroidea. **A** & **B** show brittle stars with a distinct central disc and thin, elongated arms. Plate **C** is a drawing of a sea basket. Note the webbed tangle of interbranching arms. The lower photo shows the crawling action of a brittle star.

Overall, the ophiuroids are very much like their relatives, the asteroids (sea stars), with a few notable exceptions:

(1) they lack an anus, having only a mouth as an opening to and from the digestive system;

(2) instead of having a single aboral madreporite, they have several, located orally;

(3) the tube feet lack ampulla; and

(4) ophiuroids possess special oral slits, called bursae, that permit water to circulate into the coelomic cavities for respiration and osmoregulation.

Brittle stars are truly fascinating creatures. Watching one maneuver over, under, and around tide pool obstructions is a treat for anyone who appreciates animal organization. They are highly mobile animals, using their flexible arms as levers to propel themselves through their marine habitat. Remaining secreted during the day, they emerge at night to prey on small animals, to filter plankton, or to dredge organic material from the sediments.

Ophiuroids have long tendril-like arms (usually five) that radiate from a distinct central disc (above photo). The arms are usually protected by platelike shields, having somewhat the appearance of a coat of armor. In sea baskets, the arms branch over and over to produce the tangled web so characteristic of that group (below). The basket is then used to trap food, sometimes relatively large prey, from the flowing water. The arms are segmented and easily dislodged, either accidentally or autotomously, thus, the common name of brittle star.

Sea Urchins & Sand Dollars
class Echinoidea

Members of the class Echinoidea come in two body forms: regular (globose or round) and irregular (discoid or flat and disc shaped). Globular echinoids are represented by the cylindrical shape of sea urchins and pencil urchins (photos on this page). The many spines projecting from the round body may be short, sharp or blunt or they may be long and, in some species, quite venomous.

Figure 15.8. A variety of urchins. Note the differences in spine length and shape. Some are blunt as in the pencil urchins while others are long and sharp. Some, such as those indicated by the arrows are venomous and can be quite painful.

The irregular, discoid-shaped echinoids are represented by the extremely flattened sand dollars and the more rounded sea biscuits (Figure 15.9). Short spines profusely cover the animals lending an almost furry appearance to living forms. The shortened spines are thought to be an adaptation to the burrowing habit of these animals. It's easier to tunnel if the spines are short. Tube feet are found on both the oral and aboral surface. Those on the aboral face protrude through slits arranged in the shape of petals, called **petaloids**, to serve as tiny respiratory surfaces. The madreporite is centered aborally and is surrounded by five small pores, **gonopores**, for the exit of gametes.

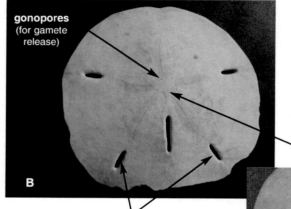

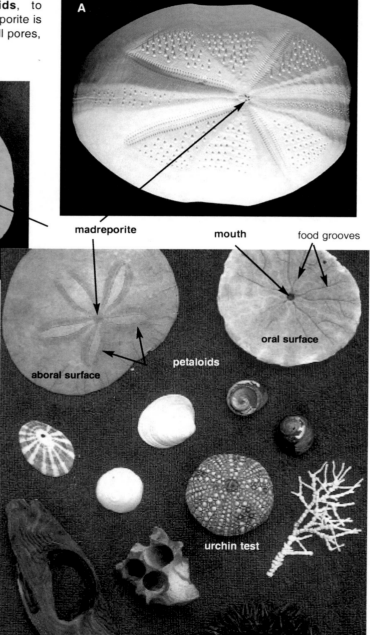

Figure 15.9. **A**, a sea biscuit endoskeleton exposing the petal-shaped petaloids and centrally located madreporite; **B**, the aboral surface of a sand dollar test. Note the large lunules. **C**, a collage of seashore items including a sand dollar (at top) showing both oral and aboral surfaces. The lunules are obscured in this photo. A live sea urchin is shown at lower right and an urchin test (or skeleton) is shown as well.

The digestive system in this group differs considerably from that seen in other echinoderms. The mouth is surrounded by an apparatus consisting of a series of hard, calcareous plates and complex muscles called **Aristotle's lantern** (so-called because Aristotle thought it looked like the valves on a lantern). The lantern is a feeding device that can be protruded from the mouth to scrape algae or to tear food into small pieces. The lantern leads to a non-protrusible stomach, which, in turn, leads to an elongated intestine where digestion and absorption occur.

At times, sand dollars feed by burying one end in sand or mud so that they "stand" erect facing the shoreline. As water flows over them, small food items are captured among the spines on the oral face and moved by cilia along food grooves toward the mouth. Open spaces in the endoskeleton, called **lunules**, allow water (we think) to pass through the animal without knocking it over— much like allowing air to pass through your open fingers if you stick your hand out of a moving vehicle.

Figure 15.10. The two body forms of the class Echinoidea. The globose form in plate **A** is a ventral view of a sea urchin revealing the feeding structure called Aristotle's' Lantern. The discoid specimen in plate **B** is a denuded sand dollar showing location of the mouth (1), anus (2), food grooves (3) and lunules (4).

Sea Cucumbers
class Holothuroidea

Sea cucumbers are soft-bodied echinoderms that, as adults, show radial symmetry and may reach three feet in length. They are rather wormlike in appearance with distinct anterior and posterior ends; however, they show no tendency toward cephalization. There is no true head nor is there a concentration of nervous tissue into a brain.

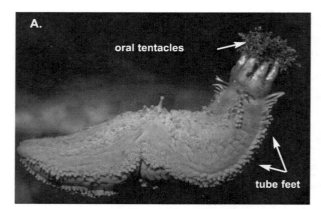

Like the previous group, the sand dollars and urchins, sea cucumbers lack arms. But instead of being cylindrical or flattened, their body is drawn out into a tube that looks very much like their cucumber namesake. Their skin has very few spines, no pedicellaria, and is usually adorned with knobby or warty growths. The endoskeleton is much reduced, limited to a few scattered ossicles buried within the skin. The water vascular system is similar to other echinoderms but the madreporite is internal rather than external. They are, however, the only member of the phylum that can lay claim to actual internal respiratory structures, called **respiratory trees**. These are a pair of tubes branching through the coelomic cavity. Water is brought into the tubular trees through the anus. It then circulates through the branching network where gaseous exchange occurs with the fluid of the coelomic cavity.

Sea cucumbers live primarily on the ocean floor where they plow through silt and sediments filtering organic debris. While they may feed on some of the same materials as urchins and sand dollars, they do so in a different manner. First of all, there is no Aristotle's lantern in this group. They capture food, instead, with elongated, feathery **oral tentacles**, which are really tube feet that encircle the mouth (Fig. 15.11)

The tentacles either have a sticky mucus to entrap food particles or are used as scoops to ingest ooze from which organic material is filtered. Tentacles with trapped food are alternately brought into the mouth where the food is cleared away, much like a person licking barbecue sauce from her fingers. The nonoral tube feet are scattered in rows along the surface of the animal in contact with the substrate. These tube feet, along with muscular contractions of the body wall, are responsible for the slow crawling locomotion of sea cucumbers.

Figure 15.11. A. A showy sea cucumber revealing the linear arrangement of tube feet and the extended oral tentacles. Plate **B** is a 10" sea cucumber trying to right itself after being placed on its back. Plate **C** is a close-up of the oral tentacles of a rather large unidentified sea cucumber. Note the feathery appearance of the tentacles that lead to the centrally located mouth. The tentacles are used to filter food from the surrounding water. Hidden among the tentacles is a well-camouflaged cleaner shrimp (arrow) gleaning the surface of the tentacles.

Many echinoderms, including asteroids (sea stars), crinoids (sea feathers), and echinoids (urchins and sand dollars) take an active interest in their young. In these cases, the larvae are protected by the brooding adults. Sea cucumbers are no exception. They, too, often brood their young by actually attaching them to the body surface (Figure 15.12).

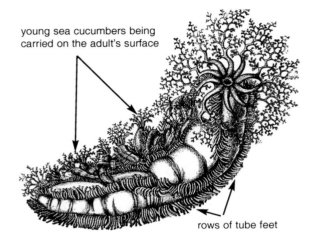

young sea cucumbers being carried on the adult's surface

rows of tube feet

Figure 15.12. *Cucumaria*, a sea cucumber brooding its young along its upper surface (top arrows). Note the rows of tube feet along the ventral surface (bottom arrows).

An echinoderm without spines seems to be a contradiction in terms. But that's exactly what sea cucumbers are. They are soft-bodied, slow-moving, naked, tube-shaped creatures. That they are good to eat is supported by the fact that, to the Japanese, dried sea cucumber skin is a delicacy known as trepang. So how does a seemingly defenseless animal protect itself? Well, while "cucumbers" may appear vulnerable, they actually have several means of self-defense. Some mimic the spiny appearance of their urchin cousins (Figure 15.13), while others exude a toxin from their skin to repel predators.

IT'S A FACT

Then there is **evisceration**. This practice, unique to sea cucumbers, is one of the most bizarre defensive maneuvers in all the animal world. Certain sea cucumbers, when harassed, may expel a stream of sticky, noxious tubules out of their anus. These threads, called **Cuverian tubules**, can entangle a predator in a sticky morass and are, apparently, quite effective. However, under extreme conditions, sea cucumbers will carry this act one step further. They may actually eject their viscera, including the digestive system, respiratory tree, and gonads, from their anus as an extra defensive tactic! This is the only known example of voluntary evisceration in the animal kingdom. The considerable regenerative ability of echinoderms enables them to regrow the lost viscera within a few weeks.

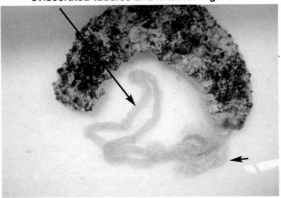

eviscerated tubules and internal organs

Figure 15.13. A. The skin of the sea cucumber *Parastichopus* lacks calcareous spines but not the mimicry of hard spines by the spiny-like soft tissue. Plates **B** and **C** show the bizarre defense process of evisceration practiced by some sea cucumbers. Cuverian tubules (arrows) and even the internal organs are emitted from the anus to repel predators.

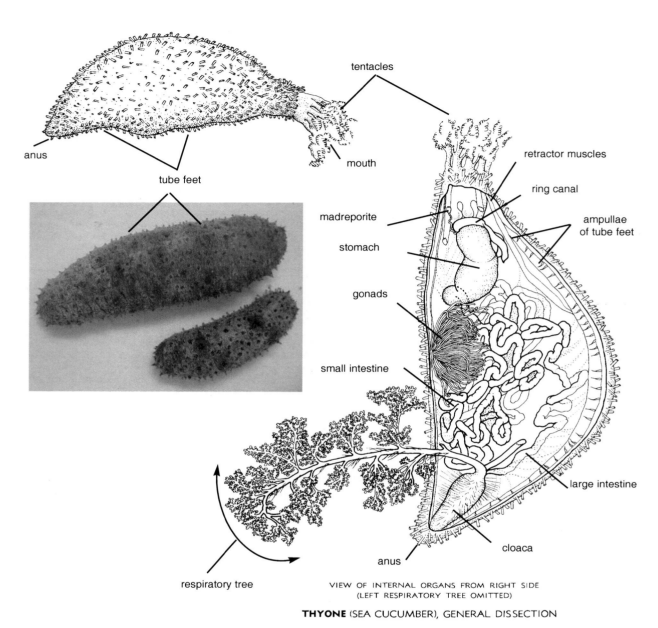

tentacles

anus

mouth

tube feet

madreporite

stomach

gonads

small intestine

retractor muscles

ring canal

ampullae
of tube feet

large intestine

cloaca

anus

respiratory tree

VIEW OF INTERNAL ORGANS FROM RIGHT SIDE
(LEFT RESPIRATORY TREE OMITTED)

THYONE (SEA CUCUMBER), GENERAL DISSECTION

© Kendall/Hunt Publishing Company

Figure 15.14. Anatomy of a sea cucumber.

Chapter 15 Summary

Key Points

1. Certain embryological features of echinoderms and chordates show their close relationship.

2. Echinoderms, for the most part, show little tendency toward cephalization, having instead an oral-aboral orientation.

3. Most echinoderms are pentamerous, although some show slight modifications of the five-rayed architecture.

4. A spiny skin covers an endoskeleton consisting of hard, interconnecting plates called ossicles.

5. The surface of the skin, in addition to possessing spines, also bears tiny clawed pedicellaria. These tiny pincers have a variety of functions ranging from feeding, to protection, to house-cleaning.

6. Echinoderms lack gills, lungs, or trachea for respiration. Instead, they use extensions of the body surface known as papulae for gaseous exchange. Tube feet may also function in respiration.

7. Echinoderms lack a circulatory and an excretory system.

8. As a group, echinoderms have separate sexes and use external fertilization (broadcast spawning). They also use fragmentation as a form of asexual reproduction.

9. The major diagnostic feature of echinoderms is the water vascular system. The WVS consists of a series of canals that transport water toward the thousands of tube feet. Muscles lining the ampulla portion of the tube foot exert pressure on the enclosed water, which, in turn, causes the tube feet to become erect. Erection of alternating tube feet then permits the slow locomotion so characteristic of sea stars and other echinoderms.

10. Tube feet are also used as respiratory surfaces, as excretory interfaces, and in feeding.

11. Echinoderms feed in a variety of ways. Some, such as the sea lilies, baskets, sand dollars, and sea cucumbers, filter food particles from the surrounding water. Others (sea stars, urchins) either scrape food from surfaces or are predatory. Sea stars can even evert a portion of their stomach into the body cavity of clams to digest the contents.

12. Echinoderms show diverse methods of self-defense. Some exude a poisonous skin secretion; sea stars have venomous pedicellaria, while many urchins have poison spines. Sea cucumbers can evert noxious threads (Cuverian tubules) or even internal viscera out of their anus to repel predators.

Key Terms

ambulacral grooves
ampullae
Aristotle's lantern
autotomy
basket star
brittle star
central disc
circular canal
cirri
Cuverian tubules
deuterostome
endoskeleton
evisceration
gonopores
lunules
madreporite
oral-aboral orientation
oral tentacles
ossicles
papulae
pedicellaria
pentamerous
petaloids
protostome
pyloric stomach & cecum
radial canals
radial symmetry
respiratory trees
sand dollars
sea cucumbers
sea lilies
sea star
sea urchins
stone canal

Points to Ponder

1. How do zoologists justify the close relationship between echinoderms and chordates?

2. Give three characteristics that all echinoderms have in common.

3. Why are echinoderms considered to be bilateral animals when they appear to be classic examples of radial symmetry?

4. Why do echinoderms have an oral-aboral axis rather than an anterior-posterior one?

5. How do echinoderms locomote?

6. How do echinoderms breathe?

7. What are pedicellaria and what are their functions?

8. How do echinoderms reproduce?

9. Describe the water vascular system and give its functions.

10. How do echinoderms feed?

11. How do brittle stars, sea stars, and sea urchins differ?

12. What is evisceration and what kind of echinoderms use it?

IV. Inside Readings

Birkeland, C. 1989. The Faustian traits of the crown-of-thorns starfish. Am. Sci. 77:154-163.

Coleman, N. 1991. Encyclopedia of Marine Animals. New York: Harper Collins Publishers.

Ellis, R. 1996. Deep Atlantic: Life, Death, and Exploration in the Abyss. New York: Alfred A. Knopf.

Lawrence, J. M. 1987. A Functional Biology of Echinoderms. London: Croom Helm.

Macurda, D. B., Jr., and D. L, Meyer. 1983. Sea lilies and feather stars. Am. Sci. 71:354-365.

Nichols, D. 1962. The Echinoderms. London: Hutchinson University Library.

Pechenik, J. A. 1996. Biology of the Invertebrates. Dubuque, IA: Wm. C. Brown Publishers.

Chapter 15

Laboratory Exercise

Phylum Echinodermata

~ the spiny-skinned animals ~

Laboratory Objectives

✓ to differentiate between the five major types of echinoderms,
✓ to identify the characteristic features of this fascinating group of animals,
✓ to dissect and identify internal (and external) features of a sea star,
✓ to microscopically examine the epidermis of a sea star,
✓ to locate and identify functions of external features of the sand dollar.

Supplies needed

✓ prepared and living specimens of the echinoderms
✓ sea stars for dissection
✓ prepared specimens of sand dollars
✓ dissecting equipment
✓ compound and dissecting microscopes

Chapter 15
Worksheet

Phylum Echinodermata
The "Spiny-Skinned" Animals

Review the preserved and live specimens on display. You should be able to recognize the different groups by their distinguishing features as described in this chapter and outlined below.

Crinoids (class Crinoidea)
✓ possess a stalk and a crown
✓ many branched, feathery arms
✓ lacking spines, madreporite, and pedicellariae

Sea Stars (class Asteroidea)
✓ usually pentamerous (five armed)
✓ spiny skin with pedicellariae
✓ aboral madreporite and anus within the central disc
✓ tube feet confined to ambulacral grooves on oral surface

Brittle and Basket Stars (class Ophiuroidea)
✓ thin, mobile arms, usually pentamerous in brittle stars
✓ many branching arms in basket stars
✓ presence of oral bursae (ventral slits to receive water for respiration)

Urchins and Sand Dollars (class Echinoidea)
✓ arms lacking
✓ urchins globular in shape, sand dollars and sea biscuits more flattened
✓ in sand dollars the aboral tube feet penetrate the skeleton via petaloid slits for respiration
✓ spines short and numerous in sand dollars, longer, often very long, in urchins
✓ sand dollar endoskeletons penetrated by openings (lunules) for stability in rushing water
✓ all forms have a feeding device called an Aristotle's lantern

Sea Cucumbers (class Holothuroidea)
✓ no arms
✓ bilateral symmetry; anterior-posterior body axis
✓ anterior tube feet forming a whorl of tentacles around the mouth
✓ remaining tube feet limited to ventral surface for locomotion
✓ no spines or pedicellariae; skin warty
✓ endoskeleton reduced to microscopic ossicles imbedded in skin

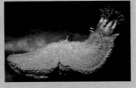

SEA STAR EXTERNAL ANATOMY
(refer to page 351, Fig. 15.6)

Before you begin your dissection of the sea star, take time to examine its exterior. Locate each of the following features:

✓ oral and aboral surfaces
✓ radial, five-part symmetry (is this primary or secondary?)
✓ central disc
✓ tube feet
✓ madreporite
✓ mouth
✓ ambulacral groove
✓ spines
✓ pedicellariae (perform microscopic demonstration; see page 364)
✓ examine endoskeletal arrangement of ossicles on preserved specimen

SEA STAR INTERNAL ANATOMY

1. Place your specimen so that the **aboral** (dorsal) surface is facing you. With a scalpel, or better yet, with a pair of scissors, cut around the **central disc** and gently lift it so that you can view the **stone canal** attached to the **madreporite**. If you are careful, you can also see, attached to the underside of the central disc, the slender **rectum**. Note the large membranous **stomach**, which has two parts, the **cardiac** and the **pyloric**. You can identify the latter as it connects to the **pyloric ducts** extending into each of the five arms. Carefully move the stomach aside as you trace the stone canal to locate the **circular** (ring) **canal**.

2. Now remove the skin from the aboral surface of ~~one~~ or two arms. The first organ in view is the large **digestive gland**, which is an extension of the pyloric ducts (discussed above). These sit in the sea star's **coelomic cavity**. If you rub the inside of the cavity, you will notice a slippery, whitish membrane...this is the **peritoneum**. Now, lift the digestive glands and notice that each arm has two depressions, one on each side of an **ambulacral ridge** running the length of the arm. Within the ridge is the **radial canal** of the water vascular system.

3. When you lift the digestive gland, you will also find a pair of **gonads** in each arm, again, one on each side of the ambulacral ridge. In a sexually mature individual, the gonads are quite large; otherwise they will be about the size of a dime or even smaller. Without microscopic examination of crushed gonads, you cannot tell if your animal is male or female.

4. Finally, examine the numerous bulbous (bubblelike) structures on the floor of each arm, radiating along each side of the ambulacral ridge. Each one is an **ampulla** and is connected to a **tube foot** that emerges orally along the **ambulacral groove**.

Checklist of sea star internal structures:

✓ ~~stone canal~~
✓ ~~cardiac and pyloric~~ stomach
✓ ~~pyloric ducts~~
✓ circular (ring) canal
✓ digestive glands (hepatic cecum)
✓ coelomic ~~cavity~~
✓ ~~peritoneum~~
✓ ambulacral ridge
✓ ~~radial canals~~
✓ ~~gonads~~ gonads
✓ ampullae
✓ tube feet
  mouth
  tube feet
  ring canal

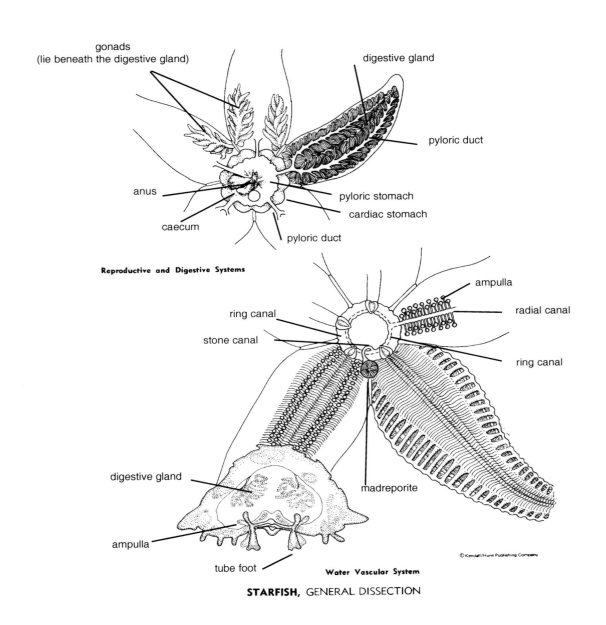

gonads
(lie beneath the digestive gland)

digestive gland

pyloric duct

anus

pyloric stomach

cardiac stomach

caecum

pyloric duct

Reproductive and Digestive Systems

ampulla

ring canal

radial canal

stone canal

ring canal

digestive gland

madreporite

ampulla

tube foot

Water Vascular System

STARFISH, GENERAL DISSECTION

© Kendall/Hunt Publishing Company

Figure 15.15. The internal anatomy of a sea star.

ray - Help with movement
madreporite - where water enter in vascular system
ampulla - Fill up water & then release in tube feet
oral surface - mouth side (ventral
Aboral surface - Anus side (Dorsal)
Ring canal - connects the stone canal, radial
canal & carries water from
stone canal to radial canal in
vascular system

Echinoderm Exercises   (also on page 607)

I. Perform a "scraping test" by shaving a small portion of tissue from the aboral surface of a sea star with a scalpel (about one-half the size of a pencil eraser). Place the tissue on a microscope slide and add a few drops of water. Swirl the tissue around until it's a thin film— the thinner the better. Now observe the mixture under the low power of a compound microscope. Identify spines and pedicellariae and sketch each below.

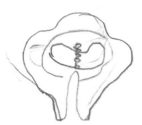

II.   In the space below, sketch both surfaces of a sand dollar. Label and give functions of:

Tiny respiration ✓ petaloids (petal-shaped aboral features with slits through which dermal branchia [papulae] protrude)
   ✓ mouth (large oral opening) SUCK UP FOOD
   ✓ anus (small oral opening) Excrete out Food
   ✓ food grooves (oral lines leading to the mouth) Direct Food into mouth
   ✓ madreporite (at junction of petaloids on aboral surface)
   ✓ lunules (five large oval openings through entire skeleton) To pass out water
Exit of gametes ✓ gonopores (five pin-sized aboral openings between bases of petaloids)

<u>Be sure to label and give functions of each structure.</u>

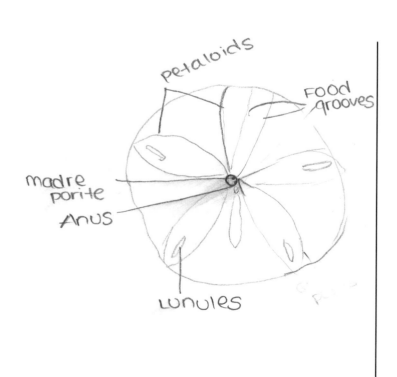

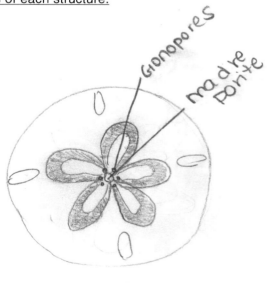

| ORAL SKETCH | ABORAL SKETCH |

16 Chapter 16

PREDATOR-PREY

The Predatory-Prey System

In natures' great game there is no right or wrong,
no book of rules...competition is the arena and survival is the only trophy.

Figure 16.1. Wolves are skilled predators. Here one protects a recent kill (a deer) from other members of the pack.

The Great Game

*P*redation, in its broadest sense, is defined as the consumption of one living organism by another. Although this sweeping definition includes herbivory, the eating of plants by animals, we will adopt the more common interpretation to include only carnivory...one animal, the **predator**, eating another, the **prey**.

Ecologically, the predator-prey system is a classic example of energy transfer. A predator uses the prey as food, an energy source. But the interplay is far from just a simple transfer of energy. Indeed, close examination reveals a complex arena where the two participants play a dangerous game of survival. The prey is ever on the defensive, trying to outwit, outrun, outmaneuver, and outbluff the predator. Not to be outdone, the predator tries just as hard to outsmart, outpace, or overpower the prey.

Each encounter can result in sudden death for one and a meal for the other. Since there can only be one victor, each is deadly serious as they grapple together in this ongoing struggle. The prey develops elaborate defenses to prevent catastrophe and the predator evolves equally intricate devices to puncture

those defenses. Through this age-old interplay they each respond genetically to the devices of the other. The prey evolves in reaction to the predator, and the predator counter-evolves as it adapts to the prey. Like living boomerangs, they bounce back and forth in a contest of genetic one-upmanship.

In evolution's brazen game, a slender advantage seems to be on the side of the prey. After all, the selective pressure is greater for them. If their defenses are inadequate, they pay the ultimate price. Mother Nature is quick to eliminate the slow, the dimwitted, the careless, the straggler; but she is just as quick to favor the fast, the smart, the careful. Evolution can be kind toward the fit.

Predators, on the other hand, usually have more than one prey species from which to choose, so pressure on them is not so urgent. As a result, the pursued seems to keep slightly ahead of the pursuer in this genetic seesaw. Even so, both lurch ahead to gain an advantage through a rich panoply of behaviors, weapons, tools, and schemes.

A Matter of Numbers

One of the most distinguishing features of the predator-prey system is that prey always outnumber predators. It's a simple matter of numbers. There has to be more flies than spiders, more shrimp than squid, more squid than sharks, more mice than barn owls, more snowshoe hares than lynx. There must be more deer than wolves and more gazelles than leopards. The very survival of the predator depends upon a plentiful and continuous supply of prey animals. In ecology, this is represented as an "energy pyramid" (below). At the pyramid's base are the vast autotrophs (the producers = phytoplankton or green plants); above them are the heterotrophs (the consumers = herbivores, omnivores, and carnivores). At the very apex of the pyramid are the top carnivores. This simple sequence is typical of energy as it moves through various trophic levels within any ecosystem. Each level outnumbers the level above it. The producers outnumber herbivores, who outnumber carnivores. Numbers decrease as you travel up the sloping pyramid with the top carnivores having the smallest population sizes of all.

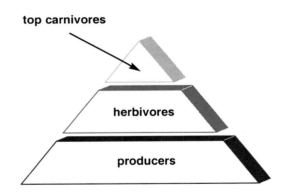

top carnivores

herbivores

producers

Figure 16.2. Population size can be depicted as an energy pyramid. The lowest levels are represented by autotrophs, primarily phytoplankton and green plants. These producers far outnumber those above, and the number of top carnivores is the smallest in any ecosystem.

How much a prey population must outnumber the predator is a product of both *size* and *selectivity*. Let's suppose that a relatively large predator such as a coyote has insects, mice, and rabbits available within its feeding range. Let's also suppose that the coyote must select only one type of prey. You can easily see that it would take more insects than mice to support a single coyote. Likewise, there would need to be more mice than rabbits. Thus, to the predator, the number of prey needed for survival depends largely on prey size. If, on the other hand, the coyote is less selective and feeds on insects, mice, and rabbits, then numbers of each prey population can be more flexible. The predator can just sample a few prey from each population. Or if the level of one population decreases, the predator can shift its efforts to the other two. But if the predator relies only on one prey species, then that population must be relatively large.

On the plains of South Africa there is a predator-prey system that well illustrates the relationship between predator survival, prey density, and prey selectivity. The aardwolf (*Proteles cristalus*) is a member of the hyena family that, unlike its ferocious cousins who prey or scavenge on large game, feeds almost entirely on a single species of insect, a termite. The snouted harvester termite (*Trinervitermes* spp) is small, even in termite terms. It lives in mounds and comes to the surface at night to forage. The habits and anatomy of the aardwolf are linked to the activity of their tiny prey. Each night the aardwolf searches out long columns of foraging termites, then laps them voraciously with a long, pointed, sticky, and mobile tongue. Imagine, a predator nearly the size of a German shepherd dog feeding entirely on 1/4"-long termites. A single meal would easily consist of thousands and thousands of termites; indeed it is estimated that over 200,000 are eaten each evening by each aardwolf. The number of snouted termites must be astronomical to support a prosperous aardwolf population.

When Prey Numbers Crash

Prey numbers, as important as they are, rarely remain constant. Instead, they tend to fluctuate—at times upward, at other times downward. Under certain circumstances, the fluctuation is remarkable. When conditions remain favorable, a prey population may increase generation after generation until incredible numbers are reached. But conditions change, and when they do...watch out! If food becomes scarce, if drought occurs, if temperature shifts, then an otherwise growing population may begin to slide. The decline may be slow and gradual or fast and precipitous. At times, circumstances may be so severe that drastic crashes occur. Since predators depend upon a plentiful supply of prey, they too inevitably suffer. The whole stack of cards may come crashing down as shown below.

Crashes can have serious consequences for predators since they rely so heavily on prey numbers for their survival. If the predator doesn't somehow respond, its population may become so deteriorated that recovery to previous status is impossible. Or worse yet it may suffer extinction. So how then does the predator respond to such crashes? It seems they do have a few options available.

(1)   When prey numbers plunge a predator may respond *by increasing its foraging range.* In fact, normal patrolling areas may be expanded significantly. For example, the usual foraging range of snowy owls (*Nyctea scandiaca*) is the Arctic tundra, reaching south to the Canadian border. But during times of prey shortages, especially in lemming populations, snowy owls may migrate far south. Under such circumstances, they have been spotted feeding on the southern shore of Lake Michigan and further south.

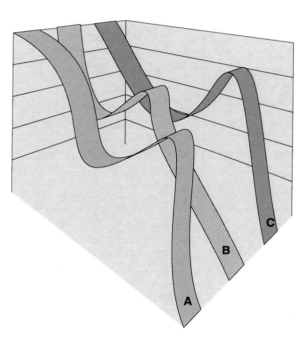

Figure 16.3. Fluctuations in animal populations often follow a predictable and well-established pattern as shown here. Curve **A** represents a food source for a population of herbivores (curve **B**). As food levels dip and then decline precipitously, so does the herbivore population. This is soon followed by a similar drop in any predator population dependent on the herbivores as indicated by curve **C**. Because of the decline in the food supply (A), both population B and C also declined.

IT'S A FACT

A common assumption is that predators, by their very nature, are themselves a major cause of prey crashes. While that is a popular idea and may at times occur, it seems to have little factual basis. For predators to induce severe population declines, they would have to suppress or somehow regulate the prey's reproductive ability. In other words, the predator must remove a large percentage of reproductive prey to cause such crashes.

However, virtually all studies indicate that predators do not select mature, healthy individuals, but rather concentrate on the weak, young, or unhealthy...what ecologists term the "doomed surplus." To concentrate their efforts on healthy prey is just too costly; it takes too much time and too much effort. Actual predation, in which non-reproductive individuals are removed, has little effect on overall prey numbers. So it seems that crashes are rarely attributed to the action of nonhuman predators. The cause of crashes is instead almost always due to nonpredatory forces such as scarcity of food, change in climate, drought, disease, and/or human intervention.

(2) Another option during bleak times is for a predator *to expand its food selection.* It is well known that predators, like all animals, have preferred prey species: snowy owls prefer lemmings, wolves prefer deer, foxes have a liking for mice. But during a time of crises, when that favored source is scarce, the predator may have no choice but to feed on a wider range of less preferred food. If mice populations decline, foxes may eat berries. When rabbit populations fall, coyotes start looking favorably on insects. In fact, adopting a more general feeding strategy is probably the most common reaction to a slump in preferred prey numbers.

Figure 16.4. When preferred prey numbers decline, a predator may respond by feeding on nonpreferred prey. A coyote, for example, may begin to eat insects instead of mice.

(3) One short-term solution to food shortages is also an extreme one. But then, extreme circumstances sometimes call for extreme solutions. When facing starvation, predators may turn on each other. **Cannibalism**, as harsh as it sounds, may actually have some positive effects. First, it provides an alternate food source for an animal facing starvation. Second, it reduces competition for what few prey might be available. And, third, cannibalism often weeds out the weak, sick, or old within the predator population itself. Thus, because of cannibalism, when prey species rebound the predator population may actually consist of younger, more vigorous individuals. In other words, cannibalism can serve to cull out the less fit among one's own species. However, over an extended period of time cannibalism is a maladaptive trait, especially if the cannibals become too aggressive or begin to feed on their own offspring. For that reason, long-term cannibalism is considered an evolutionary liability and is rarely seen except in cases of chronic stress.

(4) One other way that predators may respond to severe states of malnutrition is by *altering their reproductive patterns.* Food shortages can ultimately reduce a female's ability to become impregnated or may limit the number of offspring she produces. Those young that are issued invariably face a bleak future and suffer a high rate of mortality. There is also ample evidence that under severe stress the male or female parent will even cannibalize their own young. The effect of all of this is, of course, to limit the size of a predator population. As a result, there are fewer of one's own kind to compete for the available food, which, in turn, enhances the survival chances of those that remain.

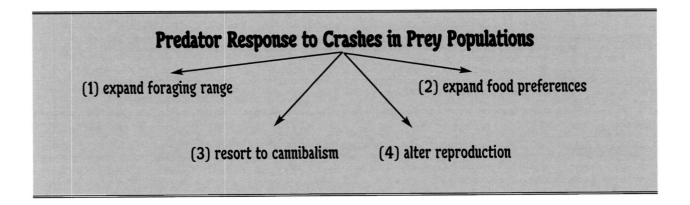

Adaptations to the Predator-Prey Relationship

*B*oth participants in the predator-prey interaction are affected by it; neither remains neutral. They both, in fact, show a number of special adaptations that promote survival. As selection pressure increases on the prey, it improves overall means of defense. It becomes faster, craftier, or more deceptive. Likewise, the predator improves its offensive course of action. Predators may become larger, stronger, faster, or more aggressive. Whatever the particular response, there is an evolutionary kinship between the two populations, a mutual give-and-take resulting in what is termed **co-evolution**.

The interesting feature about co-evolution is its reciprocal nature. As the prey improves its defenses, it causes the predator to respond likewise. Then as the predator counters by enhancing its offensive arsenal, it, in turn, elicits even better responses on the part of the prey, and so on. In essence, *co-evolution is an interplay of adaptation and counter-adaptation to ever-changing pressures brought on one population by another.* A never ending ebb-and-flow in adaptations between the two populations occurs.

An excellent example of reciprocal evolution is in the development of sensory structures. Both prey and predators must be acutely aware of what happens around them. The survival of each depends upon it. The hunted must pay meticulous attention along a road fraught with danger where misfortune waits at every turn. The hunter, too, sits on a precarious perch since meals are few and far between for the dull-witted predator. Thus, the tendency is for both to hone their awareness skills through sharp vision, a keen sense of smell, acute hearing, a delicate sense of touch, and overall vigilance. Back and forth, each continually adapts to the other through nature's ever-changing landscape.

Figure 16.6. Predators, too, have a keen sense of awareness as indicated by the cougar (above) and shark.

Figure 16.5. A group of deer stand alert sensing possible danger.

Predator Adaptations
~ how to locate and subdue prey ~

Successful predation requires several distinct and recognizable steps. To thrive, a predator must find, capture, subdue, and eat its prey. If there is a disruption in any of those steps, the predation attempt fails and the hunter goes hungry. Predators have, therefore, evolved a number of features and strategies to maximize their success.

Hunting Tactics

Predators have developed three distinct hunting patterns to locate and capture their prey: stalking, pursuit, and ambush. Each tactic has its specific advantages and disadvantages. For example, **ambush** requires that a predator spend much of its time lying in wait for a prey to appear. This is a risky practice since a suitable prey animal may not materialize for some time. However, since this is a passive process, the predator expends very little energy in doing it. Thus, the lie-in-wait mode of hunting invests considerable time, has a relatively low success rate, but involves minimal energy expenditure. The cost in time is balanced by the energy conserved. The ambush style of hunting is characteristic of many insects, spiders, frogs, toads, snakes, owls, and alligators as shown below.

At the other extreme is the **active pursuit**. Here the predator spends very little time in searching and virtually no time in waiting but exerts considerable energy in the capture itself. Therefore, the prey must be relatively large to provide enough food to offset the amount of energy spent in pursuit. In this tactic, time investment is low but the energy spent is high. Active pursuers include many fish, hawks, some sharks, a few of the big cats, wolves, hyenas, and hunting dogs.

The final hunting category seen with predators is the **stalk**. Here the hunter can pick from two strategies. First, the predator may spot a prey animal, slowly follow it, then explode in a fierce attack. Such is the practice of leopards, cheetahs, lions, and many other deliberate predators. A second stalking approach is to gradually explore or canvass a given area until a likely prey is uncovered. This is again followed by a frenzied strike. The roving stalk is typical of herons, bitterns, some raptors, and certain predacious fish. The stalk, whether trailing a victim or locating one randomly, takes considerable search time but involves little pursuit. As a result, moderately sized or even small prey can be taken.

A lone hunter must be careful not to tackle dangerous prey or those considerably larger than itself. However, predators do, in fact, often take prey that fall into this category. One highly successful way to do this is to *hunt cooperatively in packs*. Several predators working together can tire and subdue prey that a single predator could never handle. Certain fish, most notably the piranhas, wolves, hunting dogs, hyenas, jackals, African lions, and sharks often resort to group hunting.

Hunting Strategies

ambush   the stalk

active pursuit

B

A

C

Figure 16.7. The three hunting strategies of predators: **A**, the lie-in-wait tactic of the ambush is shown by this alligator; **B**, a herd of zebra are actively pursued by African hunting dogs; and **C**, a great blue heron slowly stalks fish, crustaceans, or other prey species. Wading birds use their shadow to startle prey which they quickly spear.

Deception

~ an enterprising con game ~

Many predators owe their success to trickery. An unseen or unrecognized predator is more often than not a successful one. Deception just makes the hunt, whether it's an ambush or a stalk, so much easier. A prey that doesn't spot the predator until that last rush will most likely suffer a sudden death. It is not surprising then that predators have invested much of their evolutionary energies into subterfuge. Some have concentrated on developing confusing colors or patterns, while others have opted for stripes, or spots, or background blending. Still others place success in the form of disruptive coloration, which breaks up the body's outline. Whatever the chosen route, concealment is the goal. The more invisible the hunter, the better the hunt is likely to go.

Some predators even resemble inanimate objects. There are praying mantids, for example, that mimic orchids upon which their prey feed. An unsuspecting moth, butterfly, or other insect doesn't stand a chance against this well-concealed, formidable predator. Likewise, a species of ambush bug matches the yellow florets of the goldenrod plant to surprise any careless passerby. Stonefish and scorpion fish are highly poisonous animals that blend remarkably with the rocks, sponges, and debris of the ocean floor. Unwary prey that swim too close are suddenly engulfed in the vacuum-like mouth of these voracious predators. In a similar manner, alligators and crocodiles have the appearance of floating logs, which enables them to drift unnoticed to within striking distance.

An even more incredible form of deception is for a predator to resemble the prey itself. This allows the hunter to mingle among unsuspecting quarry without detection. An example is the robber fly (*Mallophora bomboides*) that mimics bumblebees with amazing accuracy. In fact, the imposter blends so well among feeding bees that it can intermix unnoticed, and attack without warning.

A similar ploy is used by certain fireflies. As members of the order Coleoptera, fireflies are bioluminescent beetles that use species-specific flashes to signal each other. The females often lie in the grass and use their amorous flashes to lure males. The two meet and mating occurs. However, there is a predacious species that can mimic another species' code. She, too, lies in the grass to lure males, not to a romantic interlude, but to a certain death.

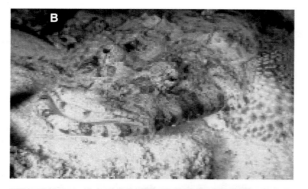

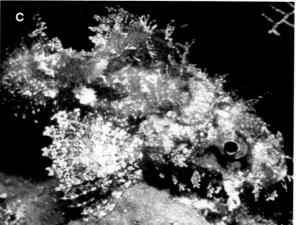

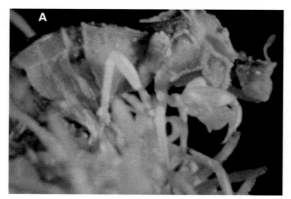

Figure 16.8. Camouflage can greatly assist an ambushing predator. For instance, plate **A** shows an ambush bug that blends perfectly with its background. Plates **B** and **C** displays the tremendous camouflage used by predacious fish to match their surroundings while awaiting their unsuspecting prey.

Weaponry
~ the predator's arsenal ~

Locating prey without being detected is but the first step toward a successful hunt. To be victorious the predator must also be able to overpower or somehow subdue the quarry. Thus, it's to the predator's advantage to be larger, faster, and more powerful than its prey. While that's what we would expect, that's not exactly what we find. It *is* true that the vast majority of predators are larger and more powerful than their prey, but they are not always faster. In fact, selective pressure on the prey has often yielded species that are speedier than their pursuers. This is especially true when the confrontation occurs over relatively long distances. So, for the predator it's that initial contact that is of utmost importance. That first burst of speed very often determines the outcome of an encounter. If the predator doesn't overpower the prey in the opening few moments, the contest may go to the hunted. That's exactly why predators have evolved ambush, stealth, and stalking as such important elements in their bag of tricks.

That initial burst of speed may allow the predator to overcome the prey, but ultimate success depends upon having the necessary weaponry to kill. And here, natural selection has outdone itself in providing tools that inflict as much damage as quickly as possible. Elongated fangs, slashing claws, sharp spines, powerful appendages, slicing mandibles, bone-crushing jaws, and powerful toxins are just some of the weapons within the predator's arsenal.

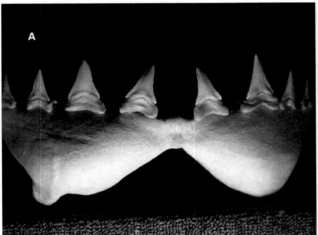

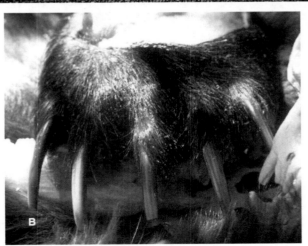

Figure 16.9. Some of the weaponry displayed by predators. **A**, shark jaws, **B**, bear claws, **C**, raptorial beak of a hawk, **D**, fangs of a tiger, and **E**, the formidable jaws of a tiger beetle.

The wide use of *chemical weapons* is another shrewd evolutionary development. From the nematocysts of the cnidarians to the powerful toxin of the blue-ringed octopus, venom is a popular weapon in the predator's armory. Spiders, centipedes, and a host of insects are among the many that can inject venom to overwhelm their prey. And, of course, no predator uses venom more effectively than snakes, over 300 species of which are poisonous.

The bounty of weapons evolved by predators is astonishing. Spiders use sticky webs to entrap their victim, while the angler fish and alligator snapping turtle use specialized structures (fins, tongue) to lure prey within striking distance. Bats and dolphins use echolocation to find their targets, and owls have feathers modified into facial discs and asymmetrical ear openings to locate quarry in virtual darkness. Many fish, especially sharks, have a lateral line system to detect weak electrical fields or subtle vibrations emitted by their prey. Electric eels, electric catfish, and torpedo rays can discharge an electric shock of several hundred volts to stun or kill their victims. The arsenal is truly impressive.

Figure 16.10. Many predators use venom to capture prey. Some, such as the anemones (top) use a passive approach, while others like the copperhead and scorpions strike with sudden death.

Figure 16.11. The facial discs of owls help to channel noises to the ears, which are positioned asymmetrically to better locate the noise source. The "ears" on top of the head are not used in prey location but are for display purposes only.

Prey Adaptations

~ how to avoid being eaten ~

*F*rom our discussion so far, it may seem that the advantage in the predator-prey struggle is well on the side of the predator. It may seem that way, but as we shall soon discover, their advantage, if any, is slight. Keep in mind that both predator and prey have co-evolved within this give-and-take relationship. As the hunter developed better ways to locate, capture, subdue, and devour a prey, so have the hunted developed a countering suite of traits to avoid being located, captured, and eaten. Thus, prey have their stockpile of tricks as well. In fact, with their remarkably diverse array of anti-predator devices to avoid becoming just another item on a predator's menu, they may just be a little ahead in this game.

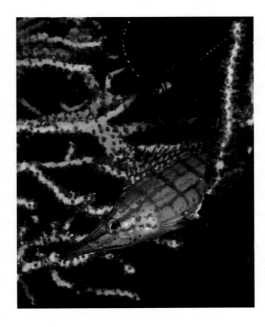

Reducing Detection

In any contest, the first line of defense is very often the best. This is absolutely true for prey animals. If they can avoid detection, they will not be eaten...it's as simple as that. If an animal is seen, heard, or scented, it is at risk. The need to remain undetected, thus, is foremost. So, it's not at all surprising that prey, like predators, have devoted a considerable portion of their adaptive effort toward deception. And the art of deception is nowhere more evident than in **camouflage**, the ability of an animal to blend with its surroundings. It enhances their chance of survival and thus increases the likelihood of transmitting these same survival traits to the next generation.

Camouflage is perhaps the most commonly used deceptive device by both predators and prey. The better an animal's coloration or shape allows it to blend with its general surroundings, the more likely it will go unnoticed. Through **substrate matching** a prey animal may blend with local soil, pebbles, dead or living tree trunks, or other features. Fish, reptiles, and many ground-nesting birds show a remarkable ability to melt into the background. To avoid detection, many prey animals have also adopted "**freezing**" as a viable option, especially when face-to-face with a large predator.

Figure 16.12. Substrate matching is a favorite ploy used by prey to avoid being seen. A fish (upper right) uses color patterns to blend with its surroundings as does the above grasshopper poised on a leaf. The flatfish to the right matches its surroundings so that it's all but invisible.

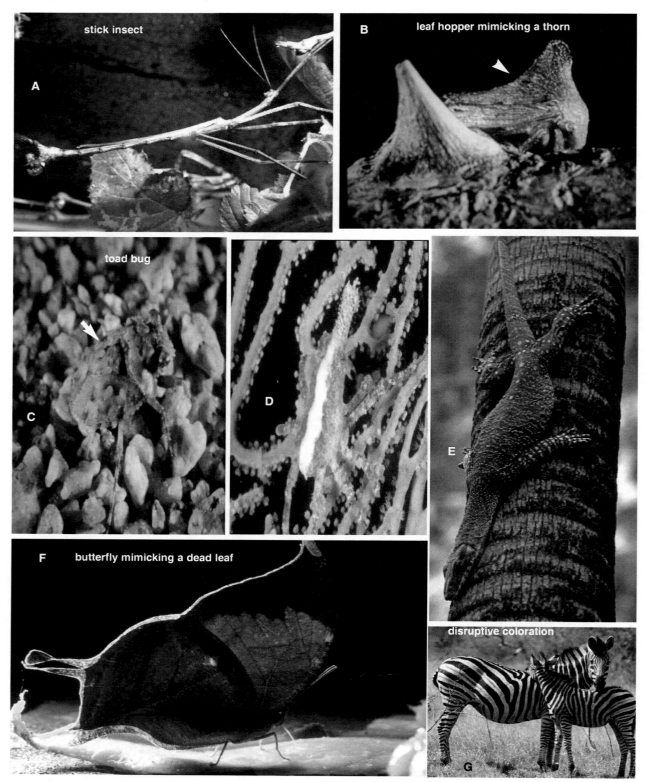

Figure 16. 13. Prey species display an incredible ability to blend into or match their background. Walking sticks (**A**) are well-known examples of animals that look like inedible features of their surroundings. But so is the leaf hopper that simulates a thorn (**B**), the toad bug that virtually disappears among the pebbles around it (**C**), the gorgonian crab in plate **D**, the lizard hugging the tree trunk (**E**) and the butterfly in **F**, are all able to blend due to their likeness to their surroundings. Another way prey use to conceal themselves is to break up the body's contour through countershading, or with contrasting stripes or spots as seen in the zebra in plate **G**.

An animal's outline poses a particular problem that must be overcome to avoid detection. The problem has to do with sunlight. Overhead light causes shadows to appear beneath an animal and, as a result, a uniformly colored animal actually becomes more conspicuous. So, many animals are darker dorsally than ventrally, which tends to counteract the effect of shadows. **Countershading** is seen in almost all camouflaged species and is especially prominent in aquatic species. When viewed from beneath, a countershaded swimmer blends with bright light beaming from above. That same animal, when viewed from above, blends with the blackness of the underlying water. This is clearly displayed by most game fish, and most predators including sharks (below) and killer whales.

countershading

Furthermore, color patterns can be used to distort an animal's shape from a distance. For example, small and large hervibores, from chipmunks to zebras, invariably have stripes or other **disruptive patterns** that break up the body's contour and make it hard for the predator to recognize its prey.

Camouflage isn't always manifested in simply matching an animal's general surroundings. There are many instances of prey resembling specific, inedible environmental features.   Leafhoppers mimic thorns, bush-crickets, some fish, and katydids have a remarkable likeness to leaves, caterpillars and walking sticks resemble twigs and sticks. Some prey species, such as caterpillars and other larval insects, may even take on the appearance of bird droppings, lumps of soil, clumps of leaves, and so on. Again, the key is to avoid detection by whatever means possible.

escaping into a burrow

One technique not yet discussed is one of the most obvious (above). Prairie dogs, when catching sight of a predator quickly dart into their protective burrows. Rabbits bolt into thick brush, birds fly off, an octopus slithers into its rocky lair, or a fish ducks out of sight into a thicket of algae, coral, or other sanctuary. Taking refuge may be the most common, and most effective means that prey have of avoiding being seen.

Reducing Capture

Should a prey animal actually be spotted, it must summon any resource available to avoid being captured. Once discovered, a favorite tactic of many prey (as mentioned above) is to take flight or find refuge in a den, burrow, crevice, tree cavity, or other shelter. A different approach, the **aggregation response**, is available to prey that occur in large groups. Many songbirds, for example, flock together to mob a marauding crow, owl, or hawk.   When threatened by a pack of wolves, muskox encircle their calves with their massive heads pointed outward. Such group defense has also been observed in meerkats, banded mongooses, capuchin monkeys, and warthogs. In each case, the group bands together to repel a predator.

Another advantage of group living is that each individual in the group can capitalize on a *collective sense of awareness*. Rather than relying on one pair of eyes scanning the horizon, or one pair of ears tuned to the slightest rustle, or one nose sniffing the air, a single prey animal can benefit from hundreds or even thousands of surveying eyes, listening ears, and sniffing noses. It is virtually impossible for predators to spring an ambush in such settings. All they can do is attack quickly with the prospects that a slow-moving animal might be caught in the confusion. However, in reality, predators generally don't waste their energy on such situations.

There is yet one more way that large aggregations of prey can protect the individual. It's called the **selfish herd concept.** In simple terms, the larger the crowd the less the chances that any one individual will be captured. One lone sandpiper feeding on a mudflat may be easy pickings for a hawk or a raiding fox. But, if that same bird is surrounded by a thousand others, its chance of capture is reduced significantly. Even if a hawk or fox is successful in catching a sandpiper, the chance that it is *our* sandpiper is minimal. In other words, the selfish herd is willing to sacrifice one individual for the sake of the pack. From the standpoint of natural selection, we might even assume that it's the slowest, dullest, or least aware that is captured, thus culling out the less fit, allowing those with more adaptive capabilities to survive and reproduce.

Figure 16.14. At times, prey occur in extremely large groupings so that each member benefits from the collective sense of awareness. Also, the likelihood that any one animal being captured is greatly reduced in what is termed the "selfish herd."

As we have seen, camouflage is one of a prey's most important lines of defense. But other aspects of color, beyond simple camouflage, can play a significant role in prey survival as well.

Distractive Coloration

Imagine that a predator, say a blue jay, has spotted a succulent butterfly. Just as the jay strikes, in full expectation of a tasty meal, the butterfly suddenly jerks open its wings to reveal two huge **"eyespots."** The startled jay stops in mid-lunge, turns, and flees. The butterfly has escaped, not because of concealment, but rather due to a precisely placed flash of color. What the blue jay apparently "saw" was a potential confrontation with an even larger animal than itself. This scenario is common in the predatory-prey scheme (see below). Eyespots are conspicuously displayed on many butterflies, moths, mantids, fish, amphibians, and snakes, among others. The spots may be large, for intimidation purposes or they may be small, and serve, instead, to deflect the predator's attention away from a vital area.

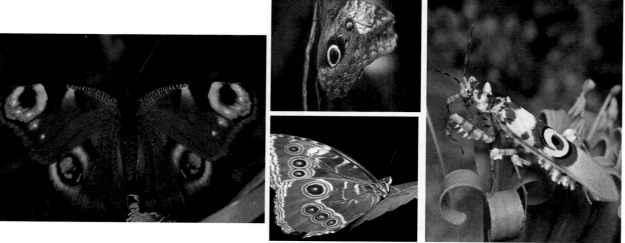

Figure 16.15. Many prey animals display conspicuous patterns resembling eyespots. These may be large to resemble a much larger animal so that when flashed they serve to repel an attacker or may be small spots scattered over the surface of the prey to simply distract an attacker. Note that in each of the cases shown above, the eyespot is located in a nonvulnerable part of the body so that if an attack does occur the individual may survive.

Distractive coloration, as this phenomenon is sometimes called, is not always manifested in eye spots. Take, for example, the rump of a white-tailed deer. When in danger, a deer will flick the tail sending a white warning signal to other deer in the area. When attacked, it again conspicuously flashes the tail, potentially diverting the attacker toward the rear and away from the head, trunk, or other vital area. Distractive coloration is evident in a large number of animals from insects, to fish, to birds, to mammals.

Figure 16.16. Contrasting patches of color (arrows) is thought to distract a predator toward a less vulnerable area. The color is often concealed until the last moment, then flashed as the prey retreats as seen in the fleeing deer.

Aposematic Coloration

Warning or aposematic coloration is used by many prey animals to alert predators of some toxic, noxious, or other unpalatable property: a sting, a venomous bite, a putrid smell, or an offensive taste. The idea is for some color combination to *clearly advertise the unpalatability*. Any bold or contrasting bright colors will do, although black or white stripes set against a yellow, red, or orange background is the usual arrangement. Rather than relying upon concealment, as camouflaged prey do, animals with warning coloration do just the opposite. They display themselves conspicuously. When approached by a potential predator, they seemingly flaunt their colors, often turning toward the intruder to show their colors to the best advantage.

For aposematic coloration to work properly two things are necessary. First, the colors must serve clear notice that the bearer is unpalatable. *The bold color pattern is intended to remind a predator of a previous unpleasant encounter with that species.* Following the initial encounter, the predator will thereafter associate the color with that episode and avoid any animal displaying that color combination. You can now see why brightness and contrasting colors are used. Second, the toxicity cannot be lethal to the predator, otherwise the whole mechanism is self-defeating.

Warning coloration is widespread among prey animals but is especially pronounced in cnidarians (sea anemones), molluscans (sea slugs and nudibranchs), insects (hornets, wasps, bees, ants, beetles, butterflies, caterpillars), fish (lionfish, zebrafishes), frogs (poison dart frogs, tree frogs), salamanders (tiger salamander, fire salamander), lizards (gila monster), snakes (coral snakes, sea snakes), and even certain mammals (skunks).

You may have noted that birds are missing from the above list; yet, many birds are adorned in brightly contrasting colors such as the goldfinch below. It seems that the ability to fly, and thereby escape predation, has left birds pretty much out of the warning coloration business. Few, if any, birds are aposematically colored simply because birds as a group lack stinging devices, poisons, or noxious smells. There are only a few records of tropical birds whose flesh is apparently distasteful and who display warning coloration. Bright, conspicuous colors are used by birds primarily in courtship displays, not as warning colors.

Figure 16.17. Aposematic coloration is used by a wide variety of animals to warn of some noxious or dangerous property. Contrasting yellow against a black background is a common pattern as seen in the poisonous flatworm (**A**) and caterpillar (**B**). An exception occurs with birds, which have very few examples of warning coloration. Above is a goldfinch (**C**), also adorned in bright yellow and black. But goldfinches are not noxious; the color is related to courtship behavior and not to warning.

Figure 16.18. Orange against black, yellow and black contrasted against a white background, red and black— these are the colors that animals have selected to convey the unmistakable message, "don't mess with me." Refer to page 530 for color photographs of additional aposematic animals.

Mimicry
~ the brilliant art of deceit ~

It's difficult to know exactly how effective warning coloration is in protecting any prey species. We can assume that the process is fairly effective, but we don't know for sure. However, what we do know is that there's a built-in problem with the mechanism: *it's a one-on-one learning experience.* Each predator must have had at least one prior harmful experience before it can connect the color pattern with the unpleasantry. This means that there must be a large number of distressful encounters for the process to work at all. Nonetheless, the fact that mimicry is used by such a large group of animals is testament to its apparent effectiveness.

But popularity is not the only way to measure the outcome of warning coloration; there's one other line of evidence available. You will recall that in the predator-prey contest deception is the passport to success. Nowhere in nature is fraud more evident than in the phenomenon of mimicry: *the resemblance of one organism to another (or to an object) for the purpose of deceiving a predator.* It comes in four forms.

(1) In **Batesian mimicry** an edible prey animal (the mimic) assumes the coloration of an inedible or unpalatable one (the model). A predator that has had an unpleasant experience with the model learns to avoid the mimic as well. It's a con game of sorts: the mimic gains from warning coloration even though it has no unpalatable properties itself.

Batesian mimicry is seen in a variety of invertebrate and vertebrate animals. Perhaps the most classic example of Batesian mimicry occurs with two snakes, one poisonous and the other not (Figure 16.19). The several species of venomous **coral snakes** all have a conspicuous pattern of red, yellow, and black stripes. A pattern that any experienced predator will soon learn to avoid. A similar color pattern occurs a variety of harmless **king snakes,** which thereby are equally well protected from predators. In Batesian mimicry, the mimic gains the same degree of protection as the model even though it's nontoxic.

(2) **Mullerian mimicry** also involves models and mimics but differs in one important respect from Batesian mimicry. Rather than one edible form resembling an inedible animal, *both the model and the mimic are toxic.* In this case, each prey population is noxious, and each population displays similar warning coloration; to a predator, they all look somewhat alike. Therefore, the predator only has to sample *one* prey animal from *one* population to learn to avoid *all* populations. Again, Mullerian mimicry differs from Batesian in that all participants are toxic and thus the learning curve for the predator is much reduced. As a result, fewer prey must pay the price of educating predators, a price that all too often results in their own death. Mullerian mimicry is

Figure 16.19. One good example of Batesian mimicry is that between the venomous coral snake (upper photo) and the harmless kingsnake (lower). The North American coral snakes have a banded pattern in which red abuts yellow (arrows), whereas red lies next to black in the kingsnake.

common among insects, especially beetles, ants, bees, wasps, moths, and butterflies.

(3) Some prey animals have a keen ability to **mimic inedible objects**. There are birds that perch in such a way that they resemble a tree knot; many insects mimic fecal droppings; other insects called tree hoppers resemble thorns; stick insects mimic twigs; beetles look like stones; there are fish that resemble leaves or floating debris; squid and octopuses that take on the appearance of algae or corals; and hermit crabs that place sponges and sea anemones on their shell. The fact is, many prey animals mirror their surroundings in such a way as to appear not only unpalatable, but inedible.

Figure 16.20. The insect above uses its own feces to disguise itself from predators. The photo to the right shows the Virginia opossum (*Didelphus marsupialus*), the only marsupial in North America. When harassed, the "possum" is known to realistically feign death, a behavior known as **akinesis**. This is a warm climate species that has moved northward. Note the tattered ears resulting from frost bite.

(4) Many predators will only eat prey they have actually killed themselves and will avoid an animal already dead. There are a number of prey species that take advantage of that through a process called **akinesis**; they pretend they're dead. Certain spiders and beetles will fall motionless when threatened, as do some chameleons. The grass snake and the hognose snake are two species that roll over on their backs with mouth agape and tongue distended when threatened. But the best actor of all is the North American opossum. When attacked, it not only assumes a deathlike posture, which it can maintain for a considerable time, but it, like the hognose snake, can even emit an odor of decomposing flesh. Most would-be opossum eaters are disgusted by this repugnant display of deception. Pretense of this sort works well for the opossum unless, of course, a hungry scavenger that relishes foul-smelling corpses comes along.

Mimicry
(resemblance of one species to another or to an inanimate object)

Batesian Mimicry - resemblance of a nontoxic animal to a toxic one,

Mullerian Mimicry - resemblance of several toxic animals to each other,

Mimicry of Inedible objects - resemblance to inedible environmental features such as twigs & leafs.

Akinesis - the ability of an animal to pretend it's dead

Preventing Consumption
~ the last resort ~

The final ploy an animal has to avoid predation is aggression...hoping in desperation that a bluff or a counterattack will permit an escape. Some animals puff up to appear large to intimidate an assailant. In fact, when threatened, humans often do likewise. But the champion "puffer" is the porcupine fish (below). It swallows air or water to get bigger and bigger until a harasser gets frustrated and just leaves. Other prey lunge at an attacker and some bare their teeth or emit loud screams in an effort to scare off an assailant.

porcupine fish
(in defensive posture)

But prey *are* inevitably captured. The nature of the game is that predators are successful, at least some of the time. If a prey animal has, indeed, been detected and caught it has but one last defensive strategy and that's to avoid being eaten. There are a variety of ways this can be accomplished.

Chemical defense. The use of toxins is a common prey tactic to gain release. Bees, wasps, hornets, yellow-jackets, some species of ants, and a variety of other insects can inflict a painful sting if provoked. Centipedes are capable of rendering a painful bite as are certain aquatic insects. Likewise, many aquatic invertebrates, such as sea anemones, urchins, and sea cucumbers, have toxic armaments to repel predators. Millipedes and centipedes can exude a defensive and poisonous chemical from special glands.

The list of prey relying on chemicals is a long, long one, and some are highly ingenious. For example, squid and octopuses are capable of releasing a numbing ink to confuse and perplex predators. Salamanders, frogs, and toads may ooze a slimy toxic skin secretion—sometimes, as in the tiny poison dart frogs, the secretion is exceedingly poisonous. The bombadier beetle releases an explosive chemical cocktail that burns upon contact. Snakes and a few lizards may inject poisons that

can cause human death. But no user of chemicals is more well known than those black and white denizens of the American woodlands: the skunks. When disturbed, they can spray a precisely aimed, nauseating chemical vapor primed to discourage the most determined of predators.

Body armor. An impenetrable covering is another one of a prey's effective safeguards against being eaten. Many animals, when threatened, simply find sanctuary within a shell, casing, carapace, or other armorlike retreat. From amoeba to armadillos, a host of prey animals find refuge in some type of protective covering.

giant tortoise
(in defensive posture)

Weaponry. When cornered, many otherwise docile prey will strike back. They may possess teeth, claws, spines, or other devices that can become effective weapons when necessary.   Porcupines and hedgehogs have sharp quills that can inflict painful wounds. The bite of a trapped groundhog, skillfully delivered, can discourage a coyote or fox. Large antlers, swinging to and fro, may repel a pack of wolves. Even a claw poking in the right place may be the remedy to an attack. And it's only the bravest of predator who will undertake an attack against a young animal when the mother is nearby. The truth is, predation can be risky business when the life of your victim is at stake.

Autotomy. One of the strangest, and at the same time most reliable, tactics is the ability of some prey to sacrifice a body part when attacked. Self-amputation of this type, called autotomy, is common among a variety of prey species. Many arthropods, such as spiders, daddy longlegs, crustacea (crayfish, lobsters, crabs), and others, will "willingly" relinquish an appendage when in a predator's grasp. Sea stars, for example, readily surrender an arm when the circumstance warrants. Certain lizards will do likewise with their tail. The sacrificed piece may even continue to move or wriggle thereby distracting the predator while the prey makes a hasty escape.

However, for autotomy to be an effective maneuver, the missing piece must be replaced. Thus, this strategy is only used by those prey species able to regenerate lost parts. In most cases, there is a fracture plane where the feature breaks off from the body. The breakage quickly closes to prevent excessive loss of blood or other essential bodily fluids.

Figure 16.21. A drastic maneuver to escape predation is to sacrifice a body part. **Autotomy** works only if regeneration is possible as in the skink (top photo: tail) and the sea star below.

Figure 16.21. Otherwise passive prey can become aggressive when facing certain death. The baboon threatens with large teeth and the porcupine defends itself with sharp quills.

Satiation. What happens to those prey species that may lack camouflage? What about those who have no toxic properties, or who cannot mimic, possess inadequate weapons, or don't engage in self-amputation? Are they without means to fend off the horde of predators? The answer is, no. There is yet one more approach open to the hunted. A preposterous, yet effective approach. They simply produce an overabundance of offspring so that any marauding predator can eat until it becomes sated. By flooding the market with offspring, they produce more than even an army of hungry predators can possibly consume. Thus, a percentage, albeit it a small percentage, of their young will escape.

Many animals, from oysters to insects, from frogs to fish, practice this strategy. Mating horseshoe crabs, for example, deposit millions and millions of eggs on sandy beaches, more than even migrating birds can devour. A world away, sea turtles do likewise (see photo below). In just a few evenings hundreds of female turtles bury batches of plum-sized eggs in sandy crypts. Months later a mass emergence of thousands of tiny turtles litter the beaches as they run a gauntlet to the sea. And even though many are lost to predators, some do invariably make their way to open water and eventual adulthood.

Rabbits, lemmings, and many other herbivores produce large litters. Even foraging ungulates such as wildebeest and caribou have a high annual output. This is all done on the premise that if enough young are issued at the same time, predators can't eliminate them all. Satiation seems to be a wasteful tactic, but it does ensure that some members of the species will carry on into the next generation.

Figure 16.22. Some animals, such as these venomous snakes, are efficient predators yet still serve as prey to larger animals.

Figure 16.21. A last-resort tactic used by many prey animals is to produce more young than can survive. Thus, predators can eat until sated without eliminating all of the offspring. Sea turtles practice such a strategy through mass egg laying.

A Dual Purpose

You have probably already realized that many predators are, at times, themselves, prey. They may prey on smaller animals but serve, in turn, as food for larger ones. Except for top carnivores, hunters are also hunted. Thus, for most animals, adaptations serve a dual purpose; they have both defensive and offensive functions. Camouflage can serve to conceal a predator from its prey as well as from its enemies. A weapon may be used to subdue a victim or to ward off an attacker. Venom can be an awesomely effective way to prepare a meal, or it can be an equally awesome means of self-protection.

The diamondback rattlesnake (*Crotalus atrox*) of the southwestern U. S. is a good example of the duality of adaptations. It is a large, highly venomous, aggressive predator. Well camouflaged, it lies in wait to ambush mice, small squirrels, and other warm-blooded quarry. But its camouflage also provides protection from potential predators (birds of prey, large mammals, other snakes, humans). Concealed within its upper jaw is a pair of large fangs, ready to deliver a lethal dose of venom into a careless rodent, or just as ready to discourage that careless backpacker. It is finely adapted with dual skills as it lives on evolution's cusp between predator and prey.

Chapter 16 Summary

Key Points

1. In the predator-prey relationship, the prey always outnumber the predators. The degree to which they exceed predator populations is a factor of their size and the selectivity of the predator.

2. The number relationship is also related to energy transfer and can be illustrated through an energy pyramid.

3. Prey numbers often fluctuate. When they do, predators have several options: they can increase foraging range, expand their food choices toward less preferred items, turn to cannibalism, or alter their own reproductive patterns.

4. Both predator and prey adapt to the relationship through the process of co-evolution as each attempts to outdo the other.

5. Predators have a number of specific adaptations, among these are

 a. various hunting strategies such as ambush, stalking, and active pursuit;

 b. deception through camouflage; and

 c. weapons such as fangs, powerful jaws, and venom.

6. Prey, likewise, exhibit several adaptations, such as

 a. reducing detection through substrate matching, countershading, disruptive coloration, and mimicking inedible objects;

 b. reducing capture through the aggregation response, selfish herd response, distractive coloration, aposematic coloration, and mimicry; and

 c. to avoid being eaten, prey may feign death, act aggressive, use what weapons they have, use chemical warfare, or resort to satiation.

Key Terms

active pursuit
aggregation response
akinesis
ambush
aposematic coloration
Batesian mimicry
camouflage
cannibalism
co-evolution
countershading
disruptive patterns
distractive coloration
energy pyramid
eyespots
Mullerian mimicry
predator
prey
satiation
selfish herd response
stalking
substrate matching
warning coloration

Points to Ponder

1. Why is it true that prey always outnumber predators?

2. What is an energy pyramid and what does it have to do with the predator/prey relationship?

3. What are the ways predators may respond to a crash in prey numbers?

4. What is co-evolution? Give two examples of this phenomenon in the predator/prey system.

5. What type of hunting strategies do predators use?

6. What other types of specific adaptations do predators have to the relationship?

7. What kinds of camouflage do prey use?

8. What is the aggregation and selfish herd response?

9. Give an example of distractive coloration and how it is used by prey animals.

10. What is aposematic coloration? Give three examples.

11. Differentiate between Batesian and Mullerian mimicry.

12. What is akinesis? Give an example.

13. What is satiation and how do prey animals benefit from it?

Inside Readings

Ernst, C. H. 1992. Venomous Reptiles of North America. Washingyton DC: Smithsonian Press.

Freeman, S. and J. C., Herron. 1998. Evolutionary Analysis. Upper Saddle River, NJ: Prentice Hall.

Owen, D. 1980. Camouflage and Mimicry. Chicago, IL: Univ. of Chicago Press.

Scholz, F. 1993. Birds of Prey. Mechanicsburg, PA: Stackpole Books.

Snyder, N. and H. Snyder. Birds of Prey: Natural History & Conservation of North American Raptors. Stillwater, MN: Voyageur Press.

Wolfe, A. 1990. Owls: Their Life and Behavior. New York: Crown Pub.

17 Chapter 17

THE CHORDATES

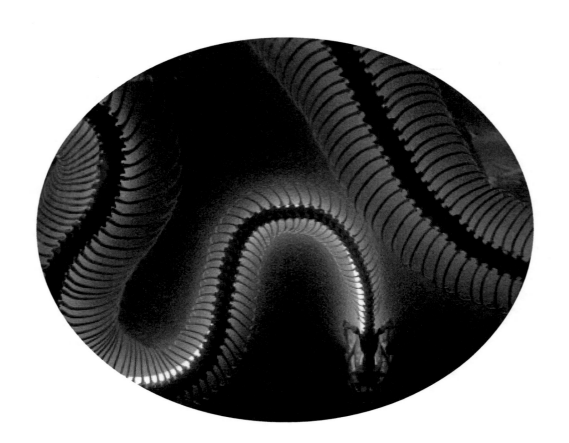

Phylum Chordata

*The sea is the land's edge also, the granite
into which it reaches, the beaches where it tosses,
It hints of earlier and other creations
the starfish, the hermit crab, the whale's backbone...
the sea has many voices.*

-Eliot

"The Whale's Backbone"

*O*ur final phylum is by no means the largest. Its members are not the most diverse, and they are certainly not the most prolific. They are, to many zoologists, not even the most interesting. But what they are, not what they aren't, is worthy of our notice. The phylum Chordata contains the most advanced and complex animals that have ever existed. It seems that evolution, in all of its contortions, all of its twists and turns, has given us two capstone groups...the arthropods on the one hand, and the chordates on the other. Each is highly successful in its own domain, each strives for mastery. The two are engaged in a cosmic contest of sorts, a contest for domination.

In the one group, the arthropods, Mother Nature has invested her energy into small size, large numbers, and most of all, into diversity...*not size, not bulk, but variety.* In the other group, the chordates, as we will soon see, she has concentrated her creative talents into larger body size (we find here some of the true giants of the animal kingdom), fewer offspring, and *most importantly, into the nervous system*...giving rise to animals with more superior brainpower than, perhaps, all others combined. *One branch led to insects with unparalleled diversity; the other to the mammals, with unsurpassed intelligence.*

But what makes a chordate a chordate? How are they different from other animals? What kinds are there? How do they use their intelligence? These are the questions this brief chapter will address. We shall see how chordates resemble each other and how they differ from any other animal on earth. What is a chordate? It's a segmented animal with a supporting, dorsal rod, a series of gill slits, a dorsal nerve cord, and a gland that processes iodine. And, oh yes, one other feature....it has a tail!

*What am I, Life?
A thing of watery salt
Held in cohesion by unresting cells.*

-Masefield

The Notochord

Every chordate, at some time in its life, has a supporting longitudinal rod passing from one end to the other. This rod, the notochord, is flexible yet rigid; it can bend but does not shorten. It is packed with gelatinous material under hydrostatic pressure. As a result, it gives the animal considerable support while, at the same time, allowing enough resilience to permit side-by-side swimming movements. The functional beauty of the notochord is that it prevents the animal from compressing (i.e., telescoping) while simultaneously allowing flexibility.

In some of the more primitive animals, such as the tiny *Amphioxus*, the notochord is the primary skeletal unit, covered by a connective tissue sheath to which muscles are attached. But in higher animals the notochord becomes incorporated into the vertebral column, the "whale's backbone" as it were.

Pharyngeal Gill Slits

As we shall see in the next chapter and beyond, some adult chordates possess functional gills borne in the neck region on special skeletal elements called pharyngeal gill arches. However, even though most adult chordates lack gills, they all, at some point in their lives, possess bilateral gill clefts or slits that open externally in the region of the neck. These are called the pharyngeal gill slits. In amphioxus, lampreys, sharks, and other fish, for example, these are apparent in adult animals and function in the important exchange of gases.

The embryos of all higher chordates (amphibian, reptiles, birds, and mammals) also show this feature but it is often obscured by the time adulthood is reached. Nevertheless, all chordates possess these slits at some time. Indeed, some human babies are born with a fistula (a slitlike opening) in the neck region, which is a remnant of the pharyngeal slits. The opening is closed surgically.

Dorsal, Hollow Nerve Tube

In those *invertebrates* that have a nerve cord (e.g., arthropods, annelids, certain mollusks, and flatworms), it is always ventrally located and is always solid. A solid nerve cord presents animals with an ominous physiological problem. Nervous tissue, by its very nature lacks blood vessels; it is **avascular**. Since blood vessels don't penetrate into nervous tissue, oxygen must diffuse from the outside. So a solid nervous structure must be limited in size and function due to a lack of readily available oxygen. In chordates, the nerve cord is in the form of a hollow tube situated dorsally, just above the region of the notochord and enclosed by the vertebral column. In vertebrates, the delicate nerve cord is protected by the "whale's backbone."

The marvelously important feature of the nerve tube is its hollow nature, which terminates, anteriorly, in a hollow brain. Although the nervous system of higher animals is hollow, it is not empty. Inside of the hollow tube is a nourishing fluid, the cerebrospinal fluid (CSF). Nerve cells, even those buried deep within the brain, can be nurtured internally by this fluid or can be nourished externally by blood capillaries. This means that the brain, being hollow, is able to grow with the needs of the animal. More importantly, it (the brain) was preadapted to expand in response to evolutionary pressures. As chordates became ever more complex, their brain kept pace, permitting even greater complexity, leading to greater brain expansion, and so on. Indeed, one of the hallmarks of chordates is this explosive expansion of the brain...leading, eventually, to animals with intellectual capacity never before seen. This we shall explore in greater detail as we travel along the chordate road.

Subpharyngeal Gland

Life depends upon the ability of cells to carry on metabolism. They must be able to acquire energy, use it in the maintenance and activity of the cell, and in the release of waste by-products. Metabolism is the term given to those total bioenergetic activities of a cell. In chordates, metabolism requires the presence of iodine. Without it, life is seriously impaired. So all chordates have, in the region of the pharynx, a gland that fixes or binds dietary iodine for metabolic use. In the "lower" chordates, this gland is called an **endostyle**. But in higher animals, such as humans, that gland is the **thyroid**.

Postanal Tail

If you tell your closest friend that he or she, as a chordate animal, possesses a tail, you may be placing your friendship in grave jeopardy. But it's true. All chordates, at some point, have a tail. Of course, this is quite obvious in many chordate animals; fish, salamanders, lizards, leopards, and elephants have a conspicuous tail. But humans? A tail is defined as an extension of the body beyond the anus; and, while adult humans have no visage of such a structure, they do have tails as embryos. And it's an impressive tail at that—at one point accounting for about one-third of the embryo's total body length! In fact, some babies are born with a small tail similar to that of a pig, which is adroitly removed forthwith.

Other Characteristics

To complete our picture of a chordate, there are just a few more pieces to add:

(1) chordates are **bilateral**;

(2) they are **deuterostomes** (the anus is derived from the blastopore);

(3) they are **segmented** (metameric);

(4) virtually all chordates have an **endoskeleton** of some sort. Whatever form it takes, the endoskeleton is always derived from mesodermal germ tissue; and

(5) almost all chordates have a circulatory pump...the **heart**. Whenever a heart is present, it's always ventral, unlike the dorsal hearts of invertebrates.

Chordate Taxonomy
~ a short edition ~

Phylum Chordata

Dispersed widely in marine, freshwater, and terrestrial environments. At some stage in their life cycle, they possess a tail, bilateral gill slits, a dorsal, hollow nerve tube, and a subpharyngeal gland to fix iodine.

Subphylum Urochordata

Marine animals called sea squirts (tunicates). They only possess a tail, notochord, and nerve tube as larvae; adults are sessile.

Subphylum Cephalochordata

Marine animals called lancelets (*Amphioxus*). They retain a tail, dorsal nerve tube, endostyle (iodine-fixing gland), pharyngeal gill slits, and the notochord throughout their lives.

Subphylum Vertebrata

Dispersed throughout aquatic and terrestrial habitats. Characterized by the presence of a vertebral column enclosing and protecting the dorsal hollow nerve tube. Tail, gill slits, and subpharyngeal gland all present, at least, in the immature stages.

Superclass Agnatha

Eel-like fish lacking jaws, scales, and appendages. Endoskeleton is made of cartilage: *lampreys, hagfish*

Superclass Gnathostomata

Jaws are present.

Class Chondrichthyes

Cartilaginous fish with paired appendages and scaled skin. Lacking a swim bladder for buoyancy:
sharks, skates, rays

Class Osteichthyes

Fish with bony endoskeleton, scales, and a swim bladder. Opercular plates are present to protect the delicate gills:
"bony fish"

Class Amphibia

Gill, lung, or cutaneous respiration; skin with mucous glands; aquatic larva metamorphose into adults. Tetrapods (having four legs):
frogs, toads, salamanders, newts, and caecilians

Class Reptilia

Skin dry and covered by epidermal scales; amniotic (cleidoic) eggs; embryos develop either within the female or deposited within the amniotic egg; lungs only; tetrapods or limbs lost secondarily:
snakes, lizards, turtles, crocodilians

Class Aves

Animals that possess feathers, a beak, most fly, endothermic, amniotic eggs:
birds

Class Mammalia

Possess hair, endothermic, mammary glands, amniotic eggs; marsupials, or placentate:
mammals

An Inside View 17.1
The Biogenetic Law

~ Evolution's Dirty Laundry ~

When early zoologists, as far back as the 1850s, began to examine embryos closely, they soon uncovered some very odd and unexpected phenomena. It seemed that the embryos of higher chordates displayed remarkably similar structures to the embryos, and even the adults, of lower chordates. They observed a number of instances in which an embryonic feature of a bird, reptile, or mammal seemed to resemble the adult feature of "lesser" vertebrates (e.g., fish and amphibians). They saw gill arches in reptiles, a tail in humans, a yolk sac in higher mammals.

They were perplexed and yet excited. How were they to interpret these shared homologies? Why, for example, do bird embryos have a well-developed gill apparatus quite like that of adult fishes? Or, why does a human embryo develop a yolk sac or a tail reminiscent of lower vertebrates even though humans have no use for yolk or a tail?

Ernst Haekel, an early German embryologist had an answer. He concluded that these situations were evidence of common ancestry among the vertebrates. To him, and many other zoologists of the time, these homologous features demonstrated an evolutionary link between the various vertebrate classes. As a bird passed through its own embryonic development, it proceeded also through a fish stage, an amphibian stage, and a reptilian stage. Likewise, as mammalian embryos developed, they, too, went through stages similar to adults of lower classes. Was *this anatomical evidence that each vertebrate class evolved from those somewhat lower on the phylogenetic scale?* In other words, did an individual's development (ontogeny) repeat (recapitulate) its evolutionary history (phylogeny)? Haekel thought so; thus, the *Biogenetic Law* was born.

Haekel even produced some elaborate drawings to prove his case. These diagrams became one of the cornerstones upon which the proof of evolution was built. Even today, textbooks use his drawings as evidence of biogenesis.

But, is it true?

Well, yes...and no!

Although Haekel's perception is no longer generally accepted, in fact, it seems that his famous drawings were not altogether accurate (some even say fraudulent; that he sketched what he presumed was there, not what really was present), there is, nonetheless, no doubt that a number of similarities do exist between embryos and adults along the vertebrate line.

So how then should *we* interpret these homologies? The current idea is that they represent nothing more than a general developmental pattern upon which class-specific structures originate. These features are present in each class simply as building blocks or as foundations for anatomical development in that particular group. Gill arches, for instance, form as a precursor to the adult pharyngeal area of, say, a reptile, bird, or mammal. Similarly, the human embryonic tail that is so prominent early in our lives is just a stage in the developmental schema of the human pelvic region.

However, these homologies are nevertheless, intriguing. Although biogenesis is no longer considered evidence of direct ancestry, it is, at least, a convincing demonstration of the anatomical relationships that exist between the various vertebrate groups.

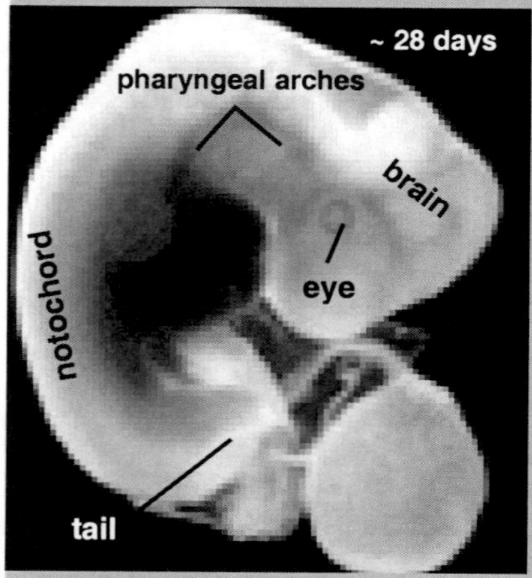

Figure 17.1. The early human embryo has the general features of all chordates including pharyngeal (gill) arches, a notochord, and a tail.

The Lesser Chordates

Subphylum Urochordata

Sea squirts (also known as tunicates) are about as unlike a tiger, a rattlesnake, or a shark as you can possibly imagine. Yet, they, in their odd way, are truly chordates. Here again, as with the echinoderms, we encounter animals that are grossly atypical because of their preferred habitat. As adults, tunicates are soft-bodied sessile animals attached to the floor of tide pools or rocky shores. It is because of their favored microhabitats that they have lost much of their chordate characteristics. They have no notochord, no nerve tube, no tail, not even a head. They look like a transparent or semitransparent bag, perhaps more closely related to cnidarians or sponges than to the fish swimming nearby.

So how do we know they're chordates? To answer that we must defer to larval tunicates. Here we find a tadpolelike animal with a notochord, dorsal nerve tube, and a tail. In other words, a true chordate. For the larva to assume an attached existence as an adult, it must transform itself into the strange-looking adults you see in Figure 17.2 & 17.3. It does so by adhering to a suitable surface and then undergoing metamorphosis. During the transformation, the larva loses its tail, notochord, and dorsal nerve cord. The only recognizable chordate feature in the adult are **pharyngeal gill slits.**

In general, the adult is shaped like a bag with two holes. One opening is an **incurrent siphon** through which water enters the animal; the other is the **excurrent siphon** whereby water exits. Entering water passes through the gill slits where oxygen and carbon dioxide are exchanged. Here also food particles are filtered from the water stream and trapped in a mucus thread formed by a grooved organ, the **endostyle.** Water then leaves via the excurrent siphon. Because the animal is so flimsy, the flowing water creates a **hydrostatic skeleton** that prevents it from collapsing. Tunicates are covered by a cellulose-like mantle, the **tunic,** from which they gain their common name.

Urochordates occur individually or in colonies. In colonial species, large numbers of individuals share a common excurrent siphon, but retain their individual incurrent siphons. They are **monoecious** (hermaphroditic) and fertilization may occur in the coelomic cavity or externally in the surrounding water. All individuals in a colony coordinate the release of sperm and eggs so that all share in the reproduction. Members can also reproduce asexually by budding off new individuals.

When disturbed, tunicates will squirt a stream of water ("sea squirts") to ward off a potential predator. However, that's not their only defense. They are also highly poisonous when eaten. As a result, many species are adorned with bright colors as a form of warning to potential predators.

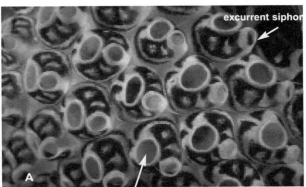

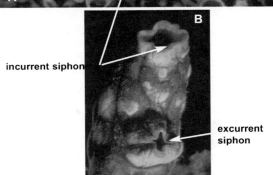

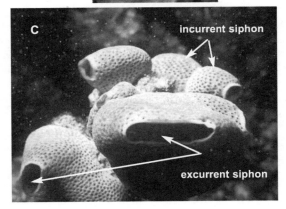

Figure 17.2. Representative tunicates, also known as sea squirts. **A,** a colonial sea squirt showing the incurrent and excurrent siphons. Plate **B** is a solitary tunicate and **C** displays another colonial tunicate (the very tiny holes are the incurrent siphons).

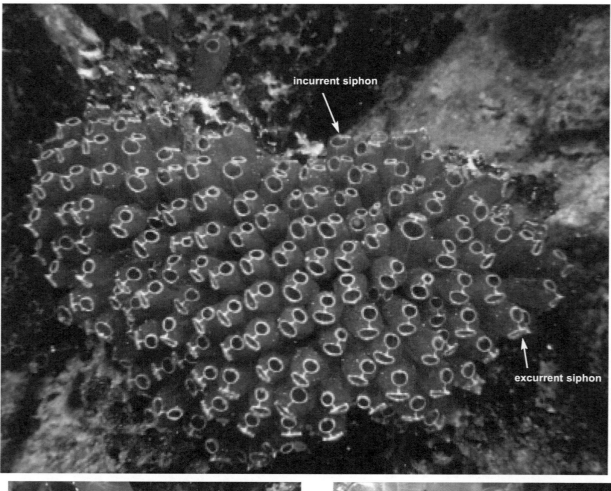

Figure 17.3. Three varieties of tunicates (sea squirts).

Subphylum Cephalochordata

Amphioxus (Figure 17.4 -17.6) is a well-known example of the subphylum Cephalochordata and is the closest relative of the vertebrates. It is a small animal, around 5 cm long (2 inches), pointed at both ends (thus, the common name of lancelet), and found along sandy shores throughout the world. The streamlined shape is an adaptation for tunneling in sandy substrates and although not adept swimmers, lancelets can thrash back and forth adequately enough to move short distances.

The subphylum derives its rather complicated formal name (Cephalochordata) from the fact that the **notochord** extends into the "head" region and, thus, runs the length of the animal (Figure 17.5). The notochord prevents compression of the animal when burrowing or swimming. Additional chordate features include a **dorsal, hollow nerve tube,** segmented V-shaped muscle blocks (myotomes), an array of **pharyngeal gill slits**, and a **postanal tail**. Although lancelets do have an anterior-posterior orientation, they lack a well-defined head and are minus a brain.

Amphioxus feeds by burrowing into the sand leaving only their head slightly exposed. In this position, they filter food particles from the flowing water. At the extreme anterior end of the animal is an **oral hood** whose entrance is guarded by fingerlike **cirri**. A water current, containing particulate matter, is created by the beating of cilia located on a **wheel organ** within the **oral cavity** (vestibule) and on the gill bars. Large items are excluded from the oral cavity by the combined action of the cirri, much in the manner if you held your fingers crisscross in front of your mouth to keep out unwanted debris. Water filters through the cirri, continues through the body cavity, passes the gill slits, enters a chamber, the **atrium**, and finally exits via an **atriopore**. As the gills are exchanging gases, food is being extracted from the water stream by the gill bars and then transported to the digestive system.

The circulatory pattern of *Amphioxus* is outlined in Figure 17.6. Note the presence of a dorsal and ventral aorta and the direction of blood flow through each: posteriorly through the dorsal aorta and anteriorly through the ventral aorta. The ventral aorta pulsates to propel blood through the various vessels and thus serves as a heart.

A cross section of *Amphioxus* (Figure 17.6) shows the central, dorsal location of the notochord and its relationship to the lateral muscle blocks, called myotomes. These muscles pull against the notochord to effect a zigzag force that propels the animal in brief spurts. The cross section also indicates the large pharyngeal cavity surrounded by the gill apparatus. Finally, note the enlarged atrium in the lower right of the diagram and the massive gonadal mass it contains.

Cephalochordates are dioecious and fertilization is external. Gonads can be seen packed in rows along the body wall. Gametes are released into the atrial cavity and gain their exit by way of the atriopore. A free-swimming larva develops, which eventually metamorphoses into an adult.

The significance of the cephalochordates lies in their close affinity to the vertebrates. The two groups share a number of features (notochord, muscle blocks, dorsal nerve tube, tail, similar sensory structures, a ventral aortic heart, and the endostyle for fixing iodine). Accordingly, lancelets are often considered to be a transitional animal between the lower chordates and those that follow.

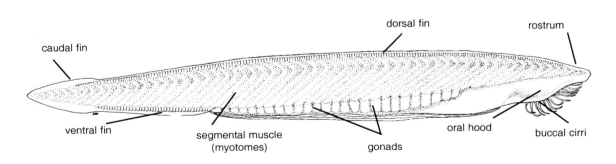

Figure 17.4. External anatomy of the lancelet, *Amphioxus*.

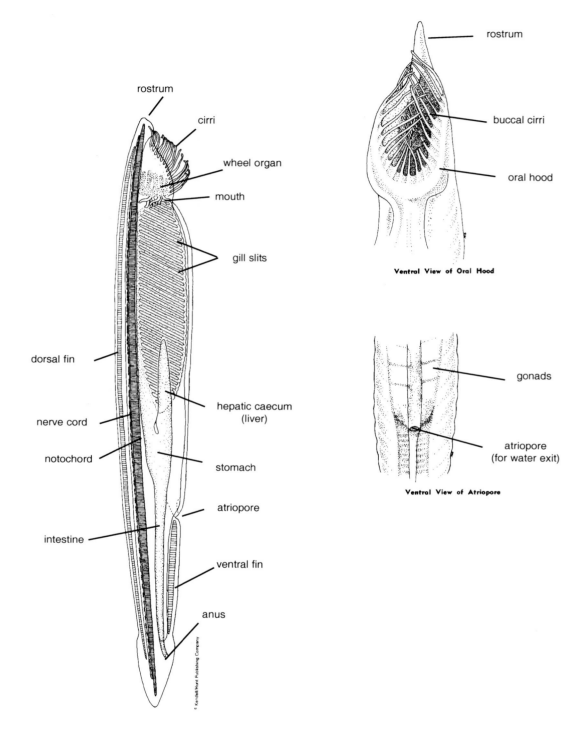

Figure 17.5. External anatomy of the lancelet, *Amphioxus*.

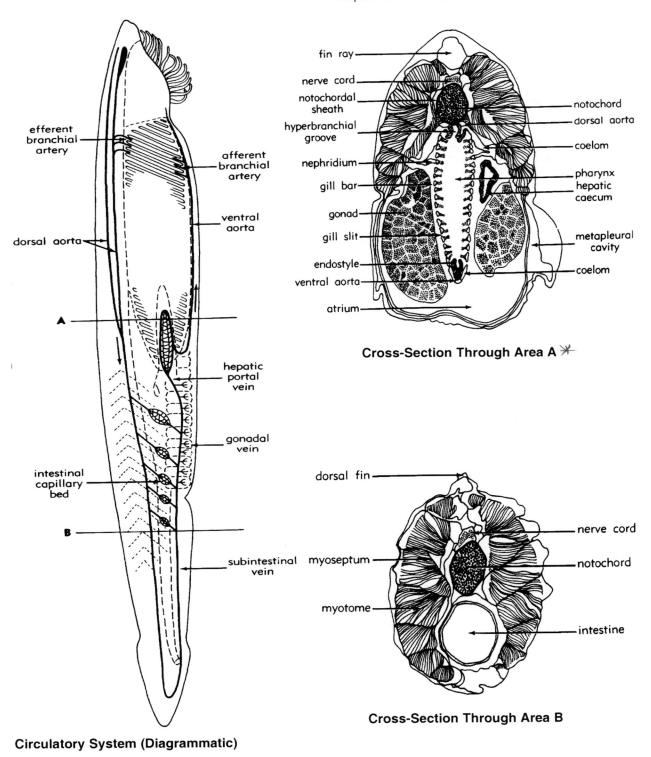

Cross-Section Through Area A ✳

Cross-Section Through Area B

Circulatory System (Diagrammatic)

Figure 17.6. The circulatory pathway and selected cross section of the lancelet, *Amphioxus*.

Chapter 17 Summary

Key Points

1. Chordates are animals that possess a notochord for support, a dorsal, hollow nerve cord, pharyngeal gill slits for respiration, a subpharyngeal gland to fix iodine, a ventral heart, and a postanal tail.

2. Biogenesis is a rather outdated concept that relates an animal's individual development to its evolutionary (phylogenetic) history.

3. Sea squirts (also known as tunicates) show their chordate affinity through the presence of pharyngeal gill slits and their larval notochord, tail, and dorsal nerve cord.

4. The small lancelet, *Amphioxus*, belongs to the small subphylum Cephalochordata. It is a tiny filter-feeding fishlike organism with a pronounced notochord, pharyngeal gill slits, and dorsal nerve cord.

5. Because of its similarity to higher chordates, *Amphioxus* is often considered a transitional form between lower chordates and vertebrates.

Key Terms

Amphioxus
biogenesis ("Biogenetic Law")
cirri
dorsal nerve cord
endostyle
excurrent siphon
incurrent siphon
myotomes
notochord
ontogeny
oral hood
pharyngeal gill slits
phylogeny
postanal tail
recapitulation
subpharyngeal gland
thyroid
wheel organ

Points to Ponder

1. What are the common features of chordates?

2. What is a notochord and what is its function?

3. What is the function of the subpharyngeal gland?

4. What is the "Biogenetic Law"? What are its strengths and its weaknesses?

5. What is a sea squirt? Why is it a chordate?

6. Describe the cephalochordate *Amphioxus*. Why is it considered a forerunner to the vertebrates?

Inside Readings

Alexander. R. M. 1975. The Chordates. London: Cambridge University Press.

Barrington, E. J. W., and R. P. S. Jeffries, eds. 1975. Protochordates. Symposium of the Zoological Society of London, No. 36. New York: Academic Press.

Forey, P. L., and P. Janvier. 1993. Agnathans and the origin of jawed vertebrates. Nature 361:129-134.

Forey, P. L. 1994. Evolution of the early vertebrates. Am. Sci. 82:554-565.

Gee, H. 1994. Return of amphioxus. Nature. 370:504.

Gee, H. 1996. Before the Backbone: Views on the Origin of the Vertebrates. New York: Chapman and Hall.

Gould, S. J. 1989. Wonderful Life:The Burgess Shale and the Nature of History. New York: W. W. Norton.

Linzey, D. 2001. Vertebrate Biology. NY: McGraw-Hill.

Jefferies, R. P. S. 1986. The Ancestry of the Vertebrates. London: British Museum of Natural History.

Jorgensen, J. M., et al. (eds). 1998. The Biology of Hagfishes. New York: Chapman and Hall.

18 Chapter 18

THE FISHES

SUBPHYLUM VERTEBRATA:
The Fishes

Fishes

*T*here are two broad categories of fishes: those with jaws and those without. Any fish **without jaws** belongs to the superclass Agnatha ("without jaws"). All other fish have upper and lower jaws and are members of the Gnathostoma ("with jaws") superclass. Jawed fishes are, in turn, divided into two subgroups: those with a **cartilaginous skeleton** (Class Chondrichthyes) and those with a **bony endoskeleton** (Class Osteichthyes).

I. **Agnatha** (fish <u>without</u> jaws)

II. **Gnathostoma** (fish <u>with</u> jaws)

Chondrichthyes
(fish with cartilaginous skeletons)
Osteichthyes
(fish with bony skeletons)

So much for the taxonomy lesson. What would really be nice to know at this juncture is...what exactly is a fish? It's a name we use all the time, but what does it mean? First of all, whatever they are, they are aquatic vertebrates. That implies they have a dorsal spinal cord encased in a series of vertebrae to protect that vital cord and to provide axial support for the entire animal. In other words, they have a vertebral column. All fish also have a head with a brain enclosed within a cranium—a braincase; thus, they are cephalic. Furthermore, since fish are aquatic, most have a stream-lined shape, pointed at both ends and fat in the middle. A streamlined body glides more easily through the water, reducing friction and making locomotion much easier. Finally, fish are bilateral and usually have two pairs of appendages, one set forward and the other toward the rear.

The problem with this description, however, is that it's too broad; it covers too much ground. In fact, the animal I just described applies equally well to a wide variety of creatures, none of which are fish: seals, walruses, whales, porpoises, sea snakes, frogs, and penguins. In reality, the term fish is too encompassing; it's too ambiguous. Applying the term fish to include everything from lampreys to sharks to sea horses is very

Bait the hook well: this fish will bite.
- Shakespeare

much like letting "bug" refer to any creeping critter: spiders, cockroaches, flies, centipedes, roly-polys, and insects.

The term is not only vague, it's taxonomically useless. What exactly is a *star*fish? Or what is a *shell-*fish? a *cuttle*fish? a *cray*fish? How about a *jelly*fish? None of those fit the description above, yet we call them fish. You see, the term has little taxonomic meaning. It has little use beyond popular jargon. However, the term is so entrenched in our common-day language that we have little choice but to continue to use it generically; but keep in mind that it has no real taxonomic significance.

Figure 18.1. The term fish applies to many kinds of animals.

The Jawless Fish: Agnathans

There is reasonable fossil evidence to suggest that the earliest fishes, which go back to the Cambrian Period some 540 million years ago, were heavily armored, jawless agnathans. Interpretation of that fossil record further supports the likelihood that these animals lived in fresh water and were bottom-feeders. Agnathans were, therefore, most likely the ancestors of all other vertebrates.

Whether they were widely distributed during that ancient time is difficult to determine. But, we do know that only two forms have survived to the present day: the lampreys and the hagfish. They are both round, slender animals that **lack jaws**, both have **naked** (scaleless) **skin**, they have a **cartilaginous endoskeleton**, and both **lack paired appendages**.

Lampreys are rather snakelike in body form and all are parasitic on other fishes. They attach with suction from a **buccal funnel** to the outside surface of a larger fish. Using horny teeth that are well placed on their tongue, they rasp a lesion in the host's flesh, secrete an anticoagulant saliva, and feed on tissue and fluids, particularly blood, quite often killing the host in the process. In fact, some years ago the sea lamprey invaded the Great Lakes and virtually decimated the commercial fishing industry. Successful control measures were instituted in the 1950s which have somewhat reduced the lamprey population and their damage.

The habitat of adult lampreys consists of large bodies of water: seas or the Great Lakes. However, they don't reproduce in open water but seek out swift-moving streams, instead. Sexually stimulated adults migrate long distances to the very stream of their origin to breed. The female uses her suckerlike funnel to attach to a solid object, such as a rock, while the male grips the back of her head using his buccal funnel. He then entwines his body around hers and the two release their gametes into the water where fertilization occurs. The female buries the eggs in the stream bottom and then the adults die. Lampreys, therefore, reproduce only once in their lifetime. The larvae hatch and spend about three years in the stream before they metamorphose into adults. At that time, they make their way to the open waters.

Hagfish, the other existing agnathan, differ considerably from lampreys. They are exclusively marine animals surviving as predators or scavengers, rarely as parasites. Another feature that sets them apart is slime. Their skin is equipped with a battery of epidermal mucus glands that exude a copious lather of slime, so much so that fishermen usually cut a fishing line rather than to try to handle a hagfish.

Hagfish have one more trait that is so curious that it warrants mention. Although they feed primarily on small invertebrates and carcasses that drift down from above, they will, if given the opportunity, bore their way through the body of a living fish, or may even enter its mouth, begin feeding, and actually devour it from the inside, leaving only a ghostly bag of bones!

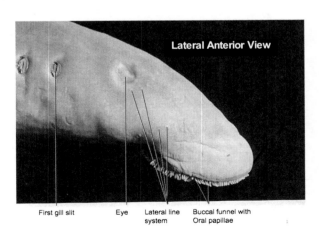

First gill slit   Eye   Lateral line system   Buccal funnel with Oral papillae

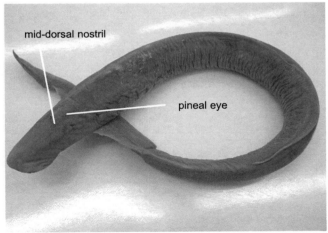

mid-dorsal nostril   pineal eye

Figure 18.2. Agnathans, such as this lamprey, are jawless, scaleless, limbless chordates. Note the external gills slits, the oval buccal funnel lined with teeth, and the mid-dorsal nostril.

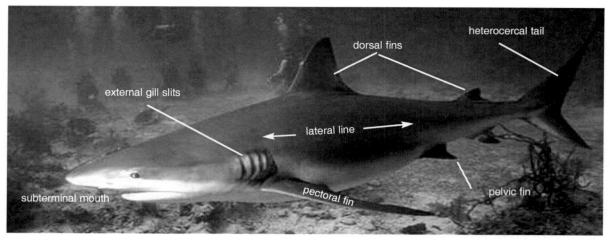

Figure 18.3. External profile of a shark. Note its streamlined shape, location of the mouth (subterminal), the two-lobed, heterocercal tail, and the countershading coloration consisting of dark above and light below.

Class Chondrichthyes

(Chondro = cartilage + ichthyos = fish)

Over the last decade or so few animals have captured the attention of the general population more so than sharks and their relatives. Almost everyone is captivated by tiger sharks, manta rays, stingray, whale sharks, and of course, the great white. It may be their incredible symmetry, their absolute efficiency as the sea's top predators, their mysterious mannerisms, or they may just evoke some primeval fear in our subconscious. Whatever the appeal, or the repulsion, we are, without question, fascinated by sharks.

Sharks and their kin are primitive in their design and have changed little since they first patrolled the Devonian seas some 375 million years ago. The group is characterized by the following set of features:

(1) They have a cartilaginous skeleton; bone is entirely missing.

(2) Unlike the agnathans with their naked, slimy skins, sharks and their kin have a scaly surface consisting of microscopic **placoid** scales. Scales of this sort have posteriorly directed barbs. This can be easily tested by first rubbing your hand along the surface of a shark toward its tail. You will feel the smooth texture that permits frictionless movement through the water. But if you caress in the opposite direction (toward the head), you will detect roughness, something akin to sandpaper. Scales provide an armor-like protection and also cancel the need to cover the body in a slimy mucus coating.

(3) The teeth are all alike, a condition termed **homodonty**. They occur in successive rows and are constantly replaced when lost or shed. The teeth are, in fact, derived from the placoid scales that cover the body.

(4) Most sharks are **fusiform** in shape (tapered at both ends) with prominent dorsal, pectoral, pelvic, and caudal fins (Figure 18.3). Note that in typical sharks the vertebral column bends upward at the tip so that the caudal fin is **heterocercal** (a larger dorsal and smaller ventral lobe).

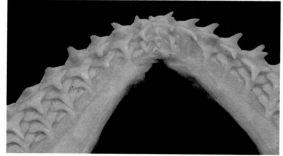

Figure 18.4. Shark teeth (bottom photo) are modified placoid scales (top) which cover a shark's surface.

(5) Members of this group lack flotation devices, such as the swim bladder so common in higher fish; sharks, instead, have a large liver that secretes **squalene** (an oily substance) that helps to keep the animal buoyant in the water. Their blood also contains substances that aide in flotation.

(6) Sharks possess a **spiral valve** within the intestine that slows the progress of food and also increases the surface area for digestion.

(7) Males possess elongated structures attached to their pelvic fins, called **claspers**, to facilitate mating.

(8) Fertilization is internal and three developmental patterns are used:

   a. **ovoviviparity** - the embryo develops within the female's uterus but is nourished by a yolk sac.

   b. **viviparity** - the embryo develops within the female's uterus and is nourished directly by the female.

   c. **oviparity** - the embryo is deposited in leathery egg cases (often called mermaid's purses, see below) where it develops outside of the mother.

(9) External gill slits open individually on the body surface (see below).

(10) The mouth of a shark is subterminal (directed ventrally), so that to feed it must twist the forward part of its body upward. Also, the upper jaw is movable to assist in this maneuver.

(11) Most sharks are active predators of fish, seals, and other aquatic prey. However, skates and rays (below) are flattened, bottom-dwelling species that either scavenge the ocean floor or feed on crustaceans, mollusks, small fish, or plankton.

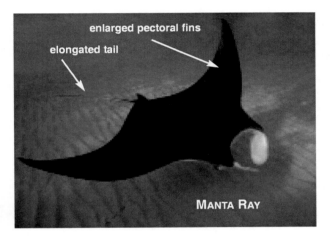

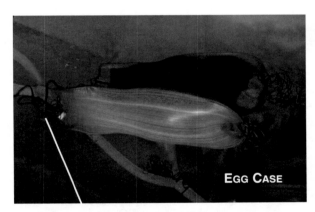

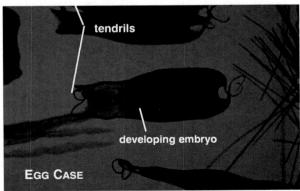

Figure 18.5. Oviparous sharks often deposit the embryo within egg cases such as the "mermaid purses" (above left). Note the tendrils at each end for attachment to vegetation or other substrate. Although most sharks are streamlined, the skates and rays (above right) are flattened bottom-dwellers with elongated tails.

(12) The shark heart consists of four compartments, but only two are actual contracting chambers (see next page).

(13) Sharks possess a unique organ, the rectal gland, which is instrumental in ion balance, primarily to regulate salt content in the blood.

(14) Sharks locate prey through vision (at close range; 25 ft. or so) or by sense of smell (distant   locators; up to a mile). They also possess special senses enabling them to detect subtle changes beneath the water surface. Most notably, sharks and their relatives possess a well-developed **electro-receptive system** for this purpose. It consists of:

    a. special organs called **neuromasts**, which are located along the side of the shark in what is termed the **lateral line system**. These mechanoreceptors have small hairs embedded in gel-filled pits. Any slight vibration in the water, such as that given off by a struggling animal, causes hairs to be displaced, which triggers an electric impulse to the brain. The shark can then determine the exact direction and distance of a wounded or struggling animal by the degree and angle of hair displacement.

    b. other sense organs, the **ampullae of Lorenzini**, are distributed over the head and snout. These function in much the same way as neuromasts but instead of detecting vibrations, they serve as unique electroreceptors to detect slight electric fields given off by potential prey.

(15) Sharks are generally **countershaded** in their coloration. With a light underbelly and a dark dorsal surface, it is difficult for prey (or predators) to see the animal either from above or from beneath.

Figure 18.6. Four sharks demonstrate the effects of countershading. Note how the animal tends to blend into the background either from above or below. The peculiar head shape of the hammer-head shark shown in the photo to the left aids in detection of prey. Electroreceptors spread along the head flanges gives precise information regarding prey location.

The Shark Circulatory System

Sharks, as with all fish, possess a *closed circulatory system* consisting of a heart, arteries, capillaries, and veins, each with its specific function. The heart's purpose, for example, is to contract and thereby provide the necessary pressure to propel blood throughout the body. Arteries carry blood away from the heart; veins return blood to the heart. Capillaries are tiny vessels connecting small arteries and veins. They serve as the site of exchange between individual tissue cells and the circulatory system. For instance, oxygen, nutrients, hormones, and other materials are transferred into cells from capillaries while carbon dioxide and nitrogenous wastes are carried away by capillaries. Since exchange can only occur by way of capillaries, *they are considered the functional unit of the circulatory system.*

The Circulatory Pathway

Even though the shark heart consists of four individual compartments, it is nonetheless considered a *two-chambered heart.* Each of the four compartments store blood but only two (the true chambers) can contract to move blood forward. Blood returning from the body first enters the heart into the **sinus venosus,** a rather thin, triangular-shaped storage compartment. From the sinus venosus, blood next passes into the **atrium,** which is another saclike area that has a thin muscular wall. The atrium contracts to move the blood into the

thick, muscular **ventricle**. It is then the job of the ventricle to provide the force to propel blood throughout the body. Blood finally exits the heart through a cone-shaped compartment, the **conus arteriosus**. The passage of blood through the shark heart thus follows this route: *(1) sinus venosus, (2) atrium, (3) ventricle,* and *(4) conus arteriosus.*

Note that it is not possible for the composition of blood to change as it passes through the heart. Whatever levels of oxygen and carbon dioxide that existed in the blood when it entered the heart remains unaltered as it leaves. The only difference is in pressure, which is increased dramatically due to the contraction of the ventricle.

Blood is carried from the heart by way of a short artery, the **ventral aorta**, which sends branches (the afferent branchial arteries) to each of the gill arches. As blood courses through gill capillaries, carbon dioxide is released and oxygen is accepted. Thus, blood leaving the gill area has changed considerably from that which entered. From the gills, blood collects into a series of efferent branchial arteries, which converge on the largest artery of the body, the **dorsal aorta.** The dorsal aorta runs lengthwise along the spinal column to give off a series of arteries to various organs and tissues. A similar series of veins returns blood from the organs to the heart.

The table below lists some of the more important arteries and the tissues they supply.

Selected Arteries*	Structures Supplied
ventral aorta	afferent branchial arteries
afferent/efferent branchial arteries	gill arches and gills
dorsal aorta	major supplier to other arteries
internal carotid artery	transports blood to the head, primarily the brain
coronary artery	supplies heart muscle
subclavian artery	transports blood to pectoral fins
celiac artery	major supplier to internal organs (gonads, stomach, and intestine)
gastric artery	supplies the stomach
hepatic artery	supplies the liver
anterior mesenteric artery	intestine (spiral valve of the ileum)
gastrosplenic artery	spleen and intestine (ileum)
posterior mesenteric artery	transports blood to the rectal gland
iliac artery	supplies the pelvic fin
caudal artery	carries blood into the tail region

* Although arteries are being emphasized, keep in mind that there is a venous system with an equal number of veins, many of which carry the same name as corresponding arteries (e.g., hepatic artery / hepatic vein).

Under Pressure: A Shark's Dilemma

One of the most important functions of the circulatory system is to deliver vital nutrients to the body's cells. These nutrients are needed to supply energy for the cell's various activities. To adequately deliver and use nutrients, two things are needed: *blood pressure* and *oxygen*.

Blood leaving the heart is under considerable pressure supplied by the beating ventricle, but it's deficient in oxygen. Blood is, therefore, shunted immediately to the gills to pick up the vital supply of oxygen. However, as it passes through the elaborate capillary network of the gills, it loses much of its pressure; up to 30% in some sharks. The problem is, and it's a major problem, the blood has yet to pass through the remainder of the body. There are literally hundreds, if not, thousands of miles of blood vessels that blood must travel through before returning to the heart where pressure can be re-established.

In higher animals, especially birds and mammals, once blood leaves the oxygen center (lungs) it returns to the heart for another pulse of pressure; but this is not the case in fish with their two-chambered hearts.

Sharks and other fish have at their disposal a number of ways to combat or compensate for their sluggish circulation. First, they exist in a medium (water) that is partially buoyant. Since the density of water counters the pull of gravity, a shark doesn't expend as much energy as a terrestrial animal does to overcome gravity's downward pull. As a result, sharks and other fish can stay suspended, just floating or resting on the bottom, for long periods without spending much of their vital energy. In addition, the squalene oil in shark livers and substances in the blood help to counter what little gravity there is and to maintain a neutral buoyancy. This suspended animation helps them conserve energy, which can then be used in explosive fashion to escape predators or to capture prey.

Second, fish, in general, are designed to move efficiently through the water. Overlapping scales, streamlined shape, and mucous skin secretions overcome friction as they swim forward. Shark's large tail fins also provide considerable force at little energy cost to propel the animal swiftly. Sharks (and other fish) also have muscular systems intended to gain the maximum efficiency for the least effort. Muscle bundles are aligned in a zigzag fashion, which gives maximal forward thrust as the muscle contracts. Finally, their muscles are proficient at using oxygen to convert nutrients into usable energy. In this regard, "a little oxygen seems to go a long way."

Figure 18.7. The streamlined shape of fish and their tendency to remain suspended for long periods help to compensate for the lack of blood pressure in their circulatory system.

Senses and the Nervous System

*A*ll higher animals, sharks included, have a nervous system divided into two parts: a central and a peripheral portion. The **central nervous system** (CNS) consists of the **brain** and **spinal cord**, while the **peripheral nervous system** (PNS) contains nerves leaving the CNS to innervate all tissues and organs.

In addition to the features indicated above, the shark also has a number of general (e.g., touch, pain, thermal) and special sensory structures. It must rely upon these senses to locate food, escape enemies, find mates, and be aware of its general environment. These special senses include organs of:

(1) **Olfaction**. For a shark, the sense of chemoreception is probably the most important and the keenest. Through olfaction, sharks can detect extremely small amounts of water borne chemicals over a considerable distance. However, that ability is not limited to sharks alone. Other fishes, such as lampreys and salmon, can relocate their streams of origin, sometimes over hundreds of miles, through chemoreception alone. Although chemical receptors are scattered over much of the animals' surface, they are especially concentrated within a pair of small chambers, the olfactory sacs. Unlike our nostrils, which help to conduct air all the way to the lungs, the nostrils of sharks lead only to these blind sacs. Water circulates through the nostrils to contact receptors in the sacs and the olfactory information is then carried to the brain where "odors" are interpreted.

(2) **Vision**. There is some debate over the importance of vision in sharks. In some species, the eyes are relatively large and surely assist the animal at medium and close distances. In other species, the eyes are much smaller and probably serve only at close range.

(3) **Hearing**, **Balance**, and **Equilibrium**. The center for all three of these functions is located in the inner ear. Obviously, sharks lack external ear lobes, thus, sound waves are "captured " by the body surface and then transmitted through the cartilaginous skeleton to semicircular canals in the inner ear. Balance and equilibrium are controlled by communication between receptors in the inner ear and the brain.

(4) The **Lateral Line** and **Electroreceptive Apparatus**. The function of these two special senses is outlined below, but for a more detailed description you may refer back to page 404. Special organs called **neuromasts** are dispersed along a shark's midlateral line from head to tail. These mechanoreceptors consist of tiny hairs, which are imbedded in a gel-like material. Slight vibrations, such as might issue from a struggling animal, cause displacement of the hairs. That, in turn, elicits electrical signals that are sent to the brain where information on prey location is processed. Additional sense organs, called the **ampullae of Lorenzini,** are concentrated on the head and snout. These function in much the same way as neuromasts, but instead of detecting vibrations they are able to discern subtle electrical fields that all animals emit. As a result of information processed by these two senses, sharks are expert at detecting far-removed or hidden prey.

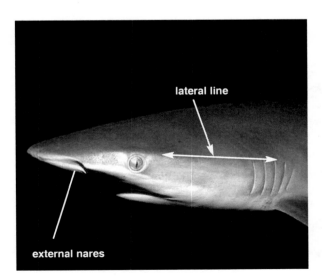

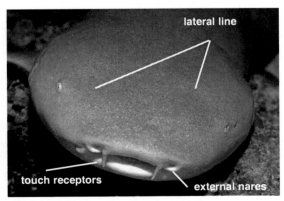

Figure 18.8. Fish have a battery of receptors available to sense their environment. Eyes are effective at medium or short range while olfactory sensors can detect odors a mile or so away. In addition, the lateral line system is sensitive to vibrations and electrical fields.

The Shark Brain

The brain of adult sharks is basically divided into three parts, each governing a separate range of functions. From anterior to posterior these are:

The Forebrain
(the Prosencephalon)

This is the anterior-most region of the brain. It consists of the olfactory bulb and olfactory tract to process incoming chemical information. These terminate on the olfactory sac where odors are detected. Immediately behind the olfactory apparatus is the small cerebrum appearing as two tiny cerebral hemispheres. The cerebrum is the center for cognitive functions in higher animals (thought, reasoning, problem solving, etc.). However, in the shark they are rather undeveloped and thus receive minimal input. A shark functions primarily on the basis of instinctive reactions; it is truly not a "thinking" animal.   Posterior to the hemispheres is a region of the forebrain known as the diencephalon. It is a regulatory center and houses the thalamus (major relay or linkage center in the brain), the hypothalamus (appetite center, temperature regulation), pineal body (see *Inside View 18.1* page 411), and the pituitary gland (major endocrine gland for hormone release).

The Midbrain
(the Mesencephalon)

The middle portion of the brain is represented by a pair of swellings, the optic lobes, which receive visual input from the eyes. These lobes also get sensory information from other areas of the brain and transmit motor (movement) impulses outward. The optic lobes thereby serve as the major integration center similar to the cerebral cortex in higher animals.

The Hindbrain
(the Rhombencephalon)

The most posterior part of the shark brain contains two noteworthy structures. Located dorsally and just behind the optic lobes is the relatively large, oval cerebellum. It is the center for muscular coordination so important to a shark. It also receives sensory information from the inner ear and lateral line system. Thus, the cerebellum is instrumental in equilibrium and balance as well as in responding to electrical and vibratory information picked up by the lateral line. The second notable structure of the hindbrain is the medulla oblongata. It is the most posterior part of the shark brain located between the cerebellum and the spinal cord. In contains major sensory and motor pathways for information to and from the spinal cord. For example, if the brain signals for a muscle, or maybe a fin, to move, that information would pass through the medulla. However, the medulla also has important basic or "vegetative" functions to perform as well, such as the control of heart rate, swallowing, and breathing.

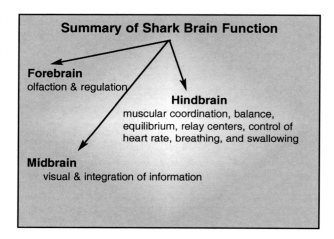

Summary of Shark Brain Function

Forebrain
olfaction & regulation

Hindbrain
muscular coordination, balance, equilibrium, relay centers, control of heart rate, breathing, and swallowing

Midbrain
visual & integration of information

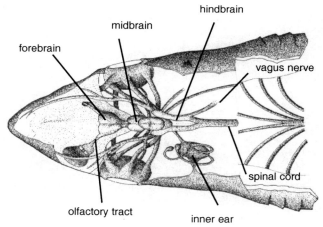

Figure 18.9. The shark brain showing the three major regions: forebrain, midbrain, and hindbrain.

Cranial Nerves

Exiting the shark brain at several points are twelve paired nerves that serve a variety of functions. Some of the major ones are listed below (the Roman numerals represent the formal number given to each respective nerve).

Selected Cranial Nerves and Their Functions

I. **Olfactory** - a sensory nerve transmitting information regarding the sense of smell

II. **Optic** - a sensory nerve transmitting visual signals

III. **Oculomotor** - a motor nerve controlling selective eye movements

IV. **Trochlear** - another motor nerve controlling selective eye movements

V. **Trigeminal** - a motor and sensory nerve responsible for jaw movement and sensation of skin in the head area

VI. **Abducens** - another motor nerve to help move the eye

VII. **Facial** - transmits sensations from the lateral line and ampulla of Lorenzini; controls jaw movement; sense of taste

VIII. **Auditory** - transmits information regarding hearing and equilibrium

X. **Vagus** - sense of touch and movement, controls heart contraction and digestive system functions

---

Class Osteichthyes

(Osteo = bone + ichthyos = fish)

The sharks and more advanced fish belonging to the class Osteichthyes have much in common. Their circulatory and nervous systems are virtually identical. Their shape, muscular systems, and body coverings help to assure smooth movement through the water. Much of their behavior is similar.

However, there are also a number of differences between the two groups. For example, all sharks have a cartilaginous endoskeleton—no bone is present. Osteichthyes, on the other hand, have at least some bone in their skeleton. In fact, the entire skeleton is ossified (i.e., bony) in most cases. Also, while both have scales, the type of scales differs considerably. The minute placoid scales of sharks are replaced by three different varieties in bony fish. The more primitive fish, such as sturgeons and gars, have a large, inflexible **ganoid** scale, whereas, advanced fish either possess a **cycloid** (e.g., bowfin and cod) or **ctenoid** type (perch, bass, and most other bony fish). A rather bizarre situation occurs in certain fish, such as flounders, that have cycloid scales on their ventral surface and ctenoid scales dorsally! Both cycloid and ctenoid scales are flexible and designed to reduce drag while swimming. They, therefore, help the fish to glide more smoothly through the water.

Another difference between sharks and bony fish has to do with their method of maintaining buoyancy. As we have discussed, sharks use a combination of liver, oil, blood chemistry, and fins to retain their position in the water. Bony fish, on the other hand, rely almost exclusively on the presence of a gas-filled chamber, the **swim bladder**, to prevent sinking and to sustain proper flotation.

Both sharks and bony fish must rely on water moving past the gill surfaces to exchange respiratory gases. Both can accomplish that by swimming with the mouth open and just allowing the water to flow over the gills. Bony fish, however, are much more adept at breathing while suspended (not moving) than are sharks. By contracting muscles in the gill chamber, they can cause water to stream across the gills. Sharks can do that, too, but not nearly as efficiently. You will also note that the gills of bony fish are protected by a shieldlike structure, the **operculum**, which is absent in sharks.

Sharks and their relatives are predominately marine organisms. There are only a few species found in freshwater habitats. Bony fish, in contrast, have mastered both fresh and saltwater environments. Since they occupy a wider range of niches, they are, therefore, much more diverse in body form than the class Chondrichthyes. There are approximately 800 known species of sharks patrolling the world's waterways, and while they vary in size and behavior, they are sparse in comparison to bony fish which number a little over 20,000 species. That's just over 25 species of bony fish for every species of shark!

sharks and bony fish are very much alike, yet very different

Figure 18.10. Some examples of the body design in bony fish, class Osteichthyes. **A,** a clown fish; **B,** an eel; **C,** a scorpion fish; **D,** a flying fish and **E,** a sea-horse.

An Inside View 18.1

"The Third Eye"

If you look carefully on the top of a lamprey head, just posterior to the entrance to the nostril, you will detect a light-colored patch of skin. This is a sensory structure known as the **pineal eye**, or by its common name, "the third eye." In the lamprey, the pineal eye plays a number of roles. We know, for example, that it is **photosensitive** (thus, its "eye" title) and, as such, helps in establishing the animal's **circadian** (daily) and **seasonal rhythms.** Through information gained about day length (photoperiod) and light intensity, an animal can adjust its behavior accordingly. For instance, it can be active at night and inactive during daylight hours, or vice versa.

The third eye also functions as a trigger for metamorphosis from a larva into an adult. Lamprey larvae, called ammocoetes, live for up to seven years buried, tail first, in their spawning stream. There they feed much like an *Amphioxus*, by filtering food substances from the water. It's due to action of the pineal eye that the ammocoete larva is eventually transformed into an adult whereupon it swims out to sea to spend its adult life.

We know, too, that the pineal eye governs the distribution of skin pigments and, as a result, can initiate color changes. The animal can, for instance, match its surroundings as a form of camouflage, an important method of defense. This is true for both the adult and larval lamprey.

The final function performed by the lamprey pineal eye has to do with the onset of sexual activity.   It seems that the third eye determines the best time of the year to reproduce. The process begins with orientation of adults toward the very freshwater stream where they, years before, began their life. The animals then migrate, often over hundreds of miles, to their spawning streams. There they mate, lay eggs, and then die. Since lampreys die shortly after their first and only mating, we can presume that migration is not a learned behavior but is genetically controlled.   Hormones secreted by the third eye influence the timing of migratory behavior.

In summary, the lamprey third eye is involved primarily with sexual maturation, rhythmic behavior, and integumental color changes.   It is certainly an organ of great intrigue. In fact, it causes me to wonder if other animals have such structures.

A scrutiny of the nervous systems of other animals indicates that many, do indeed, have a similar structure.   As a matter of fact, virtually all animals above the agnathan grade have, at least, a pineal gland, if not an actual pineal eye (only crocodiles, armadillos, sloths, and anteaters lack either). In fact, some primitive lizards (Tuataras) have a noticeable third eye sitting atop their heads, with a lenslike apparatus. It seems, then,   that practically all amphibians, reptiles, birds, and mammals have a pineal organ of some sort.

So what about us? Do humans possess a third eye?   Well, yes...and no. We are not equipped with a pineal eye, specifically, but instead we have a **pineal gland** perched on the roof of the midbrain in the vicinity where the pineal eye is located in lower animals. It's about the size of a pea and has neural connections to the "real" eyes.   In humans, the pineal gland secretes a hormone, **melatonin**, which helps to regulate   our   sleep   patterns. Scientists also suspect that it serves as an important triggering mechanism to the onset of puberty.   In other words, this photoreceptive/endocrine organ helps us to maintain our daily rhythms and sleep patterns, and to transform us into adulthood—very much the same model seen in lampreys!

Chapter 18 Summary

Key Points

1. The subphylum Vertebrata includes all remaining animals: jawless fish, jawed fish, amphibians, reptiles, birds, and mammals.

2. Although the term fish is a general term with little taxonomic validity, it is useful with reference to any streamlined, aquatic, gill-breathing vertebrates.

3. The jawless fish (agnathans) include the parasitic lamprey and the predominately scavenging hagfish. Both lack jaws, scales, paired appendages, and possess a cartilaginous skeleton.

4. The class Chondrichthyes includes the jawed, cartilaginous fish such as sharks, rays, skates, and related forms.

5. The cartilaginous fish, in addition to a skeleton of cartilage, also possess placoid scales, uniform teeth (homodonty), a heterocercal tail, an internal spiral valve, claspers in the males to grasp the female during copulation, external gill slits, a subterminal mouth, a rectal gland to regulate ion concentration, and an effective electroreceptive system using ampullae of Lorenzini and a lateral line system.

6. Sharks have a two-chambered heart that pumps blood directly to the gills. Blood pressure drops drastically as it passes through the gills and thus pressure remains low throughout the remaining circulatory system.

7. Members of the class Chondrichthyes, and sharks in particular, have well-developed senses of olfaction, vision, hearing, balance, and mechanoreception (electroreceptive system).

8. The shark brain consists of three parts: (1) the forebrain contains centers for olfaction and regulation; (2) the midbrain with areas devoted to vision and overall integration; and (3) the hindbrain, which controls muscular coordination, balance, equilibrium, heart rate, breathing, and swallowing.

9. Cranial nerves exit from the brain and innervate a number of sensory and motor activities.

10. The bony fish (class Osteichthyes) include some primitive and all advanced fish. Instead of placoid scales, they have either ganoid, ctenoid, or cycloid scales covering their bodies.

11. Bony fish have an internal swim bladder to help maintain buoyancy, and a bony shield, the operculum that protects the fragile gills.

12. Most vertebrates possess an extension of the brain, the pineal body, as a photoreceptor. It governs such things as color change, metamorphosis, and sexual maturation.

Key Terms

afferent branchial arteries
ammocoetes
ampullae of Lorenzini
atrium
buccal funnel
Cambrian Period
cartilaginous endoskeleton
central nervous system
cerebellum
cerebrum
claspers
circadian rhythms
conus arteriosus
countershading
cranial nerves
ctenoid scales
cycloid scales
diencephalon
dorsal aorta
efferent branchial arteries
electroreceptive system
forebrain
ganoid scales
hindbrain
heterocercal tail
homodonty
hypothalamus
lateral line system
medulla oblongata
midbrain
neuromasts
olfaction
olfactory sacs
operculum
optic lobes
oviparous
ovoviviparous
peripheral nervous system
pineal body
pineal eye
pituitary gland
placoid scales
rectal gland
sinus venosus
spiral valve
squalene
subterminal mouth
swim bladder
thalamus
ventral aorta
ventricle
viviparous

Points to Ponder

1. What features characterize a "fish"?

2. What are the characteristics of a lamprey?

3. How does a lamprey feed?

4. What features characterize members of the class Chondrichthyes?

5. How do sharks remain buoyant in the water?

6. What is the function of a shark's spiral valve?

7. What are ways that shark embryos can develop?

8. What is the function of the shark's rectal gland?

9. What are the parts of a shark's electroreceptive system and how does it work?

10. Trace the route of blood through a shark's heart.

11. What is the major deficiency in the shark's circulatory system?

12. What senses are important to a shark?

13. What are the three general areas of a shark brain and what is their function?

14. How do bony fish (i.e., class Osteichthyes) differ from sharks?

15. What type of scales may be found on a bony fish?

16. What is a pineal eye? Where is it? What functions does it perform?

Inside Readings

Agnathans

Alexander, R. M. 1975. The Chordates. London: Cambridge University Press.

Forey, P. L., and P. Janvier. 1993. Agnathans and the origin of jawed vertebrates. Nature 361: 129-134.

Gee, H. 1996. Before the Backbone: Views on the Origin of the Vertebrates. New York: Chapman and Hall.

Jefferies, R. P. S. 1986. The Ancestry of the Vertebrates. London: British Museum of Natural History.

Jorgensen, J. M., et al. (eds). 1998. The Biology of Hagfishes. New York: Chapman and Hall.

Linzey, D. 2001. Vertebrate Biology. New York: McGraw-Hill.

Gnathostomes

Bone, Q., and N. B. Marshall. 1982. Biology of Fishes. London: Blackie.

Bruton, M. N. (ed). 1990. Alternative Life-History Styles of Fishes. Dordrecht, Netherlands: Kluwer Academic Publishers.

Coleman, N. 1991. Encyclopedia of Marine Animals. New York: Harper Collins Publishers.

Eastman, J. T., and A. L. DeVries. 1986. Antarctic fishes. Sci. Am. 255(5):106-114.

Ellis, R. 1996. Deep Atlantic: Life, Death, and Exploration in the Abyss. New York: Alfred A. Knopf.

Evans, D. H., ed. 1993. The Physiology of Fishes. Boca Raton: CRC Press.

Hara, T. J. 1992. Fish Chemoreception. London: Chapman and Hall.

Helfman, G. S., et al. 1997. The Biodiversity of Fishes. Malden, MA: Blackwell Science.

Jameson, Jr., E. W. 1981. Patterns of Vertebrate Biology. New York: Springer-Verlag.

Klimley, A.P. 1994. The predatory behavior of the white shark. Am. Sci. 82:122-133.

Lee, D. S., et al. 1981. Atlas of North American Freshwater fishes. Raleigh: North Carolina State Museum of Natural History.

Moyle, P. B. 1993. Fish: An Enthusiast's Guide. Berkeley: University of California Press.

Nelson, J. S. 1994. Fishes of the World. New York: Wiley.

Partridge, B. L. 1982. The structure and function of fish schools. Sci. Am. 246(6):114-123.

Paxton, J. R., and W. N. Eschmeyer (eds.). 1995. Encyclopedia of Fishes. New York: Academic Press.

Radinsky, L. B. 1987. The Evolution of Vertebrate Design. Chicago: The University of Chicago Press.

Thompson, K. S. 1990. The shape of a shark's tail. Am. Sci. 78:499-501.

Webb, P. W. 1984. Form and function in fish swimming. Sci. Am. 251(1): 72-82.

Wootton, R. J. 1990. Ecology of Teleost Fishes. London: Chapman and Hall.

Wu, C. H. 1984. Electric fish and the discovery of animal electricity. Am. Sci. 72:598-607.

Chapter 18

Laboratory Exercise

The Fishes

Laboratory Objectives

✓ to identify characteristics of the subphylum Vertebrata,
✓ to locate external features of a lamprey, perch, and a shark,
✓ to dissect a lamprey to identify major internal structures and learn their functions,
✓ to dissect a dogfish shark to identify major internal structures and learn their functions,

Supplies Needed

✓ prepared and living vertebrate specimens
✓ preserved specimens of lamprey, perch, and shark
✓ dissecting equipment including bone cutters

Chapter 18
Worksheet

Subphylum Vertebrata:
The Fishes

During this lab, you will explore similarities and differences between the three major classes of fishes:

Class Agnatha ---- Class Chondrichthyes ---- Class Osteichthyes

External Anatomy

I. Class Agnatha: The Lamprey

1. Examine the exterior of a preserved lamprey. Note that it is scaleless, jawless, and limbless, although it does possess a pair of **dorsal fins** and a **caudal fin**. Locate each of these. Next, examine the **buccal funnel** leading to the mouth. The **oral papillae** on the rim of the funnel help to create a suction-like seal as the animal feeds. Note the many rows of **rasping teeth** on the funnel wall and on the large **tongue**. These are used to bore into the flesh of larger fish, which it parasitizes.

2. Along the side of the head, you will find a pair of lidless **eyes** and clusters of **pores** of the **lateral line system**. Also, there are seven pairs of **external gill slits.** The slits open into pouches bearing the gills. On top of the head, you will find a single opening. This is the **middorsal nostril** leading into the **olfactory sac.** Chemoreceptors are located in the sac to detect odors. Just behind the nostril is a whitish patch where the **pineal eye** is exposed.

3. Turn your specimen over and locate a small midventral slit about three or four inches from the tail. This is the **anus**.

4. Finally, you need to expose a section of trunk musculature. Do this by carefully removing a patch of skin (a little larger than a quarter) from the side of your specimen midway between the gill slits and the rear of the animal. Note the *W*-shape arrangement of the muscle bundles called **myotomes**. This pattern maximizes swimming movements by providing strong twisting and turning actions of the torso.

II. Class Chondrichthyes: *Squalus*, The Dogfish Shark

1. Examine the exterior of a preserved shark. First note the streamlined shape and the presence of a number of fins: an anterior and posterior **dorsal fin**, a pair of **pectoral fins**, a pair of **pelvic fins** (all four used as stabilizers during swimming), and the large **heterocercal caudal fin**, used for swimming thrust. If your specimen is a male, there will be a pair of **claspers** associated with the pelvic fins. He uses these to aid in the mating process.   In both sexes, there is an opening between the pelvic fins, the **cloaca**, for the exit of reproductive, digestive wastes, and excretory products. Both of the dorsal fins bear a **poisonous spine** for self-defense. Use a pair of bone cutters to remove both spines.

2. Moving toward the head, note the **external gill slits**. How many are there on each side? You probably counted five. But actually there are six pairs. To locate the sixth pair, you need to look behind the eyes and you will see a pair of rather large holes. These are modified gill slits called **spiracles** and are used as an alternate route for water to enter the pharynx when the mouth is closed or when the animal is feeding. Sharks have minimal ability to active-ly pump water across the gill surface, so to assure an adequate supply of oxygen they swim with their mouths open to force water across the gill surface, a process referred to as **ram ventilation.**

3. Note the **subterminal** (ventrally-directed) mouth. As you pry the mouth open to examine the teeth, be careful because they are sharp. Preserved shark jaws are on display for a closer examination. Anterior to the mouth are a pair of openings, the **external nares** (nostrils). Insert a blunt probe into one of the nostrils and note that it only passes a short distance. The nostril leads to a blind chamber, the **olfactory sacs**, for chemoreception.

4. With your bare hand, rub the surface of your shark. Note the smoothness as you caress toward the tail but the rough texture in the opposite direction. The roughness is due to the presence of **placoid scales**, which help to reduce turbulence during swimming. Examine the skin, or prepared slides, under a microscope to see the micro-structure of these scales.

5. Finally, firmly grip the snout of your specimen between your thumb and forefinger and squeeze, noting the **gel-like** material that exudes from the surface. The gel surrounds hairs in the **ampulla of Lorenzini** which is an electroreceptor that assists in detection of prey. Use a magnifying glass to get a better look at the pores leading to the ampullae.

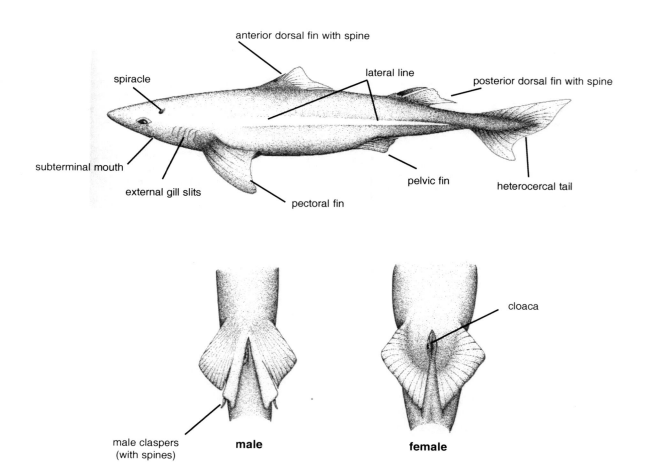

Figure 18.11. External views of the dogfish shark.

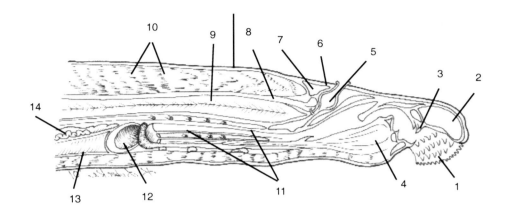

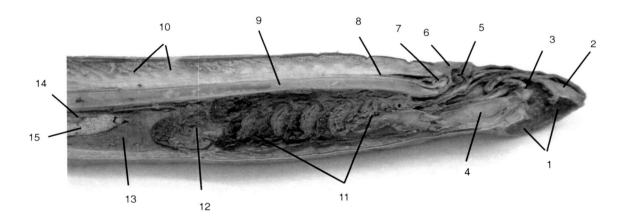

Figure 18.13. Internal anatomy of the lamprey.

1. horny teeth in buccal cavity
2. cranial cartilage
3. tongue
4. lingual muscle
5. olfactory sac
6. pineal eye
7. brain
8. spinal cord
9. notochord
10. myotome (muscle)
11. branchial basket with gills
12. heart in cartilaginous pouch
13. liver
14. kidney
15. egg mass

Internal Anatomy

I. The Lamprey

1. At this time you should make a midsagittal section through the entire lamprey specimen so that there are two equal halves. To do that place your specimen so that the dorsal surface is facing you. With a sharp scalpel, start at the buccal funnel and cut along a mid-dorsal line through the nostril and pineal eye and continuing on to the tail. It's important to stay as close to the midline as possible so that internal organs are properly exposed.

Once you have severed the specimen in two, begin to locate and examine the internal structures.

2. Lampreys have a **cartilaginous skeleton**, which can clearly be seen in the area of the buccal funnel. Find the large **tongue** and note its rasping teeth. It, too, has an internal core of cartilage, as well as powerful **lingual muscles** to provide the force for boring into flesh. The mouth is dorsal to the tongue and by using a blunt probe you can explore the **pharynx** and **esophagus**. The lamprey has no stomach; instead, the esophagus leads directly to the **intestine**, which you can find as a yellowish, somewhat flattened tube passing toward the **anus**.

3. The **gills** and the **internal gill slits** are located within an area called the **branchial basket.** Lampreys breathe by moving water through the mouth, through the internal gill slits, across the gill surface, and out the external gill slits. However, when a lamprey is attached to a host it cannot use this route, so it instead pumps water in and out of the external slits, much like a working bellows. At the posterior end of the basket is a pouch containing the muscular **heart**. Immediately behind the heart is the large arrowhead-shaped **liver**.

4. Posterior to the liver is a pair of elongated straplike **kidneys**, and hanging between them are the **gonads**. If you have a female, she may be filled with **eggs**, which have the appearance of sandlike granules.

5. Now locate the whitish **notochord** running from the head to the tail just dorsal to the branchial basket. Above the notochord is the tubular **nerve cord**. Trace it forward until it terminates in the bulblike **brain**. The whitish area above the brain is the **pineal eye**. Anterior to the pineal eye is the **olfactory sac** communicating with the external nostril.

6. Finally, dissect an eye. Cut into the outlying tissue to expose the marblelike **lens** inside. The lens is the focusing element of the eye. The black lining at the back of the eye is the **retina** where images are received. There may be some pressurized fluid inside of the eye, so be careful not to squirt yourself.

a cluster of lampreys

II. **The Dogfish Shark** - General Organs of the Body Cavity

1. Place the shark so that the ventral surface is facing you. Starting near the **cloaca**, make an incision through the body wall and into the body cavity. Be careful not to go too deep so that underlying organs are not disturbed. Continue the incision forward along the midline until you reach the region between the pectoral fins. At that point you will encounter a cartilaginous brace holding the two fins together. Do not cut through the brace at this time, but instead make two right-angle cuts toward each fin. Now return to the cloacal area and make two similar cuts there. This should allow you to fold back the body wall to expose the coelomic cavity and its contents. Note the smooth, shiny, clear lining of the body cavity, this is the **peritoneum**.

> Note - You will notice that there is a red (and maybe a blue) coloring to many of the internal structures. This is latex (rubber), injected to allow better identification of the circulatory system. We will return to that system later, however, be careful not to cut through any of the circulatory vessels.

2. At this point you need to identify the structures of the body cavity. First note the large, oily **liver**. It is divided into three lobes: right, left, and middle. Find the middle lobe; it is smaller than the other two and somewhat triangular shaped. Along its right edge a greenish sac is attached; this is the **gall bladder**. Leading from the bladder to the intestine is a tube, the **bile duct**, which you can find accompanied by a large vein passing to the digestive system. Beneath the liver is the large J-shaped **stomach**. Find it and note that it bends sharply to the right where it gives rise to the short **duodenum** of the small intestine. Continue to trace the digestive system by following the duodenum, which also makes a sharp posterior turn to become the enlarged **ileum**. The ileum has a number of parallel vessels on its surface and contains the **spiral valve** internally. The ileum leads to the short **colon** from which extends the pencil-like **rectal gland.**

3. Next open the stomach to examine its contents and do the same for the ileum. Inside of the ileum you can observe the winding series of flaps called the spiral valve. What is its function?

4. There are two major organs associated with the digestive system that you need to locate at this time. First, lying on the posterior surface of the stomach-duodenum junction is the large arrow-shaped **spleen**. It is dark colored and easy to locate. If you follow the curvature of the duodenum to the point where it joins the ileum, you will find attached to its surface the whitish, irregular-shaped **pancreas**. Actually, the pancreas occurs as two lobes. The one on the surface of the duodenum is the **ventral lobe.** Now lift the duodenum and examine its dorsal surface for the straplike **dorsal lobe** of the pancreas.

5. The **kidneys** are perhaps the most difficult of the internal organs to locate. To do so, you will need to shove all other structures to one side to expose the roof of the body cavity. The kidneys are narrow, elongated organs that lie on either side of the mid-line in the region near the vertebral column. They are darkish in coloration, straplike in form, and run most of the length of the body cavity. They are actually located beneath the peritoneum, a condition called **retroperitoneal**.

6. A male shark contains a pair of **testes** just ventral to the kidneys and beneath the liver-stomach mass. To expose them, you must lift or shove the stomach and liver to one side. You will see them as oval masses about the size of walnut or a little larger. Leading from the testes (especially in a sexually mature male) is a prominent **ductus deferens** arranged in zigzag fashion on the surface of the kidneys. At their posterior end, near the cloaca, they enlarge into a pair of saclike **seminal vesicles** where sperm may be stored.

7. A female specimen contains a pair of **ovaries** in the same location as the testes (i.e., beneath the liver and stomach mass). A mature female will show **eggs** as rounded objects on the surface of the ovary. Eggs range from pebble-size, to grapelike, to near golf ball size. Each ovary connects to a tubular **uterus**, which, in pregnant sharks may contain small **embryos**. If your specimen is pregnant, you may want to remove one of the embryos with its attached yolk sac. What is the function of the yolk?

GENERAL INTERNAL ANATOMY

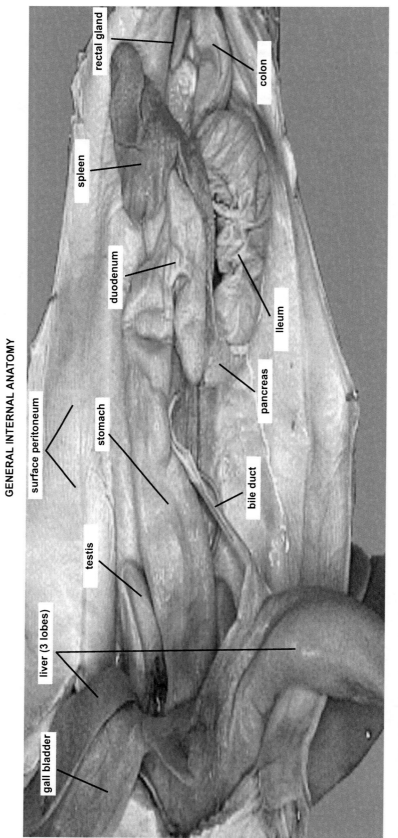

Figure 18.14. The internal anatomy of a shark.

Major Internal Organs & Associated Structures Found in Animals

stomach — stores and churns food; no digestive functions

small intestine (duodenum and ileum)— digests and absorbs food

spiral valve — slows the flow of nutrients; increases surface area for absorption

pyloric ceca* — extensions from the digestive system; increases surface area

large intestine (colon) — resorbs water from food; compacts food to form feces

cloaca — a common opening for the digestive, reproductive, and excretory systems

gall bladder — stores, concentrates, and secretes bile

heart ventricle — major pumping chamber of the heart

heart atrium — membranous chamber; propels blood to the ventricles

sinus venosus — a membranous compartment where blood first enters the heart

conus arteriosus — cone-shaped compartment from which blood exits the heart

bile duct — a passageway for bile from gall bladder to intestine

peritoneum — lining of the body cavity; derived from mesoderm, delineates a true coelom

pancreas — secretes digestive enzymes; secretes insulin

spleen — part of the immune system; provides protection from foreign substances

kidney — removes nitrogenous wastes from the body; helps regulate water balance

urinary bladder* — stores and releases urine

swim bladder* — a flotation device in many fish

gonads — testes and ovaries; produce gametes: sperm and eggs

ductus deferens — a passageway for sperm from testes to the outside

uterus — a sac in which embryos develop

rectal gland — helps to regulate ion balance, particularly salt content of the blood

liver — largest internal organ; has many functions: forms bile; removes toxins; helps with secondary
        digestion; secretes squalene (oil)

* sharks do not possess these features

Shark Circulatory System

1. We will begin our exploration of the circulatory system with the **heart**. To expose it, you will need to open the **pericardial cavity** where the heart is situated. To do that, cut through the cartilaginous support between the pectoral fins. Continue the incision forward to just below the jaw. This needs to be a relatively deep incision, but go carefully as you will not want to damage the underlying heart structures.

2. Once you have exposed the heart, identify its four compartments. The first, and most prominent, portion you will see is the muscular **ventricle**. Connected to it dorsally is a bilobed, thin-walled **atrium**. These are the two contractile chambers of the heart. However, there are two more compartments. The first, the **sinus venosus**, is connected to the atrium and can be seen by lifting the ventricle and pulling it slightly toward the shark's head. This will expose the membranous, somewhat triangular, sinus attached near the liver. The final compartment, the **conus arteriosus** leaves the ventricle anteriorly as an enlarged cone-shaped tube.

3. The remainder of the circulatory system consists of arteries and veins. For our purposes, we will only study the arteries. This is best done by starting at the heart and carefully dissecting and tracing elements one after the other. *As you do that, be sure not to sever the connections of a vessel with the organ it supplies.*

4. The first vessel is the **ventral aorta.** It is a short, whitish artery that connects the conus arteriosus with a series of arteries supplying the gills. These are the four pairs of **afferent branchial arteries**; find them as they exit the ventral aorta and pass toward the gills. After blood has coursed through the gills, it is collected by another four pairs of arteries, the **efferent branchials.** These carry blood to the largest artery in the shark, the **dorsal aorta.** The trick is to locate the efferent branchials. To do that you need to cut completely through one side of the shark's jaw at the angle where the upper and lower jaw unite. Separate the lower from the upper jaw and continue to cut posteriorly until you reach the pectoral fins; cut through the cartilage there. This should enable you to swing the entire jaw region aside to expose the roof of the shark's mouth. *The key here is to expose the roof of the mouth.*

5. Now peel away the skin from the roof. Take your time. Beneath the skin are a series of vessels, most notably the efferent branchials. Passing toward the head are the paired **internal carotids** to supply blood to the brain. Another important vessel that exits from the efferent branchials is the **coronary artery**, which courses back to supply the heart muscle. It can best be seen as spidery vessels on the surface of the ventricle.

6. Now you will follow the dorsal aorta as it courses toward the tail, identifying various vessels along the way. The first one is actually a pair of vessels that leads to the pectoral fins. These are the **subclavians arteries,** one to each fin. A short distance from these is a single, large vessel, the **celiac,** which connects to the digestive system and nearby organs. Locate this vessel and trace it to the organs it supplies.

7. As you continue following the dorsal aorta, you will next encounter three vessels that exit in the region of the small intestine. The first of these is the **anterior mesenteric artery** supplying the ileum and spiral valve. Next comes the **gastrosplenic artery** with one of its branches going to the spleen and the other to the stomach region. Trace both of these to their respective organs. The third vessel, the **posterior mesenteric artery** supplies the rectal gland.

8. There are two more vessels to locate in the caudal region of the shark. A pair of **iliac arteries** exits the aorta and passes to the pelvic fins. You may be able to find these lying superficially along the interior body wall. If not, you will need to trace them from the aorta. As the dorsal aorta enters the tail, it becomes known as the **caudal artery.** You can see this if you cut a deep section through the tail.

Questions related to the shark circulatory system

a. What structure does the subclavian artery supply? the coronary artery?

b. Blood going from the heart to the gills must pass through which two arteries?

c. How many contractile chambers are there in the shark heart? Name them.

d. What is the function of the rectal gland and from which artery does it receive blood?

e. Name, in order, the four compartments through which blood passes as it flows through the shark's heart.

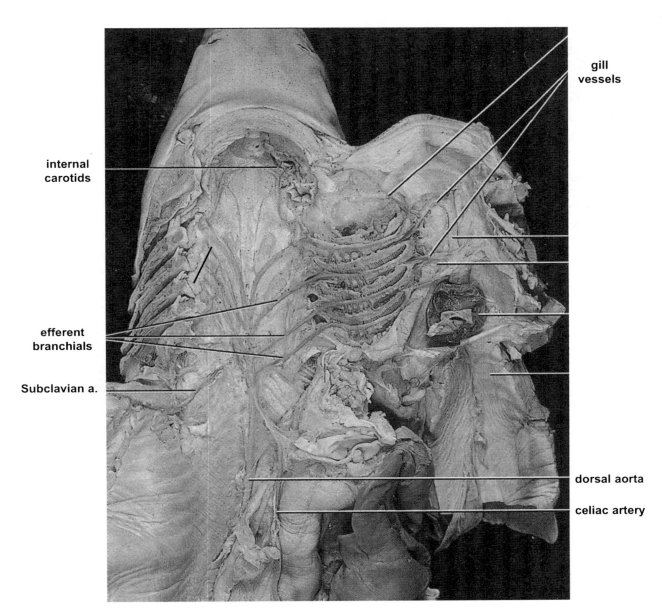

gill vessels

internal carotids

efferent branchials

Subclavian a.

dorsal aorta

celiac artery

Figure 18.15. Shark, oral cavity blood vessels

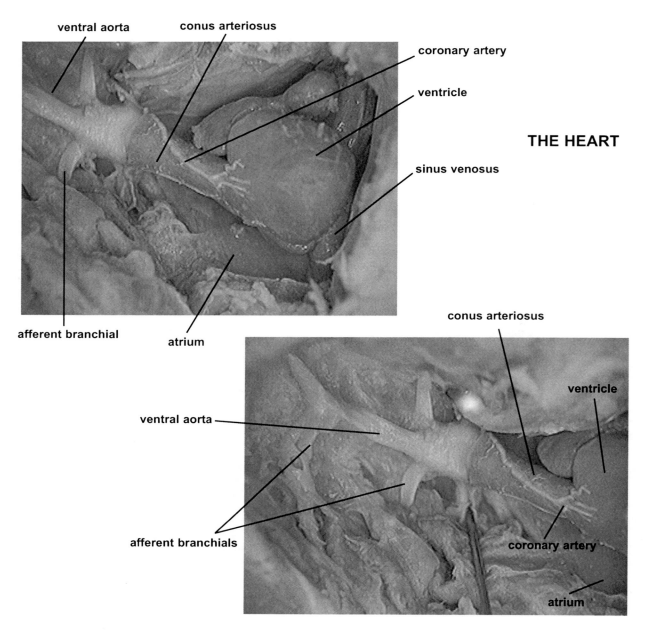

ventral aorta   conus arteriosus   coronary artery   ventricle

THE HEART

sinus venosus

afferent branchial   atrium

conus arteriosus

ventral aorta   ventricle

afferent branchials   coronary artery

atrium

Figure 18.16. Structures of the shark heart

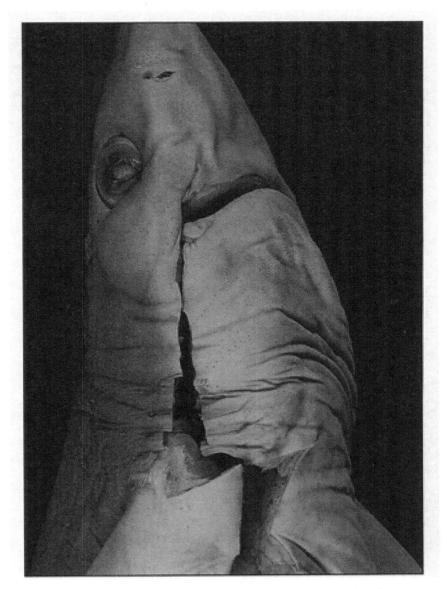

Figure 18.17. Correct incision to reach roof of mouth and oral cavity

Shark Nervous System*

1. This is the most challenging element of the shark dissection. Thus, you need to be deliberate as you reveal the brain and its components. The brain is enclosed in a cartilaginous braincase, called the **chondrocranium**, that must be partially removed. The best way to proceed is to first place the shark so that its dorsal surface is facing you. As you cut use a *shaving action* rather than to slice vertically into the cranium. Slowly shave thin layers on the dorsal surface of the head (between the eyes) until some of the brain is exposed. Then carefully continue to chip, slice, and pick away until the entire brain is revealed. The brain and its associated structures are very delicate—so proceed carefully.

2. Find the large **cerebellum** that sits atop and toward the back of the brain. What does its size tell you? We will use this as a landmark to locate the other structures. Lift the cerebellum slightly to reveal the **auricles** of the cerebellum. These are earlike structures that connect to the **medulla oblongata** that lies beneath and posterior to the cerebellum. The medulla, in turn, is connected to the **spinal cord.** These are the primary structures of the shark *hindbrain*.

3. Just anterior to the cerebellum are a pair of swellings, the **optic lobes**. Anterior to the optic lobes are two more very small protrusions, the **cerebral hemispheres,** which represent the shark *telencephalon.* What does their size tell you?

4. At the anterior-most part of the brain is the **olfactory apparatus**. It consists of a pair of **olfactory tracts**, each ending in an **olfactory bulb.** The bulb, in turn, contacts the **olfactory sacs.** The microscopic olfactory nerves pass from the bulbs to the sacs, which serve as chemoreceptors. You may need to take a little time (and a lot of patience) to properly and completely expose the olfactory apparatus.

5. Of the 12 pairs of cranial nerves, you are asked to locate and identify the following:

> **Olfactory Nerve I** - too small to see, but you should know where it is located; sense of smell

> **Optic Nerve II** - a large, whitish-colored nerve entering deep to the optic lobes; vision

> **Infraorbital trunk of nerves V** (Trigeminal) and **VII** (Facial) - the trunk is the union of nerves V & VII, it is elongated, superficial in location, and easily located. These are responsible for jaw movement, skin sensation, taste, and lateral line sensation.

> **Vagus Nerve X** - a large plexus exiting from the medulla oblongata; visceral sensations, heart beat

6. Your final dissection is to expose the paired **inner ears.** These are located lateral to the medulla oblongata encased within the cartilaginous chondrocranium. Again, you must be careful as you shave away the cartilage. Once the ear apparatus is revealed, locate the tubular **semicircular canals** and the large saclike **membranous labyrinth.** The inner ear function is to detect waterborne vibrations and it also contributes to the sense of equilibrium.

*Use the figure on the next page to help locate each of the above nervous system structures.

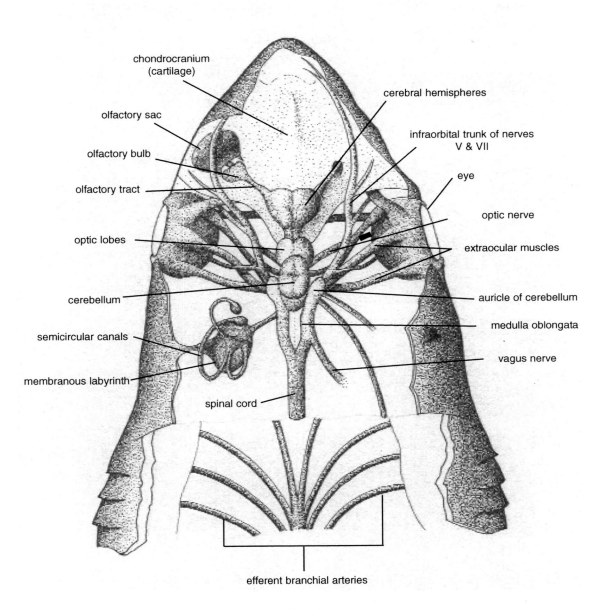

Figure 18.17. Diagram of the shark nervous system.

19 Chapter 19

AMPHIBIANS & REPTILES

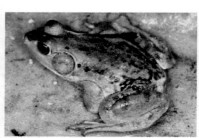

Chapter 19

Amphibians & Reptiles

Though boys throw stones at frogs in sport,
the frogs do not die in sport, but in earnest.

- Bion

Herpetology

Herpetology is the study of amphibians and reptiles and is the topic of this chapter. As an old saying goes, "frogs eat butterflies, snakes eat frogs, hogs eat snakes, and man eats hog." In this rather contracted view of natural systems, we see that reptiles prey on amphibians...or, stated conversely, amphibians provide food for reptiles. So we might well ask why would zoologists incorporate two seemingly adversarial groups into a single field of study? What do amphibians (frogs, toads, salamanders) have in common with reptiles (snakes, lizards, turtles, crocodiles)?

The answer is found in two words: **tetrapody** and **ectothermy**. Amphibians and reptiles are ectothermic tetrapods; in fact, they are the only ectothermic tetrapods. What on earth is an ectothermic tetrapod? The simplest definition of a tetrapod is that it's a vertebrate animal with four limbs, two in front and two in back. There are only four groups of animals that fit this description: amphibians, reptiles, birds, and mammals. An *ectothermic* animal, on the other hand, is one that is unable to control its own internal body temperature. It relies, instead, upon ambient (environmental) heat for **thermoregulation**. It cannot manufacture or maintain its own temperature; it's "cold-blooded." Of the four kinds of tetrapods, only amphibians and reptiles are ectotherms.

So, the binding forces that unite amphibians and reptiles into one field of study, or in our case, into one chapter in a zoology textbook, is the presence of four limbs and the inability to control internal body temperature. These two groups of animals are wrapped by the same blanket as it were—blood brothers of sort.

Figure 19.1. Frogs (upper right), salamanders (lower right), turtles, and alligators (above) are all "cold-blooded," (ectothermic) animals. They often bask in the sun to warm themselves as seen in these three photographs.

Figure 19.2. Sunning can sometimes result in overheating. Here a crocodile dissipates extra heat by panting.

Being cold-blooded has some real advantages. First of all, an ectotherm has a reliable source of body heat: the sun, or in the case of amphibians, water. It uses very little of his own energy to warm up; all it has to do is expose itself to sunlight or situate properly within a column of water. This is a cost-effective way to maintain one's internal heat. The animal expends very little of its own energy reserve to do so. All it has to do is position itself for optimum heat exposure.

Conversely, when external conditions become unfavorable, say it's too cold or too hot outside, an ectotherm just slows down. Its metabolic rate drops, its heart rate drops, respiratory rate declines, and its demand for food dissipates. The animal simply "rides out the storm" until conditions change. This is especially obvious during nighttime; when the temperature drops at night, metabolism likewise drops.

Cold-blooded animals, it seems, are able to coordinate their behavior to external conditions. This is nowhere more apparent than during extended cold periods, such as winter. It is then that ectotherms locate below the frost line and enter prolonged states of inactivity called **hibernation**. As temperatures rise in the spring, they warm and resume their activity. Some, however, use a different strategy and the result can be amazing. There are frogs of the genus *Rana* that occur in northern temperate zones (e.g., Canada) that are able to survive the coldest of winters not by relocating below the frost line, but by actually remaining exposed to frigid temperatures. They freeze. Their tissue even becomes partially crystallized. In this state of suspended animation, they remain frozen, but alive. This is possible because their blood contains a high level of glucose, which acts very much like "antifreeze" and prevents cellular death. When the temperature warms and ice melts, the frogs, too, amazingly thaw out.

In comparison to warm-blooded animals, ectotherms tend to be highly efficient energy converters. They require less food than a bird or mammal of comparable size. In fact, they use, on average, about *3% of the food needed by a similar-sized mammal.* Birds and mammals expend about 98% of their energy intake just to keep warm. That leaves very little for other functions, such as growth and repair. You will never see a bird or mammal regenerate a tail, for example! Amphibians and reptiles, on the other hand, relying on the sun for warmth, convert about 50% of their energy into new tissue. They are simply more efficient in their use of food. As a result, ectotherms can survive in situations of low food supply unsuitable to birds and mammals.

However, being an ectotherm is not all positive; it has some drawbacks as well. For instance, their geographic distribution is severely limited because of ectothermy. While they may love warm climates, only a few species can survive well in cold ones. That's why you find a superabundance of frogs and snakes in the tropics but virtually none in the tundra. There are very few snakes or toads in Alaska. They are rare on mountain tops. If you want to collect frogs or snakes, you go to a tropical rain forest; if you want to avoid them, go to Siberia.

The very fact that cool weather causes amphibians and reptiles to slow down also makes them vulnerable. If you are on a lizard collecting excursion, the best time to catch one is in the early morning while they are "sunning." Later in the day, they are just too fast or too alert to capture. So, during cool periods, such as at night, early morning, and evening, they are pretty much defenseless.

There's another side to the vulnerability story. Not only are they exposed to predators during cool temperatures, but, because they are slower, they are less likely to catch their own food. There's a real balancing act between lowered metabolism and actual starvation. Being ectothermic may offer some real advantages, but it has some serious liabilities as well.

Figure 19.3. In mornings when reptiles are sunning, they are slow and highly vulnerable. Note how this iguana is exposing his body to receive the best rays of the sun.

Class Amphibia

Amphibians include frogs, toads, salamanders, and the wormlike caecilians. As a group, they share a variety of features that separate them from reptiles.

(1) They are all **scaleless**; they have a naked skin. As a result, amphibians have serious problems retaining moisture. Water tends to evaporate from their surface presenting the threat of dehydration. To partially counter this hazard, amphibians have an integument invested with **mucous glands**. By covering their skin with the waterproofing agent mucus, the amount of water loss is greatly reduced. Nonetheless, because of their scaleless skin amphibians must stay in or near water. It's one of their primary limitations.

(2) As already discussed, amphibians are **ectotherms**. They cannot regulate their internal temperature and thus must rely on ambient (environmental) temperature for heat exchange.

(3) The majority of amphibians are **tetrapods**. The exception being the legless caecilians, which have secondarily lost their appendages as an adaptation to burrowing.

(4) All adult amphibians are **carnivorous**; there are no adult plant-eating amphibians. Most have a protrusible, sticky tongue to capture prey, which consists of a variety of invertebrate and, rarely, vertebrate animals.

(5) The skeleton of amphibians is primarily **bone**. Furthermore, in those that propel themselves by jumping, the pectoral girdle (shoulder area) is strengthened to absorb the shock of landing (see Figure 19.25, page 460).

(6) Amphibians use a variety of respiratory mechanisms. As larvae they use **gills** located in the pharyngeal area. Adults, however, rarely use gills but rely instead on a combination of **cutaneous** and **lung** respiration. Some have a supply of superficial blood vessels on the roof of the mouth where gaseous exchange can occur. This latter method is referred to as **buccopharyngeal** respiration.

(7) They have a **three-chambered heart** consisting of a ventricle and two atria. This arrangement is more efficient than the two-chambered condition of fish in that it helps to keep oxygenated and deoxygenated blood somewhat separate.

(8) All amphibians are **dioecious** and most use external fertilization. *They do not lay shelled eggs and therefore must remain linked to water to reproduce.* This is, in fact, one of the limitations to becoming truly terrestrial. Many amphibians have special **hedonic glands** in their skin that secretes a pheromone-like substance to attract mates. In frogs and toads, the simultaneous release of gametes is triggered by a mating embrace called **amplexus** (below). During amplexus the male grips the waist of a female, which causes her to release the eggs. He then simply douses them with sperm as they are shed into the water. The larval stage of a frog or a toad is a **tadpole**, which develops in water using gill respiration. Larvae then undergo an elaborate metamorphosis to become an adult.

Figure 19.4. Amplexus, the mating embrace of frogs and toads in which the male squeezes the female causing release of eggs.

Land Invasion

ALTHOUGH AMPHIBIANS WERE THE FIRST VERTEBRATES TO ATTEMPT LAND COLONIZATION, THEY WERE NOT TOTALLY SUCCESSFUL FOR TWO REASONS:

1. THEIR NAKED SKIN IS SUBJECT TO DEHYDRATION

2. THEY REQUIRE OPEN WATER TO REPRODUCE

(9) In contrast, most salamanders and all caecilians use internal fertilization and some even have internal development. Except for being aquatic and having external gills, the larval stage of salamanders closely resembles the adults. Certain salamanders have a unique developmental pattern termed **neoteny** in which larval characteristics are retained into adulthood. For example, the mudpuppy (*Necturus*) remains aquatic throughout its life and uses external gills (a larval trait). This is generally thought to represent a sexually mature individual in larval form. In some instances (tiger salamanders of the West Coast), neoteny is used only if sufficient water is available. Should the pond dry up the salamander develops lungs and breathes atmospheric oxygen. The reason for neoteny is not well understood. Some species seem to use it during times when their reproductive period is shortened or somehow limited. For instance, some mountain salamanders have adapted to shortened summers through neoteny.

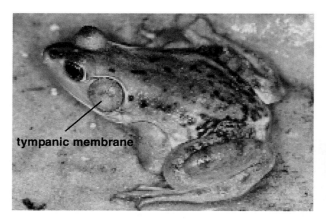

(10) The sensory systems of amphibians are rather unremarkable. They have the usual visual, olfactory, and touch capabilities. However, one system is different enough to deserve comment. Aquatic amphibians "hear" in much the same way as fishes: vibrations are transferred through external tissue to the inner ear. Terrestrial frogs, on the other hand, have two auditory routes. First, they detect airborne sounds through a large disc, the **tympanum**, located on each side of the head. Second, they (and salamanders) can detect subtle vibrations through their forelimbs. Thus they can use their front legs to "hear" the approach of a predator, an advancing prey, or an intruder. The vibrations are then transferred from the limbs to the inner ear.

(11) Amphibians have a number of ways of protecting themselves. Many jump or bound away from danger. Others change colors by altering the distribution of skin pigments to effectively match their surroundings. Frogs are especially adept at camouflaging themselves in this manner. This type of substrate matching is evident in the common grass frog, *Rana pipiens*, and others as evidenced by the tiny frog nestled in the flower below.

Figure 19.5. Photographs of amphibians showing three different characteristics. The salamander in the photo above has external gills. Although usually a juvenile feature, some adult salamanders retain gills, a phenomenon known as neoteny. The frogs in the other two photos illustrate their ability to blend into their surroundings. A tympanic membrane is also indicated in the top photo.

Perhaps the most potent defense mechanism displayed by amphibians is poison. Many frogs, toads, and salamanders have poison glands diffused throughout their skin. Some, such as the poison dart frogs of the tropics, secrete highly toxic substances.   It is said that the skin of one dart frog contains enough poison to kill 20 grown men.   Just touching these animals can be extremely dangerous. Secretions of certain toads in the Western U.S. are quite hallucinogenic and have been used by native cultures for generations to enter psychedelic states. For the most part, it is easy to tell which amphibians are poisonous and which ones aren't. The noxious ones usually advertise the trait through bold, contrasting colors: reds, oranges, black, and yellows. **Aposematic (warning) coloration** of this sort, as displayed by the salamander below, is very common in poisonous animals.

The Salamanders
(order Caudata)

Salamanders are probably the most primitive of the amphibians, at least they possess the most primitive features. For example, their **postanal tail**, which distinguishes them from all other amphibians, is itself a primitive chordate structure. The hind limbs (adapted for walking, not jumping) are splayed out to the side, another primitive feature. This means the weight of the body is supported by a muscular sling as the animal walks. Larval salamanders are aquatic but resemble adults except for **exposed pharyngeal gills** (another primitive feature).   The gills are replaced by lungs as metamorphosis transforms the animal into a terrestrial adult.

Some salamanders, however, are neotenic and retain larval characters (e.g., gills) into adulthood. Except for these few neotenized forms, adult salamanders lack gills and rely solely upon cutaneous, lung, and buccopharyngeal respiration. Because their skin serves as a respiratory surface, it must remain moist. For this reason, salamanders live near water, or at least in a moist environment. They may be found in wet woodlands or near streambeds, often concealing themselves under logs, stones, or leaf litter.

Many salamanders have poison skin glands that secrete toxins for self-defense. These glands may be located in the caudal area, skin of the head, or dispersed over the general body surface.   Some salamanders even spray their toxins from dorsal skin glands.

Figure 19.6. The red-spotted newt and the tiger salamander depicted to the left are two species that display their toxic capability through bold color patterns. The larval salamander above uses external gills for respiration.

Figure 19.7. Representative salamanders. Note the elongated tail and limbs positioned out to the side. Also, note the moist or watery microhabitats.

The Frogs and Toads
(order Anura)

Representative anurans are shown in Figure 19.8 below. Note that adult toads and frogs are characterized by the lack of a tail and the presence of large hind limbs adapted for jumping. The hind toes are clawless (except in African Clawed Frogs, *Xenopus* spp.), elongated, and webbed to facilitate swimming. The forelimbs are much shorter than the hind legs and are used as struts to support the body while at rest. They also absorb some of the force of landing following an extended jump. The skeleton is especially adapted to the leaping or hopping habit. A special elongated bone called the **urostyle**, assists the hind legs by transmitting the force of the jump throughout the body (see Figure 19.25).

Toads differ from frogs in having a rough, drier skin. They also have smaller hind legs and jumping is less of an escape mechanism than in the frogs. They have enlarged **parotid glands** just behind the eyes to secrete a variety of toxins. Both frogs and toads rely on a combination of lung and cutaneous respiration and both engage in amplexus as a means of stimulating the simultaneous release of eggs and sperm.

Larvae develop in ponds, streams, lakes, or other bodies of water, although some develop within a foamy "spitlike" substance. Metamorphosis in anurans is much more pronounced than in salamanders. The larva (tadpole) bears little resemblance to the adults and must go through a dramatic morphological change into adulthood.

Figure 19.8. Representative anurans. Note the enlarged hind limbs and elongated toes adapted for takeoff and the stout forelimbs for landing. Mating posture (amplexus) is shown in the frogs in plate **B**. Observe the size difference between male (on top) and the female—an excellent example of sexual dimorphism in amphibians. Airborne vibrations are detected through the tympanum (**A & C**) as well as through the forelimbs. Finally, note the suction discs on the toes of the tree frogs (**C** and **D**) used for climbing.

Caecilians
(order Gymnophiona)

Caecilians are rather strange-looking amphibians inhabiting tropical regions of the South American and African continents. Most are subterranean although some species are aquatic. All resemble snakes or worms more than their amphibian relatives. However, the naked, mucus-covered skin identifies them as members of the amphibian class. Caecilians are legless, having secondarily lost all four limbs as an adaptation to burrowing, and are virtually blind, also due to their subterranean existence. Their skull is thick and well fused to form a battering ram as the animal burrows.

The great majority of caecilians use **internal fertilization** and even **internal development**. The males have a copulatory organ called the **phallodeum** to aid in the transfer of sperm to the female (see photo to the right). Larvae grow within the oviducts of the female but without a yolk sac for nourishment. Therefore, larval caecilians have adopted a bizarre method of obtaining food by *scraping tissue from the lining of the oviduct with specialized teeth.*

Although most caecilians are only a few inches long, some reach a length of almost five feet. Adults feed on earthworms and other subterranean invertebrates. However, due to their secretive burrowing habits, they are rarely seen and are poorly understood.

male phallodeum

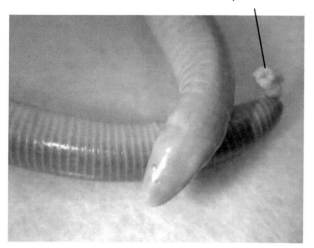

Figure 19.9. A caecilian terrestrial species of the genus *Dermophis*. Note the annular (ringed) appearance of its scaleless skin. Note also the absence of limbs, the small eyes, and stout, pointed snout—all adaptations to a burrowing habit.

Class Reptilia

*R*eptiles. The very word causes the skin of some people to "crawl" and causes others to experience sheer terror. The truth is, while some people are attracted to snakes, turtles, and lizards, many others are terrified to the point of absurdity. Some of the most irrational zoophobias that exist are in regard to reptiles. Yes, there is reason to fear some reptiles. The nile crocodile kills scores of people each year, and thousands of people die annually from snakebite (see *Inside View 20* at the end of this chapter). However, of the 2,700 or so species of snakes, only about 10% are truly dangerous—yet all snakes seem to pay the price of having venomous relatives. For instance, I have known people, intelligent people, who refuse to explore nature for a loathing of snakes. Some folks are nervous on a picnic for fear of "creepy-scaly things." I have heard stories of the aggressiveness of snakes that border on the ridiculous: "...they will bite and never let you go" ..."they will jump at you from trees"... "if a snake licks you with its tongue you will die before sundown." I once encountered a man in the woods of West Virginia who had a large revolver strapped to his hip. My inquiry as to why he was sporting a sidearm was met with the acid reply, "To kill *&%@*#* snakes." I asked, why snakes? "...they will chase you until you're exhausted, then you're a goner." I was curious about which snakes he shot. His answer was nothing less than chilling. *"All of 'em."*

On the other side of the phobic spectrum are people who adore reptiles. Reptile enthusiasts raise geckos, skinks, monitor lizards, iguanas, bearded dragons, turtles, and an endless variety of snakes as pets. Some people can't resist the temptation to turn over rocks, logs, and brush piles in search of snakes. In addition, what kid doesn't like dinosaurs? The T-Rex (Sue) exhibit at Chicago's Field Museum promises to be one of the most visited natural phenomena in history.

So, some people are attracted to reptiles and others are repulsed. Our goal in this chapter is to get a glimpse into the anatomy, behavior, and biology of this class of vertebrates, and in the process maybe reveal that of all animals, reptiles are among the most amazing.

The Invasion of Land

At the beginning of this chapter, we outlined the similarities between amphibians and reptiles. The fact is, they do have a lot in common. However, in one very important detail, they are different. In this one respect, reptiles have far outpaced amphibians on the road toward diversity. And that one difference is that reptiles are terrestrial. As a group, they have discovered the real secret of living on land and that is their great contribution to higher animals. The simple truth is that, as a group, reptiles are much better adapted to terrestrial life than are amphibians—for several reasons.

First of all, even though both groups are ectotherms, reptiles are much better at exploiting the sun than are amphibians. During periods of activity they can maintain a high and nearly constant body temperature by regulating their exposure to the sun, something that is difficult for amphibians to do. Thus, reptiles are considerably more active than their amphibian relatives.

Their high activity levels are also due, in part, to the efficiency of their circulatory system. Although the reptilian heart is three-chambered anatomically, it is virtually four-chambered functionally (in crocodilians, a true four-chambered heart does actually exist). This means that reptiles are better able than amphibians to separate oxygenated from deoxygenated blood within the heart. Thus, their muscles receive a richer, more highly oxygenated blood supply, which translates into a more active animal. If you were to observe an amphibian and, say, a lizard for a day, you would be amazed at the difference in mobility and activity levels between the two.

Figure 19.10. Terrestrial reptiles are active ectotherms.

Reptiles, like their amphibian counterparts, are tetrapods; they have four limbs. Of course, there is one big exception to this...the snakes. However, even though living snakes lack appendages, all evidence indicates that they were derived from legged ancestors. Indeed, some snakes (e.g., boas) have internal bones reminiscent of pelvic girdles (bones that support the hind limbs), and others have external spurs near the cloaca that are actual remnants of limbs.

In the majority of reptiles that possess limbs, we find that they are placed differently on the body than the legs of amphibians. Unlike the amphibian limb, which is splayed out to the side, the reptilian leg, especially in lizards, has swung into a slightly more medial placement, providing better balance. In addition, most reptiles have a relatively long limb, lifting the animal well off the ground allowing for greater mobility and speed.

A second major contribution to the reptilian conquest of land has to do with water conservation. Animals must avoid dehydration at all costs; if they dry they die. Thus, the better an animal is at conserving water, the less it will be tied to an aquatic existence. Amphibians, with their naked skin, learned this only too well and are pretty much tied to aquatic habitats. On the reptile's path toward terrestrialism, however, its skin has become one of its most formidable weapons against desiccation. Reptiles possess a nearly glandless, dry integument covered with horny **scales**. A dry, scaly skin not only provides protection against abrasion or other injury, but, more importantly, it's also impervious to water loss. It prevents the life-threatening dehydration that is so limiting to amphibians.

This means that reptiles are not linked to aquatic ecosystems but can roam far and wide without fear of drying out. But there is a price to pay for their scaly skin. Because it is impervious to water, it cannot be used as a respiratory surface. As a result, reptiles rely solely on an internal lung for respiration.

Another adaptation related to water conservation has to do with the reptilian excretory system. More primitive animals, including amphibians, remove nitrogenous wastes as ammonia or urea. This requires a fair amount of water as a flushing agent. However, reptiles excrete nitrogen, not as ammonia or urea, but as **uric acid**. Much less water is lost via this route and as a result reptiles are more efficient at water conservation than earlier animals. Thus, with a skin to prevent water evaporation, and a water-conserving kidney, reptiles are much better equipped to exploit a terrestrial environment than any animal we have encountered.

All of the factors we have discussed so far have been important, even essential, in the conquest of land by reptiles. But there is one more contributing factor...and it's more important than any of the others. By far, the greatest advantage that reptiles have over amphibians in the terrestrial contest is their method of reproduction. Reptiles practice internal fertilization and are the first oviparous vertebrates. They lay a shelled, terrestrial egg. *More than any other single factor, their exploitation of land is the consequence of a terrestrial egg.* The cleidoic (amniotic) egg allows embryonic development in a safe, nourished, and fluid environment. The shelled egg, which by the way, was also "invented" by those other terrestrial invaders, the arthropods, can be deposited well away from aquatic environments thus liberating reptiles from the aquatic-reproduction connection so apparent in amphibians. The egg can even be retained in the female's uterus in the ovoviviparous pattern of development. *The evolution of such an egg made reptiles the first chordates fully liberated from aquatic ecosystems.*

There are even a few species of lizards and snakes that forego the shelled egg and are truly viviparous. In these forms, the female provides direct nourishment to internally developing young. This is an even greater reproductive adaptation to life on land since the female can carry the embryos internally thereby assuring an even greater chance of their survival.

Reptilian success, then, as the first terrestrial vertebrates is because:

(1) they are efficient ectotherms
(2) they have an efficient circulatory system
(3) their limbs are placed more advantageously
(4) their skin prevents evaporation
(5) their kidney excretes waste as uric acid
(6) they use internal fertilization
(7) they have an efficient reproductive pattern
  ✓ internal fertilization
  ✓ the terrestrial egg
  ✓ internal embryonic development

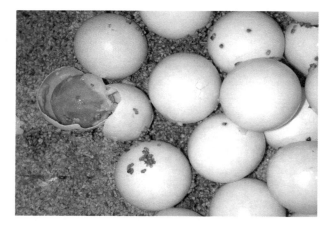

Figure 19.11. The reptilian shelled egg permitted, more than any other single feature, full exploitation of a terrestrial life.

Types of Reptiles:
turtles, lizards, snakes, crocodilians

The Turtles
(order Chelonia)

*T*urtles are among the most recognizable of the vertebrate animals. They have a keratinized (horny) beak, no teeth and are enclosed within a two-part bony shell. The dorsal portion of the shell, the **carapace,** is fused to both the vertebral column and to the ribs while the ventral portion, the **plastron,** is fused to bones of the pectoral girdle. Since the ribs are attached firmly to the carapace (see below), they cannot move to aid in respiratory movements as in other reptiles. Thus, turtles have some difficulty in breathing. To compensate, they use special muscles to press the viscera (body organs) against the lungs which expels air during exhalation. To inhale, other muscles create a partial vacuum in the body cavity, which draws air into the lungs.

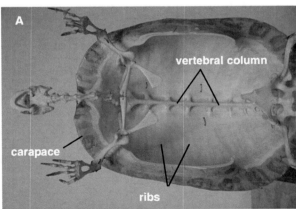

Figure 19.12. Representatives of the order Chelonia. **A**, a skeleton showing attachment of the carapace to the vertebrae and ribs; **B**, a pond turtle (slider); **C**, a terrestrial box turtle; **D & E**, sea turtles; and **F**, giant Galapagos turtles.

Turtles bury their shelled eggs in sand or soft soil and then leave them unattended. The young hatch after an embryonic period of from one to two months. In most cases, it takes several years to reach sexual maturity. Most pond turtles and small terrestrial types live for approximately 10–20 years. However, some turtles are very long-lived. The giant tortoises, for example, may well exceed 100 or more years, and there is anecdotal evidence that snapping turtles have an equivalent (or even longer) life span.

The Lizards and Snakes
(order Squamata)

Lizards and snakes make up the bulk of the known reptiles. Of the more than 7,100 species, approximately 6,850 belong to the order Squamata. And of these, about 2,700 are snakes. Thus, lizards are the most numerous reptiles in existence. They are four-legged, tailed reptiles that live in a variety of settings, from humid rain forests to barren deserts, but are, for the most part, absent from cooler, more temperate zones. They are often found near some refuge such as rocks, logs, crevices, or underground burrows into which they make a quick retreat if disturbed. Many have sticky tongues that can be launched amazing distances to capture prey. In fact, the chameleon's tongue is longer than the lizard itself! Many lizards can voluntarily lose their tail, an antipredator process called caudal autotomy. The tail is regenerated, but is usually shorter than the original. Most are oviparous, laying or burying eggs in secluded locations. Some are ovoviviparous and others are viviparous.

KINDS OF LIZARDS INCLUDE :

✓ nocturnal **geckos** have huge eyes and an ability to vocalize (some make clicking sounds while others "bark")

✓ stout-bodied **iguanas** of pet stores as well as the marine Galapagos forms

✓ **skinks** have large heads, robust bodies, and reduced legs as an adaptation to burrowing

✓ the multicolored **chameleons**

✓ **monitor lizards** and **Komodo dragons** can reach a length in excess of nine feet

✓ the nearly blind **legless lizards** live most of their lives underground and are adapted for a burrowing, subterranean existence

✓ **gila monsters** and **Mexican beaded lizards** (Heloderma) are both venomous and can inject a potent neurotoxin as they engulf their prey

B   Gila Monster

Horny Devil

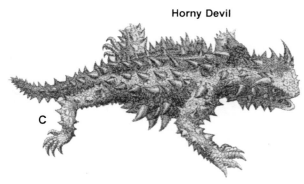

C

A

Figure 19.13. Some representative reptiles. Photo **A** shows two fat-tailed leopard geckos. Excess food, in the form of fat, is stored in the enlarged tail. **B**, *Heloderma*, the gila monster, one of only two known poisonous lizards in the world. The other is the beaded lizard, a close relative. Rather than inject venom into their prey with fangs as most snakes do, venom leaks passively into wounds as the lizard holds on with a "death grip." **C**, the Australian horny devil *(Moloch horridus)*, a horned lizard with spiny projections of the skin.

Figure 19.14. Lizard diversity. **A** and **insert**, a legless lizard;**B**, a skink...note its stout body and reduced legs for burrowing. Also note that it's regenerating its tail. **C**, a terrestrial lizard. **D - F**, iguanids.

The closest relative of lizards are the snakes. In fact, fossil records indicate that the two evolved from the same four-legged ancestor. Even so, there are three major differences between the two. First, except for a few legless species, lizards have retained their limbs while snakes have lost theirs. The fact that snakes once had appendages but have since lost them is evidenced by the remnants of pelvic limbs and external "spurs" in boa constrictors and similar species. The second primary difference is that lizards have an external ear opening, which is lacking in snakes. Finally, lizards have moveable eyelids, which snakes lack.

All snakes are carnivores, feeding on a variety of animals including insects, slugs, fish, mice, rats, birds, bird eggs, lizards, and each other. Snakes, like lizards, cannot chew their food; instead, it is swallowed whole. To accomplish that, they have a highly **kinetic skull** that allows the two jaws to disarticulate (come apart) when consuming large food items. After swallowing, the jaws are realigned with a yawnlike maneuver. However, if the mouth is full of food, say a rat or a rabbit, breathing becomes a challenge. Snakes have solved that problem by developing a tubular extension of their **glottis**. When swallowing large prey, the glottis "sneaks" out on the floor of the mouth to take in air.

The great majority of snakes (~90%) are nonpoisonous and of the 10% or so that are, few interact with humans. For a more detailed account of poisonous snakes see *Inside View 19.1* on page 446.

In keeping with their elongated shape, snakes have an extremely large number of vertebrae and ribs (photo below). Some species have as many as 400 rib-vertebrae combinations! Their left lung is reduced and the right one is elongated as are the liver, gonads, and other internal organs.

Snakes occupy a number of different habitats. Some are arboreal, such as the vine snakes or emerald tree boa, which live in trees and rarely, if ever, come to the ground. Others are exclusively burrowers (e.g., *Typhlops*, an underground snake that feeds extensively on earthworms) and seldom come to the surface. Still others, such as water snakes, anacondas, and water moccasins are semiaquatic, living an amphibian lifestyle, partly in the water and partly on land. The sea snakes, on the other hand, are strictly marine animals. However, the great majority of snakes are terrestrial, preferring to live on land (e.g., garter snakes, racers, rattlesnakes, kingsnakes, rat snakes, and dozens of desert species).

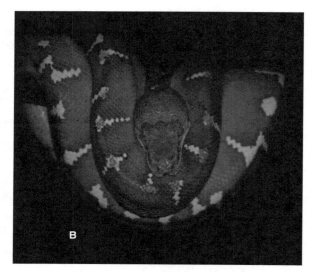

Figure 19.15. Plate **A** is an X-ray of a snake skeleton. Note the large number of rib-vertebrae connections and the flexibility they provide. A tree (arboreal) boa is shown in plate **B** and a sea snake in **C**.

Snake Senses. As a group, snakes have an advanced nervous system and an equally well-developed set of sense organs. Through the inner ear they are able to detect airborne and substrate-borne sounds. These sound vibrations may be transferred to the inner ear by way of bones of the lower jaw. Great variation exists in the visual ability of snakes. Some, such as the burrowing snakes, are virtually blind, while others are able to detect color and subtle movements.

Snakes smell their environment in a different manner than most other vertebrates. The bifurcated (forked) tongue is adapted to pick up airborne scents through constant flicking. The odor molecules are then carried to specialized grooves in the roof of the mouth and transferred to the **Jacobson's organ.** This paired organ connects to the olfactory lobe of the brain where scents are processed. Thus, by flicking their tongue snakes can detect pheromones emitted by mates or scent trails of prey.

In terms of sensory structures, the most astonishing are the **infrared (pit) organs.** These are exquisitely sensitive heat-sensing receptors that detect radiant heat given off by warm-blooded animals. These special sensors occur in two different groups of snakes. In the boas and pythons, the receptors occur as lines of small **labial pits** arranged along the scales of the lips. In the second group, the so-called "pit vipers," the sensors are consolidated into a pair of large **loreal pits** located between the nostrils and the eyes (Figure 19.16). These pits are highly refined thermoreceptors that can detect temperature changes on the order of 0.003^O C. What this means is that a pit viper can easily sense the presence of a bird or mammal by its body heat alone. They can locate prey in total darkness even if the preys temperature is close to that of the surrounding environment.

Snake Reproduction. There is little sexual dimorphism in snakes. The female may be slightly larger and heavier than the male and the male usually has a slightly longer tail than the female, but different color morphs are rare, unlike that of lizards. Snakes are unique among vertebrates, however, in that the male has a well-developed, forked copulatory device, a **hemipenes.** Although the hemipenes is paired, only one is used at a time during copulation. Developmental patterns include oviparity (egg laying) and viviparity (live birth).

Figure 19.16. Some snakes possess special infrared receptors as pit organs on their face. Pit vipers, such as the copperhead in plate **A** have two large loreal pits located between the nostril and the eyes whereas boas have similar, but miniscule, labial pits scattered along their lips (**B**). In plate **C**, two snakes are entwined in a copulatory embrace; the female is in the center.

The Crocodilians
(order Crocodilia)

AMERICAN ALLIGATOR

Of all living reptiles, the crocodilians are the most closely related to the extinct dinosaurs. And according to the fossil record, they have changed very little over the past 170 million years. They are characterized by a thick armorlike integument, a powerful laterally flattened tail, legs splayed out to the side, webbed feet, a long snout with teeth set in sockets, a four-chambered heart, and relatively large overall body size.

There are 22 species of crocodilians scattered among four principal categories: alligators, crocodiles, gavials, and caimans (Figure 19.18). However, many zoologists believe there's another living relative...birds. In fact, a number of experts consider birds to be just another type of advanced reptile of the crocodilian lineage. While the taxonomic placement of birds is an intriguing puzzle that the herpetologists and ornithologists have yet to sort out, birds are nonetheless distinct enough from other reptiles to warrant a separate class, and for us, a separate chapter.

Crocodilians are all, to some extent, aquatic animals but do not hesitate to leave the water. Using their laterally flattened tail, they are fast and agile swimmers but are also surprisingly swift on land. They have a number of adaptations to an aquatic life. For instance, their nostrils are situated at the end of the long snout to enable them to breathe while the body is partially submerged. Also, they have, along with a few birds and mammals, evolved a shelf, the secondary palate, which separates the nasal and oral passages. What this means is simply that they can continue breathing while feeding. To protect their eyes while underwater, crocodilians have a transparent membrane that "flicks" over the eye surface.

All crocodiles are efficient predators feeding on a variety of fish, birds, mammals, and whatever else they can find, including the occasional human. Some can be quite large and dangerous. For instance, Nile crocodiles

may reach a length in excess of 21 feet. An animal that large, with an unpredictable disposition, huge teeth, powerful jaws, and a slashing tail, poses quite a threat and, in fact, is responsible for about 50 human deaths per year. An alligator, or especially a crocodile, that is injured, cornered, whose young are threatened, or just hungry may attack and can be extremely dangerous. The more that humans encroach upon natural habitats of alligators and crocodiles, the more encounters will occur, some quite unpleasant.

The reproductive behavior of crocodilians differs from any other reptile. They have a fairly sophisticated repertoire of **vocalizations**, including penetrating roars, grunts, coughs, whistles, wheezes, and rumbles. These sounds are used in mating, territoriality, and as a means of maintaining togetherness of family and other groupings. There is, for instance, considerable evidence that while still in the eggs, young crocs communicate with each other and with their mother to synchronize hatching.

Crocodilians are all **oviparous**, laying up to 60 large shelled eggs. Females may pile enormous vegetation mounds around the eggs as an elaborate nest. She then patrols the area as a protective measure against nest looters. Females are also quite tender toward the young whom they guard long after hatching.

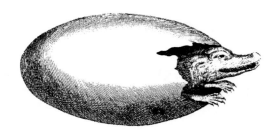

Figure 19.17. The majority of reptiles hatch from an amniotic, fluid-filled egg as seen here as a young alligator emerges. This reproductive pattern has led, more than any other feature, to the successful invasion of terrestrial ecosystems.

Figure 19.18. Diversity in crocodilians. Especially apparent is the armorlike body covering and the laterally flattened, powerful tails. The small eyes are protected by bony shields and are further safeguarded by a transparent membrane during underwater activity. Note the lateral placement of the stout limbs and the webbed, clawed toes. Finally, note the terminal placement of the nostrils on the elongated snout. The location of the nostrils and eyes are adaptations to an aquatic habitat by permitting the animal to breathe, and still view its surroundings, while almost totally submerged. **A,** a collection of Nile crocodiles; **B,** Indian gavials; **C - E,** are American alligators. One way to tell alligators from crocodiles is the appearance of the teeth when the mouth is closed. Teeth in the lower jaw of crocodiles are exposed but concealed in the alligator.

An Inside View 19.1

Venomous Reptiles

The production and use of venom by animals is widespread. The protozoan, *Paramecium*, uses poisonous trichocysts as a defensive weapon. The cnidarians (anemones, jellyfish, and their relatives) employ stinging nematocysts to subdue prey and for self-defense. There are venomous squid, gastropods, octopuses, spiders, insects, urchins, frogs, toads, salamanders, and even a mammal (the platypus). But nowhere in the world of animals has venom been more highly developed than in the reptiles, especially snakes. Of the more than 7,000 species of reptiles only about 350 (5%) are poisonous, almost all of these are snakes.   Although many of the poisonous ones are shy, secretive, or docile animals, which pose little threat to humans, there is, nonetheless, an estimated 35,000 deaths each year from snakebite. Perhaps the most dangerous snake in the world, although not the most venomous, is the cobra (*Naja naja*). This one species alone accounts for more than 10,000 human fatalities per year in India and adjoining areas.

Snakes, the vipers in particular, have evolved sophisticated venom delivery systems involving hinged fangs that inject venom much like a hypodermic syringe. As the mouth opens, the fangs automatically swing into a defensive position. The strike impales the victim and injects venom instantaneously from the large venom glands located in the roof of the mouth.

The venom itself is modified saliva contained within modified salivary glands. It is a complex mixture of over 100 chemical components, primarily powerful digestive proteins. Venom is classified into two types based on its mode of action.

Neurotoxins - these powerful enzymes attack the nervous system causing death primarily through paralysis of respiratory muscles.

Hemotoxin - this venom causes degeneration of the circulatory system, including destruction of blood and blood vessels.

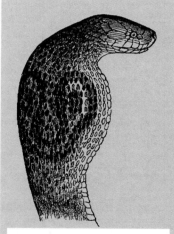

Naja naja, the common cobra

The great majority of poisonous snakes are *either* neurotoxic or hemotoxic. However, there is recent evidence that some snakes are developing the ability to deliver a mixture of the two. For instance, the venom of *Crotalus viridus*, the mojave rattlesnake of the western United States, contains both neuro- and hemotoxic venoms.

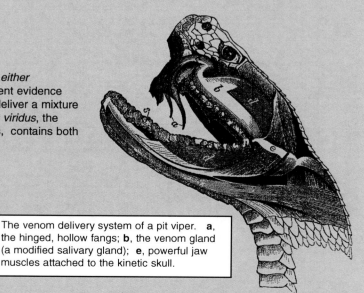

The venom delivery system of a pit viper. **a,** the hinged, hollow fangs; **b,** the venom gland (a modified salivary gland); **e,** powerful jaw muscles attached to the kinetic skull.

THE FIVE CATEGORIES OF VENOMOUS REPTILES:

(1) The **Elapids** (family Elapidae) are the most venomous snakes on earth. This group includes the coral snakes and the cobras. They have fixed (unhinged), grooved fangs along which the neurotoxic venom flows passively into the bite.

(2) The **Vipers** (family Viperidae) include the rattlesnakes, copperheads, water moccasins, puff adders, and many other Old World vipers. They possess front fangs that fold back when the mouth is closed but swing forward automatically as the mouth is opened. They inject a hemotoxic venom.

(3) The **Colubrids** (family Colubridae), such as the African boomslang, essentially allow a toxic saliva to flow into an open wound to incapacitate the prey. The chemical nature of their venom, however, is poorly understood.

(4) The **Hydrophiids** (family Hydrophiidae) are the sea snakes. They are relatively docile snakes that carry a deadly package of neurotoxic venom.

(5) The **Venomous Lizards** (family Helodermatidae) are found only in the southwestern deserts of the United States and Mexico. There are only two species, the *gila monster* and the *Mexican beaded lizard.* The venom is not injected but flows passively from salivary glands into the open wounds. They possess a potent neurotoxic venom.

IT'S A FACT

There are 21 species of venomous reptiles in the United States. Fifteen of these are rattlesnakes, two are coral snakes, and the remainder have one species each: sea snake, copperhead, water moccasin, and gila monster. Although approximately 8,000 people in the United States are bitten each year, only a few (~10) deaths result.

CORAL SNAKE

WESTERN DIAMONDBACK RATTLESNAKE

COTTONMOUTH

COPPERHEAD

Figure 19.19. The four venomous pit vipers in the United States. Although all are potentially dangerous, they are reluctant to strike unless disturbed. Their venom is used, instead, to subdue prey such as rats (lower right).

Chapter 19 Summary

Key Points

1. Herpetology is a branch of zoology dealing with amphibians and reptiles.

2. Both are ectotherms and, as such, they must rely upon the transfer of heat from the external environment to maintain their internal temperatures. Ectotherms are efficient energy converters and thus require less food than similar sized birds or mammals. However, ectothermy has some serious disadvantages as well: most notably a lessened activity under cool conditions and a consequential increase in vulnerability.

3. The class Amphibia consists of frogs, toads, salamanders, and caecilians. As a group they share a number of characteristics including moist, scaleless, mucous-covered skin; powerful hind legs often adapted for jumping (e.g., toads and frogs); carnivorous lifestyle; diverse respiration; a three-chambered heart; external fertilization of eggs (although some [e.g., salamanders] are internal fertilizers); and many have a poisonous skin secretion used in self-defense.

4. Reptiles are the first vertebrates to invade land successfully. The two features that contributed most to this is the terrestrial egg (also known as the cleidoic or amniotic egg), which emancipated reptiles from aquatic reproduction, and, secondly, their scaled skin. A scaly covering minimized dehydration, which, in turn, permitted reptiles to spread into dry habitats such as deserts.

5. Although reptiles, like amphibians, are ectothermic tetrapods, they are far more effective at using solar heat and thus are more active than their amphibian counterparts.

6. Turtles differ from other reptiles by their bony shells with attached vertebrae and ribs, the absence of teeth, and their relatively long life span.

7. Lizards are four-legged, tailed reptiles, many of which can sacrifice their tail to escape predation in a process called caudal autotomy. There are a number of different types of lizards including geckos, iguanas, skinks, dragons, legless designs, and two venomous forms, both restricted to North America.

8. Snakes are legless reptiles that have adopted a sinuous or crawling habit. As a result, their external and internal anatomy reflects this type of existence: limbless, expansion in number of ribs and vertebrae, and elongation or loss of selected internal organs.

9. Snakes differ from lizards in three ways: they are limbless, they lack movable eyelids, and they lack an external ear opening.

10. The class Crocodilia is an ancient lineage of reptiles of four general types: alligators, crocodiles, gavials, and caimans. Some authorities place the birds among the crocodilians.

11. Crocodilians are relatively large, predacious semi-aquatic reptiles characterized by a thickened, scaly skin, a flattened yet powerful tail, four-chambered heart, socketed teeth, a secondary palate, and laterally placed legs.

12. There are, worldwide, approximately 350 species of venomous reptiles although only 21 poisonous species occur in the United States. The venom is either neurotoxic and attacks the nervous system or hemotoxic with action directed toward the circulatory structures.

Key Terms

amniotic (cleidoic) egg
amplexus
autotomy
buccopharyngeal respiration
carapace
ectothermy
glottis
hemipenes
hemotoxin
herpetology
infrared (pit) receptors
Jacobson's organ
kinetic skull
labial pits
neoteny
neurotoxin
ovipary
ovovivipary
plastron
secondary palate
tetrapod
urostyle
vivipary

Points to Ponder

1. What is ectothermy?

2. What is the major advantage of being an ectotherm?

3. What is the main limitation of ectothermy?

4. What are the characteristics of an amphibian?

5. What type of skin glands do amphibians possess, and what is the function of each?

6. Contrast the importance of external and internal fertilization as it relates to the success of amphibians and reptiles.

7. How do salamanders differ from other amphibians?

8. What is a caecilian? How do they differ from other amphibians?

9. What specific features contributed to the tremendous success of reptiles as terrestrial animals?

10. What is caudal autotomy? How can reptiles afford such a drastic antipredator practice?

11. How can you tell lizards and snakes apart?

12. What is the role of the reptilian kinetic skull?

13. What role does the glottis play in snake feeding activity?

14. What is a Jacobson's organ and how is it used?

15. What are pit organs, how are they used, and what animals possess them?

16. Crocodilians have a secondary palate. Of what importance is such a feature?

17. How does reproduction in crocodilians differ from other reptiles?

18. What reptiles use venom in either capturing prey or in self-defense?

19. What type of venom do reptiles use? Which animal uses each?

20. How many venomous reptiles (and what types) are there in the United States?

Inside Readings

Amphibians

Blaustein, A. R., and D. B. Wake. 1995. The puzzle of declining amphibian populations. Sci. Am. 272(4):52-57.

Del Pino, E. M. 1989. Marsupial frogs. Sci. Am. 260(5):110-118.

Duellman, W. E. 1992. Reproductive strategies of frogs. Sci. Am. 267(1):80-87.

Duellman, W. E. and L. Trueb. 1986. Biology of Amphibians. New York: McGraw-Hill Publishing Company.

Gans, C., and G. C. Gorniak. 1982. How does the toad flip its tongue? Test of two hypotheses. Science 216:1335-1337.

Halliday, T. R., and K. Adler. 1986. The Encyclopedia of Reptiles and Amphibians. New York: Facts on File, Inc.

McClanahan, L. L., et al. 1994. Frogs and toads in deserts. Sci. Am. 270(3):82-88.

Myers, C.W., and J. W. Daly. 1983. Dart-poison frogs. Sci. Am. 248(2):120-133.

Narins, P. M. 1995. Frog communication. Sci. Am. 273(2):78-83.

Phillips, K. 1990. Where have all the frogs and toads gone? BioScience 40:422-424.

Pough, F. H., et al. 1998. Herpetology. Upper Saddle River, N J: Prentice Hall.

Souder, W. 2000. A Plague of Frogs: The Horrifying True Story. New York: Hyperion.

Stebbins, R. C., and N. W. Cohen. 1995. A Natural History of Amphibians. Princeton: Princeton University Press.

Twitty, V.C. 1966. Of Scientists and Salamanders. San Francisco: W. H. Freeman.

Zug, G. R. 1993. Herpetology: An Introductory Biology of Amphibians and Reptiles. New York: Academic Press.

Reptiles

Carr, A. 1980. The Reptiles. Life Nature Library. Alexandria, VA: Time-Life Books.

Coborn, J. 1987. Snakes and Lizards: Their Care and Breeding in Captivity. Morris Plains, N J: Tetra Press.

Eckert, S. A. 1992. Bound for deep water. Nat. Hist. 101:29-35.

Ernst, C. H. 1992. Venomous Reptiles of North America. Washington, D C: Smithsonian Institution Press.

Ferguson, M. W. ed. 1984. The Structure, Development and Evolution of Reptiles. New York: Academic Press.

Gans, C. 1970. How snakes move. Sci. Am. 222(6): 82-96.

Guillette, L. J., Jr. 1993. The evolution of viviparity in lizards. BioScience 43:742-751.

Heatwold, H. 1978. Adaptations of marine snakes. Am. Sci. 66:594-604.

Klauber, L. M. 1982. Rattlesnakes. Berkeley: University of California Press.

Linzey, D. 2001. Vertebrate Biology. New York: McGraw-Hill.

Lohmann, K. J. 1992. How sea turtles navigate. Sci. Am. 266(1):100-106.

Schwenk, K. 1994. Why snakes have forked tongues. Science 263:1573-1577.

Seigel, R. A., and J. T. Collins. 1993. Snakes: Ecology and Behavior. New York: McGraw-Hill.

Thorpe, R. S., W. Wuster, and A. Malhorta. 1996. Venomous Snakes: Ecology, Evolution and Snakebites. New York: Oxford University Press.

Chapter 19

Laboratory Exercise

AMPHIBIANS & REPTILES

Laboratory Objectives

✓ to identify characteristics of the amphibians and reptiles,
✓ to observe live amphibians and reptiles to become acquainted with some of their behaviors
✓ to dissect and identify internal structures of a frog

Needed Supplies

✓ prepared and living amphibian and reptile specimens
✓ frog skeleton
✓ preserved frog specimens for dissection

Chapter 19
Worksheet

(also on page 611)

Subphylum Vertebrata:
Amphibians and Reptiles

Herpetology

Objective:
By answering a series of questions, you will survey the anatomy, behavior, and ecological adaptations of various amphibians and reptiles. The exercise is divided into three parts.

Part I.   References
Questions posed in this section can be answered by referring to your lecture notes and/or textbook.

Part II. Observation
Questions in this portion can be answered through direct observation of living or preserved specimens.

Part III Insights
Some independent thought is required to answer each of these questions.

Part I. Text References

1. Give two <u>anatomical</u> differences between salamanders and frogs.

    a. _Salmanders- External gills, lungs, or both, Have tails_

    b. _Frogs- Tymphanic membrane, doesn4 have tail_

2. Give any three differences between amphibians and reptiles.

    a. _Reptiles circulatory more ebbicient than Amphibians_

    b. _Reptiles lungs better than Amphibians_

    c. _Reptilian jaws are modified to grab,crush prey
    Amphibians mouth is large with small teeths_

3. Give three <u>anatomical</u> differences between lizards and snakes.

    a. _lizzard -moveable eyelids Snakes- No eyelids_

    b. _____ - Ear opening _   - No Ears

    c. _____ - Tongue _   -Flick tonque
                                       that move
                                       Faster

4. Where would you find Jacobson's organ? _ROOF OF mouth_

    What is its function? _PICK chemical senses inair_

5. What is a plastron? _ventral side of turtle_

.........a carapace? _Dorsal side of turtle_

What special problem do these features cause turtles and how is it solved?
Shell → Protective
→ Restrictive breathing can only expand how
much shell allows. Solve by abdominal &
Pectoral muscle to assist breathing

6. Give three differences between a lizard and an alligator.
    a. _Alligator - 4 chambers❤️ Lizzard - 3 chamber ❤️_
    b. _Alligator - dont regenerate Lizzard - regenrate tail_ tail
    c. _Alligator - opportunistic beeders, where as_
       lizzards are not

7. Give two ways that you can differentiate anatomically between caecilians and legless lizards.
    Caecilian - Have eyes   legless lizzard - Blind
             - legless   - Have legs

8. What is neoteny and what animal on display shows this feature?
Neoteny - Larval characteristic retained in adulthood
ex: Salamanders. Retains juvenile beatures as
adult. Mudpuppy

9. What is the function of the pit in rattlesnakes?
Heat sensors, Sensory structure -

Part II. Direct Observation

10. Give three anatomical differences between frog and snake skeletons.

    a. _Frogs have no tail structure_
    b. _no legs in snake_
    c. _lots ob vertebrate_

11. In what unusual way does the alligator snapping turtle use its tongue?
     Prevents water loss

12. Observe a reptilian egg. How has this amniotic egg contributed to the success of reptiles?
     Salamander skin is a respiratory surbace
     need to live in moist environments

13. Compare the skin of a salamander to that of a gecko. What difference does this make in the life of these two animals?
     Salamanders skin is a respiratory surbace
     and need to live by moist environments
     Gecko doesn't have skin glands; can
     live in any environment

Part III. Insight

14. Arthropods periodically shed their heavy exoskeleton in order to grow, a process called ecdysis.
      Do you think that turtles, too, shed their shell (i.e., carapace and plastron) in order to grow?
      Explain how turtles are able to get larger.

Turtles don't shed their shell as their sensory nerves are attached in the carspace. When they are born they have scutes which is individual units of the shell and that grow larger

15. What are the structures just in front of the forelimbs of the mudpuppy (*Necturus*)?

External gills are in front ob borelimbs

   Is this an adult or a juvenile animal? Support your answer.

Can't tell (7-8 yrs old only live)

16. What do you think is the purpose of the large tail in the leopard gecko? (Hint: It comes in handy during periods of famine.)

Tail stores Fat, balance,

Predator avoidance

The gecko shows two types of ~~camouflage.~~ What are they?

_They can regenerate their tail so predator misconception
— can refugee in small spaces

17.   What specific type of color pattern is displayed by the tiger salamander?

Bold color patterns. That display their toxic capability through bold color patterns
Aposematic coloration - warning

18. What about a horned toad's skin tells you that it is really a lizard and not a toad?

Toads have dry skin

19. Explain how a snake's skull can permit it to swallow a prey animal much larger than its own head. Also, how does a snake breathe while doing so?

It's able to disslocated

Part IV.

Frog Dissection

I. External Anatomy

View your frog specimen from the dorsal perspective. Note the **color pattern** that helps to match the animal to its surroundings. Depending upon the substrate, the animal can change its basic color and pattern. Next, note the pair of **external nares** that leads internally to the lungs. Posterior to each eye is a **tympanic membrane** designed to detect airborne vibrations. Note the positioning of the four limbs; each is splayed out to the side, which limits the animal's overall locomotion and efficiency. Between the large hind legs is an external opening, the **cloaca**, for exit of undigested food, urine, and reproductive elements. Finally, note the webbing between the toes of the hind limbs. Webbing of this sort aids the animal in swimming.

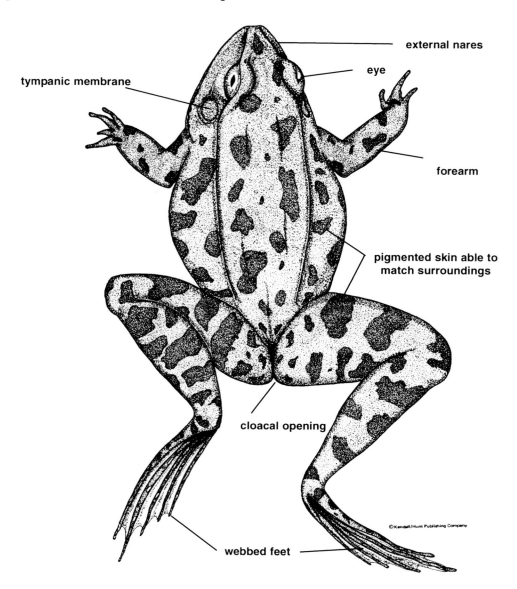

Figure 19.20. External view of the grass frog, *Rana pipiens.*

II. Oral Cavity

1. Pry open your specimen's mouth and examine the **oral cavity** as seen in the figure below and note the large **tongue**. How is it attached?   How is the attachment of the tongue related to the way in which the animal feeds?

2. A pair of **internal nares** is located near the cavity perimeter. What is their function?

3. Although frogs lack teeth in the real sense, they do possess a pair of **vomerine teeth** and a row of **maxillary teeth**. Locate and examine these.

4. At the back of the oral cavity, note the entrance into the **esophagus** and then locate the opening into the **Eustachian tubes** that lead to the inner ear. The communication between the oral cavity and the inner ear helps to equalize ear pressure in much the same way as in humans. It's why you yawn if your ears are "blocked" following an airplane ride.

III. Abdominal Organs

Place your specimen on its back and cut through the abdominal skin to expose the internal organs.

1. The most obvious organ is the large **liver**. As in the shark, there are three lobes. Locate each and find the **gall bladder**, attached (again as in sharks) to the median lobe. Locate the **bile duct** leading from the gall bladder to the **duodenum** of the small intestine. Then find the **pancreas** and note that the pancreatic duct intersects the bile duct. What is the role of:

      a. the gall bladder? _____

      b. the bile duct? _____

      c. the pancreas? _____

      d. the pancreatic duct? _____

2. Return to the area of the oral cavity and locate the **esophagus**. Trace it downward to the **stomach** and then complete your examination of the digestive system by following the **small** and **large intestine**. Note that the large intestine terminates at the **cloaca** and that the **urinary bladder** empties into the cloaca. Finally, locate and examine the **spleen**.   What is the function of:

      a. the esophagus? _____

      b. the stomach? _____

      c. the small intestine?_____

      d. the spleen? _____

      e. the urinary bladder?_____

      f. the cloaca? _____

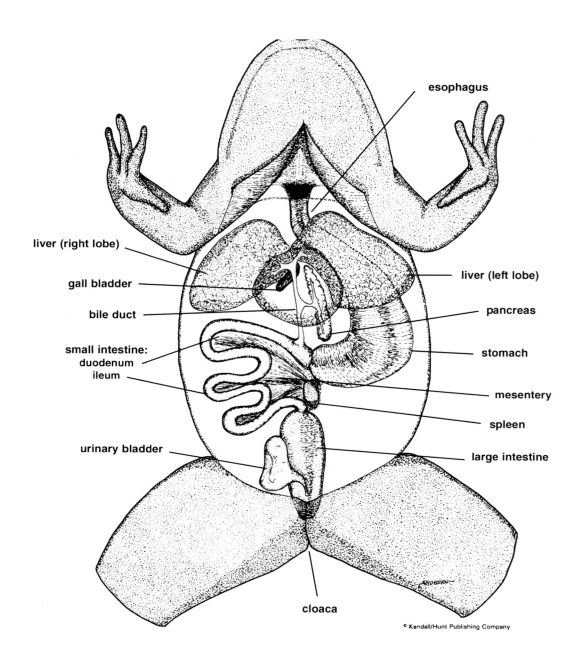

esophagus

liver (right lobe)

gall bladder

bile duct

small intestine:
duodenum
ileum

urinary bladder

liver (left lobe)

pancreas

stomach

mesentery

spleen

large intestine

cloaca

© Kendall/Hunt Publishing Company

Figure 19.21. Internal view of the frog's body cavity showing the digestive system.

IV. Thoracic Organs

1. Continue the abdominal incision forward to expose the **heart**, **lungs**, and remaining thoracic organs.

2. Respiratory System. Respiration in amphibians is accomplished through both the outer body surface (cutaneous) and through a pair of **lungs**. Examine the exterior surface of both lungs. Do they appear large or small to you? Also examine the interior of the right lung and note the branching **bronchioles** within.

3. Circulatory System. Examine the heart exterior. The more muscular portion of the heart is the **ventricle**. Located above the ventricle are the two thin-walled **atria**, above which is the **conus arteriosus**. Exiting the conus are a number of large arteries. Locate the pair that passes laterally to the lungs. These are the **pulmocutaneous arteries**. Directly above these are the two **systemic arches of the aorta**. The systemic arches unite within the abdominal cavity as the **dorsal aorta**. Finally, the most superior vessels exiting the conus are the **carotid** arches, which lead, eventually, into the head area.

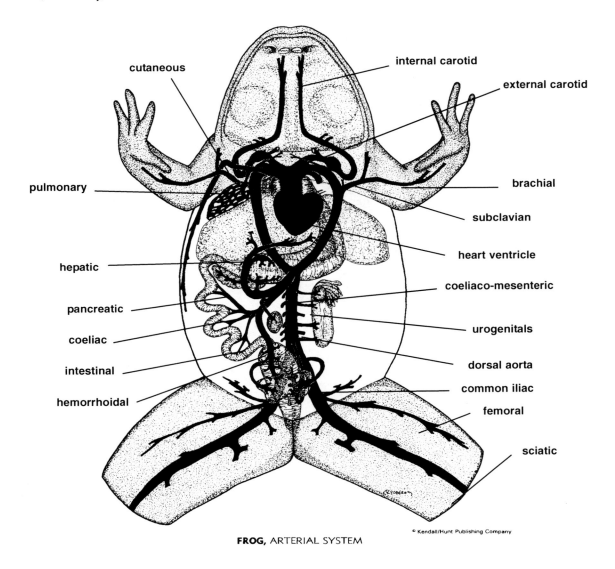

FROG, ARTERIAL SYSTEM

Figure 19.22. The major arteries of a frog.

V. Urogenital System: Female

1. If your specimen is a gravid female, the body cavity will be filled with an **egg mass**. You can locate the **oviduct** by probing beneath, and lateral to, the egg mass. Note that for eggs to be deposited externally they must pass through the oviduct and the cloaca.

2. Locate the rod-shaped **kidneys** with their attached **fat bodies**. What is the function of the fat bodies? Find the **ureter**, which carries urine to the cloaca and then to the outside.

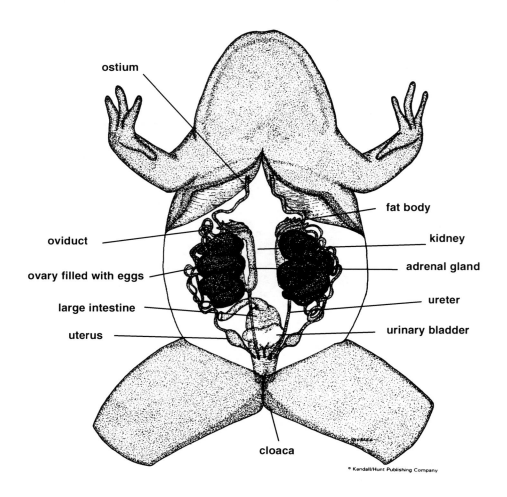

Figure 19.23. The female urogenital system.

VI. Urogenital System: Male

1. The **testis** are attached to the kidneys. Sperm travel from the testis to the kidney and then through the **urogenital duct**, to the cloaca, and then to the outside.

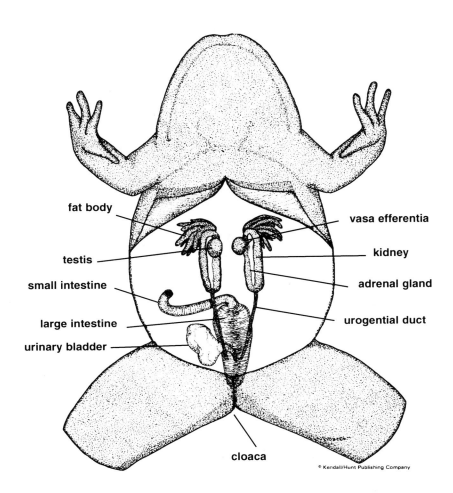

Figure 19.24. The male urogenital system.

VII. Skeletal System

1. The overall frog skeleton is adapted to a **saltatorial** (jumping) habit. Note the powerfully built, yet elongated nature of the hind legs. They, and the long toes, provide the leverage for the take-off phase of a leap. Meanwhile, the stalwart design of the forelimbs, the fused vertebral column into the **urostyle**, and the hefty pelvic and pectoral girdles are all intended to absorb the shock of landing. Even the head is fortified with a high degree of bone fusion to overcome the impact of touchdown.

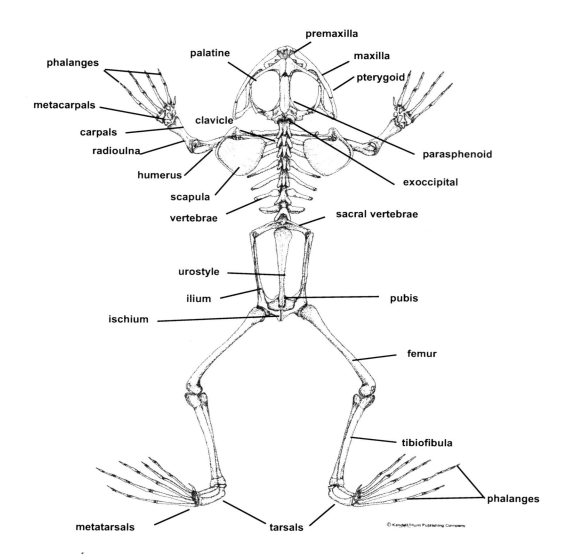

Figure 19.25. The frog skeletal system.

20 Chapter 20

THE BIRDS

20 CLASS AVES

The Birds

Staying Warm...Keeping Cool

*I*t's a cool April evening some years ago on the Anza Borrego Desert in southern California. The class I'm teaching on Vertebrate Natural History is spending a day observing wild animals. Our goal is to record any and all vertebrates seen during one 24-hour period. We arrive around noon to set up camp and I issue my usual instructions ("Always use the buddy system"..."Don't eat the cactus flowers"..."Don't lick the lizards"...and "Under no circumstances are you to pick up, molest, or handle any rattlesnake").

Exactly one day later, following a sleepless night and many hours in the field, we met to compile our data. The results reveal two worlds in one: a daytime fauna that differs significantly from that of the nighttime. During the daylight hours, although mammals are recorded, they are never plentiful. It's the reptiles that rule the daytime. For instance, on this particular visit we recorded over a dozen species of lizards, including such forms as horned "toads," chuckwallas, iguanas, skinks, whiptails, and geckos; eight different snakes (rattlers, rosy boa, shovel-nosed, lyre, gopher, and king-snakes); and even a desert tortoise.

However, the picture changed abruptly at dusk. As the temperature dropped, so, likewise, did the number of reptiles. The cooler it became the fewer snakes and lizards we encountered. In fact, the only reliable place for reptiles during a cold night was on the asphalt roadways where heat from the sun still lingered. As reptile numbers declined, they were replaced by kangaroo rats, kit foxes, woodrats, pocket mice, cactus mice, and bats. It seemed to us that, except for owls, the night was governed by mammals.

On the surface the picture reflected a rather simple dichotomy: days were given to reptiles, nights to mammals. But closer scrutiny revealed a somewhat different, and more complex, pattern. In reality, all vertebrates (reptiles, birds, and mammals) were active during the late morning and early evening hours but few were recorded during the hottest part of the day. At night, when temperatures were the coolest, reptiles were surprisingly scarce, but mammals were abundant.

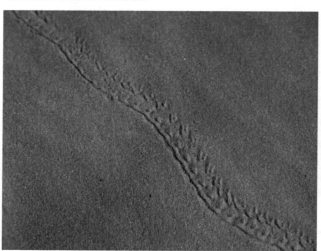

Figure 20.1. Lizard tracks in the sand (left) and a diamond-back rattlesnake, a common desert resident.

Endothermy ~ the internal furnace

Daily patterns of vertebrate animals seem to differ considerably within desert environments. To understand why that is so we need to look to their physiology, more specifically, to the ability of an animal to regulate its internal body temperature. We have already seen, in the last chapter, that reptiles are ectotherms. Their internal temperature depends upon environmental conditions. Since metabolism (cellular function) is more efficient at warmer temperatures, it stands to reason that ectotherms will be more active during the warmer part of the day.

So how then do we explain the relative inactivity of all animals in the afternoon hours, when it's really warm...hot, actually? The answer, again, is found in their physiological makeup. It's true that metabolism is more effective at warmer temperatures, but only to a point. Danger lurks if the body gets too warm. As internal temperature rises above a critical level, cells begin to dysfunction, metabolism breaks down, the body dehydrates, and death is a real possibility. So, when it's really hot on the desert, reptiles seek shelter. But so do birds and mammals. Thus, we recorded few overall animals out and about on hot afternoons. The same factors dictate reptilian behavior at nighttime. On cool nights reptilian metabolism cannot keep pace so it slows down; reptiles become lethargic and inactive, but nocturnal birds and mammals don't. Why is that? Again, one word provides the answer: **endothermy**. Birds and mammals are endothermic (warm-blooded). They can adjust and maintain their internal temperature irrespective of outside conditions. When it's warm or cool they are active, when it's hot they can still be active (but with vigilance), and when it's really cold they can be active then, too, but again, cautiously. Mammals, using their keen sense of smell have taken largely to the nighttime while birds, being primarily visual animals, have exploited the daytime.

Birds and mammals are the only animals able to self-regulate their internal temperature; they are the only endothermic animals on earth. If that's the case, there must be some advantage, some value to endothermy for these animals. The nature of that advantage is clear from the above paragraph: **activity**. *An endothermic animal is an active one.* They are not shackled to the whims of climate; their metabolism is not linked to a vicarious environment. Because they have an internal furnace that keeps them warm under virtually any circumstance, they are free to be active regardless of outside temperature. As a result, they can invade habitats not open to reptiles. Birds and mammals make up virtually all of the Alaskan vertebrate fauna. They are dominate in Siberia, Canada, Patagonia, the Arctic, and the Antarctic. They are the featured vertebrates on mountain tops, on the taiga, and on the tundra.

Endothermy, then, seems to make an animal more independent, more prosperous. That's certainly true, but it's achieved at a cost. There's a price to pay for our constant level of activity, and *the price is energy*. It takes a lot of energy to regulate our temperature and to remain active. When it gets cold outside reptiles slow down, their metabolism moderates, energy demand drops, and they cease eating. Birds and mammals, on the other hand, to remain active must keep stoking that metabolic furnace, and food is the needed fuel. As much as 98% of the energy budget of an endotherm goes to temperature maintenance. Thus, birds and mammals must keep eating. Reptiles can go days, weeks, even months without a good meal; non-hibernating birds and mammals are dead in a few days. So it's a trade-off. Remaining active is an evolutionary advantage, but endotherms must consume tremendous amounts of food to reap that prosperity.

Birds: Feathered Flight

*A*s amazingly beautiful and complex as birds are, they are defined by just two things: feathers and flight. *Among animals, only birds have feathers, and all birds have them.*

Feathers

Although feathers come in a wide variety of forms, colors, shapes, and textures, they all have one thing in common: they are all modified scales. In fact, feathers are one of the linking factors that tie birds to their reptilian ancestors. One only has to look at the feet of most birds to see scales reminiscent of those covering the surface of reptiles.

Feathers are one of nature's most versatile creations, one of her true marvels. Of relatively simple design, they are, nonetheless, rich in their diversity and utility. They provide a protective covering, and, if adequately maintained, are waterproof, strong, and durable...thus, birds spend much of their time preening, an action that helps to sustain feather vitality.

Feathers have a number of other functions. They protect the skin from dangerous UV radiation and insulate from cold and heat. Through their kaleidoscope of colors, shapes, and arrangements, feathers also serve to attract mates. They repel adversaries, are used in self-defense; they whistle, snap, boom, and flutter. They are used to line nests, they aid in species identity, and obviously, they help birds to fly. Feathers are, without question, an all-purpose innovation.

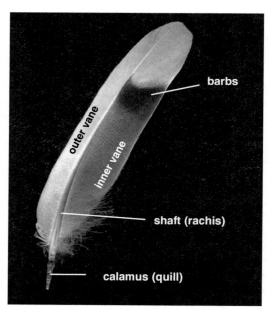

Figure 20.2. A typical primary flight feather. The narrow outer vane overlaps the inner vane of the feather in front of it, which strengthens the overall flight surface.

Flight

But birds are also defined by flight. Indeed, adaptation to flight is the singular theme in bird evolution, anatomy, physiology, and behavior. Almost everything about a bird is, in some fashion, related to flight. Although not all birds fly, the overwhelming majority do, and those that don't, apparently once did.

To fly, three things are required: **lift, forward motion,** and **elimination of friction.** Lift is the term given to the process of getting off the ground. You cannot fly if you are landlocked. However, getting off the ground, in itself, doesn't result in flight. To be a truly flying animal, you must be able to propel yourself forward. In fact, the definition of true flight (as opposed to soaring) is *self-sustained forward motion.* To fly, birds must provide the power or **thrust** for this forward motion. Finally, to maximize flight, a moving object must eliminate as much **drag** (friction) as possible. Therefore, all flying animals are **streamlined** in design. These three requirements must be met in order to fly, and birds show considerable adaptations to each.

Adaptations to Flight

I. Lift (weight-reducing) Adaptations
✓ pneumatic (hollow) bones
✓ teeth replaced by a lightweight beak
✓ tail is reduced to a small knob, the pygostyle
✓ forelimbs modified into wings
✓ no bladder, right ovary lost; no penis
✓ nonbreeding testes & ovaries atrophy (shrink)
✓ body covered by lightweight feathers
✓ no skin glands
✓ oviparous; adults don't carry heavy embryos

II. Power Promoting (thrust) Adaptations
✓ endothermy (high-energy metabolism)
✓ heat-conserving plumage
✓ high blood glucose (blood is high in energy)
✓ energy-rich diet
✓ four-chambered heart
✓ rapid heartbeat, high blood pressure
✓ efficient respiration; air sacs & flow-thru lungs
✓ efficient muscles & keeled sternum for attachment

III. Streamlining (reducing drag) Adaptations
✓ fusiform (tapered at both ends)
✓ internal gizzard replaced need for teeth
✓ liver on right/gonads on left = internal balance
✓ no pinnae (ear lobes); no tail
✓ retractable legs during flight
✓ center of gravity beneath wings
✓ use of tendons for extremities, thus limbs are light
✓ V-shaped flight patterns conserve energy

Figure 20.3. Birds spend a tremendous amount of time tending to their feathers. An American avocet and a great egret (top photos) are seen preening the important flight feathers, while the anhinga (a diving bird) exposes its wings to the drying effect of the sun. The lower left photograph shows an American bald eagle soaring. You can sense the lift given by the large wingspan (airfoil). Note also the slotting between the primary flight feathers on the outer edge of each wing. Slotting causes the air to move more smoothly over the wing to prevent stalling. Finally, the lower right photo is of a peregrine falcon, the fastest of all birds and perhaps the fastest animal on earth. Note how the overlapping feathers provide a smooth surface for air to flow.

Advantages of Flying

One of the most remarkable and dramatic steps taken in forward locomotion was winged flight. It has so many advantages that the evolutionary pressure to develop flight must have been considerable. In fact, it's surprising that so few animals have taken to the air; only insects, birds, and bats fly. Maybe, the energy demand is so high that few animals have been willing to pay the price. Nonetheless, the benefits are incredibly great. Perhaps the most obvious one is the easy escape from predators. This is so evident that birds are about the only animal, with the exception of some insects, that can afford to flauntingly expose themselves without fear of predation. Just think about the many birds each spring that find an exposed perch from which to sing, posture, dance, cavort, and display. They seem to do so with an almost flagrant disregard for any threat that might happen along.

Other advantages afforded by flight include the exploitation of otherwise unavailable niches and the ability to escape undesirable ones. Food that is out of reach of a land-bound animal is easily reached by a flying one. The same is true with nesting sites. Birds are not compelled to nest only on the ground, although many do. They can, instead, build their nests high and out of harms way. It is no accident, therefore, that most birds construct nests in trees, on cliffs, or in other inaccessible locations. If one site is unacceptable, or if food sources dissipate, relocation to a new site is always possible for a flying animal. If the climate changes, then distant relocation, too, is not out of the question, as bird migration amply attests.

Figure 20.4. The advantages of flight are many. Birds seem to almost blatantly expose themselves to danger but can easily escape most enemies by simply flying off (plate **A** & **B**). Flight also opens food resources not available to land-bound animals (**C**) as well as relatively inaccessible nesting sites as with the eagle in plate **D**.

The Senses

A flying animal depends upon its senses as much as, if not more than, any nonflying animal. A constant stream of accurate information is mandatory if a bird is to fly skillfully. The senses of **vision** and **hearing** are, thus, particularly well developed in birds. Their eyes and corresponding optic lobes are huge as is the cerebellum (for muscular coordination and orientation). Many birds have their eyes placed laterally on their heads, which gives them an almost 360° field of vision. This is important in ground-feeding birds, for instance, that must constantly survey their surroundings for predators. Other birds, such as hawks, eagles, and owls, have eyes more frontally placed. This gives them a lesser overall visual field but greater binocular capabilities. They, thus, sacrifice a wide field of vision for the greater depth perception necessary when searching for, or diving on, a prey animal.

Most birds have an acute sense of hearing and are alerted to the slightest noise or disturbance. This, combined with their sensitive eyesight, places them among the most vigilant of animals. After all, flight allows a bird to escape only if it can accurately sense its environment. However, among all birds, the auditory specialists are the owls. They live in a world of sound beyond what we could ever imagine. Few animals can match them when it comes to hearing, which is a distinct advantage for an animal that hunts at night. An owl's auditory apparatus is anatomically unique in that, in most species, the two ear cavities are placed asymmetrically on the skull. One ear is larger than the other and placed higher on the head. This enables owls to pinpoint the most delicate of noises or stirrings made by their prey even in total darkness.

As much as vision and hearing are enhanced, olfaction, in birds, is reduced. They just don't have much of an ability to detect airborne chemicals. The exception occurs in vultures and other carrion feeders. Most vultures can to detect a dead animal from considerable distance merely through the sense of smell. This explains how a soaring turkey vulture can locate a carcass in dense foliage or how a California condor can find a freshly killed animal from several miles away. But, overall, a bird's sense of smell is quite limited, probably somewhere in the same range as in humans

Figure 20.5. The placement of eyes differs between prey and predator. Ground-dwelling birds, such as the turkeys (**A**) tend to have eyes placed to the side, which gives them a wide field of vision. Predators, on the other hand, tend to have eyes placed in front of the head, as in the owl (**B**), or situated so that they can survey the ground surface while soaring (**C**). Also, in owls the ears are asymmetrically placed, one being slightly higher than the other to offer a better sense of sound direction. Except for vultures (**D**), birds have a weak sense of smell.

The Power for Flight

Birds, being endothermic, must constantly maintain a heightened metabolic rate. This requires high-energy food, coupled with efficient respiratory, circulatory, and muscular systems. Although birds feed on a variety of substances, they prefer foods high in energy such as nectar, seeds, and insects. The energy demands, for instance, of a hummingbird with its wing-beat of 80 per second are extreme. It feeds on high-energy nectar and, when not feeding, can enter into a semi-dormant, **heterothermic**, state to help conserve its all-important energy.

Unlike their reptilian ancestors, birds don't have teeth. Most biologists relate the loss of teeth and large jaws as adaptations to flight. However, even though teeth are absent, food is, nonetheless, broken down mechanically and mixed with digestive enzymes by an internal, muscular **gizzard**. Seed-eaters usually have stones or pebbles in their gizzard to help further crush the food. By the way, another link between birds and dinosaurs is the presence of such stones in many fossilized dinosaur remains.

The relationship between a high-energy diet and the power for flight is nowhere better seen than in the avian respiratory system, perhaps the most efficient in all of the animal kingdom. Bird lungs are so designed that air moves in only one direction across the respiratory surface. In mammalian lungs, by comparison, air moves in and out of alveoli (tiny air sacs) in a tidal-like action. As a result, stale air is always present in the lung of a mammal. This is not the case in birds with their "flow-through" respiration. Incoming "'fresh" air bypasses the lungs to be stored in a series of posterior air sacs.

When the bird exhales, the fresh air then passes forward through the lung to be stored in other sacs located anteriorly. Air is thus constantly in contact with the respiratory surface enabling maximum extraction of oxygen. The respiratory apparatus of birds is considered the most effective among all animals, even though their lungs are comparatively small.

Flying requires powerful muscles, which, in turn, requires a steady and abundant supply of oxygen. So it's no surprise that, along with the respiratory system, the circulatory system of birds is immensely efficient. For instance, birds were the first animals, as a group, to develop a four-chambered heart. The importance of such a heart is that oxygen-depleted and oxygen-rich blood are completely separated by the double-pump mechanism. There is no mixing of oxygen-poor and oxygen-rich blood. In addition, a bird heart beats at a pace much higher than a comparably sized mammal, which places blood under relatively high pressure. Your heart beats about 70 times each minute. The heart of a robin, on the other hand, may beat as fast as 600 beats per minute and a hummingbird may reach rates in excess of 1200 bpm!

The oxygen-carrying capacity of bird blood is also beyond what we might expect. Bird blood has an inordinately high level of hemoglobin, the substance to which oxygen clings as it travels throughout the body. As a combined result of a bird's high-energy food and its efficient respiratory and circulatory systems, the power for flight is achieved.

IT'S A FACT

The bird respiratory system is perhaps the most effective known. This is because air passes through the lungs to collect in large air sacs located within the body cavity. Thus, there is never any stale (mixed) air in the bird lung. As a result the avian respiratory system is extremely efficient.

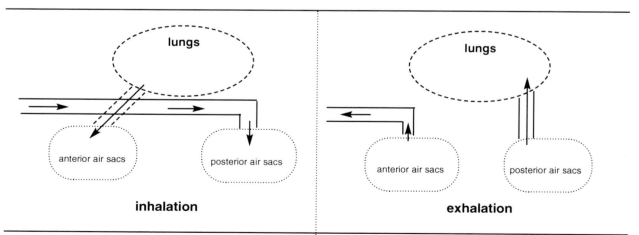

Figure 20.6. Respiration in birds, as shown above, is one of the most efficient mechanisms known. Air that is drawn in during inhalation bypasses the lungs to be stored in large posterior air sacs. As that is happening, "used" air from the lungs passes into large anterior air sacs. During exhalation, the fresh air from the posterior sacs passes into the lungs where gaseous exchange occurs. Meanwhile, the used air from the anterior sacs leaves the body through the trachea and mouth. Note that air passes through the lungs instead of accumulating as stale air, which happens in mammalian (human) lungs.

Bird Reproduction

Birds are animals of instinct. Their behavior follows a well-scripted process dictated by their genes. Although some relatively high-level learning can, and does certainly occur, their behavior overall is predominately innate, consisting of ingrained, preset patterns. Some of the more advanced species, such as crows and jays, have a great ability to mimic, and even to "solve" minor problems. But, on the whole, birds are instinctive animals. Nowhere is that more evident than in their reproductive activity. In general, the sequence of bird reproduction involves six highly instinctive steps.

1. territory establishment
2. courtship
3. pair bonding
4. nest construction
5. egg laying and incubation
6. brooding and care of young

(1) **Establishment of a Territory.** Virtually all birds are territorial. The size and types of territories may differ, but most birds possess one, especially during the breeding process. A territory is defined as *any space defended against others of the same species.* In other words, a territory owner attempts to exclude, not all birds, but only those like himself. In most cases, it's the male that establishes and defends his territory against other males of the same species. Those best able to secure and guard a territory will, in general, have the best territories on which to raise young.

A territory that is successfully defended provides a number of positive functions. For example, by dividing a habitat into defensible parcels, local resources will not be overused; thus territories help to conserve and apportion natural resources. Also, by spreading populations out over the habitat, the impact of disease and parasites is, likewise, reduced.

Ownership of a territory confers upon the owner two other advantages. First, he becomes familiar with the area. He knows where food may be found and where shelter is located. Survival is thereby enhanced. Second, owning a territory gives the owner the edge in defense. On his territory, he essentially becomes invincible against other males. Birds, and other territorial animals, become quite belligerent when their territory is invaded. Anyone who has disturbed a nest full of baby birds can attest to the parent's indignation.

Figure 20.7. Male birds establish a territory by displaying, posturing, dancing, and singing. Once established, a territory is defended from any male intruder of the same species, and potential predators.

(2) **Courtship**.   The most important function of a breeding territory is that it helps to attract potential mates. A male who owns a territory will advertise that fact. He signals his availability by singing, posturing, dancing, and otherwise displaying his various attributes. If a female is present, he will escalate his efforts, often to a frenzied pitch. This process is termed **courtship** and is initiated by the male through song and ritualized behavior. Males sing incessantly at the start of the breeding season to call attention to themselves and their territory. They also display any color, patterns, or special features they possess, all in an attempt to get the attention of potential mates. They will even offer gifts in the form of food and/or nesting material.

Figure 20.8. Bird courtship wears many costumes.

(3) **Pair Bonding.** The female listens and watches. If what she hears and sees is acceptable, she will select that male. That selection process forms a bond between the two and leads to copulation, which in birds lasts only a few seconds. Most males lack a penis and must transfer sperm to the female by means of cloacal contact ("cloacal kiss").

Females have only the left ovary and oviduct; the right organs have been eliminated. Similarly, the right testes is reduced in males. The gonads of both sexes enlarge greatly during the breeding season; in fact, a male's testes may increase as much 1,500 times when reproducing.

If a female chooses well, her young will have adequate resources and a male able to defend them and the breeding area. Such a bond can last a lifetime, as in some geese, swans, eagles, and others, or it may last only the time it takes to copulate (e.g., hummingbirds). Most likely, however, the bond will last throughout a single breeding season.

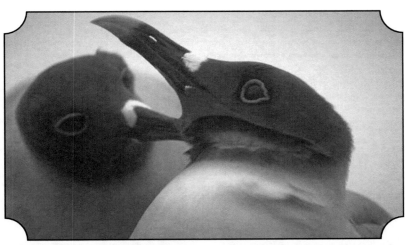

Figure 20.9. For most birds to mate successfully, they must establish strong pair bonds. This is often accomplished by physical contact between the male and female, or by dancing, mutual posturing, and other reciprocal behaviors.

However long it might last, pair-bonding in birds results in strong reproductive relationships. The most common type of bonds are **monogamous,** with one male and one female raising young as a family unit. Until recently, ornithologists believed that upwards of 90% of birds are monogamous and that the relationship was unbreakable. A new study, however, has leveled some doubt on the strength of bird monogamy. What we thought was a faithful, one-on-one relationship may not be so unfailing after all. It seems that, in reality, some birds are untrue to their partners.

In the study, genes of nestlings (newly hatched birds) were compared to their parents through DNA analysis. The idea was to determine if all of the young were from the original pair. The results unexpectedly showed that about 75% of the nests had offspring not sired by the owner of the territory. In some cases, in fact, nests had young from several fathers! It seems that while "dad" was off singing, dancing, and defending the territory, mom was entertaining neighboring bachelors. Apparently her deception went undetected as the original male seemed to have no insight into the chicanery. He continued to defend the territory and, in many cases, to protect, feed, and care for the young—some of which were fathered by another male!

So ornithologists have adjusted their view of bird monogamy. Yes, 90% of birds are monogamous, *socially.* They establish and defend a territory together. They build the nest and raise the nestlings as a pair. They appear in public as a faithful couple. However, beneath the facade, it appears that they are far from monogamous. In these cases, the brood is a mixture of genes from more than one father. Genetically, they, or shall I say she, is rather indiscreet.

There are two other bonding patterns in birds. One is **polygamy.** The males of many bird species have more than one partner. This pattern is termed **polygyny** and is seen, for example, in red-winged blackbirds. On the other hand, a female may have more than one partner (**polyandry**) in which case the parental roles are reversed; the male builds the nests, incubates the eggs, and raises the young while she defends the territory. This is a rare condition among birds, occurring in such species as button quail and phalaropes.

The final, and least common, reproductive pattern is **promiscuity.** Here, males and females seem to mate indiscriminately. The pair-bond has a short duration, lasting essentially only as long as copulation lasts. The male then goes his way and the domestic chores are left entirely to the female. Hummingbirds are examples of promiscuous birds.

Figure 20.10. Most birds form pair-bonds that last for the length of a breeding season (e.g., one summer). However, a number of species, such as the swans shown here, mate for life.

(4) Nest Building. Nests have evolved as devices to hold, protect, or cradle the fragile, shelled eggs during embryonic development. The great majority of birds build a nest of some type (see below). However, there are exceptions. For instance, owls do not build nests, preferring to rely on vacated cavities or deserted nests of another bird. Then there are the brood parasites who use "foster" parents to raise their young. In the United States, the brown-headed cowbird (*Molothrus ater*) lays its eggs in nests of other birds, usually warblers. The foster parents (warblers) then incubate and care for the young cowbird, often at the expense of their own offspring. Once the young cowbird is able to fly it leaves the foster parents to join other cowbirds.

However, most birds construct their own nests as outlined in the *Inside View* on the next page.

Figure 20.11. Birds have adopted a variety of nesting patterns. Some nest on the ground, building scrapes or platform nests (**A** & **B**). Others prefer to take over abandoned rodent or squirrel burrows, such as the burrowing owl in plate **C**. Woodpeckers (**D**), however, excavate their own cavity nests. Many large birds build elevated platform nests as shown by the osprey in panel **E**.

An Inside View 20.1

Bird Nests

There are four basic types of bird nests:

1. A **scrape** is a simple depression in the ground that birds scratch out to receive the eggs. Many shorebirds (killdeers, sandpipers, turnstones) use this type of nest. The eggs are cryptically colored to blend with sand, pebbles, or other substrate.

2. Many birds build a **platform nest,** either on the ground or elevated. Here vegetation is piled into a large mass to hold the eggs. Ground platform nesters include geese, ducks, pelicans, loons, and many others. Elevated platform nests are built by hawks, eagles, herons, and cormorants, among others. Platform nests can become huge; for example, eagles continue to add to their platform nests year after year until the nest may exceed several meters in diameter and over a ton in weight.

3. A favorite nest type of many birds is the **cavity nest**. A hollow is excavated in soil, trees, or other material to hide the eggs. Birds such as woodpeckers and kingfishers excavate their own cavity each year. However, many other birds (owls, starlings, nuthatches) occupy abandoned cavities. Eggs of cavity nesters, being hidden from view, are usually white.

4. Most birds build a **cupped nest** of some sort in which the eggs are cradled in a sinklike depression. The most common cupped nest (**statant**) is supported from below. Examples are many: robins, warblers, cardinals, sparrows, and hummingbirds. Some birds (swallows and swifts) glue nests to a vertical surface such as a building, bridge, or rock face (**adherent nest**). Another cupped type is the **pendulous** nest of orioles and weaver birds in which the eggs are suspended at the bottom of a long saclike container.

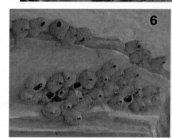

Key to Nest Photos

1. scrape	6. cupped (adherent)
2. elevated platform	7. scrape
3. ground platform	8. elevated platform
4. cupped (pendulous)	9. cavity (excavated)
5. cupped (statant)	

(5) Egg Laying and Incubation. The next phase in the avian reproductive cycle is egg laying and follow-up care. The number of eggs laid by a bird varies with the species. The smaller birds such as robins, sparrows, and warblers, usually have four to six eggs per clutch, while the larger ground-nesting birds (geese, ducks, pheasants) generally have larger clutches, up to 24 eggs, or so. However, there are many exceptions. Pelicans, hawks, condors, eagles, and penguins, for example, are large birds that produce only one to three eggs per year. On the other hand, hummingbirds, the smallest of birds, lays only one or two eggs per clutch.`

The size of eggs depends, of course, upon the size of the female. The smallest known eggs are from the smallest known bird, the Cuban bee hummingbird. The egg is pea sized, although it constitutes 6% of the adult's weight. That's about the same ratio seen in humans (e.g., a 125-lb. woman carrying a 9-lb. embryo). Of course, the woman doesn't have to carry her baby while flying! The largest egg is from the ostrich (grapefruit sized > 3 lbs.). In fact, it is the largest single cell in existence.

The color and pattern of eggs depends, for the most part, on its degree of concealment. The more exposed or unprotected eggs are usually cryptically colored to match the surroundings. Conversely, the more concealed eggs, or those actively guarded, are usually white, or some pale color. The actual egg color is due to pigments laid down in the oviduct as the fertilized egg passes through. Pigments are primarily derived from bile; thus, the colors are usually in the blue, bluish-green, to brown spectra. If the egg rotates as it passes through the oviduct, it will be streaked; if not, it will be spotted or unmarked.

For the embryo to develop within the egg, it must be kept at a constant, warm temperature. In fact, it must be regulated within fairly narrow limits, depending upon the species. To accomplish that, most birds incubate the eggs by sitting on them. This allows warmth from the body of the incubator to pass to the eggs. Incubation is a chore that usually falls to the female, although males of many species participate as well. Eggs also require an optimum humidity for embryos to develop properly. Many birds, such as ducks and songbirds, carry water on their feathers, which is then sprinkled on the eggs to maintain the appropriate humidity.

Incubation periods vary considerably. Smaller birds generally hatch within 7–14 days, but large birds can go up to two months. An ostrich egg, for example, takes six weeks to hatch. A bird sitting upon eggs is highly vulnerable. As a result, incubating birds (i.e., the female) are usually camouflaged to blend with the nest and immediate surroundings. When both sexes incubate, the male, too, is usually cryptically colored.

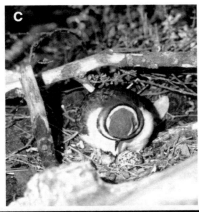

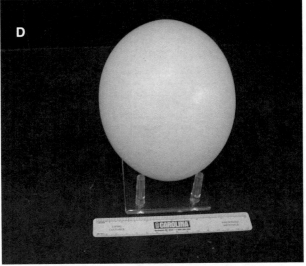

Figure 20.12. Eggs deposited in ground nests are typically mottled or streaked to blend with the background (**A**) although some birds (ducks and geese) have white eggs (**B**) which the cover with downy feathers when they leave the nest. Incubating adults are most often cryptically colored as with the plover above (**C**). The ostrich egg (**C**) is the largest of all bird eggs weighing about 3 pounds.

(6) Hatching. In most cases, birds emerge from the egg unassisted, often only with the aid of a sharp projection, the "egg tooth" on top of their upper beak. Evidence from a number of sources indicates that embryos often emit calls from the egg just prior to hatching. Behavior of this type could serve to synchronize hatching among all embryos or may serve to alert the adults that hatching is imminent.

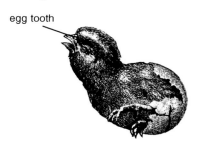

egg tooth

There are two categories of hatchlings, **altricial** and **precocial**. Those birds that hatch naked, or nearly so, unable to feed themselves, and unable to walk are termed altricial. Most songbirds fall into that category, as do hawks, eagles, pelicans, and a large number of other species. Altricial young demand ongoing care from the adults. At hatching they are unable to control their internal temperature so adults continue to transfer body heat to the nestlings, a process called **brooding**. As feathers develop, endothermy is attained and the brooding process gradually ends.

At hatching, precocial young, unlike their altricial counterparts, are covered with downy feathers, can walk on their own, and require less attention from the parents. Brooding may or may not be needed, often depending upon the weather. Precocial young in a single nest often hatch synchronously and leave the nest *en masse* within hours of hatching. Geese, ducks, shorebirds, gulls, and chickenlike birds (pheasants, quail, grouse) are all examples of precocial birds.

Even after young birds fledge (leave the nest), parental duties are not necessarily complete as most young birds continue to receive attention for some time. They are usually fed and protected as they gradually learn the process of finding and securing food on their own. However, eventually, young birds withdraw from their parents to fend for themselves. That frees adults from their reproductive duties, although many immediately repeat the entire breeding process. Some will remain together until migration when the pair-bond is dissolved. Of course, those birds that pair for life remain together constantly.

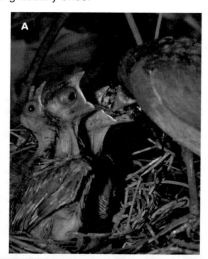

Figure 20.13. Hatchlings are either altricial and lack feathers and mobility at hatching (as the cardinals (**A**) or herons (**B**), or precocial (**C - E**). Precocial young possess feathers and are relatively mobile at hatching, although they still require some parental care. Note the young loon riding on its parent's back and the goslings protected by their mother's wing.

Species Recognition
"Who Am I?"

*H*ow does an owl know it's an owl and not a raven, a woodpecker, a duck, or a sparrow? Then, once it knows it's an owl, how does it know it's a barn owl rather than a snowy or great horned owl? Birds must have some way of knowing who they are, otherwise reproduction would be a chaotic mess. Can you imagine a sparrow who thinks he's an eagle, or a hummingbird who aspires to be a peacock? The truth is, if birds are confused as to their identity they will be unable to form pair-bonds and to carry out the reproductive process...resulting in complete disorder. So how do they do it? It so happens that they learn their identity very early in their lives through a precise process called **imprinting**.

Imprinting is *a genetically determined learning process in which a young animal forms an irreversible impression or attachment to an object*, usually its mother. It only occurs during a *critical, sensitive period* (e.g., immediately after hatching); it happens rapidly, and without positive reinforcement. As a result, powerful social bonds are established and the animal "learns" species-specific signals such as sounds, color patterns, and behaviors. From that time on, it recognizes and responds to its own kind.

Imprinting is a nice, fail-safe method of species recognition and individual identity. However, it's not altogether infallible. In fact, the whole process can be easily sabotaged. Normally the hatching bird imprints on the first large moving object it sees, usually its mother. But what if it sees something else, something very different from its mother; what then? As an undergraduate college student, I was part of a research project that examined the process of bird imprinting. Our biology department operated a field station that had a resident flock of semi-wild Canada geese. I, and several other students, collected goose eggs and artificially incubated them. We timed it so that the first view a hatching gosling had was of one of us. The result was astounding. Each gosling became imprinted on its "foster mother" (a college student) and for the first month or so of its life, it followed us wherever we went. It was not uncommon to see biology majors trudging across campus followed by one or more fuzzy goslings.

It was fun for us, but ultimately not so much fun for the geese. These birds grew up confused. Sure, they looked like geese, they acted like geese, they even sounded like geese, but they preferred the company of humans. I didn't fully realize the strength of the imprinting process until these birds became sexually mature. What a mess. They ignored any and all courtship overtures of other geese. They spurned the head tosses, the cheek waggles, and the incessant honking "come-ons" of mature geese. But at the same time, they were driven by their hormones to reproduce...but not with a goose. No, they were bound and determined to mate with the imprinted image they had formed a few years earlier. That made springtime at the biology station one curious place to be. Geese chasing after students, students determined not to be caught! Imprinting is indeed a powerful force in the life of birds.

Figure 20.14. Most birds develop irreversible attachments to the first moving object they see upon hatching. This is usually their mother with whom they establish strong bonds, almost as though they were physically tied to her. **Imprinting** of this sort helps to maintain species identity throughout life.

Bird Sounds

Birds produce sounds through vocal and nonvocal means. Nonvocal sounds may be made by bill clapping, vibration of the feathers, foot stomping, or even through the use of foreign objects. True vocalization in birds, however, is produced by a bronchial structure called the **syrinx**. The syrinx, while similar to our vocal apparatus (the larynx), differs in its location and design. As a result, many birds have intricate and versatile vocal patterns. Although some birds lack a syrinx, and thus are mute (storks, pelicans, and vultures), most birds produce some vocalization. Indeed, certain birds make sounds that reach a high degree of tonal complexity.

For the most part, birds sing, or make other sounds, due to an *ingrained genetic program.* They repeat the same series of notes over and over. Thus, bird vocalization is highly stereotyped and ritualized. As a result, it is easily understood, clear, and unambiguous.

There are two types of true vocalization. A **call** is a simple acoustic structure of no more than four or five notes without a clear pattern. It usually concerns group coordination and is rarely sexual in function. Calls are given as alarm signals, feeding summons, as a rally during migration, or as a vocal threat. For example, the peeping of a newly hatched chick is a call to maintain contact with the parents. True **bird song**, on the other hand, consists of a more detailed series of varying notes to form a recognizable melodic pattern. The identity of the singer is easily recognized. Advanced bird-watchers, for instance, can identify most birds just by hearing their song.

In addition to species identity, bird song has a number of other functions:

Functions of Bird Song

✓ to proclaim sex of the singer
✓ to indicate dominance
✓ to advertise for a mate
✓ to establish sovereignty on a territory
✓ to synchronize sexual behavior
✓ to strengthen the pair-bond
✓ to rally a flock to collective action
✓ to intimidate enemies
✓ to discharge nervous energy

Figure 20.15. Some birds, such as the marabou storks in **A** above can do little more than grunt. However, most birds, such as the mew gull in plate **B**, use calls, or have more elaborate song patterns as seen in the cactus wren (**C**) or the house finch (inset).

Bird Migration

One of the prevailing mysteries in the field of zoology is animal migration. The annual to-and-fro movement of insects, mammals, and especially birds has puzzled and perplexed scientists for centuries. There has been few natural phenomena more studied and less understood than bird migration.

Although not all birds make yearly migrations, most do. For those birds in the northern hemisphere, autumnal migration follows a southern pathway (called **flyways**). The opposite is true for birds of the southern hemisphere; they tend to follow northern treks. The distances traveled varies by species and depends to a great extent on the location at the start of the migratory process. Birds nesting in the northern-most part of their range will travel farther than one nesting in the southern regions.

Migratory pathways are specific for each kind of bird. A common migratory journey is exemplified by a warbler nesting in southern Canada. Its migration would carry it as far south as Louisiana, and perhaps even across the Gulf of Mexico to South America. However, the all-time champion migrator is the Arctic tern. This medium-sized bird makes a one-way trip each year of 8,500 miles from breeding grounds in the Arctic to feeding grounds near the Antarctic. It remains south for a few months and then makes the return trip. Arctic terns may live as long as 30 years. If that is so, then a single bird could travel over 500,000 miles during its lifetime!

There must certainly be advantages to migration, otherwise birds wouldn't expend the massive amounts of energy required. The first and most obvious is the escape from inclement weather. This seems only logical for a bird breeding in temperate zones. As cool weather approaches, birds become restless and are soon ready to head south. To many ornithologists, escaping cold weather is considered the initial and primary pressure for migration in birds. Other benefits include the ease of locating and establishing new territories, obtaining new food sources, and a safeguard against overuse of habitat and the subsequent depletion of resources. Migration may also prevent an excessive buildup of predators and disease organisms, which could occur if dense bird populations remained in one place.

Figure 20.16. A high sky filled with migrating birds: a herald of spring and harbinger of winter.

Many factors interact to initiate the act of migration. *A primary force is the genetic pulse.* Innate signals automatically cause a bird to become restless as internal hormonal changes occur. Other possible factors include the depletion of food supply, the degree of fat deposited on the body, and environmental changes, such as falling temperatures, wet seasons, or dry seasons. However, the changes in **photoperiod** (i.e., day length) is considered to be the most influential and important migratory cue for birds. In other words, *dwindling daylight is the major trigger that sets the migratory urge into motion.*

The big question, and the one that has yet to be fully answered, relates to the precise mechanism. Just how do birds accomplish such an incredible feat? Keep in mind that we are talking about a bird, for instance, a house wren in my backyard, that begins a month-plus journey each autumn from a Chicago suburb to Argentina. There it remains until similar cues send it north in the spring. This tiny wren, smaller than a mouse, ends its multi-thousand mile trip where it started, in my backyard. Notwithstanding accidents, it will repeat the journey each of its four or five years of life.

Over the past 40–50 years, the migration of thousands of birds have been tracked using leg bands. Details of their flights has been carefully recorded so that we now know with certainty that long trips of this type are the rule, not the exception. So, again, how do they accomplish this overwhelmingly difficult exploit? The answer is, we don't know for sure. But our best educated guess indicates that birds use two general migratory mechanisms.

The Truth Behind V-Flight Patterns

Have you ever looked skyward at flocks of birds during their annual migrations? If you have, then you most likely witnessed the famed V-shaped flight patterns so common among certain large migrating birds. Birds fly in this arrangement for a very good reason...it's easier. The V-pattern actually conserves energy. During V-flights the lead bird, the one in front, is subjected to the most friction as it flies forward. Those birds trailing behind fly off to the side and slightly above the bird in front. Thus, each bird benefits from "drafting", much as race cars do. Every so often the lead bird peals off to take a position near the rear and is replaced by another in the flock. In this way all birds take a turn at leading and, thus, all benefit from the V-formation.

How Birds Migrate

(1) <u>Route-based</u> <u>Navigation</u> is a process wherein the migrating bird keeps track of landmarks (visual or auditory) on an outward journey so that those landmarks can be used in a reverse sequence on the return trip. The traveler views lakes, mountains, ridges, ravines, waterways, and maybe even man-made artifacts. It may also use infrasound (low-frequency sound waves) emanating from various landmarks. The theory is that the bird hears rivers, seashores, wooded areas, prairies, jet streams, and who knows what else as directional aids. It then combines all of these visual and auditory elements into a memory bank for the return journey.

(2) <u>Location-based</u> <u>Navigation</u> is a process developed even before the trip begins. In this type of migration, birds establish the direction of travel from information available at the journey's *origin*. It incorporates the use of navigational aids such as "sun compasses," polarized light, celestial cues, stars, and/or the earth's magnetic field. Apparently birds have the ability to "read" magnetic pulses that serve as longitudinal pathways. It is also thought that those birds using celestial clues have some sort of internal clock that helps regulate the timing of migration.

These two explanations, as logical as they may seem, still leave a number of unanswered questions. For example:

a. How can young birds that have yet to make the trip use such sophisticated mechanisms?

b. How can a bird that "memorizes" landmarks on the first leg of the trip recall the same landmarks several months later, and in the reverse order?

c. How can a bird the size of mouse travel so far and yet find its way to the same nest site used the previous year?

d. What exactly does a bird use to read the earth's magnetic field? No special organ has been identified to do so—at least we haven't found one. Their nervous system shows no special structure or pathway designed to detect magnetic pulses, yet there must be some dimension of the brain that provides this special skill. The   exact way birds are able to read the magnetic compass remains an unrevealed ornithological mystery.

Truly there is a lot to be discovered regarding the mechanism of bird migration. Perhaps the only thing we know for sure is that year after year, for thousands of years, birds have been doing it!

Bird Taxonomy

Birds are the best-known taxonomic group of animals with approximately 10,515 known species distributed among 40 living orders listed in the table below. There are very few undiscovered birds and, on average, only two or three new species are identified each year.

There are two basic reasons for our degree of familiarity with birds. First of all, they are easy to locate. Because of their tendency to sing from exposed perches and to fly, birds is among the most obvious of any animal type. The second factor to account for our knowledge of birds are the number of amateur bird watchers. No other group of animals has anywhere near the legions of admirers as birds. There are literally millions and millions of bird watchers throughout the world. These are individuals dedicated to the art of bird identification and avian biology. Amateurs are often extremely knowledgeable and have added much to our understanding of birds and their ways.

Birds are identified primarily through a combination of external features, vocalizations, behavior, and habitat. For example, beak shape, feet characteristics, plumage coloration, song, specific habits, and where a bird is encountered all aid in its identification. Beaks, especially, are identification aids as the photographs on the next two pages confirm.

Living Orders of Birds

Order - bird type (# of species)

Tinamiformes – tinamous (47)
Struthioniformes – ostrich (2)
Rheiformes – rheas (2)
Casuariformes – cassowaries and emus (6)
Apterygiformes – kiwis (5)
Anseriformes – (waterfowl) ducks, geese, swans, screamers (173)
Galliformes – (upland gamebirds) pheasants, turkeys, partridges, grouse, quail, jungle-fowl, curassows, guinea fowl, megapodes (299)
Gaviiformes – loons (5)
Sphenisciformes – penguins (19)
Podicipediformes – grebes (23)
Procellariiformes – albatrosses and petrels (136)
Phoenicopteriformes – flamingos (6)
Phaethontiformes – tropicbirds (3)
Ciconiiformes – storks (19)
Suliformes – cormorants, anhingas, darters, frigatebirds, gannets, boobies (56)
Pelecaniformes – pelicans, ibises, spoonbills, herons, egrets, bitterns (113)
Accipitriformes – eagles, hawks, vultures, buzzards, ospreys, secretary bird (263)
Falconiformes – falcons, caracara, kestrels (66)
Otidiformes – bustards (26)
Mesitornithiformes – mesites (3)
Cariamiformes – seriemas (2)
Eurypygiformes – sunbittern, kagu (2)
Gruiformes – cranes, rails, trumpeters, limpkins, coots, moorhens, swamp hens (174)
Charadriiformes – sandpipers, plovers, avocets, stilts, snipes, curlews, phalaropes, jacanas, gulls, terns, skuas, jaegers, oyster catchers, skimmers, puffins, guillemot, murres (383)
Pteroclidiformes – sandgrouse (16)
Columbiformes – pigeons and doves (330)
Psittaciformes – parrots, macaws, parakeets, cockatoos, cockatiels, lorikeets, lovebirds, budgerigars (386)
Opisthocomiformes – hoatzin (1)
Musophagiformes – turacos (23)
Cuculiformes – cuckoos, anis, roadrunners (148)
Strigiformes – owls (221)
Caprimulgiformes – poorwills, whip-poor-will, nightjars, oilbirds, frogmouths, potoos, nighthawks (118)
Apodiformes – swifts, swiftlets, needletails, sicklebills, racquet-tail, and hummingbirds (458)
Coliiformes – mousebirds (6)
Trogoniformes – trogons and quetzals (44)
Leptosomiformes – cuckoo-roller (1)
Coraciiformes – kingfishers, bee-eaters, todies, motmots, rollers, ground-rollers, kookaburra (158)
Bucerotiformes – hornbills, ground-hornbills, woodhoopoes, hoopoes (73)
Piciformes – woodpeckers, flickers, toucans, barbets, jacamars, honeyguides, tinkerbirds, puffbirds (432)
Passeriformes – flycatchers, larks, swallows, shrikes, waxwings, wrens, mockingbirds, thrushes, warblers, gnatcatchers, tits, nuthatches, treecreepers, sparrows, buntings, finches, tanagers, honeycreepers, vireos, blackbirds, starlings, bowerbirds, birds of paradise, crows, magpies, jays, and the nutcracker (6267)

Figure 20.17 Representative bird beaks. **A & B**, recurved beaks of avocets; **C**, stout chickenlike beak of a barnyard rooster; **D**, massive beak of a toucan; **E**, raptorial beak and naked head of a vulture; **F**, spatulate beak of a roseate spoonbill; **G**, parrotlike beak of a macaw; **I**, straight beak of a great blue heron; **J**, generalized beak of a blackbird; **H & K**, decurved beaks of a flamingo and an ibis; (opposite page) **L & M**, conical beak of a house finch and a house sparrow; **N**, straight beak of a heron; **O**, raptorial beak of a red-tailed hawk; **P**, terete beak of a hummingbird; **Q, & T**, gular sacs of pelicans; **R & S** flattened beaks of a barnyard goose and a male mallard duck.

Bird Beaks

Chapter 20 Summary

Key Points	Key Terms

Key Points

1. Endothermy plays an important role in the lives and ultimate success of birds.

2. Birds are uniquely identified by the presence of feathers.

3. Birds are also characterized by flight, which affords a number of advantages. To fly, birds must overcome gravity in an act called lift; they must provide forward motion (i.e., thrust); and they must eliminate drag. Birds are designed to fly through specific adaptations to all three requirements.

4. The sequence of bird reproduction demonstrates their overall instinctive behaviors. The six stages of bird reproduction normally involve the establishment of a territory, specific courtship behaviors, the establishment of a pair-bond, building of a nest, laying and incubating eggs, and the brooding and care of young.

5. Since birds are instinctive animals, imprinting plays an important part in their identity and overall existence.

6. Vocalizations are significant in the life of birds. Song may consist of a complex series of sounds meant to provide a number of benefits or advantages.

7. Most birds migrate as a way to escape inclement weather, relocate to new habitats, find new sources of food, and prevent the overuse of habitat.

8. There are two general theories as to the way in which birds migrate: route-based and location-based navigation.

Key Terms

altricial young
bird call
bird song
brooding
brood parasites
drag
endothermy
fusiform
gizzard
heterothermy
imprinting
infrasound
lift
location-based navigation
monogamy
ovipary
pair-bond
photoperiod
platform nest
pneumatic bones
polyandry
polygamy
polygyny
precocial young
promiscuity
pygostyle
route-based navigation
scrape
syrinx
thermoregulation
thrust

Points to Ponder

1. What is the major advantage of endothermy?

2. What is the relationship between endothermy and food intake?

3. What specific functions do feathers serve?

4. What are specific adaptations that birds show to lift (i.e., weight-reducing adaptations)?

5. What are specific adaptations of birds to thrust (i.e., power-promoting adaptations)?

6. What are specific adaptations of birds to prevent friction while flying (i.e., adaptations to streamlining)?

7. What are the six stages of bird reproduction?

8. What are specific advantages of flying?

9. What types of pair-bonds do birds form?

10. What is imprinting and what role does it play in the life of birds?

11. How do birds sing, and what role does singing play in the life of birds?

12. What advantages are provided by bird migration?

13. Describe the two theories of bird migration.

Inside Readings

Alcorn, G. D. 1991. Birds and their Young. Harrisburg, PA: Stackpole Books.

Black, J. M. (ed). 1996. Partnerships in Birds: The Study of Monogamy. New York: Oxford University Press.

Davies, N. B., and M. Brooke. 1991. Co-evolution of the cuckoo and its hosts. Sci. Am. 264(1):92-98.

Feduccia, A. 1995. Explosive evolution of Tertiary birds and mammals. Science 267:637-638.

Harrison, C., and A. Greensmith. 1993. Birds of the World. DK Publishing, Inc.

Heinrich, B. 1986. Why is a robin's egg blue? Audubon 88(4):64-71.

Jellis, R. 1984. Bird Sounds and their Meaning. Ithaca, NY: Cornell University Press.

Johnsgard, P. A. 1997. North American Owls. Washington, D C: Smithsonian Institution Press.

Kerlinger, P. 1995. How Birds Migrate. Mechanicsburg, PA: Stackpole Books.

Knudsen, E. I. 1981. The hearing of the barn owl. Sci. Am. 245(6):112-125.

Marshall, L. G. 1994. The terror birds of South America. Sci. Am. 270(2):90-95.

Mock, D. W., et al. 1990. Avian siblicide. Am. Sci. 78:438-449.

Myers, J. P., et al. 1987. Conservation strategies for migratory species. Am. Sci. 75:18-26.

O'Connor, R. J. 1984. The Growth and Development of Birds. New York: John Wiley and Sons.

Pettingill Jr., O. S. 1970. Ornithology in Laboratory and Field. Minneapolis: Burgess Publishing Company.

Proctor, N. S., and P. J. Lynch. 1993. Manual of Ornithology. New Haven: Yale University Press.

Rahn, H., et al. 1979. How bird eggs breathe. Sci. Am. 240(2):46-55.

Seymour, R.S. 1991. The brush turkey. Sci. Am. 265(6):108-114.

Sanders, J. 2000. Internet Guide to Birds and Birding. Dubuque, IA: McGraw-Hill Publishing Company.

Scholz, F. 1993. Birds of Prey. Mechanicsburg, PA: Stackpole Books.

Shipman, P. 1998. Taking Wing. New York: Simon & Schuster.

Terborgh, J. 1992. Why American songbirds are vanishing. Sci. Am. 266(5):98-104.

Waldvogel, J. A. 1990. The bird's eye view. Am. Sci. 78:342-353.

Wellnhofer, P. 1990. Archaeopteryx. Sci. Am. 262(5):70-77.

Welty, J. C. 1975. The Life of Birds. Philadelphia: W. B. Saunders Company.

Zeigler, H.P. and H. J. Bischof. 1993. Vision, Brain, and Behavior in Birds. Cambridge, MA: MIT Press.

Chapter 20
Laboratory Exercise

The Birds

Laboratory Objectives

✓ to relate bird beaks to feeding habits
✓ to identify external characteristics of birds
✓ to identify various birds from external features

Needed Supplies

✓ selected bird museum skins
✓ selected items from birds

Chapter 21

SUBPHYLUM VERTEBRATA
Class Mammalia

Hair, Canines, and Mother's Milk

*H*air and mammary glands are uniquely mammalian features; all mammals have them and no other animal does. Morphologically, hair is an epidermal outgrowth that covers the animal as a coat or **pelage**. The coat usually consists of two types of hair: coarse, elongated *guard hairs* that overlie and protect the shorter *underfur* (fur), which can be extremely dense as in northern animals (beaver, fox, otters, mink). The densest fur of any mammal is that of the sea otter (~100,000 hairs per cm^2). For centuries, wild mammals have been heavily trapped for their valuable, dense underfur. Fortunately, recent advances in captive breeding of fur bearers (e.g., mink farms) have relieved some of the pressure on wild populations. Of course, wool (fleece) has been harvested from sheep, llamas, and goats for centuries.

In many mammals, hairs are attached to a small **arrector pili** muscle buried within the dermis. Contraction of this tiny muscle causes the hair shaft to become erect. Erection of hairs has two functions. First, it can trap air and thus assist in thermoregulation by warming or cooling down the animal; and, second, pilo-erection can make the animal appear larger and more ferocious, thus serving as a deterrent to potential enemies. In humans, pilo-erection causes the skin surrounding hairs to form a tiny mound resulting in "goose bumps." Although this may have been of insulating value at one time, it no longer has that or any other known function. An interesting note here is that when a person is frightened or in danger, body hair stands erect. This is especially true of males, where pilo-erection of this sort reveals a link to a past when erect body hairs promoted a savage appearance.

Although all mammals have hair, not all are completely covered by it; indeed, certain large tropical animals and some aquatic forms hardly have any hair at all. Whales, dolphins, elephants, hippopotamuses, rhinoceroses, naked mole rats, and, of course humans, all have a much reduced hair covering. Some adult whales, in fact, only have a few oral whiskers, the rest of their body being completely naked.

Hair, like feathers, evolved primarily to insulate a warm-blooded animal. Insulation prevents excess heat loss and, at the same time, excludes cold air. Hair, however, does more than just insulate. It also helps an animal to perceive its surroundings. For example, modified guard hairs, the whiskers (**vibrissae**) around the mouth and eyes of cats, dogs, and most other mammals, are sensitive to touch and vibration. The same is true of elongated guard hairs scattered over the body of many other mammals. Even in humans, body hair can detect such things as a soft-blowing wind, pressure on the skin surface, a slight caress, to an insect crawling up our legs.

Hair has a number of other functions: it protects the tender skin from damaging UV radiation; it's used in defense, such as stabbing devices (porcupine quills). Hair also serves as camouflage (e.g., zebra stripes), in sexual display (lion's mane), for communication (through color patterns and display), to provide warning signals (tail flash of white-tailed deer), and even in feeding as with the filtering whalebone of baleen whales.

The feature for which the class is named is the mammary gland. Like hair, this too is a distinctly mammalian feature. Only female mammals nourish their young by milk secreted from these special glands. The mammary gland is a modified sweat gland that begins to form milk, under hormonal control, shortly after pregnancy begins. Milk is then released, again under the influence of hormones, as the newborn begin to suckle.

An interesting side note is the observation that both male and female mammals possess mammary glands. Of course, they are nonfunctional in normal males. Nonetheless, it raises the intriguing question of the origin and reason for such glands in the male of the species. The answer seems to lie in the timing of embryological development. During development, both sexes travel the same gender route until a genetic signal heralds one along the female path and the other toward maleness. Apparently, by the time the genetic signal is issued, both sexes have traveled beyond the point of mammary gland formation so that both sexes are born with these unique glands. Of further interest is the fact that if males are given sufficient female hormones the mammary glands can even become functional!

The length of time that young mammals nurse depends upon the species, but in all cases, it is a prolonged period. Extended nursing is made possible, in part, by the late eruption of mammalian teeth. These first "milk teeth" are later replaced by permanent teeth, a condition known as **diphydonty**, consisting teeth of different types (**heterodonty**) such as incisors, canines, and molars.

Figure 21.1. All mammals are invested with hair, although some, such as the elephant in the above photo, are only sparsely covered. From piglets to puppies, mammalian young are also nourished with secretions from mammary glands.

The Chewing Vertebrate

Unlike other vertebrates, mammals can chew their food. Fish, amphibians, reptiles, and birds either swallow their food whole or, if too large, rip, twist, or tear it into small enough pieces to swallow. No other animal masticates food within the oral cavity as mammals do. Chewing food in the mouth cavity accomplishes two things. First, it macerates the food into smaller pieces that can be more easily swallowed and, thus, more easily digested. Second, it allows the food to come into contact with saliva. This, of course, lubricates the food mass for swallowing but, more importantly, permits certain digestive enzymes to begin their work. Thus, as food is swallowed it is already undergoing partial digestion. Human saliva, for example, contains amylase, an enzyme that digests carbohydrates. The potato you swallow is being predigested as you chew it.

To chew effectively, a number of morphological changes must occur in the jaws, oral cavity, and teeth. Let's look at teeth first. True mastication requires a grinding action that would be impossible if teeth interlocked when the jaws were closed. Imagine an alligator trying to grind its food. It simply cannot be done. Thus, the first change is to have the crowns of the upper and lower teeth contact each other when chewing. In other words, a chewing animal, especially a herbivore, needs **occluding teeth.**

Having one type of all-purpose tooth is inefficient for a varied diet. Therefore, many mammals are **heterodonts**, they have developed teeth of four different types for different purposes. **Molars**, for instance, have large surfaces that can be used to pulverize plant material or seeds. To bite off pieces of plants, the front teeth, the **incisors**, are modified into nippers. Grinding teeth are not so important for a flesh eater, so carnivores evolved teeth adapted to impaling prey and shearing off chunks. Enlarged, sharp **canines** are used to stab the prey while bladelike **carnassial** surfaces on rear teeth shear off smaller pieces. Omnivores, such as raccoons and primates, have a combination of all four types.

The teeth of some mammals have been modified for special purposes, and the result can be amazing. Consider, for example, the narwhal. It only has two teeth, both of which are nonfunctional. In the male, one (the left incisor) emerges as a lance like tusk extending up to ten feet! Although the precise role for this exaggerated tooth is uncertain, it most likely plays some sexual function or as a sign of male dominance. Or consider the elephant or mammoth. They, too, have exaggerated incisors used for a variety of purposes from lifting, digging, defense, and display. The tusks are much larger in males and grow throughout their lives. Tusks average about 130 lbs. each and those from old bulls have been weighed in at 290 lbs. apiece. Walruses and saber-toothed cats have overgrown canines that project as two formidable pikes used in defense, predation, in displays of male dominance, and, in the case of the walrus, to help haul the animal onto ice floes.

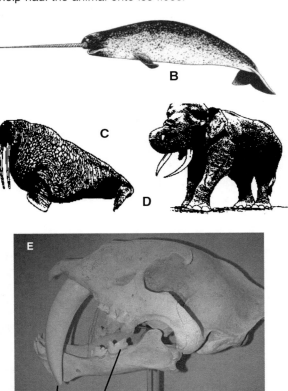

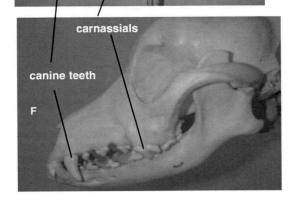

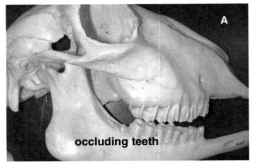

Figure 21.2. Representative teeth of mammals. **A,** a herbivore showing the occluding teeth necessary to chew plant material; **F,** a carnivore with the heterodont condition resulting in a variety of tooth types used for feeding on flesh; **B,** a narwhal; **C,** walrus; **D,** an elephant; and **E,** the extinct saber-tooth cat, all with overgrown canines or incisors.

Final Exams, SATs, and IQs

We spend a tremendous amount of time and energy thinking about, and worrying about, how smart we are. Others test us and we test ourselves. How many math tests have you endured? How many history, anthropology, and science examinations have you written? Life seems to be one test after another. There are even driver's exams, psychological profiles, Civil Service exams. You can't get into the army without being tested. Some of the most popular television programs have to do with how much we know about this or that, trivia they call it. We have even developed measuring sticks of how smart, or dumb, we are in comparison to everyone else: an IQ test we call it. We are truly obsessed with our intellect.

Other animals don't escape our curiosity about intelligence, either. Animal behaviorists want to know how smart animals are. They test everything from planaria, squid, cockroaches, mice, and elephants. They put them through mazes, teach them encoded messages, and test them on color, word, and shape discrimination. Biologists even seek to discover if other animals have a "self-identity," if they can solve problems, have reasoning ability, or even if non-human animals have a moral sense. Perhaps we will find out that some mouse somewhere is supersmart, the all-time maze champion. Perhaps we will uncover the Einstein among dolphins, or maybe we will find a chimp that belongs in college (he or she would, of course, have to take the SAT, or ACT).

So, what's all this fuss about intelligence anyway? Well, the fact is...mammals are smart. Among all animals that have ever existed, mammals have evolved the greatest degree of intelligence. The contest isn't even close. While insects dominate because of size, reproductive prowess, and instinctive behavior, mammals dominate because of intelligence. Birds have concentrated their evolutionary gifts into flight. Their brains reflect it: large optic lobes for acute vision and a large cerebellum for muscular control and coordination. Mammals, on the other hand, have invested their evolutionary heritage into a large cerebrum...the reasoning center, the problem-solving center, the communication center, and the thinking center. *A huge cerebrum is the hallmark of mammals.* It, as much as hair, mammary glands, or occluding teeth, is their distinction.

Non-human mammals have complex behaviors, emanating from their large cerebrum, that enhance survival. From birth they are **inquisitive**. They explore every inch, every nuance of their terrain. They, more than any other animal on earth can, learn. Much of their learning is Pavlovian or **conditioned**, that is when a normal response to one stimulus becomes associated with a new stimulus. But for higher mammals, especially primates, learning can be of the highest order, due to **insight**. There is even some convincing evidence that chimpanzees have some awareness of self.

However, the crowning achievement of intellect is the human. The human mammal survives, not because of its size, speed, strength, or natural weapons. We survive because we have been given the bonus of vast cerebral hemispheres...

we survive because we are smart.

Reproduction

If we compare the reproductive patterns of lower animals with that of higher ones, a singular trend is apparent...*the reduction in the number of offspring produced.* Some lower forms, such as nematodes and oysters, literally produce hundreds of millions of zygotes per year. Yet, during the same time frame, a mammal, such as a primate or an elephant, may produce but a single offspring, or even fewer.

The difference then has to do with how reproductive energy is spent. Lower forms expend most of their energy store on the *number* of zygotes, whereas higher animals place their energy at the other end of the reproductive spectrum, on the *care* of their offspring. This is accomplished through three means: (1) a decline in broadcast spawning in favor of internal fertilization; (2) a trend toward internal development of the embryo (viviparity); and (3) increased effort spent on postnatal development and parental care of young.

Lower animals, in their front-loaded system, expend their reproductive energy producing large numbers of zygotes, but spend little if any effort toward their care. Conversely, mammals allocate much less of their reproductive energy on the number of offspring produced, choosing, instead, to invest their effort primarily on postnatal care. To do so, mammals have evolved three reproductive patterns.

(1) The most primitive members of the class are not viviparous but are, instead, oviparous. The monotremes lay reptilelike, amniotic eggs from which the young hatch. Although the females lack normal mammary glands, they do, nonetheless, produce milk to suckle the young.

(2) Marsupials lack such an egg but the young are born at a very immature stage. They then attach to nipples in the pouch (**marsupium)** where they complete their development.

joey in pouch

B

joey in pouch

C

(3) The great majority of mammals use a placenta to nourish young, which are born at a relatively advanced stage. **Placentate** mammals have an extended postnatal stage in which the young learn survival skills from adults.

D

Figure 21.3. Reproductive patterns fall into one of three categories. Only a few animals are egg layers, such as the platypus above (**A**). The marsupial approach is a little more common in which young are confined for a period within a pouch (**B & C**). However, most mammals are placentate (**D**) wherein a development is assisted by the nourishing placenta and later by adults.

Special Adaptations

*E*ndothermy very likely evolved in mammals as they adapted to an insectivorous, nocturnal lifestyle not available to ectothermic reptiles. This is because early mammals probably maintained their body temperature close to the ambient nocturnal range. Later, they most likely became more active during the daytime (diurnal) and maintained body temperature at a higher level. This enabled them to eventually survive in a wide variety of habitats, including dry, hot deserts, tropical rain forests, and temperate zones, as well as in extremely cold arctic environments. This led to a more active and agile animal with a number of specific adaptations.

(1) Limbs are no longer splayed out to the sides as in many reptiles but, instead, became rotated beneath the body for more efficient balance and locomotion. Mammals are easily the fastest and most nimble of all land animals.

(2) A more active lifestyle requires a more efficient supply of oxygen. Lung surface area, thus, is enlarged and a muscular **diaphragm** is used to alter pressure inside of the thorax, which, in turn helps to ventilate the lungs while breathing.

(3) The circulatory system improved to better deliver oxygen to the cells. The four-chambered heart is perfected to supply a constant high level of blood pressure, which, in turn, assures timely blood supply through the thousands of miles of capillaries to the trillions of cells.

(4) Likewise, the digestive system underwent improvement through the addition of fingerlike extensions called villi and microvilli. These carpet the small intestine to increase surface area for digestion. Certain ruminants (e.g., cattle, sheep, bison, goats) have developed a multi-chambered stomach that permits them to re-chew their food and to help with the digestion of the massive amount of plant material consumed.

(5) The teeth of mammals and the chewing apparatus have also undergone extensive modification as discussed earlier in this chapter.

(6) Kidneys are designed to remove an increased volume of urea, which is stored in an expandable urinary bladder. The kidneys also are well adapted to regulate the body's water content.

(7) Finally, mammals possess three important skin glands, each with a specific purpose:

  a. sweat glands help to regulate body temperature.

  b. sebaceous glands secrete sebum, an oily substance that serves to waterproof and soften the skin.

  c. apocrine glands are modified sweat glands that secrete a sexual attractant, much like a pheromone.

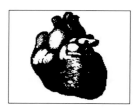

four-chambered heart

A Summary of Mammalian Characteristics

Three landmark features:
* *hair*
* *mammary glands*
* *intelligence*

Other features include:

✓ occluding teeth (heterodonty & diphydonty)

✓ aquatic, arboreal, subterranean & terrestrial habitats

✓ endothermy

✓ large cerebral hemispheres

✓ advanced olfactory, visual, and auditory receptors

✓ extended postnatal development

✓ active and exploratory lifestyle

✓ insulated by hair and subcutaneous fat

✓ apocrine, sweat, and sebaceous glands

✓ pinnae (external ear)

✓ non-nucleated red blood cells

✓ larynx

✓ secondary palate

✓ diaphragm

✓ effective heart, lungs, and kidneys

✓ limb placement for speed and agility

Living Orders of Mammals

Order - mammal type (# of species)

Afrosoricida (30)
Artiodactyla (240)
Carnivora (286)
Cetacea (84)
Chiroptera (1104)
Cingulata (21)
Dasyuromorphia (71)
Dermoptera (2)
Didelphimorphia (87)
Diprotodontia (142)
Erinaceomorpha (24)
Hyracoidea (4)
Lagomorpha (91)
Macroscelidea   (15)
Microbiotheria   (1)
Monotremata (5)
Notoryctemorphia (2)
Paucituberculata (6)
Peramelemorphia (21)
Perissodactyla (17)
Pholidota (8)
Pilosa (10)
Primates (376)
Proboscidea (3)
Rodentia (2277)
Scandentia (20)
Sirenia (5)
Soricomorpha (407)
Tubulidentata (1)

Kinds of Mammals

Mammals, especially marsupials and placentates, underwent an extensive radiation to produce a number of distinctive **taxonomic orders**. Some of these are outlined below.

Monotremata
(duck-billed platypus, echidna)

✓ oviparous (lay shelled eggs)
✓ incubate eggs yet suckle their young
✓ fur covered
✓ toothless
✓ reptilelike limbs
✓ testes confined within the abdomen
✓ poison spurs on hind legs of male platypus

echidna - "spiny anteater"

duck-billed platypus

Diprotodontia
(kangaroos, sugar gliders, koalas)

✓ prehensile tail
✓ marsupium
✓ premature birth

adult red kangaroo

koala

Chiroptera
(bats)

- ✓ flight
- ✓ echolocation
- ✓ heterothermy
- ✓ nocturnal or crepuscular

Rodentia
(beaver, mice, woodchuck, squirrels)

- ✓ largest group of mammals
- ✓ cosmopolitan (distributed worldwide)
- ✓ constantly growing incisors
- ✓ herbivores/insectivores

Figure 21.4. **A**, a flying fox (a type of bat; order Chiroptera) hangs suspended by its hind legs; **B**, a red squirrel; **C**, a rairie dog; **D**, a porcupine; **E**, a house mouse; and **F**, a chipmunk (B - F are all members of the order Rodentia).

Carnivora
(wolves, bears, foxes, cats, raccoons)

✓ predators
✓ enhanced olfaction
✓ most are speed adapted
✓ specialized dentition

Cetacea
(whales, porpoises)

✓ marine
✓ limbs modified into flippers
✓ blowhole for breathing
✓ either homodonty or no teeth at all

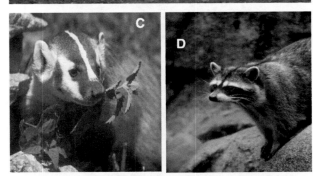

Figure 21.5. Members of the orders Carnivora (A - F) and Cetacea (G - H). **A**, lioness on kill; **B**, grizzly with fish; **C**, badger; **D**, raccoon; **E**, red fox; **F**, cougar; **G**, killer whale; and **H**, dolphin.

Artiodactyla
(pigs, hippos, deer, elk, moose, camels, alpacas)

✓ elongated legs; cornified hooves
✓ foot usually with two toes
✓ usually possess horns
✓ many are ruminants (i.e., chew their cud)

elk

moose

musk oxen

warthog

Primates
(apes, monkeys, lemurs, humans)

✓ arboreal/terrestrial/agile
✓ stereoscopic vision
✓ grasping hand; generalized limb
✓ enlarged cerebral hemisphere
✓ omnivores
✓ nails instead of claws
✓ reduced olfaction
✓ bipedalism
✓ advanced communication
✓ intelligence/inquisitiveness

gorilla

orang-utan

Homo sapiens

Chapter 21 Summary

Key Points

1. As with birds, endothermy plays an important role in the lives of mammals.

2. Mammals are warm-blooded animals characterized by the presence of hair, mammary glands, and occluding teeth. They also reach the peak of intelligence among all animals.

3. As a group, mammals use three types of reproduction. As a result, we see monotremes, marsupials, and placentates.

Key Terms

arrector pili muscle
canine teeth
carnassial surface
diaphragm
endothermy
guard hairs
heterodonty
incisors
marsupial
molars
monotreme
occluding teeth
pelage
placentate
secondary palate
thermoregulation
underfur

Points to Ponder

1. What are the three primary features of mammals?

2. What role does hair play in the life of mammals?

3. In what ways are teeth important to mammals?

4. What is the relationship between mammalian survival and intelligence?

5. Discuss the three reproductive strategies seen in mammals.

6. Describe the features of the respective mammalian orders: Monotremata, Marsupialia, Chiroptera, Rodentia, Carnivora, Cetacea, Artiodactyla, and Primata.

Inside Readings

Altringham, J.D. 1996. Bats: Biology and Behavior. New York: Oxford University Press.

Augee, M.L. (ed). 1992. Platypus and Echidnas. Mosman, New South Wales: The Royal Zoological Society of New South Wales.

Austad, S. N. 1988. The adaptable opossum. Sci. Am. 258(2):98-104.

Bauer, E. A. 1997. Bears: Behavior, Biology, Conservation. New York: Voyageur Press.

Bongaarts, J. 1994. Can the growing human population feed itself? Sci. Am. 270(3):36-42.

Bronson, F. H. 1989. Mammalian Reproductive Biology. Chicago: The University of Chicago Press.

Burney, D. A. 1993. Recent animal extinctions: Recipes for disaster. Am. Sci. 81:530-541.

Clark, W. C. 1989. Managing planet Earth. Sci. Am. 261 (3):46-54.

deWaal, F. M. B. 1995. Bonobo sex and society. Sci. Am. 272(3):82-88.

Feduccia, A. 1995. Explosive evolution of Tertiary birds and mammals. Science 267:637-638.

Griffiths, M. 1988. The platypus. Sci. Am. 258(5):84-91.

Hoage, R. J. ed. 1985. Animal Extinctions: What Everyone Should Know. Washington: Smithsonian Institution Press.

Honeycutt. R. L. 1992. Naked Molerats. Am. Sci 80:43-53.

Linzey, D. 2001. Vertebrate Biology. New York: McGraw-Hill.

Macdonald, D. (ed). 1999. The Encyclopedia of Mammals. Barnes & Noble, Inc.

Murray, J. D. 1988. How the leopard gets its spots. Sci. Am. 258(3):80-87.

Nowak, R. M. 1999. Walker's Mammals of the World. Baltimore, MD: The Johns Hopkins University Press.

O'Shea, T. J. 1994. Manatees. Sci. Am. 271(1):66-72.

Seyfarth, R. M, and D. L. Cheney. 1992. Meaning and mind in monkeys. Sci. Am. 267(6):122-128.

Thewissen, J. G. M., and S.K. Babcock. 1992. The origin of flight in bats. BioScience 42:340-345.

Tuttle, R. H. 1990. Apes of the World. Am. Sci. 78:115-125.

Vaughan, T. A. 1986. Mammalogy. Philadelphia: W. B. Saunders Company.

Whitehead, H. 1985. Why whales leap. Sci. Am. 252(3):84-93.

Wilkinson, G.S. 1990. Food sharing in vampire bats. Sci. Am. 262(3):108-116.

Wilson, D. E., and D. A. Reeder, eds. 1993. Mammal Species of the World. Washington: Smithsonian Institution Press.

Wursig, B. 1989. Cetaceans. Science 244:1550-1557.

Chapter 21

Laboratory Exercise

The Mammals

Laboratory Objectives

✓ identify the mammalian characteristics
✓ identify various mammals from external features
✓ identify the external anatomy of a pig
✓ dissect and identify internal features of a pig

Needed Supplies

✓ a variety of mammal museum skins
✓ selected items from various mammals

Chapter 21
Worksheet

(also on page 619)

Subphylum Vertebrata
Mammals

INSTRUCTIONS: During this exercise you are asked to give the common name of 34 mammals. Use the field guides to identify each museum study skin and be sure that your answer corresponds to the numbered tag attached to each specimen.

1. Little brown myotis
2. American wink
3. common musrat
4. Eastern Fox squirrel
5. Eastern gray squirrel
6. Eastern cottontail
7. Northern racoon
8. Red Fox
9. woodchuck
10. Virginia possum
11. Stripped skunk
12. Eastern chipmunk
13. Thirteen-lined ground squirrel
14. American beaver
15. Black bear
16. Nine-banded Armadillo
17. White-tailed deer

18. cougar
19. least weasel summer male
20. South california longtail weasel
21. Mule deer
22. _____
23. _____
24. _____
25. _____
26. _____
27. _____
28. _____
29. _____
30. l
31. _____
32. _____
33. _____
34. _____

22 Chapter 22

EXTINCTION

Chapter 22

Extinction

Nature is often hidden; sometimes overcome; seldom extinguished.
-Francis Bacon

Looking Back...Looking Forward

We began, in chapter one of this book, exploring the world of animal diversity. Since then we have seen that animals blanket the earth with an amazing tapestry of color, shape, and design. We have marveled at their simplicity and their complexity, at their strangeness and their beauty, their clumsiness and their inventiveness. The ways they've adapted to life in every crack or crevice is nothing short of astonishing. With the help of evolution's sculpturing hand, they have invaded every available ecological niche, resulting in about 40 million different kinds. If that's so, then there's little doubt that more animals exist now, and in greater variety, than at any time in the history of the world. From sponges to squirrels, the success of animals is incredible. We can take heart that in the midst of the modern world more animals are out there than ever before.

While that is true...there *are* more animals than ever...they are not on safe ground. In fact, I would be remiss if I didn't point out, here at the end of our journey, that animals are, in fact, engaged in a struggle of mammoth dimensions. Not since the time of the dinosaurs have they faced such a bleak future, and few are exempt. Frogs are in trouble, so are butterflies, snakes, whales, and yellow-headed blackbirds. Pandas are under assault, as are owls, corals, and sharks. The threatening cloud enshrouds snails, lobsters, and shrikes alike. Not many escape the entwining reach of extinction's tentacles. A war is being waged and the very survival of animals, and perhaps our own, lies in the balance.

To truly understand what is now happening to animals we need first to know something of extinctions. For example, are they a common phenomenon? Are they a recent thing? What causes them? What can we expect in the future? Is there anything we can do? Addressing these questions is the purpose of this final chapter.

First the history. Scientists estimate that over 50 billion kinds of organisms have inhabited the earth since that first spark of genesis some 4.5 billion years ago. Fifty *billion* different species have lived and thrived here on our little planet. Well, if that ancestral figure is anywhere near accurate, and if our estimate of 40 million living animals is also correct, then that means that well over 99% of the species that once lived and prospered are no more. That's a sobering thought. The truth is, *extinction is normal*; each species appears, thrives, and then disappears. On average, a species has a life span of about four million years, then it's gone. Extinction, it seems, is the fate of every species. It is a process that has occurred on a slow steady pace over the eons. It has even occurred in massive episodes wiping out virtually all existing life-forms in but a blink of time's tired eye. In fact, we believe that at least five global extinctions have occurred, the last about 65 million years ago, the so-called K-T event, which eliminated approximately 70% of living forms, including the dinosaurs. *Moreover, we are convinced that the sixth such episode is now under way.*

But Mother Nature is a persistent sculptor... always creating with a slow, inexorable hand. And the tool she uses in her creative effort is genetic diversity. If the fossil record is being read correctly, then great extinctions have offered great opportunities. After each major episode, when the slate was just about wiped clean, new life-forms evolved and in greater variety than before. If nothing else, natural selection is expedient, taking advantage of any new void. Extinction then is nature's way of turning corners on a map-less road—a way to start off in a new direction, leaving behind many of the past efforts, even successes. When one species exits the scene, genetic variety makes it possible for another to take its place. The process is wondrous in its simplicity. As long as there are survivors, there is raw material for new life-forms. Replacements are just waiting on the sidelines, waiting in the genetic fabric of the DNA molecule; waiting in mutations yet to occur.

So, if extinction is the rule in nature's celestial game, why should we be so concerned? After all, it's a natural process and nature will take care of herself...she will create new and perhaps even more wonderful species. Why be concerned? Why?...for two reasons. First, natural selection is slow, real slow. Replacement takes millions of years, far beyond our limited ability to just wait it out. Second, and more foreboding, is that as wondrous as natural selection is, it's just as heartless. It cares not for what disappears, nor for what reappears. Perhaps in our zeal to have dominion over the earth, we will erase the very diversity upon which *we* depend. Perhaps in the chaos, we will become one of the losers, to be replaced by some unknown, maybe by some unthinking, uncaring new species. That's why we should watch very carefully what's happening to our earth's biodiversity.

Causes:
How does extinction occur?

There are really only two possible causes of extinction: that which occurs as the result of human actions and that which has other causes, unrelated to us. We will call one *natural* and the other *man-induced*. Keep in mind, however, that we are certainly part of the natural order and thus not really separate from nature's forces. But, for purpose of comparison we will use these two rather artificial, but real categories.

Natural Forces

If you think about it, virtually all of earth's acts of extinctions have nothing to do with humans. At least for the first five major episodes, we were not the culprit; they all occurred long before we came on the scene. We can't be blamed for the disappearance of trilobites. Nor should we have any guilt over the demise of dinosaurs. We didn't have a hand in the passing of archaeopteryx. We weren't there, we weren't involved...don't blame us.

So, what were the causes, then? Something was surely responsible for the loss of all those billions of species. What was it? Well, there are four or five possibilities. One is **predation**. A predator may be so efficient that it devours every member of a given prey species. We can surmise that this surely happened here and there, but we can be equally certain that it didn't happen on a large scale. The predator-prey relationship is such that the two co-evolve, thereby reducing the likelihood that prey extinction *en masse* would result. **Parasites**, too, may have caused local extinctions, but not likely in great numbers.

A more likely possibility is **disease**. Diseases can be so invasive, so intrusive, so virulent that an entire species could succumb. There's little doubt that epidemic diseases have taken their toll throughout the ages. The extent, however, is again unknown, although it's doubtful that disease, by itself, has spawned extinctions on the grand scale seen in the fossil record.

A better candidate for large-scale extinctions is **competition**: both interspecific and intraspecific types. Competition for limited resources can be a compelling force in the fight for survival. Those species or those individuals better equipped in the race for resources will survive, often at the expense of others. This is without questions one of the prime forces in what's called **background extinctions**—that is, in the slow, gradual extinctions that occur constantly between the major episodes.

Figure 22.1. The likelihood that predation of one species on another, such as in this lioness feeding on the zebra, can cause large-scale extinctions is questionable. A much more likely culprit is competition. Whether it's interspecific (as between the hyenas and vultures above) or intraspecific (between the individual hyenas or between the individual vultures), competition is a potent force in the race for survival.

If predation, parasites, disease, and competition are only responsible for background extinctions, what then has been the cause of mass extinctions of the sort suffered by the dinosaurs? Our best guesswork to date is that massive disappearance of life-forms was caused by **cosmic forces**: volcanic activity, temperature shifts, asteroids, and meteors. Each of these alone can have grave and widespread effects on living organisms. If two or more happened simultaneously, the aftermath would be catastrophic. A volcano on an island can virtually extinguish all life within a few hours or days. A large one or a meteor the size of a few football fields, however, will have global consequences. The impact would most likely produce a shock wave and a poisonous cloud that will suffocate most land animals. The resulting temperature change would wipe out whole continents of animals. Sea level may drop, bodies of water may completely vaporize, photosynthesis may cease. Surely, the result would be cataclysmic.

Figure 22.2. Cosmic forces, such as volcanoes, meteors, and temperature changes, can cause havoc on entire populations of living organisms.

Human Causes

Then along came man, a newcomer on evolution's playground. The best estimates place our emergence at about 500,000 years ago. We burst on the scene as a new species, *Homo sapiens,* a species destined to dominate earth as none had before, or since. Humans entered the extinction equation and changed the rules. We, with our huge brains, our sense of self-awareness, our unique concept of morality, our immense desire to consume, have altered the very face of our planet.

Just like a massive asteroid, we have collided with earth. And we are only now beginning to comprehend the weight of that impact. Like a horrendous meteor, we have issued forth our poisonous cloud of pesticides, hydrocarbons, and acid rain. We have raised and lowered water tables, sent shock waves of atomic explosions, denuded the prairie, slashed the forests, and we have even, it seems, altered climate through global warming. What will be the price other animals must pay for *our* time here?

Based on all the evidence available to them, scientists estimate that, on average, 10 species should suffer extinction each year. At the end of the 20th century, that number reached about 100 per day! *That's nearly 4,000 times what we would expect.* A staggering change had taken place and has catapulted us head over heels into the sixth major episode of worldwide extinction. But are we really to blame?

(1) **Habitat depletion.** Since our appearance on earth, 50% of wetlands and over 90% of forests are gone. Millions of acres of grasslands and prairies have vanished; rain forests, those genetic safety nets where most of the planet's diversity resides, are disappearing at the rate of hundreds of acres each day.

(2) Our air, waterways, and soils are **polluted** with deadly toxins. Tankers break open and oil covers sea stars and sea birds, agricultural runoff sours our streams, garbage heaps choke our groundwater, dams flood pristine habitats, and acid rain showers ponds and lakes with their deadly mists.

(3) In our effort to control pests, to enhance domestic livestock, or even just to diversify habitats, we have intentionally (and sometimes accidentally) introduced new species into areas where they often out-compete natural flora and fauna. As a result, introductions of "exotics" have sometimes had disastrous consequences.

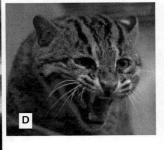

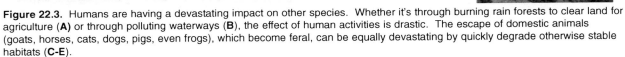

Figure 22.3. Humans are having a devastating impact on other species. Whether it's through burning rain forests to clear land for agriculture (**A**) or through polluting waterways (**B**), the effect of human activities is drastic. The escape of domestic animals (goats, horses, cats, dogs, pigs, even frogs), which become feral, can be equally devastating by quickly degrade otherwise stable habitats (**C-E**).

(4) In the name of sport, food, clothing, or other "comforts," we have **exploited** animals without regard. Passenger pigeons vanished because of our taste for pickled pigeon, the dodo was clubbed out of existence, and alligators nearly paid the ultimate price for shoes or handbags. Whales have suffered, fish populations have declined, otters have given their fur for fashion, egrets for feathered head wear, elephants sacrificed tusks for piano keys, and rhinos donated their horns for sexual pleasure (ground rhinoceros horn is thought in some cultures to be an aphrodisiac). The list is a long one; so many other species are paying the price of our pleasures.

feathers

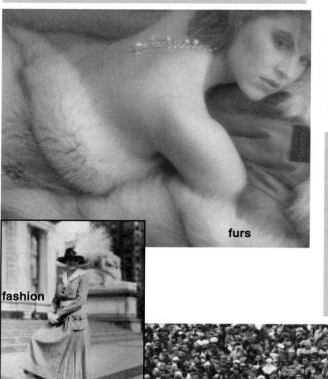

furs

fashion

(5) Finally, there's the issue of **human population size.** Our species has just passed the six billion count. That's a lot of people wandering about our planet, each one hungry, thirsty, and in need of a little space. And it's about to get worse. Experts claim that our numbers will double every 30 years or so. With food, drinkable water, and space being scarce right now, what will the future hold for us and for those other species sharing this planet? There's bound to be a squeeze. Which species will suffer most when the squeeze is applied? Your guess is as good as mine...but I'm afraid we've already seen a glimpse into that future as the number of animals nearing extinction grows dramatically day by day.

poaching

population

Figure 22.4. From feathers to furs, animals continue to be exploited for our comfort. As our numbers continue to expand, what price will future animals pay?

Chapter 22   *Extinction*

So What?

Let's face it, animals will suffer extinction no matter what we do. So what's the big deal? Why should we care? What's a species worth anyway?

Let me suggest a few ways to address such questions. First of all I, for one, don't want to live in a world without giraffes, pandas, or alligators. I would rather not see owls, butterflies, koalas, and sea anemones vanish. I would much rather share this planet with caecilians, cougars, and eagles than just with cockroaches, rats, and starlings (although I guess they have their place, as well). The truth is, I relish nature's diversity, the more the better. And I would guess that I do not stand alone; you might very well share my preference for variety.

But there's another reason to care, and this one affects us all. Do you realize that over 50% of current **medicines** are derived from living organisms? Aspirin comes from willow, penicillin from a fungus, a northwestern tree has given us an ovarian cancer drug, and clawed frogs may very well ooze the next class of antibiotics from their slippery skin. After all, plants and animals have developed protective measures against disease over millions of years—measures that we are now just beginning to tap. But if a large proportion of wild species vanish, they may carry with them the secrets to solve today's diseases and those yet to surface.

Another way to answer the "so what?" question is the **cave-in principle.** I suspect that if certain key members of an ecological community are lost, they may just bring down the entire pyramid with them. There may be a domino effect when a critical species disappears, leaving others unable to adjust; if they vanish, the whole ecosystem may collapse.

And there are some other things to consider when you ponder the value of a wild animal species:

...other benefits from animals

✓ **genetic diversity** - Each organism's DNA is unique and once lost cannot be recovered... *extinction is,* indeed, *forever."*

✓ **environmental monitors** - Many species can warn of environmental decline (e.g., eagles and other large raptors warned us of the biomagnification effect of DDT; the death of pine trees signal acid rain, which can stagnate an otherwise productive lake).

✓ **agricultural benefits** - Plants and animals are our prime source of food. As demand for new and better food grows, we will need to look to wild animals as a potential new source. But once they pass, we have lost their potential.

✓ **educational** -We have an inherent tendency to seek information about the natural world. As animals disappear, the intricate ecological web is diminished and lost to our understanding.

✓ **recreation** - Eco-tourism, bird-watching, hiking, camping, hunting, fishing, nature photography, and insect and shell collecting are all pleasures afforded by our natural world.

✓ **moral imperative** - Many people believe that preventing extinction is "the right thing" to do. We are not alone here on earth and it is only ethically right to protect those species with whom we share this planet.

✓ **survival** - The more we understand about animals and their survival, the more likely it is that we can prevent or delay their extinction, and by extension, perhaps our own.

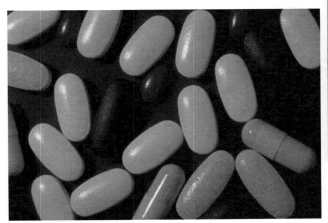

Figure 22.5. From medicines to food, plants and animals have contributed immeasurably to human welfare.

Solutions:
What can be done?

In 1973 the United States passed legislation to curb the alarming increase in animal and plant extinctions. Called the **Endangered Species Act**, it is a powerful tool to both identify and protect wildlife at risk. Since its enactment, thousands of species have been studied and hundreds have been "listed" as either threatened or endangered. It's difficult to gauge the success of such legislation; however, most scientists believe it to be an essential part of the war against needless extinctions. In fact the American bald eagle and the American alligator, two populations in trouble 30 years ago, have experienced substantial improvement to the extent that both species have recently been "delisted."

There's more...
...additional solutions include:

(1) reducing the size, or at least the rate of growth, of the human population, perhaps the greatest current threat to animal survival;

(2) setting aside large tracts of protected land as wildlife sanctuaries;

(3) reducing animal exploitation, including the control of animal harvesting for sport or commercial profit;

(4) controlling the use of biocides by intensifying efforts to use integrated pest management in our battle against agricultural and other pests, and

(5) being ever mindful of the possible damage done by the introduction of exotics into new environments.

I concluded the first chapter in this book with this statement: *"animals enrich the earth with their complexity, their matchless beauty, and their fascinating diversity."* It would be a tragedy for that matchless complexity and amazing diversity to disappear because of our failure to understand, or heed, the needs of natural systems.

Chapter 22 Summary

Key Points

1. Even though there are more kinds of animals alive now than at any time in the past, their survival is in jeopardy.

2. The 40 million living animal species represents less than 1% of the 50 billion total species thought to have lived on earth.

3. Each species is thought to have a life span of about 4 million years; thus, extinction is the norm.

4. On average, about 10 species will become extinct each year (the so-called background extinction); however, the fossil record tells us that at times the rate accelerates into massive events when most species vanish. There has been at least five of these global events in the past, the last occurring some 65 million years ago (the K-T event) when the bulk of dinosaurs disappeared.

5. The causes of background extinctions include a combination of predation, parasitism, disease, and competition.

6. Cosmic forces, such as volcanoes, temperature change, and meteors, are considered the prime cause of mass extinctions.

7. We are currently in the throes of the sixth global extinction event. This one is caused, to a great extent, by the action of humans.

8. The current extinction of ~ 37,000 per year is thousands of times above the background rate.

9. Man-induced causes include habitat depletion, pollution, introduction of exotic animals into intolerable habitats, exploitation, and the overall impact of the rising tide of human numbers.

10. There are many reasons for us to value living species. Among them are:

> ✓ the source of numerous medicines

> ✓ to prevent the domino effect whereby the loss of one species may spur the loss of others, even the entire ecosystem

✓ the loss of genetic diversity

✓ the service many animals perform as monitors of habitat decline

✓ the loss of food and other agricultural benefits

✓ because of their educational value

✓ because of their recreational value

✓ it's the ethical thing to do

✓ our survival may be tied to the survival of other animals with whom we share this plant

11. The Endangered Species Act and other such legislations have helped to curb the rate of extinction in the United States and throughout the world.

12. Other possible solutions include (a) reducing human population growth; (b) forming as many wildlife sanctuaries as possible; (c) reducing the degree of animal exploitation; (d) reducing the use of biocides and increasing the reliance upon integrated pest management; and (e) reducing the introduction of exotics.

Key Terms

animal exploitation

background extinction

cave-in principle

cosmic forces

Endangered Species Act

exotics

extinction

global extinctions

habitat depletion

integrated pest management

K-T event

Points to Ponder

1. What is background extinction? What causes it and how many species are expected to disappear each year due to background extinction?

2. What are mass extinctions? What are their causes?

3. What are some of the ways that humans are involved in animal extinctions?

4. In what ways are natural populations of animals beneficial to humans?

5. What are some of the things humans can do to alleviate pressures on natural wildlife populations?

Inside Readings

Ackerman, D. 1995. The Rarest of the Rare. New York: Random House.

Bongaarts, J. 1994. Can the growing human population feed itself? Sci. Am. 270(3):36-42.

Burney, D. A. 1993. Recent animal extinctions: Recipes for disaster. Am. Sci. 81:530-541.

Clark, W. C. 1989. Managing planet Earth. Sci. Am. 261 (3):46-54.

Hoage, R. J. ed. 1985. Animal Extinctions: What Everyone Should Know. Washington: Smithsonian Institution Press.

Hubbell, S. 1999. Waiting for Aphrodite. Boston: Houghton Mifflin & Company.

Raup, D. M. 1991. Extinction: Bad Genes or Bad Luck? New York: W. W. Norton & Company.

Simon, N. 1995. Nature in Danger: Threatened Habitats and Species. New York: Oxford University Press.

Souder, W. 2000. A Plague of Frogs. New York: Hyperion.

A Color Portrait – Part II
...FROM SEASTARS TO WHALES...AND MORE

ECHINODERMS

TUNICATES
"sea squirts"

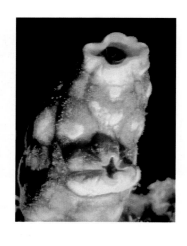

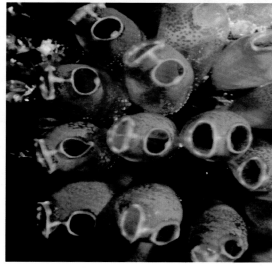

Cartilaginous Fish
Class Chondrichthyes

All of the animals shown here have two things in common, they possess a skeleton, not of bone, but of cartilage and they are beautifully designed. Hammerhead sharks are shown in plates **A** and **B**. A beautiful, well-camouflaged tiger is depicted in **C.** The huge whale shark (**D**) is contrasted to the long-tailed, wide-bodied manta ray in **E**, and an ominous predator patrols the oceans in **F**.

Bony Fish
The Osteichthyes

Mother Nature reveals her creative talents in the architecture of the bony fish. From the plain coloration of the bony fish in plate **A** to the elaborately disguised scorpion fish (**B**), or the strange shape of the sea horse (**C**), evolution has outdone itself. The lion fish in plate **D** displays bold colors warning of its poisonous spines while the beautiful clown fish (**E**) makes its home among the stinging tentacles of a sea anemone.

Amphibians

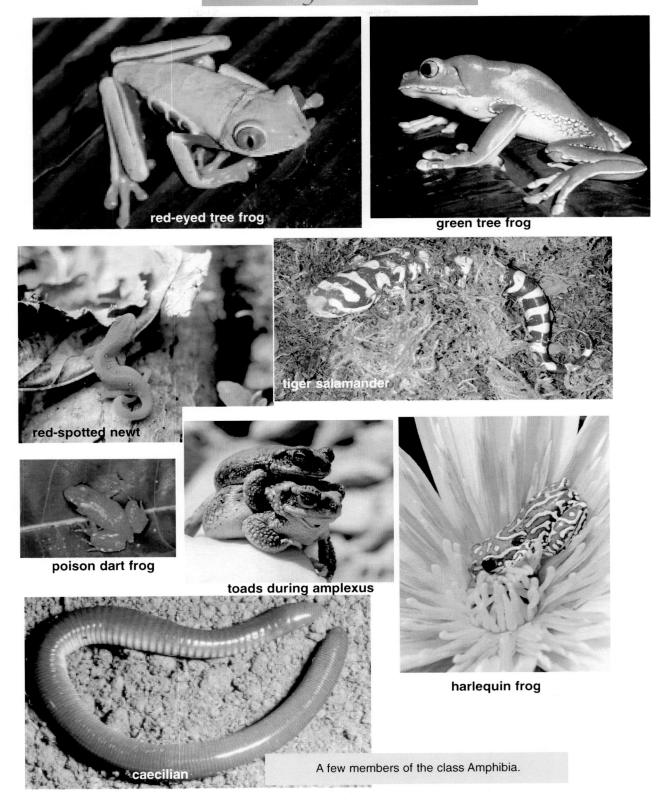

red-eyed tree frog

green tree frog

red-spotted newt

tiger salamander

poison dart frog

toads during amplexus

harlequin frog

caecilian

A few members of the class Amphibia.

gavials

American alligator

box turtle

giant tortoise

green sea turtle

sliders

Reptiles

Some representative reptiles. All have a dry, scaley skin, internal lungs, and reproduce by way of a terrestrial egg.

legless lizard

iguana

western diamondback rattlesnake

coiled viper

emerald tree boa

copperhead (venomous)

gecko (ventral view)

Birds: Class Aves

scarlet macaw

vulture

toucan

long-eared owl

male hummingbird

A few of the over 9,000 species of birds. Members of this diverse group all have feathers, a cornified beak, hollow bones, and, like reptiles, a terrestrial egg. Most are capable of flight. The wings of the hummingbird (lower right) can exceed 80 beats per second!

blue-footed booby

American bald eagle

crowned crane

sage grouse

male frigate bird

male mallard duck

common peafowl

More birds, showing some of their elaborate and colorful displays.

avocet in process of preening

Mammals: Class Mammalia

red squirrel

male lion

elk stag (male)

orang-utan

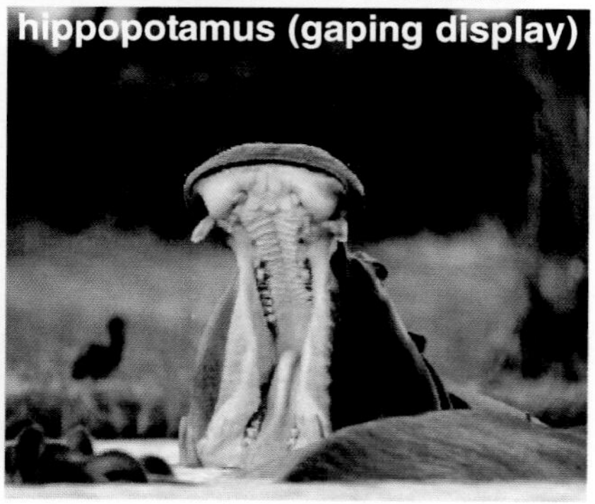

hippopotamus (gaping display)

raccoon

killer whale (*Orca*)

tiger

Fur, hair, mammary glands, and a highly inquisitive nature characterize the mammals. They are a diverse and colorful group of chordates, as the above photographs reveal.

fruit eating bat

giraffe

sea otter

cottontail rabbit

dolphin

gorilla

red fox

chipmunk

zebra

male red kangaroo

gray wolf

532

Camouflage

butterfly camouflaged as a leaf

fish matching its background

toadfish

grasshopper on leaf

head

toadbug among pebbles

rattlesnake

cicadas against tree trunk

giant water bug with eggs

brittle star against a rock

leopard well camouflaged in grass

Both predator and prey species rely heavily upon concealment. On this page and the next, note how each of the animals shown above uses its color and, at times, its behavior to blend with its surroundings.

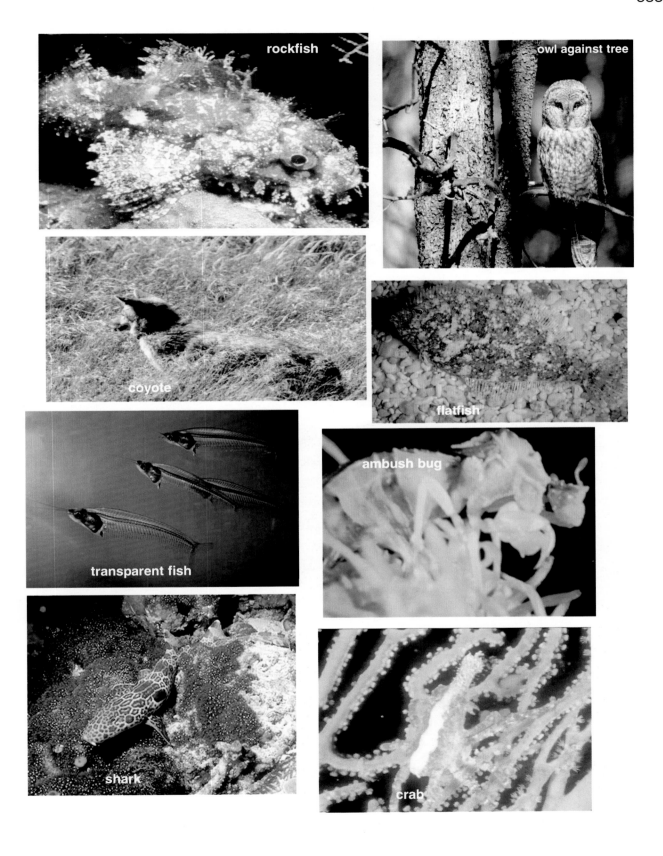

Warning Coloration

honeybee

toxic flatworm

toxic grasshopper

toxic nudibranch

toxic flatworm

lion fish

strawberry poison-dart frog

monarch butterfly

toxic millipede

black widow spider (female)

Many animals advertise their toxic properties through bold patterns called warning or aposematic coloration. A predator quickly learns to avoid such animals after a distasteful encounter.

Mimicry

BATESIAN MIMICRY
the mimicry of a toxic animal by a nontoxic one as shown by the two snakes below

kingsnake (harmless)

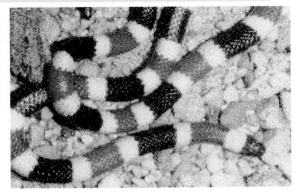

coral snake (venomous)

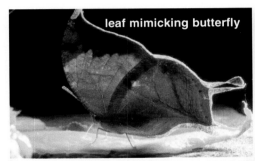

leaf mimicking butterfly

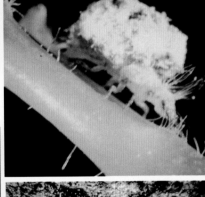

feces mimicking insect

mantid with an "eyespot"

stick insect

"Eyespots" found on many animals may either scare off a predator or may help to deflect the attention of a predator toward a less vulnerable part of the prey's body.

fish with an "eyespot"

thorn mimicking insect

Akinesis = the ability of an animal to "play" dead.

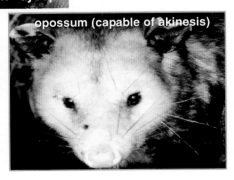

opossum (capable of akinesis)

Weapons

Both predator and prey animals have a number of formidable weapons at their disposal

Predator

Prey

Glossary

A

abdomen - The part of the body that houses the internal organs. In arthropods it is the region posterior to the thorax. In vertebrates it is the area between the thorax and the pelvic girdle.

abducens nerve - One of the motor cranial nerves that aids in the movement of the eyes.

aboral surface - The region of an animal (e.g., echinoderms or cnidarians) opposite from the mouth.

Acanthocephala - An uncommon phylum of animals known as the spiny-headed worms. They are related to the roundworms.

acoelomate - The condition in which an animal lacks a body cavity.

acontia - Threadlike structures that extend from the mouth of sea anemones. They are studded with stinging cells used in defense.

active pursuit - One of the hunting strategies used by predators usually involving a high-speed chase. An energy demanding strategy.

adaptation - Any beneficial feature that improve an animal's chance of surviving and reproducing.

adaptive radiation - When a large number of species evolves from a single ancestral form.

adductor muscles - Muscles in bivalves (clams) that serve to close opposing shells.

adherent nest - A type of cupped bird nest that is built by attaching it to a vertical surface, (e.g., swallows).

afferent branchials - Arteries in the shark that carry blood to the gills.

African sleeping sickness - A human disease caused by the protozoan, *Trypanosoma*. It's transmitted to humans through the bite of *Glossina*, the tsetse fly. The disease occurs in the blood and when it reaches the brain it induces a comatose state, leading to death.

African clawed frog - A common laboratory frog of the genus *Xenopus*, one of the few anurans with clawed toes. Originally from Africa, it was brought to the U.S. as a laboratory frog which has subsequently been released into the wild and is now a pest in California.

aggregation response - A strategy used by prey species to avoid capture. In this strategy, the prey occur in large numbers, thus decreasing the likelihood that any one member would fall victim to a predator.

Agnatha - A superclass of vertebrates characterized by the absence of jaws and paired appendages. They have a cartilaginous endoskeleton and a prominent notochord. Lampreys and hagfish.

ahermatypic - Corals that do not form reefs.

air sacs - A series of membranous extensions of a bird's respiratory system. They are instrumental in the flow-through pattern of bird respiration. There are usually four anterior and four posterior air sacs.

akinesis - A defensive strategy used by some prey species in which they feign death.

alligators - Reptiles belonging to the order Crocodilia and characterized by a broad snout and unexposed lower teeth when jaws are close. Common in southern United States.

alternation of generations - A life pattern in which an animal alternates between an attached, asexual stage, and a free-swimming sexual phase. Seen in the cnidarian *Obelia*.

altricial young - A type of animal born or hatched in a rather helpless state, (compare to precocial).

ambulacral grooves - Oral grooves radiating from the mouth of a sea star. The grooves contain portions of the water vascular system.

ambush - a hunting strategy used by predators whereby they lie in wait until a victim comes within striking distance.

ametabolous development (= simple development) - Arthropod development in which the young resemble the adults and go through continuous molts throughout their lives (compare to hemi- and holometabolous development).

ammocoetes larva - The larval stage of a lamprey.

amniotic egg - A shelled egg that contains a fluid-filled sac enclosing the embryo (e.g., birds).

amoebic dysentery - An intestinal flu-like disease caused by *Entamoeba histolytica*.

amoebocytes - A type of sponge cell responsible for the formation of spicules, reproduction, and nutrient transfer.

amoeboid movement - Locomotion by cells (e.g., ameba) in which the protoplasm slowly streams through foot-like extensions called pseudopodia.

amoeboid sperm - Spermatozoa that lack the typical flagellum for locomotion, using pseudopodial flow instead.

Amphibia - One of the seven classes of the phylum Chordata. Its members are characterized by a slimy, mucus covered skin that serves as a respiratory surface. Development usually involves an aquatic stage and metamorphosis into the adult. Frogs, toads, salamanders, and the legless caecilians.

Amphiopoda - An order of crustacean known as sand fleas.

Amphioxus - A primitive chordate of the subphylum Cephalochordata. Known as lancelets, these small marine animals are pointed at each end and display a number of primary chordate characteristics such as a notochord, tail, dorsal hollow nerve tube, and an endostyle.

amplexus - The mating "embrace" of toads and frogs in which the male positions himself on the back of the female and grips her waist. When he squeezes, she responds by emitting a stream of eggs which he then fertilizes.

ampullae of Lorenzini - Part of the electroreceptive system of fishes, especially sharks. They consist of gel-filled pits with sensory hairs that are displaced by subtle electrical fields generated by other animals. Used to help find prey.

ampullae - A bulb-shaped structure as in the ampullae of Lorenzini or the dorsal part of the echinoderm tube foot.

Animalia - The kingdom of organisms known as animals. Characterized by multi-cellularity, having eukaryotic cells, and heterotrophy.

Annelida - A phylum of invertebrate, segmented, wormlike animals (earthworms, leeches, and polychaete worms).

Anoplura - An insect order consisting of ectoparasitic, blood-sucking lice.

antennae - A threadlike sensory structure extending from the head of arthropods.

antennal gland - A pair of excretory glands in the head of certain crustaceans, such as lobster and crayfish. See also green gland.

antennules - A pair of small antennae emanating from the head of crustaceans.

anterior - The head end of an anima. Usually the part that first meets the environment, opposite of posterior.

anterior intestine - A portion of the flatworm intestine that extends toward the head, as in planaria.

anterior ganglion - A collection of nerve cells toward the head region of an animal, the earliest type of a brain in animals.

Anthozoa - A class of the phylum Cnidaria containing the sea anemones and corals.

anticoagulant - A chemical that prevents or delays the clotting of blood. Found in the saliva of many blood-sucking parasites such as mosquitoes, leeches, and ticks.

antler - A bony growth on the head of certain ungulates (deer, antelopes, moose), usually present only during the breeding season.

anus - The termination of the digestive system in most animals. The exit for undigested waste material (feces).

aortic arches - Blood vessels in annelids that transport blood from the dorsal to the ventral artery. Sometimes referred to as "hearts".

Apicomplexa - A protozoan phylum characterized by an apical complex used to penetrate host cells. Examples are *Plasmodium*, *Toxoplasma*, and *Monocystis*.

apocrine sweat gland - glands located in the armpit, around the nipple, and in the groin area that secrete an odorous substance. The purpose is not fully under stood but is thought to function as pheromones for sexual attraction

aposematic coloration - Boldly contrasting color patterns (usually yellow, red, orange, and black) that warn of an animal's dangerous or noxious properties. Warning coloration of this type is seen in many animals from Cnidarians, Mollusca, Arthropoda, and Vertebrata.

appendicular skeleton - The bones or cartilage of the upper (arms) and lower (legs) extremities. Includes elements of the pelvic and pectoral girdle.

aquatic emancipation - The adoption of a terrestrial existence by an animal group. It was accomplished twice in the history of animals—by the arthropods and the reptiles. The shelled (land) egg played a major role in this process.

Arachnida - A class of chelicerate arthropods with four pairs of legs. They breathe by way of book lungs or trachea. Spiders, mites, ticks, scorpions, and harvestmen.

Aristotle's lantern - A beaklike feeding apparatus of echinoid echinoderms.

arrector pili muscles - Small integumental muscles attached to hair follicles that cause hair to become erect, either indicating fear or used for defensive purposes.

artificial selection - Human control of animal reproduction to produce certain favored genetic traits.

Artiodactyla - A taxonomic order of two-toed mammals characterized by cornified hoofs, antlers or horns, and elongated limbs.

Ascaris lumbricoides - The human roundworm. The largest intestinal nematode infesting humans.

Asteroids - The class of echinoderms containing the sea stars and characterized by an indistinct central disc, a distinct aboral madreporite, and usually five relatively immobile arms.

asymmetry - A body form with no predetermined shape, no matching parts along a central line or axis. For example, an ameba or sponge.

atoll - A coral reef formed as a ring around a spent volcano.

atrium - (10 A small muscular chamber in the heart. It pumps blood into a ventricle. (2) Also, the central chamber in a sponge.

auditory nerve - Cranial nerve number VIII which transmits sensory information related to hearing and equilibrium. Also known as the vestibulocochlear nerve.

Aurelia - A common jellyfish known as a moon jelly.

auricles - Any ear-shaped structure such a the external pinna or the extensions of the cerebellum.

autoinfection - When a final host becomes reinfected with the same parasite, for example, human pinworms reinfecting a child.

autotomy - A defensive strategy used by many prey species in which they are able to sacrifice a body part (e.g., a tail or leg) in order to escape predation. Self-amputation.

autotrophs - An organism (e.g., a plant) able to manufacture its own food from inorganic sources such as through photosynthesis.

Aves - The class of vertebrates containing birds. Animals characterized by feathers, air sacs, hollow bones, endothermy, and amniotic eggs.

B

background extinction - The low-level rate of extinction that occurs between major, widespread extinction events.

barnacles - A type of crustacean belonging to the class Cirripedia. These animals are highly modified in that they have no head, no gills, and a greatly reduced abdomen. Many have assumed a parasitic lifestyle.

barrier reefs - Coral reefs that are separated from the mainland by lagoons or channeled waterways.

basal disc - The aboral structure used to attach to the substrate in animals such as *Hydra* and sea anemones.

basket stars - Echinoderms belonging to the class Ophiuriodea. Characterized by long, many branched arms.

Batesian mimicry - When a nontoxic animal resembles the color, shape, or pattern of a toxic animal thereby providing a measure of protection. When one species resembles another noxious species and is, itself, protected by aposematic (warning) coloration.

bats - A flying mammal belonging to the order Chiroptera. Usually nocturnal and uses sonar for navigation.

beaver - A mammal belonging to the order Rodentia. Characterized by enlarged and constantly growing incisors, thick, valuable underfur, and a flattened tail.

benthic - The bottom substrate of any body of water, especially the oceans, seas, or large lakes.

bilateral symmetry - Body plan in which only one midsagittal section will result in two mirror images. Bilateral symmetry is characteristic of active animals with a definite anterior head and a posterior tail.

Bilateria - The general rank given to all bilaterally symmetrical animals.

bile duct - A tube leading from the gall bladder to the small intestine to transport bile for fat digestion.

binary fission - Asexual reproduction in which mitosis results in two separate organisms.

binomial nomenclature - A system of naming organisms in which each type of organism has two names consisting of the genus and species (e.g., *Homo sapiens*)

biocontrol - The control of pests, such as insects, using natural means rather than through chemical pesticides. Includes such means as natural diseases and predators.

biodiversity - The variety of organisms living in a defined area such as an ecosystem.

biogenesis - (Biogenetic Law) An observation by Ernst Haeckel that an animal's embryonic life reflects its evolutionary history. In other words, ontogeny recapitulates phylogeny. The "Law" is now discounted as direct evidence of evolutionary ancestry.

biogeography - The study of life's geographic distribution.

biology - The study of life.

biomass - The living part of an ecosystem.

biological species concept - The view that a species is a group of similar organisms reproductively isolated from all other such groups.

biophilia - The concept that all persons have an affection for living things.

bipedal - Walking upright on the hind legs; e.g., humans.

biramous appendage - Appendages that have a base attached to the body and two distal processes.

bird clutch size - The number of eggs that accumulate in a nest before hatching. It is usually is fixed for each bird type.

Bivalvia - A class of mollusks whose members have two shells attached by a hinge ligament, possess a prominent foot, but lack a radula.

black widow spider - A poisonous spider (*Latrodectus mactans*) characterized by a black body with a red hour-glass shape on the ventral abdomen.

black fly - A group of biting flies, some of which are vectors of human disease; the black fly, *Simulium,* transmits the blinding worm, *Onchocera volvulus.*

bladderworm - A cyst stage of a tapeworm.

blastocoel - A fluid-filled cavity in the embryonic stage of development called the blastula.

blastula - A stage of embryonic development at the end of cleavage in which the embryo consist of a hollow ball of cells.

blinding worm - The common name given to *Onchocerca volvulus*, a dangerous human parasitic roundworm.

blood flukes - Trematodes that parasitize the bloodstream.

bonding - A process in which two animals, usually male/ female, or female/ young, form lasting, strong relationships.

bony fish - Common name given to members of the class Osteichthyes in which the endoskeleton is primarily bone.

book gills - A series of external flattened plates formed from the exoskeleton and used as respiratory surfaces.

book lungs - Similar to book gills (above) but appear as internal rather than external plates for respiration.

box jellies - A relatively large jellyfish characterized by a box-shaped bell and long trailing extremely poisonous tentacles.

Branchiopoda - A class of tiny marine crustaceans with leaf-like appendages and a pair of chitinous valves. *Daphnia* is a common example of the phylum.

branchial - Referring to gills.

branchial hearts - supplemental hearts in squid to aid in pumping blood through the gills.

Branchiura - A class of parasitic crustacea known as fish lice.

brittle stars - An echinoderm of the class Ophiuroidea characterized by long, tendril-like arms, tube feet lacking ampullae, no anus, and numerous slit-like madreporites.

broadcast spawning - The free release of gametes (eggs and sperm) into the water where fertilization occurs. This process is used by many fish and other aquatic animals.

brood parasites (= social parasites) - Birds such as brown headed cowbird and European cuckoos that lay eggs in another birds nest. The foster parents then incubate, hatch, and care for the cowbird nestlings.

brooding - The process of caring for young. In birds, it refers to the transfer of body heat from the parent to the young.

brood patch (= incubation patch) - A vascularized area of abdominal skin that adult birds (e.g., ducks and geese) use to transfer heat to the eggs during incubation.

brown recluse spider - One of the two dangerous spiders in the United States (the other is the black widow).

brown-headed cowbird - A brood parasite (see above) that uses another bird as a foster parent.

buccal funnel - The hoodlike area surrounding the mouth. For example, in a lamprey it is used as a suction device to attach to large prey.

buccopharyngeal respiration - The exchange of respiratory gases through moist, vascularized tissue in the roof of the mouth and pharyngeal area of amphibians.

budding - A process of asexual reproduction in which a young develop as small growths (buds) on or in the adult.

C

caecum - A saclike extension of the digestive system to increase area for storage, digestion, and absorption of food.

caecilian - A rarely seen, snake-like, legless, subterranean amphibian belonging to the order Gymnophiona.

calamus - The part of the feather shaft buried in skin.

calcareous cups - Depressions in the skeleton of corals caused by the secretion of calcium carbonate by individual coral polyps. Corals withdraw into the cups as a means of escape.

calcium carbonate test - The protective shell of several types of amoeboid organisms. Also known as calcareous tests.

Cambrium period - The geological period from 530 to 525 million years ago when most of today's life forms came into being.

camouflage - Blending with an organism's surroundings.

canines - Long, sharp teeth used to rip flesh. Also a term used to refer to dog-like animals.

capillary - A thin vessel located between arteries and veins through which materials pass into and out of the circulatory system. It is the functional unit of the circulatory system.

carapace - A dorsal covering or shield over the cephalothorax of crustacea or the dorsal plate of turtles.

cardiac stomach - The first of two chambers in the stomach of echinoderms. It is the portion of such a stomach that can be everted for external digestion as in feeding on clams.

carnassial surfaces - The sharp, shearing surfaces of molars in flesh-eating animals.

carnivore - A member of the mammalian order Carnivora; also, any flesh eating animal.

carpet mites - Tiny air-borne arachnids (*Dermatophagoides pteronyssimus*) that feed on dead skin and float in dust along with other debris.

carrying capacity - The maximum number of living organisms that an environment can support.

cartilaginous skeleton - An endoskeleton composed of cartilage rather than bone.

caste systems - The arrangement of eusocial insects into various distinct forms such as queen, worker, drones, soldiers, etc.

caudal - In reference to a tail or toward a tail.

caudal autotomy - The voluntary sacrifice of a tail to escape predation; often seen in lizards or snakes.

Caudata - The order of amphibians to which salamanders belong.

cave-in principle - The concept that the extinction of one key species in an ecosystem may cause many others to follow, and perhaps even cause the collapse of the entire ecosystem.

cavity nests - A nest that is located within a chamber that may be dug or excavated in wood or earth; for example, woodpeckers, kingfishers, burrowing owls.

cellular differentiation - When cells within an organism become specialized to perform different functions.

centipedes - A group of dorso-ventrally flattened uniramians characterized by multiple segments each bearing a pair of legs, reduced eyes, and maxillipeds modified into venomous fangs.

central disc - The somewhat circular area where the arms of a sea star meet; houses the digestive system and part of the water vascular system.

central nervous system (CNS) -The portion of the nervous system consisting of the brain and spinal cord.

centromere - The location on a chromosome where spindle fibers attach and where two sister chromatids are held together.

cephalic - Having to do with or toward the head.

cephalization - The process of concentrating sensory and nervous structures into the anterior portion of an animal: formation of a head.

Cephalochordata - A subphylum of chordates characterized by a number of chordate features such as a tail, notochord, gill slits, and dorsal nerve tube. Commonly referred to as lancelets.

Cephalopoda - The class of mollusks that contain the squid, cuttlefish, octopus, and nautilus.

cephalothorax - The combined head and thorax in crustacea and certain other arthropods.

cercaria larva - A developmental stage of a fluke; a larval digenetic trematode formed asexually from a sporocyst or redia usually within a snail intermediate host.

cerci - Antennae-like posterior appendages on certain arthropods (e.g., centipedes, insects).

cerebellum - A portion of the hind brain that controls coordination and equilibrium.

cerebral ganglia - A collection of nerve cells into a primitive brain.

cerebral hemispheres - The part of the forebrain that controls higher functions (thought, reasoning, problem solving) and voluntary muscle movement. The largest part of the mammalian brain.

Cestoda - The taxonomic class of flatworms that contains the tapeworms.

Cetacea - The order of mammals that contains the whales and dolphins.

chameleons - A type of lizard noted for its long extrusible tongue and its ability to change body color.

chelicera - A pair of anterior appendages in arachnids modified for food handling or for piercing and sucking.

Chelicerata - The subphylum of arthropods containing the scorpions, spiders, mites, and ticks. The body is arranged into a prosoma and an opisthoma and the first two pair of appendages are the pedipalps and chelicerae.

chelifores - Feeding structures of sea spiders equivalent to the chelicerae of arachnids.

cheliped - The first walking leg of a some arachnids and crustaceans modified into a pincer-like offensive and defensive structure.

Chelonia - The order to which turtles belong characterized by two shells, a dorsal carapace and a ventral plastron, and lack of teeth. (another name for the order is Testudines).

chemoreceptors - Specialized receptors designed to respond to chemical stimuli; an example would be odors detected by the olfactory apparatus.

chewing mandibles - Insect mouthparts adapted for eating plant material.

Chilopoda - Class of arthropods containing the centipedes; characterized by flattened body, one pair of appendages per segment, reduced eyes, and maxillipeds modified into venomous glands.

chipmunk - A type of mammal belonging to the order Rodentia; usually striped with contrasting patterns.

Chironex - The genus of the poisonous box jelly.

Chiroptera - The order of mammals to which bats belong; characterized by winged flight and sonar locomotion.

chisel-like beak - A type of bird beak possessed by woodpeckers and adapted to drill into hard substances such as wood.

chitin - The material (polysaccharide) that constitutes the material of the arthropod exoskeleton.

chitinous exoskeleton - The hard covering of arthropods.

chitons - A type of mollusk belonging to the class Polyplacophora; characterized by eight overlapping dorsal plates and a prominent ventral foot with which they attach to the substrate.

chloroplasts - An photosynthetic organelle in plant cells.

chlorophyll - The substance in green plants, located within chloroplasts, responsible for photosynthesis.

choanocytes - A type of cell in sponges responsible for creating a subtle water current from which they trap food particles. Also known as collar cells.

Chondrichthyes - Class of vertebrates to which sharks, rays, skates, and chimeras belong. They have a cartilaginous skeleton, paired pectoral and pelvic fins, and lack a swim bladder.

Chordata - The phylum of animals characterized by a notochord, a dorsal hollow nerve cord, pharyngeal gill arches, and a postanal tail. Fish, amphibians, reptiles, birds, and mammals.

chromosomal replication - The duplication of chromosomal material (DNA) prior to cell division.

chrysalis - The pupal case or cocoon of a butterfly in which metamorphosis takes place.

chyme - a semifluid mixture of partially digested food and digestive secretions formed in the stomach.

cicada - An insect of the order Homoptera known for its loud buzzing sounds and for its prolonged underground development.

cilia - Hairlike extensions of certain cells.

Ciliophora - The protistan phylum characterized by the presence of cilia for locomotion or feeding, multinuclei (macro-and micronuclei), heterotrophic feeding, and complex cell structure: *Paramecium, Stentor, Vorticella.*

circadian rhythms - A biological pattern (rhythm) that lasts for a period of about one day.

circular (ring) canal - Part of the water vascular system of echinoderms located within the central disc, it joins the stone and radial canals.

cirri - (1) A tuft of fused cilia used for locomotion by some protists (2) Organs used for copulation by certain invertebrates. (3) Tufts of hair-like setae on polychaete parapodia. (4) Barnacle feeding appendages (5) The arms in feather stars.

Cirripedia - The crustacean class to which barnacles belong.

cladogram - A tree-like representation of the evolutionary relationships of groups of organisms.

claspers - Spine bearing devices on the pectoral fins of male sharks used during copulation.

cleavage - The mitotic division of a fertilized egg (zygote).

cleidoic egg - The shelled egg of arthropods and chordates capable of being deposited on land (the terrestrial egg).

clitellum - A glandular ring on an earthworm that secretes mucus during copulation and forms the cocoon to protect the developing embryos.

cloaca - A common cavity or opening for the digestive, reproductive, and excretory systems.

cloacal kiss - The term given to bird copulation; since most birds lack a penis, the two cloacae must come into contact for sperm transfer.

cloacal spurs - Poisonous spines on the hind legs of the duck-billed platypus.

Clonorchis sinensis - The human liver fluke.

closed circulatory systems - A circulatory system in which blood is confined to vessels (arteries, veins, and capillaries).

clown fish - A small fish that has a symbiotic (mutualistic) relationship with sea anemones. The fish either fails to elicit the release of the anemone's stinging cells or is otherwise resistant to the poison.

Cnidaria - The phylum containing animals characterized by radial symmetry, diploblastic tissue arrangement, stinging cells (nematocysts), and a gastrovascular cavity (GVC).

cnidarian bladder - The top floatation device in a jellyfish, Portuguese Man-O-War, or other cnidarian.

cnidocytes - The cell that houses the stinging cell (nematocyst) of a jellyfish or other cnidarian.

cobra - A highly dangerous snake characterized by small fangs and a potent neurotoxic venom. Worldwide, cobras cause more human deaths than any other snake.

cocoon - Any protective case for a developing embryo; especially in annelids and arthropods.

coelom - A cavity that contains the visceral organs and is lined by a mesodermally derived peritoneum.

coevolution - The inter-relationship between two species such that each exerts a strong evolutionary influence upon the other. As one evolves it causes the other to evolve in response. An example would be the predator/prey relationship.

coiling - The process of shell formation in many gastropods resulting in a twisted or coiled shape. Thought to aid in overall balance for the animal.

cold-blooded - A common reference to an ectothermic condition in which an animal must rely upon external (environmental) temperature to regulate its own temperature.

Coleoptera - The order of insects to which beetles belong.

collagen - the fibrous protein constituent of bone, cartilage and connective tissue.

collar cells - A cell type in sponges responsible for creating a water current from which it filters food. Also known as a choanocyte.

colony - A collection of organisms.

colonial theory - A hypothesis that multicellular animals were derived from colonial protists.

colubrids - A large family of snakes, some of which are highly venomous.

column - The body stalk of a hydra or of certain echinoderms.

commensalism - A relationship between two species in which one benefits and the other is unaffected.

commissures - An interconnection between parts of the nervous system.

common housefly - *Musca domesticus*.

common ancestry - When two or more species are derived from a common evolutionary source.

comparative anatomy - The study of groups of adult animals in an attempt to discern evolutionary relationships.

comparative embryology - The study of animal embryos in order to determine similarities.

competition - The interaction between two or more animals for the same resource.

competitive exclusion law - The notion that two animals cannot occupy the same ecological niche at the same time.

complete digestive system - A digestive tract consisting of a mouth, a digestive tube, and an anus so that food follows a one-way path.

complete metamorphosis (development - The dramatic transformation of arthropods that proceeds from a larva to a pupa to an adult that differs significantly from the immature forms. Also known as holometabolous development.

complex eye - The eye of higher animals consisting of a focusing lens and a highly innnervated retina.

compound eye - An arthropod eye in which the lens consists of numerous individual units called ommatidia.

conical beak - A powerful type of bird beak that is cone-shaped and adapted to feed on seeds. Possessed by such birds as sparrows, finches, cardinals and grosbeaks.

conjugation - A type of sexual reproduction in ciliates in which mating types exchange genetic material in the form of haploid micronuclei.

contractile vacuole - An osmoregulatory organelle in protists that collects and removes excess water.

conus arteriosus - A noncontractile compartment in the shark heart that leads from the ventricle to the ventral aorta.

convergent evolution - When two unrelated species evolve similar structures.

cooties - A vernacular term for human head lice.

Copepoda -A class of tiny crustaceans characterized by a single median eye, no carapace, and a special egg carrying device, the ovisacs.

copulation - The transfer of sperm into a female by a male.

coral reefs - Diverse assemblages of anthozoan cnidarians each depositing a calcium carbonate skeleton. Over time the skeleton "grows" into vast underwater forest of corals.

countershading - Color patterns on many animals in which the ventral surface is white and the dorsal surface is dark. When viewed from below by predators or prey the animal blends with the light-colored sky, likewise, when viewed from above it blends with the dark water or other surface.

coxa - The portion of an insect leg closest to the body.

coxal gland - The excretory organ of a spider which exits near the base (coxa) of the leg.

crepuscular - Being active at sunset or sunrise.

Crinoidea - A class of echinoderms containing the feather stars and sea lilies.

Crocodilia - The reptilian order to which alligators, crocodiles, caimans, and gavials belong.

crop - An enlargement of the digestive system used for food storage. In annelids and insects, the crop is the foregut, in birds it is an expansion of the esophagus.

cross fertilization - The process in which two copulating organisms trade sperm, thus, fertilizing each other.

cross-section - A plane of bisection on the long axis that results in an anterior and a posterior segment. Also known as a transverse section.

Crustacea - A subphylum of arthropods characterized by two pair of antennae, a pair of mandibles, two pairs of maxillae, and biramous appendages. Lobsters, crabs, shrimp, crayfish.

cryptic coloration - Color patterns that cause an animal to blend with its environment. A form of camouflage. See substrate matching.

cryptobiosis - An inactive state usually induced by dehydration or other adverse environmental conditions. Allows an animal to be dormant and to therefore escape unfavorable conditions.

ctenoid scales - A type of fish scale in which the free surface has comb-like projections. Most bony fish have this type of scale (e.g., bass, bluegill, perch).

cupped nest - The most common type of bird nest in which the egg receptacle is shaped like a cup. Examples: robin, cardinal, warblers, sparrows).

cutaneous respiration - Exchange of respiratory gases (O_2 and CO_2) through the body (skin) surface.

cuticle - The non-cellular covering of an animal derived from the epidermis. In nematodes it forms a protective outer layer for the animal. It is a term that refers to the outer layer of skin in higher animals.

cuttlebone - An internal, vestigial shell used for flotation (buoyancy) in cuttlefish, a type of mollusk.

Cuverian tubules - Whitish, sticky threads that sea cucumbers eject from its anus to ward off predators.

cycloid scales - A type of fish scale found on certain bony fish (bowfin, cod) in which the free surface is rounded.

cysticercosis - Infection with the bladderworm or cyst (cysticerca larva) of a tapeworm.

cysticercus - The larval stage or bladderworm of a tapeworm in which the head (scolex) in inverted in a fluid-filled sac.

cytoplasm - The contents of a cell between the nucleus and the cell membrane.

cytoplasmic bridges - Strand of cytoplasm that extends from one mating type of paramecium to another. Genetic material travels across the bridges form one animal to the other.

cytostome - An opening in many protozoa through which food passes.

D

daily torpor - An animated state that some animals (e.g., bats and hummingbirds) experience on a daily basis.

dauer larva - A larval form of some nematodes that can resist changing conditions by becoming dormant.

DDT - A widely used organochlorine pesticide that has now been banned worldwide due to its environmental and health hazards.

Decapoda - The order of crustaceans to which the crabs, shrimp, and lobsters belong; all have ten walking legs.

decurved beak - A bird's beak that curves downward; used to sift invertebrates from water or to probe into soft substrate.

definitive host - The host where parasites reach adulthood or undergoes sexual reproduction. Also known as the final host.

dehydration - The state of severe water loss leading to death.

Demodex - The generic name of the human follicle mite.

deposit feeding - A process of gaining nutrients from sediments of soft bottom habitats such as mud or sand. The substrate is ingested and food extracted from it.

dermal branchiae - Thin tissue projections that extend onto the body surface of echinoderms for respiratory and osmoregulatory exchange.

Dermaptera - The insect order to which earwigs belong.

Deuterostomata - On e of the two major branches of animal evolution leading to chordates. When the blastopore forms the anus.

diapause - Many insects enter an inactive state prior to the onset of adverse environmental conditions such as cold or extreme aridity.

diaphragm - An arched respiratory muscle situated between the thoracic and abdominal cavities in mammals.

diencephalon - Part of the forebrain where regulatory structures such as the thalamus, pineal body, and pituitary gland reside.

differential survival - Variation in animal survival due to differences in genetic fitness.

diffusion - The inherent movement of molecules from an area of higher to lesser concentration.

digenetic flukes - Parasitic trematodes with one or more intermediate hosts in their life cycle. Example: *Fasciola hepatica, Clonorchis sinesis*.

digestion - The physical and chemical process whereby large food molecules are broken down into smaller ones that the digestive system can absorb.

digestive gland - A large organ in a variety of organisms (e.g., crayfish, sea stars) that serves much like a liver and a pancreas by storing nutrients, secreting digestive enzymes, and absorbing nutrients.

dioecious - When an animal can form only one type of gamete, either sperm (male) or ovum (female). Contrast to monoecious.

diploblastic - Characterized by two tissue layers derived from the embryonic endoderm and ectoderm, for example the cnidarians with their epidermis and gastrodermis.

diploid - A state in which chromosomes occur as homologous pairs (2n) as opposed to occurring singly in the haploid (n) condition.

Diplopoda - The class of arthropods to which the millipedes belong; characterized by a rounded rather than flattened body and having two pairs of legs per diplosegment.

diplosegments - The fusion of two adjoining segments in millipedes resulting in the condition of two pairs of legs per visible segment.

Diptera - The insect order to which flies, mosquitoes, and their relatives belong.

direct life cycle - When a parasite passes from definitive (final) host to definitive host without using an intermediate host. Example: human pinworm and hookworms.

direct deposit feeding - Ingestion of a substrate (e.g., soil) and removing the organic component as the material passes through the digestive system, e.g., earthworms.

directed locomotion - Purposeful movement toward or away from an object as opposed to random or drifting movement.

disruptive coloration - Patterns that break up, or mask the outline of an animal making it difficult to see, e.g., zebras, tigers, chipmunks.

distal - The location of a structure away from an organism's midline.

distractive patterns - Coloration, or color patterns, that diverts attention away from a vulnerable area. For example, the white patch on a deer's tail that distracts a predator away from the head.

diurnal rhythms - Activities that occur on a 24-hour basis.

diversity - The variety of organisms living in an area.

diverticula - Small branches of the digestive system such as seen in trematodes.

division of labor - When the demands of an organism are divided among different cells that each perform a different function.

dominance hierarchy - A social "pecking order" in which one animal physically dominates others in an organized fashion. The dominant or alpha animal thus has priority in mating, feeding, and other group interactions.

dorsal -The back of an animal opposite of the ventral surface; the same as posterior for bipedal animals.

dorso-ventrally flattened - A animal that is flattened from top to bottom (pancake-like) rather than side to side. Examples: trematodes, centipedes, stingrays.

drag -The friction caused by an animal moving through air. For example when a bird is flying drag is a resistance to forward flight.

drag line - A silk strand emitted by spiders and used for anchorage.

drone - The male honeybee or other insect.

ductus deferens -A tube designed to transmit sperm from the testes to the penis or to the outside.

E

ecdysis - The periodic shedding or molting of the arthropod exoskeleton to accommodate growth. May also refer to the shedding of reptilian skin.

ecdysone (= molting hormone) - A steroid hormone of arthropods that stimulates molting.

echinoderms - Members of the phylum Echinodermata characterized by a spiny skin, endoskeleton, secondary radial symmetry, pentamerous body organization, and tube feet. Sea stars, brittle stars, sand dollars, sea cucumbers, sea feathers, and lilies.

Echinoidea -The class of echinoderms that contains the sea urchins and sand dollars.

echolocation - Using sound and its reverberation to determine the location of prey, as in bats.

ecological niche -The role an animal plays in its environment, including, but not limited to, its habitat, its feeding habits, its behavior, and its interactions with other animals.

ectoderm - The embryonic tissue (germ layer) that gives rise to the outer covering of an animal (epidermis) and to the nervous system.

ectoparasite - A parasite that infests the outside of its host; e.g., a flea, tick, louse, mosquito.

ectoplasm - The outer layer of cytoplasm in a protist. More viscous than the endoplasm.

ectothermy - Using heat from the environment to maintain internal temperature ("cold-blooded").

efferent branchials - Arteries that carry blood away from the gills in fish. As opposed to the afferent branchials that transport blood to the gills.

egg tooth - A projection on the top of the upper beak in a bird used to break through the shell when hatching. Some reptiles have a similar process.

electroreceptive system - A series of special receptors on the surface of many fish able to detect weak electrical fields and waterborne vibrations.

elephantiasis - A mosquito-borne chronic human disease of the lymphatic system caused by the roundworm *Wuchereria bancrofti*; results in gross enlargement of extremities and other parts of the body. This disease is confined for the most part to the tropics.

endocytosis - The process by which materials pass through the plasma membrane and into the cell.

endoderm - The innermost of the three embryonic germ layers (mesoderm and ectoderm are the other two) giving rise to much of the digestive and respiratory systems.

endoparasite - A parasite infecting the interior of the host; e.g., tapeworms, roundworms, flukes.

endoplasm - The innermost portion of the cytoplasm of a protist, usually more fluid than the ectoplasm.

endopodite - The branch of a crustacean biramous appendage that projects medially (toward the body mid-line).

endoskeleton - An internal skeleton that lies under the epidermis. It may be bony or cartilaginous as in chordates, or made of calcium carbonate as in echinoderms.

endostyle - A mucus secreting groove in the floor of the branchial chamber of some chordates used to entrap food particles.

endothermy - The ability to regulate one's own body temperature without reliance upon ambient heat. Compare to ectothermy.

energy pyramid - A representation of trophic levels within an ecosystem. The pyramidal shape depicts the energy losses that occur at each successive trophic level.

entomology - The study of insects.

environmental resistance - The limitations placed on a population by a number of environmental forces, such as weather, temperature, food availability, shelter availability, and so on.

Ephemeroptera - The insect order containing mayflies.

epicuticle - The outer, usually waxy, covering of the arthropod exoskeleton.

epidermis - The outermost layer of cells that cover the surface of an animal.

epithelio-muscle cells - Special contractile cells of cnidarians that integrate epithelial tissue covering the animal's surface with underlying muscles for movement.

epitoke - The reproductive form of certain polychaete annelids that differs from the rest (non-reproductive form) of the animal.

erratic parasite - A parasite that is in, or on, the proper host but in the wrong location.

ethology - The study of animal behavior that stresses processes and the impact of evolution on natural habitats.

euglenoid movement - Locomotion using a flagellum.

eukarote - A cell in which the genetic material is confined within a membrane-bound nucleus.

eutelic growth - Growth of an animal due to enlargement of individual cells rather than by the increase in cell number.

eusocial - A group of organisms genetically programmed to work together for the betterment of the entire group. For example, termites, ants, and honeybees.

evisceration - The loss of internal organs.

evolution - Genetic change in a population over time.

excurrent siphon - The tube-like opening through which water exits a clam or a tunicate.

exopodite - The lateral branch of a biramous appendage.

exoskeleton - The outer skeleton of an animal as in arthropods.

exotic - A species of animal introduced (usually because of favorable features) into a new environment, often lacking the normal checks and balances.

external fertilization - The union of sperm and egg outside of the adult organisms.

extinction - The complete or irrevocable loss of an entire species.

extracellular digestion - A process in which food is digested outside of individual cells through the use of digestive enzymes.

eyespot - A photosensitive organelle in a protist, also called a stigma.

F

facultative interaction - Relationships between animals that isn't necessary for survival but will be used, given an opportunity.

feeding polyps - A member of a cnidarian colony that specializes in gathering food for the entire colony.

femur - The portion of the vertebrate leg closest to the body. In insects, it is the third segment between the trochanter and tibia.

feral - An animal that was once domesticated but is now wild. Feral goats, pigs, dogs, cats, and so on

filter feeder - An animal that obtains its food by entrapping small particles from the surrounding air or water.

final host - The parasitic host on which the parasite reaches maturity, also known as the primary or definitive host.

fission - A type of asexual reproduction in which a cell mitotically divides into two cell, as in binary fission. In other cases, the cell can undergo multiple fission resulting in a number of individual cells. In each instance, the daughter cells become individual organisms. Example: most protists.

flagellum - A single hair-like cell process that is usually used for locomotion. Example: *Euglena* or a sperm cell.

flame cell - Part of the osmoregulatory structure in the protonephridia of flatworms. The flame cell consists of a ciliary tuft that waves similar to a flickering flame.

flattened beak - The beak of ducks, geese, and swans. The beak is flattened dorsoventrally and is used to sift food from water, mud, or other substrate. Also known as a compressed beak.

flatworm - Any member of the phylum Platyhelminthes. Animals are flattened dorso-ventrally: Planarians, flukes, and tapeworms.

fluke - (1) A member of the flatworm class Trematoda. Many are internal parasites of invertebrate and vertebrate hosts. (2) The lateral extension of a whale's tail.

food chain - The transfer of energy through an ecosystem from one organism to another in a rather linear series.

food vacuole - An internal, saclike organelle formed during phagocytosis and used in intracellular digestion.

food web - The transfer of energy through an ecosystem by way of a network of interrelated organisms rather than in a simple linear sequence.

foot - The lower portion of a vertebrate leg that contacts the substrate; (2) the muscular locomotor organ of a mollusk.

footprint pheromone - A chemical secreted by a queen bee that inhibits the production of other queens.

foraminiferan - A type of ameba of the phylum Sarcodina that forms an outer shell of silica or calcium carbonate.

fossil - Any trace of an organism from a prior geological age.

fouling - A term referring to the mixture of digestive wastes (feces) with incoming food. Seen in certain mollusks, flatworms, cnidarians, and other animals.

founder's groups - When a small segment of a population becomes isolated and colonizes a new habitat. Since a small group will not contain a representative sample of the various genes of the wider population, the gene frequency changes, often radically, reflecting only that of the founder's group. The result is known as the founder's effect.

fragmentation - A form of asexual reproduction in which an animal breaks into various pieces, each piece capable of becoming a new individual.

fringing reefs - Coral reefs located along rocky shorelines.

frontal section - A plane of bisection that divides an animal into dorsal and ventral portions.

fusiform - The shape of an animal or structure that tapers at each end.

G

Galapagos Islands - A group of islands about 600 miles off the coast of Ecuador. It was observations of this archipelago's flora and fauna that helped Charles Darwin formulate his theory of evolution by natural selection.

gall bladder - A saclike organ used to store and concentrate bile from the liver. It releases bile into the small intestine to aid in the digestion of dietary fats.

gamete (= sex cell or germ cell) - A haploid cell (sperm/ovum) resulting from meiosis and used in sexual reproduction.

ganoid scales - A relatively large and inflexible fish scale covering more primitive teleosts, such as sturgeons, gars, and bowfins. Contrast to cycloid and ctenoid scales.

gastric mill - A grinding device within the stomach chambers of certain crustaceans, such as crayfish and lobsters.

gastrodermis - The tissue that lines the gastrovascular cavity in cnidarians.

Gastropoda -The molluscan class that contains snails, slugs, sea slugs, and other sea shells; characterized by torsion and a coiled shell (if a shell is present).

Gastrotrich - A small phylum of pseudocoelomate animals that lives in sediments of marine environments.

gastrovascular cavity (GVC) - The mouth In cnidarians and a few other invertebrates leads to this large chamber that is used for food digestion. It is part of an incomplete digestive system.

gastrozooid - A feeding polyp in certain colonial cnidarians, such as the Portuguese man-of-war and *Obelia*.

gastrula - A stage of early embryonic development in which the organism encloses two germ layers (the ectoderm and the endoderm) and possesses a central cavity.

gastrulation - An embryonic process during which cells migrate to transform a blastula into a bi-layered gastrula.

gecko - A type of lizard (usually nocturna) characterized by the presence of hairlike setae projecting from the feet that provides excellent climbing ability.

gemmules - An over-wintering reproductive stage of freshwater and certain marine sponges. It is a sack filled with mesenchymal cells, each capable of becoming a new individual.

gene flow - Changes in the frequency of genes due to immigration of organisms from a larger population.

gene pool - The various genes contained within a population or species of organisms.

generalized beak - A type of bird beak that can be used for feeding on a wide range of materials, such as plants, seeds, and insects.

generalists - Animals that are rather broad in their adaptations and, as a result, can survive in a variety of different settings.

genetic code - The sequence of nitrogenous bases in the DNA molecule that signals the formation of various proteins.

genetic recombination - An event that occurs during meiosis in which homologous chromosomes exchange genetic information through crossing over. A major source of genetic variation in a population.

genetics - The study of the transmission of genes from one generation to another and the subsequent expression of those genes.

genetic uniformity - The result of mitosis whereby organisms are genetically identical to each other.

genetic variation - The result of meiosis whereby sexually reproducing organisms have a wide range of characteristics so that no two members (except for identical twins) are exactly alike, genetically.

genital operculum - A reproductive, platelike appendage on the ventral surface of a horseshoe crab.

genotype - The genetic makeup of an organism. Opposed to the phenotype which is the physical expression of an organism's genes.

geographical distribution - The deployment (i.e., placement) of plants and animals on the earth.

geophagy - The desire of an animal to consume soil.

georeceptors - Sensory nerve endings that react to the pull of gravity.

germ layers - The three layers of cells that form during early embryonic development; the ectoderm, endoderm, and mesoderm.

gestation - The period of embryonic development in viviparous animals from fertilization to birth.

Giardia - A protozoan flagellated parasite of the human digestive system.

gila monster - One of the two venomous lizards in North America; found in southwestern U.S. and northern Mexico.

gill - A respiratory surface in aquatic animals.

gill slit - An opening in the body wall of certain chordates allowing water to pass from the pharynx, over the gill surfaces, and to the outside. Oxygen and carbon dioxide are exchanged and, in some forms, food may be filtered from the water stream.

gizzard - A muscular portion of the stomach capable of grinding or mashing hard foods such as plants, and seeds. Occurs in earthworms and birds.

glochidia - A larval stage of certain freshwater bivalves (Mollusca) that lives as an external parasite on fish gills.

glottis - A flap of tissue in the larynx of mammals that closes off the trachea. In snakes it is a tube-like extension of the trachea used to breathe while the snake is feeding.

Gnathostomata - A vertebrate superclass in which hinged jaws and paired appendages are present.

gonad - A gamete-producing organ; testis in the male and ovary in the female.

gonopod - The first one or two pairs of swimmerets in male crayfish (and lobsters) used to transfer sperm to the female during copulation.

gonopores - Openings for the release of gametes; located on the aboral surface of sand dollars and their relatives.

gradual development - Insect metamorphosis in which the young have a slight resemblance to the adults and gradually attain adult characteristics through a series of molts. Also known as hemimetabolous or incomplete metamorphosis.

gradualism - Cumulative, minor evolutionary changes over time that eventually results in the appearance of a new species.

gravid - An egg carrying or pregnant female.

green gland - The excretory gland in crayfish; also known as the antennal gland.

grub - A name given to the larval stage of a beetle.

guard hairs - The elongated coarse hairs that cover the denser fur in certain mammals.

gular pouch - A type of beak in which the gular (throat) area is expanded into a saclike structure used to capture fish; for example, in pelicans.

gustation - The sense of taste.

Gymniphiona - The order to which the legless amphibians (caecilians) belong.

H

hagfish - A jawless fish (agnathan) related to lampreys.

hair - An slender epidermal projection characteristic of mammals.

halteres - The stubby, club-shaped remains of a fly's hind wings (order Dipetra) used for balance while flying.

haploid - The condition in which chromosomes appear as single entities not paired as in the diploid condition.

heart ventricle - A muscular chamber of the heart usually located between the atrium and the aorta, although in fishes it is sandwiched between the atrium and the conus arteriosus.

heart atrium - A muscular chamber of the heart, usually located between the vena cavae and the ventricle, although in fishes it is situated between the sinus venosus and the ventricle.

Heloderma - The gila monster, a venomous lizard.

hemimetabolous development (= gradual or incomplete metamorphosis) - A type of insect metamorphosis in which the immature forms (nymphs) differ from adults and gradually gain adult features through several molts.

hemipenes - One of two copulatory organs of snakes and lizards.

Hemiptera - The insect order to which the true bugs belong.

hemocoel - Large, internal blood-filled cavities of animals with open circulatory systems. Example: mollusks and arthropods.

hemolymph - The blood of an animal with an open circulatory system.

hemotoxin - The venom of certain poisonous animals that attacks the circulatory system. Example: rattlesnakes.

herbivore - An animal that eats plants.

hermaphrodite - The monoecious condition in which a single individual can produce both sperm and egg.

hermatypic - Reef-making corals.

herpetology - The study of amphibians and reptiles.

heterocercal tail - A tail whose terminal vertebrae bend upward giving the caudal fin an elongated dorsal lobe and a smaller ventral lobe. Example: most sharks.

heterodonty - Having teeth of different types, each specialized for a different function. For example, incisors for biting, canines for tearing, and molars for crushing or grinding.

heterothermy (= cold-blooded) - A state in which an animal cannot control its internal temperature but must rely upon an external source, usually the sun, for body warmth.

heterotroph - An organism that obtains its nutrients from inorganic and organic elements within its environment. Contrast to an autotroph.

hinge ligament - The structure that holds the two valves of a bivalve together. When adductor muscles relax the hinge ligament causes the valves to gape.

hirudin - An anticoagulant in the saliva of leeches to prevent its host's blood from clotting.

Hirudinea - The annelid class to which leeches belong.

holdfast - A structure used for attachment to a substrate. Example is the aboral disc of a hydra.

holometabolous development - see complete metamorphosis.

homeostasis - The condition wherein the internal environment of an animal remains in a state of equilibrium or within narrow limits.

homodonty - The condition in which teeth are all similar and non-specialized in their function.

homologous - Structures that have a common ancestry such as the wing of a bird and the wing of a bat.

homologous chromosome - Chromosomes that are similar in shape and that carry genes for the same traits.

hookworm - A roundworm parasite that bores directly through the skin of a person and then lives in the human intestine.

horseshoe crab (= king crabs) - A chelicerate arthropod of the class Merostomata. It lacks antennae or mandibles, its body occurs in two parts (cephalothorax and abdomen), and it possesses specialized appendages, the chelicera and pedipalps.

host - The recipient within a symbiotic interaction, particularly within a parasitic relationship.

host specificity - When parasites are specialized to infest only one type of host.

human elephantiasis - The disease caused by the filarial worm, *Wuchereria bancrofti* that blocks lymphatic passages, thus causing drastic and sometimes grotesque swelling of a person's extremities.

hydatid cyst -A large, bladderlike capsule filled with tiny larvae of the dog tapeworm, *Echincoccus granulosus*. The cyst develops when a person swallows the larval stage. Such a cyst can be fatal.

hydrostatic skeleton - A fluid-filled body cavity which, when under pressure, becomes rigid, thus providing support for the body wall.

hydrostatic pressure - A measurement of the pressure contained within a fluid-filled body cavity.

Hydrozoa - A class of cnidarians whose members are mostly polymorphic, having both polyp and medusoid body forms. Examples: *Hydra, Obelia, Physalia*

Hymenoptera - An insect order to which ants, bees, wasps, hornets, and yellow jackets belong. Members often have stinging structures. The body is attached to the abdomen by a narrow, waist-like pedicle. Members of this order may form highly social groupings.

hyperparasitism -The condition when a parasite, itself, is host to a parasite.

hypertonic - A solution that causes a cell to lose water and shrink (crenate). Usually a solution that has a higher salt content than the cell's cytoplasm.

hypodermic impregnation - Fertilization that occurs when the penis of some flatworms is inserted through the body wall of a sex partner to inject sperm, much in the manner of a hypodermic syringe.

hypothalamus - A part of the forebrain responsible for regulating appetite, body temperature, and other metabolic processes.

hypothesis - An explanation of a scientific phenomenon based upon observations.

hypotonic - A solution (usually with low salt content) that causes an influx of fluid into a cell so that is swells.

I

imago - An adult insect.

imprinting - A process that occurs early in life in which a young animal develops a strong, irreversible bond with another animal (usually its mother) or other object.

inbreeding - Reproduction between closely related animals.

incidental parasite - A parasite located in the appropriate tissue but in a nonpreferred host.

incisor - A type of tooth in mammals designed for cutting.

incomplete digestive tract - A digestive system with only one opening that serves as both mouth and anus.

incomplete metamorphosis (= gradual or hemimetabolous development) - Insect development through a series of molts that gradually transforms an immature into an adult.

incubation - The process in which an adult (usually the mother) transfers body heat to eggs.

incubation period - The length of time that bird eggs are incubated prior to hatching.

incurrent siphon - An opening in mollusks or tunicates through which water enters an animal.

indirect life cycle - A parasitic life cycle that includes at least one intermediate host.

infrared receptors - A sensor adapted to receive light within the infrared spectrum.

infrasound - Sound waves with frequencies below the level of human hearing.

inner vane - The larger of the two feather vanes.

insectivore - A plant or animal that feeds on insects. Also, any member of the mammalian taxonomic order Insectivora.

instar - The stage between successive molts in insects and other arthropods.

integrated pest management - The control of pests through a variety of means, including chemicals, natural predators, diseases, and sterilization.

integument - The outer surface or body covering of an animal.

intermediate host (= secondary host) - The host in a parasitic life cycle that harbors the immature stage of the parasite.

internal fertilization - The union of egg and sperm (i.e., syngamy) that occurs within the body of an animal, usually the female.

interspecific competition - The struggle between different species for the same resources.

interspecific diversity - Variety in a given area as measured by the different kinds of species.

interspecific relationships - The interactions between members of different species.

intracellular digestion - The digestion of food that occurs within a single cell. For example, phagocytosis of food particles by an ameba.

intraspecific competition - The struggle of common resources between members of the same species.

intraspecific diversity - The differences that occur between members of the same species.

intraspecific relationships -Interactions between members of the same species.

intromittent organ - Any structure used to inject sperm into another organism.

island biogeography - The distribution of plants and animals on islands.

isopods - A crustacean of the order Isopoda. For example, the sow bugs (roly-polys).

Isoptera - The insect order to which termites belong.

isotonic - The equal distribution of materials (e.g., water and salt) inside and outside of a cell. As a result, the cell does not shrink nor does it swell.

J

Jacobson's organ - An olfactory receptor in the roof of the mouth of many reptiles. Together with the tongue, it is used to analyze airborne chemicals.

Johnston's organ - A receptor for the sense of hearing (mechanoreceptor) located at the base of the antennae in certain male dipterans, such as mosquitoes and midges.

jet propulsion - Locomotion accomplished by the forceful expulsion of water (or air) causing an object to move in the direction opposite the thrust. Examples are squids and octopuses.

juvenile -An immature stage of an organism, especially of insects.

K

keeled sternum - The shape of the sternum of birds in which the ventral surface is prolonged into a blade or keel.

keratin - A fibrous, waterproof protein replacing the cytoplasm of cells within the skin's epidermal layer. It is a prominent constituent of hair, feathers, claws, beaks, nails, horns, talons, and hoofs.

keystone species - A dominant species within an ecosystem; often influencing the very characteristics of that ecosystem. An example would be an alligator within a swamp ecosystem.

kidney - An organ designed to perform dual functions: the removal of nitrogenous wastes (e.g., urine) and to aid in the balance of water within the body.

kinesis - Movement in response to a stimulus. Example, response to light = photokinesis. Compare to taxis.

kinetic skull - The state of certain reptiles (esp. snakes) in which the lower jaw can disarticulate from the upper jaw to accommodate a particularly large meal.

kingdom - The highest taxonomic classification of life. Five kingdoms are currently recognized.

krill - A common name given to a number of small marine crustacea. Serves as an important food source for a number of oceanic animals including squid and toothless whales.

K-T event - A major extinction episode that occurred at the junction of the cretaceous and tertiary period, some 65 million years ago. The cause is thought to be the impact of a comet or asteroid that collided with the earth off the coast of Mexico's Yucatan peninsula. The resulting fireball, acid rain, sulfur dioxide cloud, earthquakes, volcanoes, tidal waves, and global cooling caused the elimination of up to 80% of the world's species including the dinosaurs and marine reptiles.

L

labial palp - (1) A chemosensory appendage located on the lower lip (see labium) of insects. (2) A flap of tissue guarding the entrance to the mouth in bivalve mollusks.

labial pits - Heat receptors located in the lip area of certain snakes, such as boas, to detect warm-blooded prey.

labium - The "lower lip" mouthpart of insects

labrum - The "upper lip" mouthpart of insects.

lactation - The production and release of milk by the mammary glands.

Lamarckism - A rejected concept, forwarded by Lamarck, that claimed acquired characteristics are inherited. Also known as the *use and disuse principal.*

lamprey - A jawless, scaleless, eel-like fish belonging to the groups of chordates known as agnathans.

lancelet s - A common name for the primitive chordate, *Amphioxus.*

large intestine - The lower part of the digestive tube.

larva - An immature animal that differs in form from the adult. For example, the immature stage of a holometabolous insect.

larval instars - The immature, feeding stage of a holometabolous insect, the stage between molts.

larynx - The vocal apparatus of amphibians, reptiles, and mammals. Contrast to the syrinx of birds.

latent learning - Learning that occurs as a result of exploratory behavior without immediate rewards.

lateral - Away from the midline of a bilateral animal.

lateral line system - A series of mechanoreceptors along the side of fishes (esp. sharks) capable of detecting water vibrations.

lateral undulation - A mode of locomotion in which an animal propels itself in a side-to-side action as it pushes against lateral supports ("pegs"). Example: fish and snakes.

law of competitive exclusion - The notion that two animals cannot coexist in the same ecological niche.

leech - A type of annelid belonging to the order Hirudinea. Many are blood-sucking ectoparasites of various vertebrates.

Lepidoptera - The insect order to which butterflies and moths belong.

lethal mutations - Alternations in the genetic code that are incompatible with life, thus are fatal.

leukocyte - Any of five types of nucleated cells in blood. A white blood cell.

lift - The ability of an animal, such as a bird, bat, or flying insect to launch itself from a surface. To begin flight.

limiting factor - An environmental resource that is in short supply and thus restricts an animal's ability to survive and/or reproduce.

Limulus - A genus of horseshoe crabs.

linear descent - Evolutionary development along a single line rather than in a radiating fashion.

lipid - Fat; adipose tissue.

littoral zone - A shallow area along the shoreline of a lake, pond, or other body of water.

liver - An internal organ responsible for a number of functions, including secondary digestion of carbohydrates, proteins and fats, blood formation, production of bile, and detoxification of blood. The largest internal organ of humans.

location-based navigation - A method of bird migration in which the animal uses information gathered at the point of origin. Celestial cues, polarized light, and the earth's magnetic field are all thought to contribute to the migratory process.

longitudinal muscle bundle - Muscle fibers in nematodes and other invertebrates that run parallel to the body axis, used to bend the animal from side to side.

longitudinal nerve cords - A nerve cord that runs along the length of the animal. In invertebrates it is located ventrally, while in vertebrates it is dorsal.

loop of Henle - A U-shaped portion of the kidney tubule that is used to conserve water.

loose connective tissue - A flexible type of connective tissue in which cells and fibers are widely scattered. Examples are adipose (fat) tissue, membranes, and areolar connective tissue.

lunules - A moon or crescent-shaped opening such as those occurring in the skeleton of a sand dollar. They are thought to serve as channels for water passage to prevent the animal from tumbling while in an upright, feeding position.

lumen - The central opening or cavity of an organ or tube such as that of blood vessels or the digestive tract.

lymphoid organ - Any tissue or organ that produces protective lymphocytes (e.g., tonsils, spleen, thymus, lymph nodes).

lymphatic system - A series of tubules and intervening lymph nodes that returns lymph fluid to the circulatory system. It also filters bacteria and other foreign material from the fluid.

M

macroevolution - The evolutionary process that leads to the formation of new species of plants or animals.

macrophage - A phagocytic white blood cell.

macronucleus - The larger of two nuclei found in ciliates> The macronucleus controls cellular metabolism and asexual reproduction.

madreporite - Part of the water vascular system of echinoderms. The madreporite is a scablike protrusion on the animals surface that serves as an entrance through which waters enters the animal.

maggot - The larval stage of dipterans (flies).

magneto-receptor - A sensor in many animals (insects, reptiles, birds, and some mammals) that is able to detect magnetic fields.

Malacostraca - The class of crustaceans to which crayfish, lobsters, shrimp, hermit crabs, true crabs, and wood lice belong.

maladaptive mutations - Changes in the genetic code that make an organism less fit for a particular environment. See also lethal adaptation.

malaria - A disease caused by a *Plasmodium* protistan and transmitted to humans through the bite of an *Anopheles* mosquito.

malpighian tubules - Fingerlike excretory tubes that project from the midgut of insects and certain other arthropods.

mammary gland - The milk-producing gland (breast) of mammals.

mandible - The lower jaw in vertebrate animals. Also, refers to the tearing and biting mouthpart of insects and other arthropods.

mantle - Part of the visceral mass of mollusks. It is the fleshy outer layer that often encloses the internal organs and secretes the shell.

marsupium - The pouch of marsupial animals where young undergo postnatal development.

mass extinction - A global or near-global incident that results in the widespread elimination of many plants and animals.

massive beak - The large beak of certain birds such as toucans.

Mastigophora - The protozoan subphylum to which flagellates belong.

mating type - A group of ciliates (such as *Paramecium*) that are able to mate (conjugate) with each other but not with members of other mating types.

maxilla - A type of mouthpart located posteriorly to the mandibles of many arthropods. Example: crayfish.

maxilliped - A mouthpart appendage posterior to the crustacean mandible.

mechano-receptor - A sensor capable of detecting movement, tension, vibrations, or changes in pressure.

medial - Any structure located toward the midline of an animal.

median (parietal) eye - A photoreceptor located on the dorsal surface of many vertebrate heads. Sometimes called the third or pineal eye.

medulla oblongata - The portion of the brainstem located anterior to the spinal cord. It serves as a vital relay center as well as controlling certain "vegetative" functions, such as heart rate and breathing.

medusa - One of two body forms of cnidarians. The medusoids have a jellyfish-like shape and are the sexual stage in the cnidarian life cycle.

meiosis - The process whereby sexual reproduction is accomplished due to the reduction of the diploid number of chromosomes to the haploid condition. Meiosis is the formation of gametes, generally in the form of sperm and eggs.

melanin - A dark pigment found in skin, hair, feathers, and other body coverings.

melatonin - A hormone secreted by the pineal body (gland) that helps to regulate daily rhythms and sleep patterns.

Merostomata - The arthropod class to which horseshoe crabs belong.

merozoite - An infective stage of the malarial parasite, *Plasmodium* that results from multiple fission.

mesencephalon - The middle part of the brain that contains the visual centers.

mesenchyme - Tissue arising from mesoderm that has the capability of forming a variety of structures. Also a gellike layer of tissue found in certain invertebrates where it is known mesohyl or mesoglea.

mesentery - A thin sheetlike tissue derived from mesoderm that helps to suspend various abdominal organs.

mesoderm - The basic embryonic layer that gives rise to certain tissues (such as the muscles, circulatory system, and skeleton). Mesoderm is located between the endoderm and the ectoderm

mesoglea - The gellike layer located between the outer epidermis and the inner gastrodermis in cnidarians. The "jelly" of jellyfish which provides buoyancy.

mesohyl - Middle gelatinous layer in a sponge, also called mesenchyme.

metabolism - The total chemical activities of a cell or of an organism.

metacercarial larva - An infective stage in the life cycle of certain parasitic trematodes. It often assumes a dormant state.

metamere - A single segment in a metameric (segmented) animal.

metamerism (= segmentation) - Arrangement of the body into repeated units or segments that may be ringlike as in an earthworm.

metamorphosis - The development of a larvae into an adult. It often involves dramatic transformations such as from a caterpillar to a butterfly.

metanephridium - An excretory structure of many invertebrates. It consists of a long tube, with a fun nel at one end draining the body cavity, and a pore, leading to the outside, at the other end.

Metazoan - A multicelled animal.

Metridium - A genus of sea anemone.

Mexican beaded lizard - One of only two venomous lizards in North America.

microclimate - The temperature, humidity, atmospheric pressure, salinity, and so on in a restricted area such as a microhabitat.

microevolution - Change in gene frequency that produces variety in a population but does not result in the formation of a new species.

micronucleus - The smaller of two nuclei in ciliates. It is responsible for reproduction, as in conjugation.

midbrain - The portion of the brain devoted to vision; also known as the mesencephalon.

migratory flyway - The path taken by migrating birds.

milk teeth - The first set of teeth in a diphyodontic animal.

millipedes - Arthropods belonging to the class Diplopoda, characterized by two pairs of legs per diplosegment.

mimicry - The resemblance of one animal or species to another often affording some form of protection.

miracidium larva - An immature trematode fluke.

mitosis - Nuclear division resulting in daughter cells with the identical genetic makeup.

molars - A flat, grinding tooth of heterodontic animals.

molecular mimicry - An instance in which a parasite imitates the molecular makeup of its host, thereby bypassing the host's immune system.

mollusk foot -The fleshy, muscular mass of a mollusk used for locomotion, burrowing, and sometimes in killing prey.

mollusk visceral mass - see visceral mass

mollusk mantle - see mantle

mollusk radula - see radula

molt inhibiting hormone (MIH) - A hormone secreted by the X- gland of crustaceans that prevents molting from occurring.

molting hormone (MH) - A hormone, such as ecdysone, secreted by the Y-gland of crustaceans causing molting to occur.

monoecious (= hermaphroditic) - The state in which one organism can produce both types of gametes: sperm and egg.

monogamy - A pair bond between a single male and a single female. The condition of having but a single mate at a time.

monophyletic - A group of organisms that have evolved from a single ancestor. Contrast to polyphyletic.

morphogenesis - The development of shape or form as in the morphology of a single organ or a group of organs.

motor pathway - The portion of the nervous system that conveys impulses to elicit some response, usually by stimulating muscular reaction.

mucous gland - A cell or gland that secretes the viscous, lubricating material called mucus.

Mullerian mimicry - When two or more toxic species resemble each other, thus rendering each some protection.

multicellular - An animal consisting of more than one cell.

multiple fission - A form of asexual reproduction in which a cell, or its nucleus, divided into a number of viable units.

musculo-nervous interaction - The inter-relationship between the muscular and nervous systems so that the two work together.

mutagen - Any agent, such as temperature, a chemical, or radiation that may cause a mutation.

mutual copulation - The reciprocal exchange of sperm between two animals.

mutation - A permanent change in the genetic makeup (DNA) of an animal.

mutualism - An interaction between members of two different species in which both are benefitted.

myotomes - An embryonic block of mesoderm from which muscles are formed. (2) Also refers to the zigzag arrangement of muscles in lampreys and sharks.

N

nacre - The inner lining of a mollusk shell from which pearls and mother-of-pearl are formed.

naiad - An aquatic, immature stage of a hemimetabolous (gradual metamorphic) insect.

naris (pl. = nares) - The opening into the nasal cavity.

natural selection - The primary mechanism by which evolution occurs in natural populations as advanced by Charles Darwin and Alfred Wallace.

negatively phototrophic - Repelled by light.

nematocyst - A stinging apparatus of cnidarians.

Nematoda - Phylum to which the roundworms belong.

neo-Darwinian theory - The current explanation of evolution that combines elements of Darwin's original concept with modern knowledge of genetics and molecular theory.

neoteny - The attainment of sexual ability by an immature animal.

nephridiopore - The point of exit of the annelid nephridium.

nephridium - (1) The embryonic structure that gives rise to the adult kidney. (2) The excretory apparatus of an annelid.

nephrostome - A funnel-shaped opening into the annelid nephridium through which fluid enters.

nerve cell - see neuron.

nerve net - The primitive nervous system of cnidarians consisting of a diffuse network of nerve cells.

neuromast - A receptor cell in the lateral line system of vertebrates. It is designed to detect vibrations.

neuron - A nerve cell capable of transmitting an electrical impulse.

neurotoxin - Any venom that attacks the nervous system.

niche - The role an organism assumes in its environment, including all of the biotic and abiotic factors.

nidamental gland - structures in the female squid that forms a gelatinous coating around a mass of eggs.

nictitating membrane - A thin, transparent "third" eyelid of certain amphibians, reptiles, and birds.

nitrogenous waste - The form in which metabolic nitrogen is removed from the body; usually as ammonia, uric acid, or urea.

non-directional mutation - The idea that mutations are random, spontaneous, and non-predictive.

non-lethal mutation - A mutation that does not result in death.

nonrandom elimination - The theory that natural selection eliminates nonfit individuals from an evolving population.

notochord - A rod-shaped, stiff, supportive structure along the midline of all larval, and many adult chordates. The feature after which the phylum Chordata was named.

nuclear dimorphism - A cellular condition in which there is more than one kind of nucleus as in the Ciliates: For example, *Paramecium*.

nucleus - The membrane-bound portion of a cell that houses the genetic material.

nutrient - A substance that provides nourishment (energy) for an animal.

nuptial pad - A thickened region of skin on the thumbs or other areas of the body that male amphibians use to grip the female during mating (amplexus).

nymph - The immature stage of those insects using gradual metamorphosis (hemimetabolous development). A nymph resembles the adult but lacks wings and is sexually immature.

O

obligatory relationship - An interaction between two organisms that is required for survival of one or both.

occluding teeth - When the cusps of upper and lower teeth come into contact as the mouth is closed.

ocellus (plural = ocelli) - A simple eye.

oculomotor nerve -A cranial nerve that aids eye movement.

Odonata -The order of insects to which the dragonflies and damselflies belong.

oil gland (= uropygial gland) - Gland at the base of a bird's tail that secretes oil used in preening.

olfactory - Of, or having to do with, the sense of smell.

oligochaete - Members of the annelid class Oligochaeta, such as earthworms.

ommatidia - The individual sensory subunits of the arthropod compound eye.

omnivore - An organism that feeds on both plant and animal material.

onchocercis larva - A dormant larval stage of a tapeworm.

ontogeny - The development of an individual organism.

oogenesis - The meiotic development of an ovum (egg).

ootheca - A protective casing containing fertilized eggs.

open circulatory system - A circulatory system in which blood enters cavities (sinuses) to bath tissues.

operculum - A protective cover or shield. Example are the flap over the gill chamber of fishes and the plate over the opening to a snail shell.

opisthosoma - The posterior body portion of a horseshoe crab or an arachnid.

optic - Relating to the eyes or to vision.

optic nerve - The large nerve that carries visual information from the eye to the brain.

oral - Of, or relating to, the mouth.

oral-aboral axis - Body orientation based upon location of the mouth. The oral surface is next to the mouth and the aboral surface is opposite to it.

order - The taxonomic level between class family.

organ - A collection of tissue designed to perform a specific function.

organogenesis - The embryonic development of an organ or of an organ system.

Orthoptera - The insect order to which grasshoppers, crickets, praying mantids, walking sticks, and cockroaches belong.

osculum - A large opening through which water exits a sponge.

osmoregulation - The maintenance of proper solute concentration (especially water and NaCl) within a cell or an organism.

osmosis - The diffusion of water through a biological membrane due to concentration gradients.

osphradia - Sense organs located in the mantle of certain mollusks, such as clams and snails, to monitor the quality of incoming water

ossicles - (1) The individual plates in an echinoderm endoskeleton. (2) The small bones of the middle ear.

ostium (plural = ostia) - Any opening. For example, the small pores through which water enters a sponge.

ovary - The female gonad capable of producing eggs.

oviduct - A tube for the passage of eggs once they have left the ovary.

ovigers - Specialized legs of a male sea spider (Pycnogonia) used to carry and brood fertilized eggs.

oviparous - The reproductive strategy of an animal, such as an insect or bird, that deposits external eggs in which young develop.

ovipositor - A reproductive device of females (e.g., insects) used to deposit eggs into a substrate such as the ground, a tree trunk, or another animal.

ovoviviparous - The development of young within the body of the mother in which nourishment is derived from the egg (usually in the form of yolk) but not from the mother.

ovulation - The release of eggs from the ovary.

ovum (plural = ova) - An egg.

P

paedomorphosis - See neoteny.

pair bonding - The establishment of a close relationship between two members of the same species, such as a mating male and female or a female and her young.

paleontology - The study of earth's early lifeforms.

palp - A small fleshy appendage near the mouth usually involved in feeding.

pancreas - An internal organ responsible for the secretion of digestive enzymes and the hormone insulin.

papula (pl. = papulae) - A fleshy projection of an echinoderm's integument used for respiration and excretion.

parapodium - Segmental, fleshy appendage in polychaetes (e.g, the sandworm) used for swimming, crawling, and burrowing.

parasitism - The interaction between two organisms in which one benefits at the expense of the other, usually without causing extensive damage.

parasitoid - A parasitic animal that lays eggs within the body of its host. The larvae then feed on the living tissue of the host, eventually killing it.

parenchyma - Undifferentiated cells used as packing material around internal organs.

parturition - The process of giving birth.

parthenogenesis - The development of an organism from an unfertilized egg.

patent - Being open or inflated.

pecking order (= dominance hierarchy) - A social order in animals that determines the priority rank for feeding, mating, and other activities.

pectinate organs - Comblike sensors located ventrally on scorpions used to detect vibrations.

pectoral girdle - Skeletal elements that attach the forelimbs to the axial skeleton.

pedicel - A narrow stalk, such as the "waist," connecting a spider's abdomen to its thorax.

pedicellaria - Tiny clawlike appendages on the surface of echinoderms (e.g., seastars) used for cleansing, protection, and feeding.

Pediculus humanus capatis (= "cooties") - Human head lice.

pedipalps - Small appendages, near the mouth of arachnids, used in feeding.

pelage - The coat of hair or fur in mammals.

pellicle - A noncellular, thin, but tough covering of many protists.

pen - A reduced shell imbedded dorsally in the mantle of a squid.

pendulous nest - A bird nest in which the eggs are held at the bottom of a bulblike, swinging receptacle.

penis - The male organ used to insert sperm into a female.

penis fencing - a reproductive behavior in flatworms whereby sexual partners stab each other with a stylet-like penis.

pentamerous - Having to do with structures that occur as five, or as multiples of five.

pericardium - A membranous sac that encloses the heart.

periodic parasite -A parasite that only occasionally visits its host, for example a tick or mosquito.

periodicity - A condition in which a parasite synchronizes its behavior to that of a vector .

periostracum - The outermost, often dark colored, layer of a molluscan shell.

peristalsis- Movement of material through a tube due to rhythmic contraction of its muscular walls.

peripheral nervous system (PNS) - The nervous system excluding the brain and the spinal cord.

peritoneum - A mesodermal membrane enclosing the coelomic cavity.

peritonitis - Inflammation of the peritoneum, may be a fatal condition.

permanent parasite -An animal that depends upon its parasitic lifestyle in order to survive.

petaloid - A petal-shaped object.

phagocytosis - The process whereby a cell surrounds and engulfs a substance, such as a food item.

pharyngeal gill slits - Openings in the wall of the pharynx of chordate animals.

pharynx -The part of the digestive system located between the oral cavity and the esophagus.

phenotype - The physical characteristics of an organism due to its genetic expression.

pheromone - An airborne, hormone-like substance secreted by an animal that elicits a specific behavior on the part of another animal of the same species. Example: many female insects release pheromones that attract males for mating.

phoresis - A symbiotic relationship in which one animal uses another for transport only.

photoperiod - The varying length of daylight during a given 24-hour period. Also, may refer to the effect of daylight on an organism.

photophore - An organ that emits light, especially prominent in bioluminescent organisms.

photoreception - The ability to perceive visible light.

photoreceptor - Any light detecting sensor.

photosynthesis - The ability of organisms (e.g., green plants) to convert light energy into chemical energy. See autotrophs.

phylogenetic tree (= dendrogram) - A diagram showing the proposed evolution of a group of organisms.

phylogeny - The evolutionary history of an organism or a group of organisms.

Physalia - The genus name for the Portuguese Man-O-War, a venomous cnidarian.

physiology - The study of function.

phytoplankton - Minute algae and plants that are suspended in water. Often, along with zooplankton, it forms the basis of ecological food pyramids.

pinacocytes - The thin, flat cell type that covers the external surface of a sponge.

pineal eye (= third eye) - An organ beneath the surface of the dorsal head skin that serves as a photoreceptor in many vertebrates such as lampreys, sharks, amphibians, and reptiles.

pineal body (= pineal gland) - A small mass in the midbrain, derived from the pineal eye, that secretes the hormone melatonin and is involved in biological rhythms.

pinworm - A parasitic roundworm of the human large intestine. The smallest and most prevalent intestinal nematode in the United States.

pit organs - Heat sensing devices located on the facial area of certain snakes.

pituitary gland - The "master" endocrine gland of higher organisms. It is attached to the ventral surface of the brain and is responsible for secreting a large number of hormones.

placenta - A structure derived from both the mother and the embryo through which exchange of oxygen, nutrients, and wastes occur.

placoid scales - The body scales of sharks.

planula larvae - A ciliated, free-swimming cnidarian larvae.

Plasmodium - The protist that causes malaria.

plasma - The liquid portion of blood.

plastid - The plant cell organelle containing chlorophyll and involved in photosynthesis.

plastron - The ventral shell of turtles.

platform nests - Broad, somewhat bulky nests built either on the ground (loons, grebes, some pelicans) or elevated above the ground (herons, hawks, eagles.

pleopods (= swimmerets) - Small, paired appendages on the abdomen of certain crustaceans, such as crayfish.

pneumatic bones - Hollow bones of birds.

podium (pl = podia) - A tube foot of echinoderms used in locomotion and controlled by the water vascular system.

poison glands - Glands capable of secreting a venomous or toxic substance.

poikilotherm (= cold-blooded) - An animal unable to regulate its own body temperature.

polyandry - A female having more than one male mate.

polychaetes - The class of annelids characterized by pronounced segmentation, the presence of paired parapodia, and a large number of setae.

polygamy - The condition in which an animal has more than one mate at the same time.

polygyny -The condition in which a male has more than one female mate.

polymorphism - A condition in which a population has a variety of genetically determined traits. An example would be certain cnidarians in which a polyp and a medusoid body form exist.

polyp - The attached, asexual stage of a cnidarian.

polyphyletic - A group of organisms which have evolved from more than one ancestral line.

polyspermy - The fertilization of an egg by more than one sperm. An unviable (lethal) condition in most cases.

Porifera - The phylum to which sponges belong.

porocyte - A tubular cell type in sponges that forms channels for the entrance of water.

Portuguese Man-O-War - A venomous colonial cnidarian belonging to the genus *Physalia*.

postanal tail - The extension of the body beyond the anus. One of the unique features of chordates.

posterior - The region of an animal opposite from the head. The portion that last meets the environment.

preadaptation - A pivotal evolutionary concept stating that a genetic structure or feature is present in a population prior to its selection. One of the primary mechanisms of phyletic evolution.

precocial young - Individuals that have reached a relatively high degree of development at hatching or birth. They are capable of locomotion, feeding, and limited self-care. Compare to altricial young.

predation - A relationship in which one animal preys upon another, usually for food.

preening - Care of feathers by birds. It involves passing feathers through the beak to restore or maintain a functional state.

prehensile - A structure adapted for grasping.

pressure receptors - Sensors able to detect tension.

primary host (= final or definitive host) - Organism (host) in which a parasite reaches sexual maturity (adulthood).

primary producer - Organism, such as green plants, capable of converting light energy into a chemical form, thus making energy available to itself, and ultimately to other organisms.

primitive - Ancestral. A feature or trait that has led to other (advanced) features.

prismatic layer - The middle layer of a mollusk shell.

procuticle - The inner, thicker layer of the arthropod exoskeleton that hardens after a molt.

proglottid - A section or subunit of the tapeworm's body adapted for reproduction.

prokaryote - A primitive organism or cell lacking a nucleus and other organelles; mainly bacteria.

promiscuity - A type of pair bond in which mates are selected randomly or indiscriminately.

propagation- To increase or multiply as by reproduction.

prosencephalon - The anterior most part of the vertebrate brain including the cerebrum, olfactory apparatus, and other features.

prosoma - The body part of an arachnid encompassing the head and thorax.

prostomium - The fleshy lobe anterior to the mouth in earthworms and other annelids.

protists - Members of the kingdom Protista consisting of unicellular or colonial protozoans.

protonephridia - A simple osmoregulatory structure of flatworms consisting of a closed tubular system containing flame cells.

protopodite - The basal portion of the crustacean biramous appendage.

protostomes - A lineage of animals whose mouth originates from the embryonic blastopore.

proximal - Toward the midline or the point of attachment of a structure on an animal.

pseudocoel - A condition in which the body cavity is not derived from mesoderm, and thus not lined by a peritoneum. Characteristic of the phylum Nematoda.

pseudocoelomates - Animal groups that possess a pseudocoel.

pseudopodia - Variable cytoplasmic extensions of amoebas used in feeding (endocytosis) and for locomotion.

punctuated equilibrium - An interpretation of the evolutionary process that differs from the traditional gradualistic explanation. The punctuated equilibrium concept proposes that evolution occurs as a pattern in which long periods of relative inactivity (stasis) is interrupted by fairly rapid changes taking place over thousands of years rather than the millions proposed by the gradualistic theory. See gradualism.

pupa - An immature, nonfeeding stage of a holometabolous insect leading to adulthood.

puparium - A case or housing in which a pupa transforms into an adult.

pupation - The transformation of pupae into adults.

Pycnogonida - The class of arthropods consisting of the sea spiders.

pygidium - The terminal segment of certain animals such as earthworms.

pygostyle - The knoblike, reduced "tail" of a bird.

pyloric cecum - A blind sac leading from the digestive system used to increase digestive surface.

Q

quadruped - An animal the walks on four limbs.

queen - The female of certain eusocial insects such as bees, ants, and termites.

queen's substance (= queen mandibular pheromone) - A secretion of a queen bee that prevents the development of other queens.

quill - (1) The hollow interior portion of a feather shaft. (2) A rigid, defensive, spine-like hair such as pos sessed by hedgehogs or porcupines.

R

rachis - Solid portion of a feather shaft.

radial canals - A portion of the echinoderm water vascular system that extends into each arm.

radial symmetry - A form of body structure in which any plane passing through the oral-aboral (longitudinal) axis results in mirror images.

radial cleavage - A pattern of embryonic cell division in which the daughter cells are aligned with the central axis of the entire cluster.

radiolarians - A type of sarcodine (ameba) that forms hard protective shells.

radioles - A series of branched, ciliated tentacles usually near the mouth of tube worms (annelids).

radula - A tongue-like rasping device of many molluskans used to scrape food, such as algae, from a hard surface.

ram ventilation - A pattern of breathing in many fish in which water passes over the gills as the fish swims with its mouth open.

raptorial feeding - Seizing live prey with feet, talons, or tentacles and bringing it to the mouth where it's torn into pieces for swallowing. Characteristic of squid, octopuses, birds of prey, and others.

raptorial beak - A bird beak adapted for tearing apart meat or captured prey.

ratite - Referring to flightless birds that lack the keeled sternum of flying birds. Examples: ostrich and emu.

recapitulation - The repetition of an animals evolutionary past during its embryonic development.

rectal gland - An osmoregulatory organ of sharks and their relatives that removes excess ions (esp. NaCl) from the blood and empties it into the cloacal area.

rectum - The terminal section of the digestive system.

recurved beak - A bird's beak that is upturned at the end for sifting food from mudflats or other shallow water.

redia larva - A larval stage of a digenetic trematode.

regeneration - Regrowth of lost parts.

regional specialization - The functional organization of a area of the body such as the head, thorax, or abdomen.

releaser - A stimulus that triggers a behavioral response.

renal - Of, or relating to, the kidney.

renette - An excretory device of some roundworms.

reproductive isolation - The inability of animals to mate even though they may occupy overlapping niches.

repugnatorial glands - Organs that secrete a noxious substance such as the malodorous emissions of centipedes and millipedes.

reservoir host - Where parasites can multiply (usually asexually) without necessarily damaging the host.

respiratory pigment - A metallic organic compound such as iron or copper with which oxygen can combine.

respiratory tree - Internal branching tubules radiating from the rectum of a sea cucumber to function in gas exchange (respiration).

retina - The portion of the eye that bears photoreceptor cells to transform light into electrical signals that are relayed to the brain.

retroinfection - Reinfection of a host by one of its current parasites.

rhabdites - Mucous secreting cells concentrated on the ventral surface of certain flatworms .

rhombencephalon - The portion of the brain called the hindbrain that houses the pons, cerebellum, and medulla.

ring canal - The part of the echinoderm water vascular system that encircles the mouth between the stone canal and the radial canals.

rod cell - A photoreceptor cell adapted for vision in low light and for color vision.

royal jelly - A sugary substance fed to larva by worker bees to induce development of a new queen.

roly-polys - A common name for any animal that rolls into a protective ball, especially members of isopod crustaceans such as pillbugs and woodlice.

rostrum - A beaklike projection from the head.

roundworms - Any member of the phylum Nematoda.

route-based navigation - Navigation by the use of landmarks along a migratory pathway.

rumen - The first part of the stomach in cattle, sheep, and deer.

rumination - The regurgitation of food from the stomach for chewing.

S

sacrificial reproduction - The price of breeding in many animals, often involving some sacrifice, even death.

sagittal section - A plane of bisection along an animal's longitudinal axis to produce right and left halves.

saliva - Secretions of the salivary glands that contain enzymes, mucous, and anti-bacterial agents.

salt gland - A gland in many marine vertebrates located near the eye or nostril that excretes excess salt.

saprozoic nutrition - Gaining nourishment by feeding on dissolved nutrients.

satellite male - A male frog that positions himself near a courting male and intercepts females attracted by the courtier.

satiation - A defensive strategy of a prey species in which more young are produced than predators can consume, thereby assuring that some prey will survive.

scavengers - Animals that feed on dead animals.

schizogony (= multiple fission) - Cell reproduction involving multiple nuclear divisions resulting in numerous genetically identical individuals.

scientific method - The process or approach that scientists use to identify and solve problems. Involves asking questions, making observations, forming hypothesis, testing, and making conclusions.

scolex - The bulb-shaped holdfast organ at the anterior end of a tapeworm.

scrape -A type of bird nest consisting of a shallow depression on the surface of the ground.

scrotum - An external pouch that contains the testes in mammals.

scute - A horny plate or scale such as that on turtles or some lizards.

sebaceous oil gland - A skin gland that secretes an oily substance called sebum.

secondary eyes - Specialized arachnid eyes adapted for night vision.

secondary palate - A bony plate that forms the roof of the oral cavity and separates it from the nasal cavity. Occurs in mammals and some reptiles.

secretion - Liberation of a substance from a gland.

sedentary - Attached or not moving, as a barnacle.

segmentation (= metamerism) - Division of the body into repeating units called segments or metameres.

selective deposit feeding - The process of filtering food from a muddy substrate such as on the floor of the ocean.

self-fertilization - The union of sperm and egg from the same animal.

selfish herd response - A defensive strategy of prey animals using large numbers to assure individual survival.

self preservation - The tendency of all animals to stay alive.

semen - The fluid that contains sperm.

seminal receptacle - A sac-like organ of female reproductive systems used to gather and store sperm received during copulation. Found in annelids and many insects.

seminal vesicle - (1) Organs in the male reproductive system of annelids that store sperm prior to its release. (2) Organs in male mammals that add a nutrient-rich substance to semen.

senescence - The aging process.

sensory neuron - A nerve cell that relays signals from receptors to the central nervous system.

sepia - Inky substance that is secreted by squid, cuttlefish, and octopuses for defensive purposes.

septum - A layer of tissue that divides two compartments or chambers such as in the heart or between adjoining segments in an annelid.

sequential hermaphroditism - A condition whereby an animal is one sex (e.g., female) during a part of its life cycle then later becomes the opposite sex (i.e., male).

serial homology - A series of structures having different functions but all derived from a common source. For example, the appendages of crustaceans.

sessile - Attached to a substratum.

seta (plural setae) - A hairlike, chitinous bristle on many invertebrates, including annelids and arthropods.

sexual reproduction - Reproduction accomplished through meiosis whereby two haploid cells unite to form a new, genetically unique offspring.

sexual selection - Choosing mates based upon differences in structure, displays, or other behavior.

shivering thermogenesis - The generation of body heat through quivering.

silica - A white or colorless crystalline compound (silicon dioxide) often occurring as sand or quartz. It is sometimes used by animals in the making of casings, shells (the tests of foraminiferans), or skeletal elements (spicules of sponges).

simple development -See ametabolous development.

simple eye - A photoreceptor that detects light but does not form a distinct image.

Simulium - The genus of black flies that serves as a vector for *Onchocerca volvulus*, a roundworm parasite known as the blinding worm.

simultaneous hermaphrodites - Animals that transmit sperm to each other synchronously.

sinus venosus - A heart compartment leading to the atrium.

sinusoidal locomotion - Forward movement in a side-by-side, or S-shaped, pattern.

siphon - A funnel-shaped or tubular structure found in many mollusks that permits water to enter and leave the mantle cavity.

siphonoglyphs - Cilia-lined grooves in sea anemones that helps to produce hydrostatic pressure.

sister chromatid - One of the pair of replicated chromosomes joined by a centromere.

slit organs - Ventral sensory structures in certain arachnids used to detect vibrations.

small intestine - The portion of the digestive system between the stomach and large intestine that is responsible for the bulk of digestion and food absorption.

social parasitism - See brood parasites

soft corals - Fleshy corals that lack a skeletal or rigid support.

solute - Any substance that is dissolved in another substance.

solution - The liquid in which a substance is dissolved.

solvent - A liquid, such as water, that dissolves another substance

somites (= segments = metameres) - Mesodermal blocks in the embryonic wall appearing as a series of repetitive units.

specialization paradox - A concept that specialization to one's environment is an important route to success but at the same time over-specialization may ultimately lead to extinction. An animal must be able to balance generalization with specialization to avoid the paradox.

562

specialists - Animals that have become closely adapted to use or exploit specific features in their environment.

speciation (= macroevolution) - The process of forming a new species.

species (singular and plural) - (1) For sexually reproducing forms: an ongoing group of interbreeding natural populations reproductively isolated from all other groups. (2) For asexually reproducing forms: groups of organisms similar to each other but recognizably different from other groups.

species perpetuation - An ongoing activity of all animal populations to reproduce and thereby pass on genetic traits.

sphincter - A ring of muscles capable of constricting an opening or passageway.

spicules - Needle-shaped skeletal elements of sponges.

spinnerets - Fingerlike tubules on the posterior abdomen of spiders through which silk is released.

spiracle - An opening used for ventilation of the respiratory surface in arthropods. A narrow opening in the wall of the exoskeleton that leads into the tracheal respiratory system.

spiral valve - (1) A corkscrew membrane in the intestine of sharks used to increase surface area for digestion. (2) A helical valve separating oxygenated and deoxygenated blood in amphibians.

spiral cleavage - A method of development in which daughter cells twist in a helical pattern as more cells are added to the growing embryo. It is characteristic of most invertebrates other than arthropods and echinoderms.

spleen - An organ of the immune system.

spongin - A fibrous protein that makes up the endoskeletal framework of some sponges.

sponging mouthparts - A feeding apparatus of certain insects (e.g., housefly) that sops up liquids.

spongocoel - The internal cavity of a sponge.

spore - An inactive form of a zygote, often resistant to environmental conditions.

sporogony - Process of producing spores (e.g., in *Plasmodium*, the malarial parasite)

sporozoite - An infective stage of the malarial parasite, *Plasmodium*, formed from a spore.

squalene - A low-density fat or oily substance produced by shark livers to help maintain buoyancy.

squid pen - See pen.

stabilizing selection - An evolutionary pattern which favors the norm in a phenotypic range at the expense of the extremes. Results in a narrowing of the phenotypic range.

statant - A type of cupped bird nest that is supported from beneath, such as the nest of robins, cardinals, and warblers.

statocyst - A hollow organ in many invertebrates responsible for balance and equilibrium. The statocyst contains granules (statoliths) that stimulate sensory hairs which detect the direction of gravity.

statoliths - Granules (sand, or dried mineral mass) within a statocyst that deflect sensory hairs in response to gravity.

stereoscopic vision - three-dimensional vision accomplished by the placement of eyes laterally on the head (e.g., hawks, eagles, and other raptors).

sternal keel (= carina) - A long bladelike process connected to the ventral side of a bird sternum to which powerful flight muscles attach. See keeled sternum.

stigma (plural, stigmata) - A photoreceptor in *Euglena* and other protists.

stoma (plural stomata) - A mouth, or mouthlike opening.

stomach - A food storing pouch within the digestive system.

stone canal - A tubular part of the echinoderm water vascular system connecting a madreporite to the ring canal.

straight beak - A stabbing type of bird beak possessed by herons and others in which the contact between the upper and lower beak (i.e, the commissure) is in a straight line.

streamlining adaptations - Features in birds that overcome drag or friction while flying.

strobila - The body of a tapeworm, posterior to the neck, that contains the proglottids.

subpharyngeal gland - An iodine-fixing gland of chordates located ventral to the pharynx.

substrate (= substratum) - Any surface upon which something rests, such as the ocean bottom, the surface soil, ground, rock, and so on.

substrate matching - The ability of some animals to physically match their surroundings through colors, patterns, or shape.

subterminal mouth - The mouth of some fish, such as sharks, that is directed ventrally.

sun compass - A navigational aid to some migratory birds in which the sun is used as a tool to determine direction.

suprapharyngeal ganglia - Nerve masses located ventral to the pharyngeal region. A type of primitive brain.

surperparasitism - A condition in which a host harbors more than one type of parasite.

suspension feeders - The removal of food from the surrounding water, usually through a filtration device.

swim bladder - A gas-filled sac in fishes that aids in maintaining buoyancy.

swimmerets (= pleopods) - Abdominal appendages of certain crustaceans (e.g, crayfish) to which eggs may be attached in the female.

sycon sponge - A sponge body-form in which choanocytes do not line the central cavity but line radial canals instead.

symbiont - The initiator of a symbiotic relationship.

symbiosis - An intimate relationship between members of different species. Four common types occur: phoresis, mutualism, commensalism, and parasitism.

symmetry - A correlative arrangement of body parts on either side of a central axis so than an animal can be bisected into similar halves.

sympatric - Relating to overlapping habitats. Applies specifically to evolution that occurs between two populations even though they occupy the same geographical region.

synapsis - The coming together or alignment of homologous chromosomes during meiosis.

syncytium - Any tissue consisting of large cells with several nuclei, most likely derived from the fusion of several cells.

syngamy (= fertilization) - The union of egg and sperm nuclei resulting in the formation of a zygote.

syrinx - The vocal apparatus of birds. Contrast to the larynx of mammals.

systematics - The study of the evolution and classification of organisms.

T

tactile receptor - A touch receptor.

tagma (plural tagmata) - Fusion of neighboring body segments to form a distinct region such as a head or thorax.

tagmatization (= tagmosis) - Process of forming distinct body regions (tagmata) in segmented animals.

tapetum (= tapetum lucidum) - A reflective layer near the retina to enhance night vision in nocturnal animals.

tarsus -(1) The distal segmented part of an insect leg. (2) The portion of the vertebrate foot between the leg and the metatarsus.

taxis - The movement of an animal in response to an environmental stimulus.

taxon - A group of genetically related individuals.

taxonomy - The naming and classifying of organisms.

tegument - The external covering of tapeworms and flukes.

teleology - Inappropriate claim as to cause or purpose of a process.

telson - (1) The spinelike posterior projection of a horseshoe crab (2) The flattened posterior appendages of decapod crustaceans used as a paddle or flipper.

tentacle - An elongated, unsegmented, often fleshy, protrusion usually associated with the mouth.

tergum - The dorsal surface of an arthropod.

test - The hard casing of certain sarcodinians (amoebas) and urchins (Echinodermata).

testis - The primary (sperm producing) reproductive organ of a male animal.

terete beak - A bird beak that is somewhat spikelike as in hummingbirds.

terrapin - A semi-terrestrial aquatic turtle (chiefly British usage)

terrestrial egg - See cleidoic or amniotic egg.

terrestrialization - See aquatic emancipation.

territory - Any area actively defended against others of the same species. It often provides the space to acquire food, shelter, or reproductive partners.

tetrapod - A general term referring to any vertebrate having four limbs such as amphibians, reptiles, birds, and mammals.

thalamus - A area of the brain that links the cerebral hemispheres to each other and to associated regions.

theory - A general description or statement that explains a body of facts.

thermal stratification - Distinct layering within lakes due to temperature differences.

thermogenesis - The generation of heat by an animal.

thermo-receptor - A sensor sensitive to temperature changes.

thermoregulation - The ability to control internal body temperatures.

third eye - See pineal eye.

thigmotaxis - Movement of an animal in relation to direct tactile (touch) stimulus. Many animals select a habitat where they can be in direct contact with some surface.

thorax - The tagma between the head and abdomen of arthropods and other animals. It typically bears the locomotor appendages.

threshold - The level that a stimulus must reach to evoke a response.

thrust - The force generated by a flying animal to overcome friction or drag.

thyroid - An endocrine gland located in the neck region that is involved in iodine fixation and overall body metabolism.

tibia - (1) The larger of two bones of the human lower leg. (2) The fourth division of an insect's leg located between the femur and the tarsus.

tick - A blood-sucking arthropod parasite belonging to the class Arachnida.

tissue - A group of similar cells that coalesce to perform a particular function.
Examples: muscle, skeletal, nervous.

tissue specificity - A condition in which a parasite favors a certain tissue within its host.

top carnivore - The carnivorous animal within a particular ecosystem that is not preyed upon by another animal.

torpor - State of inactivity induced by lowered temperature.

torsion - A process in which the visceral mass of a gastropod embryo is twisted into a U-shape.

tortoise - A general term for any terrestrial turtle.

touch receptor - A sensor that responds to tactile stimuli.

toxin - Any substance produced by one organism that has some negative or harmful effect upon another.

trachea (plural tracheae) - (1) The respiratory tube that connects the nasal or oral cavity with the bronchi. (2) The interconnecting tubes leading from spicules throughout the body of certain arthropods, especially insects.

tracheal respiration - The respiratory mechanism in many arthropods, especially insects. It consists of slits in the body wall (spiracles) which allow air to enter a branching network of tubules (tracheae) distributing oxygen throughout the body.

tracheole - Small tubules of the tracheal respiratory system.

transverse section (= cross section) - A plane of bisection dividing an animal into anterior and posterior parts.

trematode - Parasitic flatworms belonging to the class Trematoda.

trepang - Certain sea cucumbers of the southern Pacific and Indian Oceans or the prepared food derived from them.

trichinosis - The disease caused by infection with the roundworm, *Trichinella spiralis*, characterized by fever, muscular pain or spasms, and tissue swelling.

trichocyst - A hairlike defensive device expelled from the pellicle of certain ciliates.

trigeminal nerve - Cranial nerve number five, involved in jaw movement and head skin sensations.

tripartite intestine - The three-part intestine of many flatworms (esp. Turbellarians).

triploblasty - The state in animals in which their bodies are derived from three embryonic tissue layers: ectoderm, mesoderm, and endoderm.

trochanter - The second, proximal division of an insect leg.

trochlear nerve - Cranial nerve number four involved in controlling eye movements.

trochophore - A free swimming, ciliated larvae of many mollusks, annelids, and other protostome animals.

trophic levels - The nutritive, or feeding, level of an organism within an ecosystem. Autotrophs occupy the producer trophic level whereas animals occupy the consumer level.

tsetse fly - The insect vector of African sleeping sickness.

tube feet - Blind-ended muscular sacs of the water vascular system of echinoderms used in locomotion, feeding, respiration, and attachment.

tubules of Cuvier - Threadlike, sticky or noxious strands extruded from the anus of sea cucumbers for defense.

tumor - A cellular mass resulting from uncontrolled cell division.

tunicate - Members of the subphylum Urochordata, also known as sea squirts.

tympanic membrane (= eardrum) - A membrane capable of detecting airborne vibrations. Prominent externally on many insects and amphibians.

typhlosole - A straplike longitudinal infolding of the earthworm intestine that increases surface area for absorption.

U

ultraviolet (UV) radiation - Wavelengths of radiation on the border of X-rays. May be damaging to living tissue.

umbo - The oldest portion of a bivalve shell, consists of a rounded prominence or bulge located at the anterior hinge.

underfur - The dense, soft pelage that lies under the more coarse, outer guard hairs of certain mammals.

unicellular organization (= cytoplasmic organization) - The pattern in which all of life's activities are conducted within the confines of a single cell. Protists are the only unicellular organisms.

Uniramia - A subphylum of arthropods whose members are characterized by possessing one pair of antennae, one pair of mandibles, and unbranched (uniramous) appendages.

uniramous - Unbranched as in the uniramous appendages of insects.

upwelling - The rising of colder, nutrient-rich water due to wind, current, temperature differential, or climactic conditions.

urchin test - The endoskeleton or shell of a sea urchin.

urea - The form in which metabolic wastes, such as nitrogen and carbon dioxide, are excreted in certain animals, especially mammals.

ureotelic - Excretion of nitrogenous waste in the form of urea.

urethra - The tube carrying excretory wastes (e.g., urine) from the body.

uric acid - The major nitrogenous waste in egg-laying terrestrial animals (insects, birds, reptiles).

uricotelic - Excretion of nitrogenous waste in the form of uric acid.

urinary bladder - A saclike organ that temporarily stores urine.

Urochordata - The chordate subphylum in which are enclosed in a tunic. The sea squirts or tunicates.

urogenital aperture - Opening through which urine and gametes (sperm and eggs) are released in certain fishes.

uropod - Terminal pair of appendages in decapod crustaceans (e.g., crayfish) which, together with the telson, forms a paddlelike tail fin.

uropygial gland (= oil gland) - A gland at the base of a bird's tail that produces oil for preening. Also known as the preening gland.

urostyle - The pronglike fused vertebrae of a frog.

use and disuse (= Lamarckism) - The refuted theory that acquired characteristics can be passed to offspring. It was once incorrectly thought to be a prime mechanism in the evolutionary process. Also known as the theory of acquired characteristics or LaMarckism, after its major proponent.

uterus - The hollow, muscular organ of female mammals where the embryo develops.

V

vacuole - Any small bubblelike space formed within a cell.

vagina - A tubular organ that leads from the uterus to the outside in the female reproductive tract. May be used as a receptacle for the penis during copulation.

vagus nerve - Cranial nerve number ten that helps to control heart contractions and certain functions of the digestive system.

valves - (1) The two halves of a bivalve (mollusk) shell. (2) Devices that control the flow of fluids through a tube or chamber by preventing back-flow.

vane feather (= contour feather) - The primary type of bird feather in which there is a vane on either side of a central shaft.

vas deferens (= sperm duct) - Tubule through which sperm pass on their way to the outside.

vasoconstriction - Arterial constriction due to contraction of smooth muscles.

vasodilation - Arterial dilation due to smooth muscle relaxation.

vector - Any organism that actively transmits a parasite or disease organism from one animal to another.

vegetative brain functions - Activity of the brain concerned with basic functions such as digestion, circulation, excretion, and respiration.

vein - (1) Any blood vessel that returns blood from capillaries to the heart. (2) Tubule within an insect wing that contains blood vessels and nerves.

veliger larva - A free-swimming larval stage of many mollusks. It is derived from the trochophore larva and undergoes metamorphosis into the adult.

vent - The narrow external opening into the cloaca.

ventral - Referring to the belly of an animal or to the surface that faces the substratum. The same as anterior for bipedal (upright) animals.

ventral aorta - A relatively short ventral artery leading from the heart of fishes.

ventral nerve cord - A longitudinal mass of nerves that runs along the ventral surface of an invertebrate animal. Contrast to the dorsal nerve cord of vertebrates.

ventricle - (1) The muscular chamber of the heart that delivers blood to an artery. (2) A brain chamber filled with cerebrospinal fluid.

vermiform - Any structure shaped like a worm.

vertical migration - Daily movement of aquatic organisms (especially zooplankton) upward at night and downward during the day.

vesicle - Any bubblelike, membrane-bound structure that contains material entering the cell by endocytosis or leaving the cell by exocytosis.

vestigial structures - Any structure that was functional in the past but is now useless. These structures are often used as evidence for evolution.

vibrissa (plural vibrissae = whisker) - A long, stiff facial hair of a mammal used as a touch receptor.

viper - A venomous snake belonging to the family Viperidae.

virulence - The capacity to cause a disease.

vitalism - A belief that living organisms have unique powers.

viscera - The internal organs of an animal.

visceral mass - A saclike area of a mollusk that contains the viscera.

viviparity - The condition in which embryos develop within the body of the mother and are nourished by her.

vasodilator - Any substance that causes blood vessels to dilate (expand).

W

wafting - The movement of a substance in fluid across a ciliated surface.

waggle dance - The behavior of a scout honey bee by which it transmits information to worker bees about the direction and distance of a food source.

walking legs - Jointed, walking legs of arthropods.

warning coloration - Highly visible color patterns of one animal that communicates a dangerous or noxious condition. Serves to deter predation.

water vascular system (WVS) - A branching network of fluid-filled canals in echinoderms used for locomotion, respiration, and excretion.

web - A network of silk formed by spiders and used for a variety of purposes such as catching prey, reproduction, and locomotion.

weight reducing adaptations - Features in birds that promote lift for flight.

whiskers - See vibrissa.

withdrawal reflex - A simple automatic action that pulls a limb or other structure away from injury.

work - The exertion needed to accomplish something.

worker jelly - The sugary substance fed by bee workers to larvae.

X

X gland (= X organ) - The crustacean gland that secretes molt inhibiting hormone (MIH).

xanthophore - A cell that contains yellow, red, or orange pigment.

Xenopus - See African clawed frog -

Xenotransplantation - The use of animal parts or organs in humans.

Y

Y-gland (= Y-organ) - Crustacean gland that secretes a molting hormone (MH).

yolk - The nutrient component in an egg.

yolk glands - Structures that secrete yolk for eggs.

yolk sac - A membrane that encloses yolk in embryos.

Z

zooid - A single individual in a colony.

zoology - The study of animals.

zoonosis - A disease that's transmitted from animals to humans.

zooplankton - Aquatic animals suspended in water.

zooxanthallae - Mutualistic photosynthetic protists living within another animal such as a sponge, coral, or clam.

zygote - The cell resulting from fertilization of an egg by a sperm.

Index

A

aardwolf, 367

Abducens nerve, 409

Aboral surface, 97

Acanthocephala, 140

Acoelomate, 119, 141

Active pursuit, 371

Acontia, 102

Adaptation, 12

Adaptive radiation, 16, 23

Adductor muscles, 184

Adherent nest, 476

Advantages of flying, 468

Afferent branchial arteries, 405

African Sleeping Sickness, 73, 78

Aggregation response, 377

Agnatha, 391, 400, 401, 415

Ahermatypic coral, 104

AIDS, 78

Air sacs, 470

Akinesis, 383

Alternate evolutionary explanations, 30

Alternation of generations, 97

Altricial young, 478

Ambulacral grooves, 349

Ambush, 371

American Alligator, 447

Ametabolous development, 324 - 325

Amniotic egg, 440

Amoeba, 69 - 70

Amoebic dysentery, 71, 78

Amoebocytes, 86 - 88

Amoeboid movement, 69

Amoeboid tests, 71

Amphibia, 391

Amphioxus, 395 - 397

Amplexus, 9, 433, 437

Ampullae of Lorenzini, 404, 407

Ampullae (sea star), 349

Anemones, 97, 115

Animals defined, 42

Annelida

    characteristics, 218

    circulatory system, 220

    digestive system, 219

    excretory system, 220

    reproduction, 220

Antennae, 288, 294

Antennal gland, 288, 292

Antennules, 288, 294

Anterior ganglion, 121

Anthozoa, 102

Ants, 258, 326

Anura, 437

Aortic arches, 220

Apical complex, 77

Apicomplexa, 77

Aposematic coloration, 380, 435

Aquatic emancipation, 259

Arachnida, 273

Aristotle's lantern, 355

Arrector pili muscle, 23, 495

Arthropods, 258

    characteristics, 263

    osmoregulation, 260

    reproduction, 261

    respiration, 260

    senses, 260

    thermoregulation, 260

Artificial selection, 27

Ascaris lumbricoides

    anatomy, 143, 149, 150

    life cycle, 159

Asexual reproduction, 54, 56 - 58

Asteroids, 350, 351, 361

Asymmetry, 45, 69, 117

Atoll, 104

ATP, 49

Atrium, 395, 405

Auditory nerve, 409

Aurelia, 102

Auricles, 121

Australian horny devil, 442

Autoinfection, 158, 162

Autotomy, 291, 385

Autotrophs, 42, 69

Aves, 391, 464

B

Bald Eagle, 33

Barnacles, 155, 290, 297

Barrier reefs, 104

Barriers to terrestrialism, 261

Basal disc, 97

Basket stars, 352

Batesian mimicry, 382

Bath sponges, 86

Beef tapeworm, 129

Bilateral symmetry, 45, 117

Bilateria, 262

Bile duct, 423

Binary fission, 54, 76

    longitudinal, 54

    multiple, 54

    transverse, 54

Binomial nomenclature, 32

Biocontrol, 327

Biodiversity defined, 3, 4

Biogenesis, 392

Biogenetic Law, 392

Biological species concept, 4

Bioluminescence, 195 - 196

    extrinsic, 195

    intrinsic, 195

Biophilia, 30

Biramous appendage, 288, 291, 295

Birds, 464

    air sacs, 470

    beaks, 484 - 485

    brood parasites, 475

    clutch size, 477

    courtship, 472

    flight, 466 - 468, 470

    migration, 481

    pair bonding, 473

    nests, 475 - 476

    radiation, 24

    reproduction, 471

    respiration, 470

    senses, 469

    sounds, 480

    taxonomy, 483

    territory, 471

Biting-chewing mouthparts, 322

Bivalves, 184

Black bear, 19

Black widow spider, 6, 274, 275

Blackflies, 164

Bladderworm, 128, 161

Blastocoel, 142

Blastula, 262

Blinding worm, 164

Blood flukes, 125

Blue-ringed octopus, 177

Body

    armor, 384

    directions, 46

    organs, 423

    shapes, 46

Bonding, 7, 8

    birds, 473

Bony fish, 409

Book gills (lungs), 271, 272

Box jellies, 106

Branchial hearts, 191

~ Index ~

~ Index ~

~ Index ~

~ Index ~

~ *Index* ~

~ Index ~

~ Index ~

~ Index ~

~ Index ~

Zoology:
An Inside View of Animals

Worksheets

Lab Notes

Chapter 1   Worksheet

_____ _____
(name)   (lab section)

Instructions: Use the dichotomous key to identify each "unknown" specimen. Give its phylum and common name in the space provided.

Animal Identification

Specimen #	Phylum	Common Name
1.	_____	_____
2.	_____	_____
3.	_____	_____
4.	_____	_____
5.	_____	_____
6.	_____	_____
7.	_____	_____
8.	_____	_____
9.	_____	_____
10.	_____	_____

Lab Notes

Chapter 2 — Worksheet
this exercise was provided by Prof. Sandy Kreiling

Body Form & Function
_____
(name)

I. Body Surfaces

1. Make a model of an animal of your choice out of "Play Dough". Just so we're sure which end is which, use the probe to draw a mouth and an anus on your animal.

2. Use the probe to mark the midline on your model between the anterior and the posterior end. Mark both the dorsal and the ventral surface.

3. Sketch your animal in the space below. Label anterior, posterior, dorsal and ventral.

II. Sections (to view the internal structures of an animal, it is often cut in half or a thin slice is cut from it)

1. Draw a line on each of the animals below to indicate a **transverse** and a **sagittal** cut.

(a transverse cut produces a **cross section** and a sagittal cut produces a **longitudinal section**)

TRANSVERSE

SAGITTAL

2. Using a scalpel, cut your animal in half making either a transverse or a sagittal section. Your partner should make the other cut. How would your view of the interior of the animals differ in the two cuts?

III. Radial Symmetry

1. Recycle your animal and use your Play Dough to make a model of a radially symmetrical animal with several appendages.
   a. where would you put the mouth?
   b. is there an anterior and a posterior end?
   c. does it matter where you draw the midline?

IV. Body Design (sack-like vs. tube-within-a-tube designs)

1. Use 1/2 of the Play Dough to make a simple sac-like body with only one opening.

2. Use the other 1/2 of the Play Dough to make a tube-within-a-tube body with an opening at each end and a continuous tube connecting the two openings.

   a. if you were to feed these two animals, where would the food enter? where would wastes exit?

3. In the space below draw each of your body plans. Use arrows to indicated the flow of food and wastes.

Lab Notes

Chapter 3   Worksheet

_____
(name)

Subphylum Sarcodina
"the amoeboids"

1. Observe live specimens of *Amoeba proteus*.
    Use the illustration in Figure 3.2 on page 70 to locate the following structures:

    - ✓ cytoplasm (ectoplasm and endoplasm)
    - ✓ pseudopodia
    - ✓ nucleus
    - ✓ contractile vacuole
    - ✓ food vacuole

How to make a wet mount to observe living protozoans:
1. Use a pipette to collect the specimen.
2. Place 1 or 2 drops on a depression slide.
3. Place a cover slip over the specimen.
4. View through the low power on the microscope.

2. Observe locomotion in the living specimen and describe it below:

AMOEBA   © Kendall/Hunt Publishing Company

3. Are these organisms autotrophic or heterotrophic?

_____

4. On a prepared slide of *Amoeba proteus*, locate the same structures as in the living specimen listed above.

5. View the slides of radiolarian and foraminiferan tests. Sketch a few of these below:

6. What is the function of these tests in the living organism?

586

Subphylum Mastigophora

"the flagellates"

7. View a prepared slide of *Euglena*.
   Although these organisms are very small, try to locate the following structures:
   (refer to Figure 3.4, pg 72 and the diagram to the right)

   ✓ chloroplasts
   ✓ nucleus
   ✓ flagellum
   ✓ stigma (eyespot)

EUGLENA

8. Observe live specimens of *Volvox*. (refer to Figure 3.4)

9. Explain why *Volvox* is considered colonial and not multicellular.

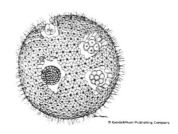

10. What structure does *Volvox* use to locomote? _____

11. Notice its green color. What does that tell you about its energy source?

VOLVOX

12. View prepared slides of *Trypanosoma*. (Refer to the photograph below and note the many parasites scattered among the tiny red blood cells.)

13. What human disease is caused by *Trypanosoma*? _____

What is the name of the insect that transmits this parasite to humans? _____

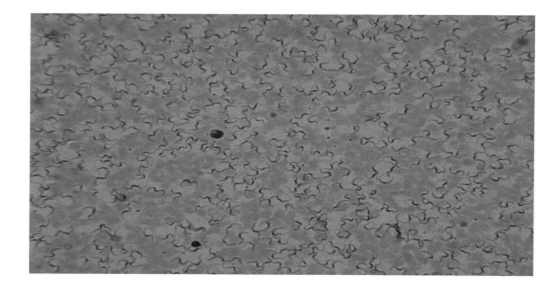

Phylum Ciliophora
"the ciliates"

14. Observe live specimens of *Paramecium* .

15. Use Figure 3.8, pg 76 and the diagram to the right to help locate the following structures on the live specimen:

    ✓ cilia
    ✓ nucleus
    ✓ contractile vacuole
    ✓ food vacuoles
    ✓ oral groove ("gullet")

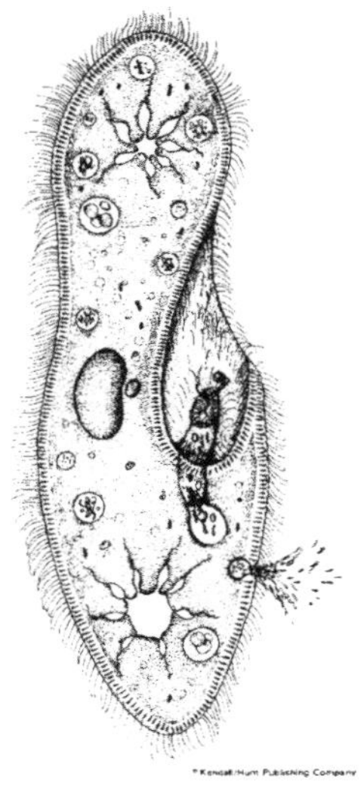

16. In the space below, describe locomotion in the living *Paramecium*.

PARAMECIUM

17. View a prepared slide of *Paramecium* "trichocysts." This slide is stained to reveal the layer of rodlike trichocysts as a single-file row just beneath the pellicle and cell membrane.

    What function do trichocysts serve? _____

18. What two methods of reproduction does *Paramecium* use?

    a. _____

    b. _____
         (Study prepared microscope slides of *Paramecium* fission and conjugation.)

19. In the space below, make two sketches: one showing a *Paramecium* undergoing fission and another illustrating a pair during conjugation.

        fission   conjugation

20. Observe a prepared slide of *Plasmodium*. Look for infected red blood cells which will appear as darkened cells with a granulated appearance. What human disease does this parasite cause?

Lab Notes

Chapter 4	Worksheet

_____
(name)

Phylum Porifera

1. In the box to the right, sketch and label a simple sponge. On the diagram include ostia (pores), an osculum, the spongocoel, and the holdfast.

2. On your sketch use arrows to show the route of water through the sponge. Then list (below) items that may be transported by the current of water. Which cell type creates the water flow? _____

3. Observe prepared slides of *Grantia* (=*Scypha*) and note the presence of ostia perforating the outer wall and the centrally located spongocoel. Refer to diagram on page 89.

4. View a sponge skeleton (i.e., a bath sponge) using a dissection microscope. Note the abundance of pores and the spongin that make up the specimen.

5. Observe a prepared microscope slide of spicules. Then sketch several individual spicules in the space below as well as the network of spicules that make up a sponge skeleton.

6. If a live specimen is available, scrape a small area from its surface. Keep the tissue wet and then tease the material apart with a needle probe. Sketch the spicules and compare them to the ones you saw above.

7. Observe a prepared slide of a gemmule and sketch one in the space below.

Lab Notes

Chapter 7 | Worksheet

_____
(name)

Phylum Nematoda
Ascaris lumbricoides External Anatomy

A. Observe the external anatomy on specimens of both a male and female _Ascaris_. Use a dissecting microscope to find the mouth of each. This is done by locating the three anterior lips surrounding the mouth (Figure 7.1). Also refer to Figure 7.1 to help locate the copulatory spurs (spicules) of the male. Finally, find the female genital pore, which is where sperm enters the female during copulation. This is a small ventral slit located about one-third of the way from the mouth.

Give three _external_ ways that you can distinguish males from females.

1. _____

2. _____

3. _____

Ascaris Dissection
(be certain to read the cautionary note below)

A. After the external examination, you will need to dissect a female _Ascaris_. First note the superficial, thin, almost transparent cuticle covering the animal and its underlying epidermis. Using a sharp scalpel or a needle probe, cut through both the cuticle and the epidermis along the entire length of the animal. Do not go too deep or you risk disturbing the underlying organs. Now pin open the worm so that the internal structures are exposed. Add a little water to keep the specimen moist. Note that the organs are freely suspended in an open space inside of the animal.

What is this space called?

B. Now use Figure 7.1 to locate the following features on the female _Ascaris_.
- ✓ mouth _____
- ✓ anus
- ✓ intestine
- ✓ ovary/ oviduct (you do not need to differentiate between these two)
- ✓ uterus*
- ✓ vagina
- ✓ genital pore

*Note: The uterus forms an inverted Y. The tail of the Y points toward the mouth.

C. What is the normal food of an _Ascaris_? _____

(you may also wish to refer to page 158 in chapter 8 for a further discussion of _Ascaris_).

Caution

Ascaris eggs can withstand adverse conditions: high temperatures, low temperatures, desiccation, and very likely, formaldehyde preservatives. Therefore, as a precaution against the possibility of infection by eggs that might have survived the preservation process, be sure to observe the following:

1. Do not put your hands near your mouth during lab.
2. There is to be no food or beverages during today's lab.
3. Gloves are to be worn during your dissections.
4. Thoroughly wash your hands with soap and water after today's lab.

Microscopic Anatomy

A. Examine a c.s. microscope slide of an *Ascaris* female (Figure 7.2). Locate the following:

✓ pseudocoel
✓ uterus and eggs
✓ ovary/oviduct (you do not need to tell them apart)
✓ intestine
✓ nerve cords (ventral and dorsal)
✓ longitudinal muscle
✓ cuticle and underlying epidermis

1. Approximately how many eggs/day can a female *Ascaris* produce? _____

2. How does a human become infected with *Ascaris*? _____

3. On the cross-section slide of a female *Ascaris,* locate the uterus. Then increase magnification so that the eggs within the uterus are in view. Note their scalloped edges.   Sketch several eggs in the space below:

B. Examine an *Ascaris* male microscope slide — cross section (Figure 7.2). Locate the following:
✓ pseudocoel
✓ testis
✓ intestine
✓ nerve cords (ventral and dorsal)
✓ longitudinal muscle
✓ cuticle and underlying epidermis

4. Where in the female *Ascaris* does fertilization occur? _____

5. What is the function of the testis? _____

6. What is the function of the male copulatory spicules? _____

7. How does the sperm of *Ascaris* differ from other animals? _____

Chapter 8	Worksheet

_____
(name)

Symbiosis

Mutualism

1. Define mutualism and give two examples among those animals that we have studied.

Commensalism

2. Define commensalism and give two examples among those animals that we have studied.

Parasitism

3. Define parasitism and give two examples among animals other than humans.

4. Examine a microscope slide of the encysted form of larval *Trichinella spiralis* (the trichina worm).

   a. Note the spiraled position assumed by the many larvae on this slide.

   b. How is this stage reflected in the parasites' name?

   c. In what type of tissue is the cyst embedded? _____

   d. What disease does this organism cause in humans? _____

5. Next, select a slide showing male and female hookworms, *Necator americanus*.

   > **Note**: The parasite is essentially identical to another species of hookworm, *Ancyclostoma duodenale*. In fact, the names are sometimes interchanged. For our purposes, we will use *Necator* as the hookworm example as it is the more common parasite of humans.

   a. How is *Necator* transmitted to humans?

   b. Why is it more common in Atlanta than in Chicago?

   c. Note the large anterior teeth (the source of its common name). What do you think is their function?

   d. Finally, examine the male hookworm and locate the posterior, flared copulatory bursa. How do you think he uses these structures?

6. The final microscope slides to examine are of male and female *Enterobius vermicularis*, the human pinworm.

   a. Note the very small size of this parasite. Which is larger, the male or female? _____

   b. Describe how this parasite may be transmitted to humans.

Instructions: The table below contains information about selected human parasites; a separate parasite for each row. The table is partially complete in that enough information is contained in each row to identify that parasite. Using lecture notes and text information, fill in the missing data.

Parasite sci. & common name	Human Location	Intermediate Host	Transmission	Pathology & Diagnosis
Taenia solium (pork tapeworm)				mild intestinal upset, loss of appetite, diarrhea eggs & gravid proglottids detected in stool
			eggs are swallowed, larva migrate to lungs then up trachea and are again swallowed to mature in small intestine	female lays 200,000 eggs per day
				most widespread human roundworm in U.S., female lays up to 16,000 eggs/day; flashlight and scotch tape tests to detect worms that crawl out of anus at night; familial infection
			filariaform larvae penetrate exposed skin from soil and migrate through blood vessels to lungs and then to intestine	
				trichinosis; may cause severe muscular fatigue
			bite of the blackfly *(Simulium, spp.)*	
				gross enlargement of the extremities

Chapter 9	**Worksheet**	

Mollusca II

_____
(name)

***Loligo* (squid) Dissection**:

Squid: External Anatomy
(Figures 9.20–9.24)

1. Examine the head of a preserved squid and note the large **eyes** (you will dissect one later), the anterior **arms** and **tentacles**. How many are there of each? Closely examine an arm and a tentacle noting the **suckers** used to grasp prey. View a sucker under a dissecting microscope. Pull the arms apart to locate the **mouth** surrounded by the **periostomial membrane**. You will dissect the contents of the mouth area later.

Describe how a squid feeds:

2. Now examine the rest of the exterior. Note that the "skin" of the squid is really the **mantle**. The squid has two large lateral **fins** along its posterior border.

What is the function of the fins? _____

3. Now find the anterior **collar,** which encloses the internal organs. Place a finger (or a probe) between the mantle and the internal organs into the **mantle cavity**. Now locate the funnel-shaped **siphon**.

Describe the role of the siphon, the mantle cavity, and the collar in squid locomotion. Explain how a squid makes a **right** turn.

Squid: Internal Anatomy
(Figures 9.21–9.24)

1. Place the squid so that its ventral side is facing you. The siphon will be exposed in this position. Use scissors or a scalpel to open the specimen along the ventral midline. Do not cut too deeply or you will destroy underlying tissue. Cut all the way to the posterior end of the animal. Now make one or two lateral cuts on each side of the animal in the region of the fins and away from the midline. This will allow you to pin back the mantle easily to expose the internal organs.

2. Do not cut into the organs until instructed to do so. You should at this point be able to locate the **pen** as a reddish colored supporting rod lying just beneath the dorsal mantle.

3. First determine if your specimen is a *male* or *female*. If it is a female you will see two large glands located in the midregion of the specimen. These are **nidamental glands** used to produce a gelatinous coating for the eggs. Also, if the specimen is female you should see masses of small **eggs** in the **ovaries** located in the posterior region of the animal. Finally, to complete the female reproductive system, locate the **oviduct** on the animal's left side. Note its opening through which eggs pass on their way to the mantle cavity.

4. If your specimen is a mature male, you should first locate the small tubular **penis** on the animal's left side. Then find the centrally positioned **testes** and bulb-like **spermatophoric gland** situated on the animals' left side. Sperm are packaged into bundles and stored in the spermatophoric gland until passed to the female when mating. The male uses a modified (hectocotylized) arm for this purpose. Although the male does have a penis, it is far too small to be used as a true copulatory organ.

5. Locate the **ink sac** lying beneath the **rectum** (actually it's dorsal to the rectum, since your animal is on its back). The ink sac may be black or somewhat metallic in color. Be careful not to rupture the sac as the ink (sepia) can be rather messy. Note that both the **anus** and ink sac empty into the siphon.

> Explain why the ink sac and rectum both empty into the siphon:

6. Next locate the large, straplike **retractor muscles**, one attached to each side of the siphon.

> What do you think is the function of the siphon retractor muscles?

7. Examine the large, lobed **gills** located on each side of the animal. Note their large surface area. Why are they so big?

8. Locate the **systemic heart.** To do that, you will need to carefully remove overlying membranous tissue between the base of the gills. The heart is easy to recognize as a fairly large, triangular, muscular mass with several vessels emanating from it. Now locate the two **branchial hearts** situated at the base of each gill.

> Explain the function of the systemic and branchial hearts.

9. The squid's **kidney** can be located as a rather diffuse whitish mass located just anterior to the systemic heart.

10. There are three more internal items you need to examine, all associated with the digestive system. First, find the very large **digestive gland** located beneath the ink sac and retractor muscles. Then locate the **stomach**. This is a little tricky since you will have to dig around somewhat to expose it. It is thumb shaped and located on the animal's right side just posterior to the gills. It's usually about the size of your little finger, but may be larger if the squid has recently fed. If food is present, you may want to open the stomach to determine its contents. The final internal structure to locate is an extension of the stomach, the **cecum**. The squid doesn't have a true intestine, so the cecum is used as a large auxiliary sac for digestion. It is often rather gel-like in texture.

* Quite often there are tapeworms in the squid caecum. Search your specimen for the white, ~1/4-inch-long parasites.

11. Now you will dissect the head contents. To do so, place the squid so the dorsal surface faces you. Then make a sagittal incision through the base of the arms, between the eyes, to expose a round muscular body about the size of your thumb. This is the **buccal bulb.** Notice a tube, the **esophagus**, leading posteriorly from the bulb to the stomach.

12. Open the bulb with a scalpel. Inside locate the large parrotlike **beak**. Pry open the beak to expose the tongue shaped **radula** inside equipped with bristlelike teeth.

Why is the buccal bulb so large and tough?

13. Next dissect one of the eyes. Be careful when you open it as fluid may squirt out. The squid eye is constructed very much like yours; in fact, the squid eye and the human eye are classic examples of *parallel evolution.* There is an opening, the **pupil,** that allows light to enter the eye proper. Light passes through the round, marblelike **lens** for focusing and then strikes the **retina**, a dark brown to blackish layer that transforms light into the visual signal sent to the brain. If cow eyes are available you may wish to dissect one and compare its anatomy to the squid eye. See page 235 for instructions on how to dissect a mammalian eye.

14. Your final dissection is to expose the **brain**. Continue the cut posteriorly that you made to dissect the buccal bulb. Cut deeply between the eyes until you see a large white mass, which is the brain. Squids have the largest eyes and brain of any invertebrate animal. Next locate the pair of stellate ganglia (fig. 9.21). Just pull the body mass to one side which will expose these star-shaped nervous structures. They function as relay centers to control the powerful mantle muscles responsible for jet propulsion.

Can you think of any relationship between predation and the squids' tentacles, siphon, eyes, and brain?

You need to be able to locate and give functions of the following squid features:

✓ arms	✓ tentacles	✓ fins
✓ mouth	✓ periostomial membrane	✓ ink sac
✓ mantle	✓ mantle cavity	✓ kidney
✓ collar	✓ siphon	✓ siphon retractor muscles
✓ nidamental glands	✓ eggs	✓ ovaries
✓ oviduct	✓ penis	✓ testes
✓ spermatophoric gland	✓ buccal bulb	✓ beak
✓ radula	✓ esophagus	✓ stomach
✓ cecum	✓ rectum	✓ anus
✓ digestive gland	✓ gills	✓ systemic heart
✓ branchial hearts	✓ brain	✓ stellate ganglia
✓ eyes (lens, pupil, retina)	✓ pen	

Lab Notes

Chapter 10	Worksheet

_____
(name)

Phylum Annelida
Class Polychaeta

One of the common polychaetes of seashores is the sandworm *Nereis*. It is a predator of the intertidal zone where it feeds on a variety of invertebrates. *Nereis* is an active swimmer and can crawl quickly in search of prey or to escape danger.

I. *Nereis*, external anatomy

1. Place a sandworm specimen so that the dorsal side is facing you. Observe its basic shape and external features. Note the pronounced segmentation and the paired, lateral appendages, the **parapodia**, on each body segment. View these under a dissecting microscope and observe the bundle of bristlelike **setae** emerging from each parapodium.

What are three functions of parapodia?

2. Sandworms have a well-developed head and pharynx, although the pharynx is often withdrawn into the animal. If your specimen has an extended pharynx, view its numerous appendages: eyes, tentacles, palps, and anterior jaws. If not, then observe the specimen on display.

*Do not dissect the sandworm.
When you are finished with the external examination, return the specimen to its container.

II. Display Specimens

Study the various other polychaetes on display. Examine the specimen labeled *Amphitrite*. The feathery structures extending from the anterior segment are called radioles.

What are the functions of the radioles?

Subclass Oligochaeta

Perhaps the most common biological specimen is *Lumbricus*, the common earthworm. They are used in virtually all beginning lab courses. However, even if you have already dissected one (or more), keep in mind that earthworms possess a number of significant annelid structures.

III. Living Earthworm, *Lumbricus*

1. Place a live earthworm on a moist paper towel. Determine the dorsal, ventral, anterior, and posterior surfaces of your specimen. (**Note**: The dorsal side is darker and setae are located only on the ventral surface.) You can feel the setae (bristles) by grasping the worm in one hand and pulling your fingers first toward the tail and then toward the head. In what direction do they point?

How do you explain this?

2. What are the two functions of the setae?

3. Now observe your worm's locomotion. This type of movement is called **peristalsis**.

How do muscles and coelomic compartments combine to produce this type of locomotion?

4. Does the earthworm have a preferred orientation? Place it on its dorsal surface; does it right itself? Virtually all animals will behave in the same manner.

How do you explain this apparent "need" to remain upright?

5. Describe your animal's response to:

    a. **Touch** - gently touch the animal with a soft object, then poke it with a needle probe. Record your observations by comparing the worm's reactions to these two stimuli.

    b. **NaCl** - place a small amount of salt directly on the worm. What is the response?

> Explain at the cellular level.

6. Farmers consider earthworms to be highly beneficial. Why is that?

IV. Preserved Earthworm

External Anatomy

    Examine a preserved earthworm. Note its general shape and distinctive segmentation. Examine the anterior end to find the mouth located between the most anterior segment, the prostomium, and the first true segment. Likewise, find the anus on the pygidium (the last segment). Next locate the ventral opening of the vas deferens on the 15th segment and the small oviduct openings on segment 14. Finally, examine the glandular, ringlike **clitellum**. In the space below, describe the function of the clitellum.

V. Earthworm Dissection
Internal Anatomy
(Figures 10.10 - 10.13)

1. Place the earthworm so the dorsal surface is facing you. Using a sharp scalpel or a needle probe, make a shallow incision through the epidermis from the prostomium (head) back to about 2 inches posterior to the clitellum. Be sure not to cut too deep as you will damage the underlying structures.

2. Use dissection pins to spread the body wall so that the internal structures are adequately exposed. However, avoid pinning the most anterior segments as this may obscure the brain and related structures.

3. Note that each internal compartment is separated by a thin tissue, the **septa**, so that each segment has its own internal coelomic cavity.

4. Now examine the circulatory system. Find the **dorsal blood vessel** lying atop the digestive system. You will see this as a dark line running the length of the animal. Trace it forward until it terminates at the series of five **aortic arches**. These are identified as small, blackish, fingerlike objects that project ventrally. The arches join the **ventral vessel**, another black line that can be located along the ventral surface of the intestine.

5. Next locate and examine the nervous system. It consists of a pair of **suprapharyngeal ganglia** (brain) that appear as tiny white "pinheads" at the anterior end of the pharynx. These form a loop around the pharynx to unite below as the **ventral nerve cord**. The cord runs the length of the animal beneath the ventral blood vessel. Find the whitish-colored cord by gently moving the intestine aside.

6. The earthworm has both male and female reproductive organs. The large whitish objects lateral to the aortic arches are the **seminal vesicles**. Buried within the vesicles are the small testes, which to observe must be carefully dissected. Exposure of the testes is optional. The ovaries, likewise, are too small to view without the aid of a microscope (finding them, too, is optional). The remaining female structures are the small paired **seminal receptacles** situated antero-lateral to the vesicles. Sperm from the mating partner are stored in the seminal receptacles until needed.

7. The most complicated apparatus in the earthworm is the digestive system. Carefully examine it and locate each of the following features. First find the muscular **pharynx**. Note how it is attached to the body wall by a series of stringy muscles. Posterior to the pharynx is the elongated **esophagus**, which leads to the saclike **crop**. The crop is a soft enlargement of the esophagus. Behind the crop is the muscular **gizzard** leading to the **intestine**, which passes to the **anus**.

8. Carefully open the intestine to reveal the straplike **typhlosole** running the length of the intestine.

> What is the function of the earthworm's typhlosole?

VI. Earthworm, Cross-Section Slide

Select a microscope slide of the earthworm in cross section. Starting at the outside (the outer edge of the specimen) and working your way internally, find the following structures:

- ✓ **epidermis** - most exterior tissue
- ✓ **circular muscles** - fibers lying just beneath the epidermis
- ✓ **longitudinal muscles -** fibers facing the coelomic cavity
- ✓ **coelomic cavity** - the open space in each segment that houses the internal organs
- ✓ **intestine** - the elongated, centrally located tube
- ✓ **typhlosole** - the round tube within the intestine
- ✓ **dorsal blood vessel** - lying above the intestine
- ✓ **ventral blood vessel** - lying beneath the intestine
- ✓ **ventral nerve cord** - lying beneath the ventral blood vessel

Chapter 13	**Worksheet**

_____
(name)

Phylum Arthropoda
Subphylum Crustacea

1. Examine the various crustaceans on display. Try to locate the following crustacean features

  ✓ the presence of two pairs of antennae
  ✓ body divided into two tagmata; the cephalothorax and abdomen (sometimes covered by carapace)
  ✓ biramous appendage
  ✓ mandibles
  ✓ presence of gills
  ✓ thick, hardened exoskeleton

Class Malacostraca
The Crayfish: External Anatomy
(Figures 13.9-13.11)

2. Examine the external features of both a male and female crayfish and locate:

  ✓ eyes and eyestalks
  ✓ antennules
  ✓ antennae
  ✓ rostrum
  ✓ mandibles
  ✓ maxillae & maxillipeds
  ✓ carapace
  ✓ abdomen
  ✓ telson
  ✓ anus
  ✓ chelipeds (first pair of walking legs, modified into a large pincer)
  ✓ walking legs
  ✓ gonopods (on male only)
  ✓ swimmerets

3. Use the features below to identify specimens as either male or female.

  Males: large chelipeds used to control female during mating, narrow abdomen, large first swimmeret called gonopods, openings from the two male sex ducts located at the base of legs # 5.

  Females: chelipeds proportionally smaller than males, broad abdomen, small or missing first swimmeret, remaining swimmerets relatively large to carry eggs, opening into the seminal receptacle is centrally located between legs # 4 & #5, the two openings from the oviducts are located at base of walking legs # 3.

Notice how hard the exoskeleton is in your specimens. What four functions does it perform?

_____

_____

_____

_____

The primitive crayfish appendage is **biramous**, consisting of a base and two branches. The embryonic appendage is modified into 10 adaptive, functional variations. This arrangement is an example of serial homology in which a basic form (here, the biramous appendage) is modified into a series of functional structures. From one of your specimens, extract one of each general type of appendage, sketch, and give its major function below.

FUNCTION	**SKETCH**

antennules: _____

antennae: _____

mandibles: _____

maxillae: _____
(either one of two)

maxillipeds: _____
(any one of three)

chelipeds: _____
(pincers)

walking legs 2 - 5: _____
(any one)

male swimmeret #1 _____
(gonopod)

remaining swimmerets: _____

uropod: _____

Chapter 14   Worksheet

_____
(name)

Insect Identification

Instructions: Use the insect keys on the previous two pages, as well as the description of the various orders to identify the 10 unknown insect specimens. <u>Identify</u> to **order** only

Taxonomic Order	**Common Name**
1. _____	_____
2. _____	_____
3. _____	_____
4. _____	_____
5. _____	_____
6. _____	_____
7. _____	_____
8. _____	_____
9. _____	_____
10. _____	_____

Lab Notes

Chapter 15   **Worksheet**

_____

Echinoderm Exercises

I. Perform a "scraping test" by shaving a small portion of tissue from the aboral surface of a sea star with a scalpel (about one-half the size of a pencil eraser). Place the tissue on a microscope slide and add a few drops of water. Swirl the tissue around until it's a thin film— the thinner the better. Now observe the mixture under the low power of a compound microscope. Identify spines and pedicellariae and sketch each below.

II.   In the space below, sketch both surfaces of a sand dollar. Label and give functions of:

✓ petaloids (petal-shaped aboral features with slits through which dermal branchia [papulae] protrude)
✓ mouth (large oral opening)
✓ anus (small oral opening)
✓ food grooves (oral lines leading to the mouth)
✓ madreporite (at junction of petaloids on aboral surface)
✓ lunules (five large oval openings through entire skeleton)
✓ gonopores (five pin-sized aboral openings between bases of petaloids)

<u>Be sure to label and give functions of each structure.</u>

ORAL SKETCH

ABORAL SKETCH

Lab Notes

Chapter 18	Worksheet

_____
(name)

Fish Characteristics

I. Give the organism, anatomical location, and function of the following:

Animal   **Location &   Function**

_____1. Pineal eye _____

_____2. Spiral valve _____

_____ 3. Heterocercal tail _____

_____ 4. Spiracle _____

_____ 5. Claspers _____

_____ 6. Mid-dorsal nostril _____

_____ 7. Operculum _____

_____ 8. Buccal funnel _____

_____ 9. Lateral line system _____

II. Define

1. Peritoneum:

2. Viviparous:

3. Ovoviviparous:

4. Oviparous:

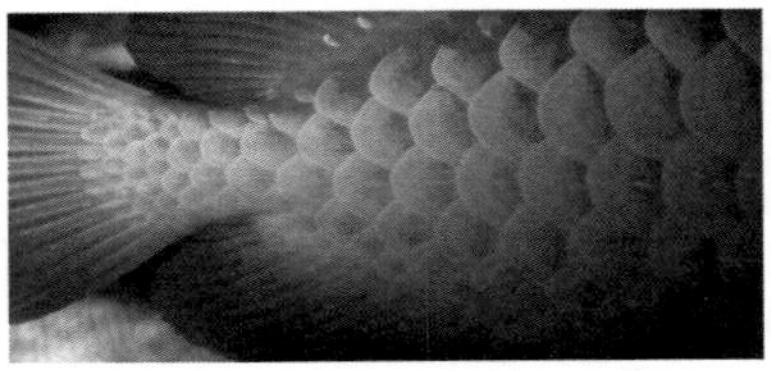

III. In the space below, sketch the four types of fish scales as observed via the microscope.

placoid   **ganoid**   **cycloid**   **ctenoid**

Which of the four scales does the perch possess? _____

Lab Notes

Chapter 19   Worksheet

Subphylum Vertebrata: _____
                                                (name)
Amphibians and Reptiles

Herpetology

Objective:
By answering a series of questions, you will survey the anatomy, behavior, and ecological adaptations of various amphibians and reptiles. The exercise is divided into three parts.

Part I. Text References

1.
> **Part I. References**
> Questions posed in this section can be answered by referring to your lecture notes and/or textbook.
>
> **Part II. Observation**
> Questions in this portion can be answered through direct observation of living or preserved specimens.
>
> **Part III Insights**
> Some independent thought is required to answer each of these questions.

Give two <u>anatomical</u> differences between salamanders and frogs.

   a. _____

   b. _____

2. Give any three differences between amphibians and reptiles.

   a. _____

   b. _____

   c. _____

3. Give three <u>anatomical</u> differences between lizards and snakes.

   a. _____

   b. _____

   c. _____

4. Where would you find Jacobson's organ? _____

   What is its function? _____

5. What is a plastron? _____

   ....a carapace? _____

   What special problem do these features cause turtles and how is it solved?

6. Give three differences between a lizard and an alligator.

   a. _____

   b. _____

   c. _____

7. Give two ways that you can differentiate anatomically between caecilians and legless lizards.

8. What is neoteny and what animal on display shows this feature?

9. What is the function of the pit in rattlesnakes?

Part II. Direct Observation

10. Give three anatomical differences between frog and snake skeletons.

    a. _____

    b. _____

    c. _____

11. In what unusual way does the alligator snapping turtle use its tongue?

12. Observe a reptilian egg. How has this amniotic egg contributed to the success of reptiles?

13. Compare the skin of a salamander to that of a gecko. What difference does this make in the life of these two animals?

Part III. Insight

14. Arthropods periodically shed their heavy exoskeleton in order to grow, a process called ecdysis. Do you think that turtles, too, shed their shell (i.e., carapace and plastron) in order to grow? Explain how turtles are able to get larger.

15. What are the structures just in front of the forelimbs of the mudpuppy (*Necturus*)?

   Is this an adult or a juvenile animal? Support your answer.

16. What do you think is the purpose of the large tail in the leopard gecko? (Hint: It comes in handy during periods of famine.)

   The gecko shows two types of camouflage. What are they?

17. What specific type of color pattern is displayed by the tiger salamander?

18. What about a horned toad's skin tells you that it is really a lizard and not a toad?

19. Explain how a snake's skull can permit it to swallow a prey animal much larger than its own head. Also, how does a snake breathe while doing so?

Lab Notes

Chapter 20	Worksheet

_____
(name)

Bird Beak Identification

Much of the biology and behavior of a bird can be determined through examination of its beak.   For example, its habitat and food source are directly linked to beak type. In this exercise, you are asked to identify each of the beak types on display (according to the numbered tags) and to indicate a possible food source for each bird.

Beak Type   **Food Type**

1. _____   _____

2. _____   _____

3. _____   _____

4. _____   _____

5. _____   _____

6. _____   _____

7. _____   _____

8. _____   _____

9. _____   _____

10._____   _____

11._____   _____

12._____   _____

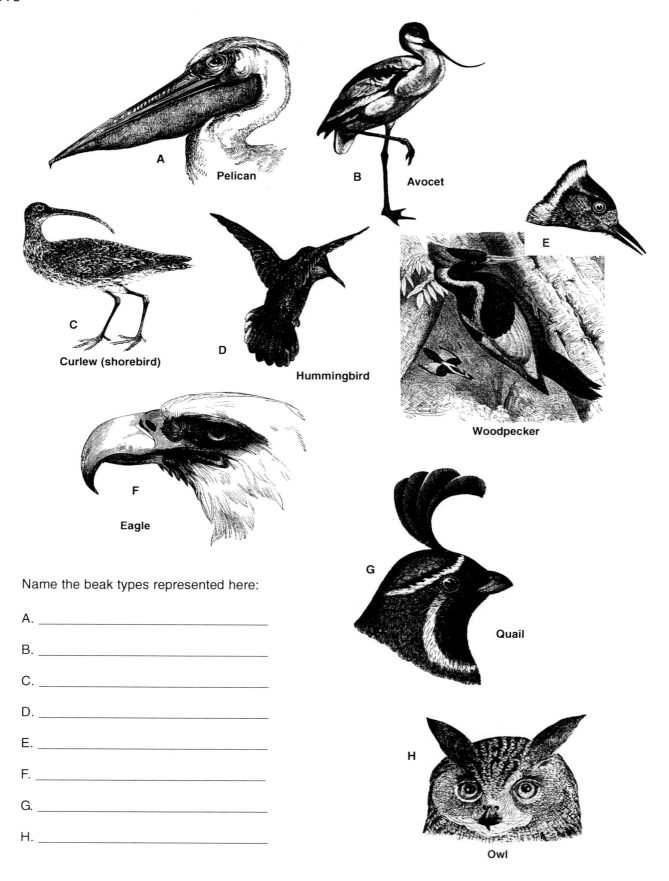

A Pelican

B Avocet

C Curlew (shorebird)

D Hummingbird

E

Woodpecker

F Eagle

G Quail

H Owl

Name the beak types represented here:

A. _____

B. _____

C. _____

D. _____

E. _____

F. _____

G. _____

H. _____

Chapter 20 Bird Identification

_____
(name)

INSTRUCTIONS: During this exercise you are asked to give the common name of 34 birds. Use the field guides provided by your instructor to identify each museum study skin and be sure that your answer corresponds to the numbered tag attached to each specimen.

1. _____

2. _____

3. _____

4. _____

5. _____

6. _____

7. _____

8. _____

9. _____

10. _____

11. _____

12. _____

13. _____

14. _____

15. _____

16. _____

17. _____

18. _____

19. _____

20. _____

21. _____

22. _____

23. _____

24. _____

25. _____

26. _____

27. _____

28. _____

29. _____

30. _____

31. _____

32. _____

33. _____

34. _____

Lab Notes

Chapter 21	Worksheet

_____
(name)

Subphylum Vertebrata
Mammals

INSTRUCTIONS: During this exercise you are asked to give the common name of 34 mammals. Use the field guides to identify each museum study skin and be sure that your answer corresponds to the numbered tag attached to each specimen.

1. _____

2. _____

3. _____

4. _____

5. _____

6. _____

7. _____

8. _____

9. _____

10. _____

11. _____

12. _____

13. _____

14. _____

15. _____

16. _____

17. _____

18. _____

19. _____

20. _____

21. _____

22. _____

23. _____

24. _____

25. _____

26. _____

27. _____

28. _____

29. _____

30. _____

31. _____

32. _____

33. _____

34. _____

Lab Notes